MARINE ORGANIC CHEMISTRY

Evolution, Composition, Interactions and Chemistry of Organic Matter in Seawater

Elsevier Oceanography Series, 31

MARINE ORGANIC CHEMISTRY

Evolution, Composition, Interactions and Chemistry of Organic Matter in Seawater

Edited by

E.K. DUURSMA

Delta Institute for Hydrobiological Research,
Yerseke,
The Netherlands
and

R. DAWSON

University of Kiel,
Kiel,
Federal Republic of Germany

ELSEVIER SCIENTIFIC PUBLISHING COMPANY
Amsterdam — Oxford — New York   1981

ELSEVIER SCIENTIFIC PUBLISHING COMPANY
335 Jan van Galenstraat
P.O. Box 211, 1000 AE Amsterdam, The Netherlands

Distributors for the United States and Canada:

ELSEVIER/NORTH-HOLLAND INC.
52, Vanderbilt Avenue
New York, N.Y. 10017

Library of Congress Cataloging in Publication Data
Main entry under title:

Marine organic chemistry.

   (Elsevier oceanography series ; 31)
   Includes bibliographies and indexes.
   1. Chemical oceanography. 2. Organic geochemistry.
I. Duursma, E. K. II. Dawson, R.
GC116.M37   551.46'001'54   80-17046
ISBN 0-444-41892-X

ISBN: 0-444-41892-X (Vol. 31)
ISBN: 0-444-41623-4 (Series)

(Photograph: P.J. van Boven)

*Easy is the descent into the dark deep, but to retrace your way
and issue into space — there is the toil and there is the task.*

(Vergil's Aeneid)

GENERAL ACKNOWLEDGEMENT

During the course of editing this book we have certainly put a great number of people under stress for which we now offer our sincere apologies.

Our respective institutes, The Delta Institute for Hydrobiological Research, Yerseke, The Netherlands, of the Royal Netherlands Academy of Arts and Sciences, and the Sonderforschungsbereich 95 at the University of Kiel, Federal Republic of Germany, carried the burden of our activities throughout.

Our particular thanks are extended to Gerd Liebezeit for all manner of help throughout, to Barbara Heywood for her patience in deciphering and correcting many of the "multi-lingual" texts, to Kenneth Mopper for the many fruitful discussions and to our colleagues and friends for their encouragement.

LIST OF CONTRIBUTORS

M. AUBERT   CERBOM, 1 Avenue Jean Lorrain, 06300 Nice, France.

W. BALZER   Institut für Meereskunde, Düsternbrockerweg 20, D-2300 Kiel-1, Federal Republic of Germany.

G. CAUWET   Centre Universitaire, Avenue de Villeneuve, 66025 Perpignan, France.

R. DAWSON   Sonderforschungsbereich 95, Universität Kiel, Olshausenstrasze 40/60, D-2300 Kiel-1, Federal Republic of Germany.

E.K. DUURSMA   Delta Institute for Hydrobiological Research, Vierstraat 28, 4401 EA Yerseke, The Netherlands.

W. FENICAL   Scripps Institution of Oceanography, La Jolla, Calif. 92093, U.S.A.

R.B. GAGOSIAN   Woods Hole Oceanographic Institution, Woods Hole, Mass. 02543, U.S.A.

M.J. GAUTHIER   CERBOM, 1 Avenue Jean Lorrain, 06300 Nice, France.

K.A. HUNTER   University of East Anglia, School of Environmental Sciences, Norwich NR4 7TJ, Great Britain.

C. LEE   Woods Hole Oceanographic Institution, Woods Hole, Mass. 02543, U.S.A.

G. LIEBEZEIT   Sonderforschungsbereich 95, Universität Kiel, Olshausenstrasze 40/60, D-2300 Kiel-1, Federal Republic of Germany.

P.S. LISS   University of East Anglia, School of Environmental Sciences, Norwich NR4 7TJ, Great Britain.

M.D. MacKINNON Environmental Affairs Department, Syncrude Canada Ltd., 10030-107 Street, Edmonton, Alberta T5J 3E5, Canada.

R.F.C. MANTOURA Institute for Marine Environmental Research, Prospect Place, The Hoe, Plymouth PL1 3DH, Great Britain.

P.H. NIENHUIS   Delta Institute for Hydrobiological Research, Vierstraat 28, 4401 EA Yerseke, The Netherlands.

A. SALIOT   Laboratoire de Physique et Chimie, Université Pierre et Marie Curie, Tour 24-25, 4 Place Jussieu, 75230 Paris Cedex 05, France.

A.W. SCHWARTZ   Department of Exobiology, Catholic University, Toernooiveld, 6525 ED Nijmegen, The Netherlands.

B.A. SKOPINTSEV Institute of Biology of Inland Waters, Academy of Sciences of the U.S.S.R., Borok, Yaroslavl Region, U.S.S.R.

R.G. ZIKA   University of Miami, School of Marine and Atmospheric Sciences, 4600 Rickenbacker Causeway, Miami, Fla. 33149, U.S.A.

VIII

CONTENTS

Chapter 1

INTRODUCTION

E.K. DUURSMA

The story of organic matter in the oceans and its conversion from and into inorganic substances is almost as old as the sea itself. The origin of life was the organic molecule. Life expanded over all the oceans from the tropics to the poles and into its eternal, abyssal depths. Is it not surprising that life should endure and thrive up to the present day? This question is rarely posed although organic matter itself may condition environments where the prerequisites for life were absent.

Since the earliest oceanographic studies, organic matter has been related to oxygen. The old techniques used oxygen consumption as a measure for organic matter. Equally, the oxygen minimum in the oceans has fascinated our predecessors in oceanography.

Although the chances in geological time scales could have favoured a Black Sea condition, our oceans are still aerobic and not anoxic below the euphotic zone. Why should this be? The explanations so far are to be found in exchange patterns between deep and surface waters. Little exchange coupled with rich, productive, euphotic zones should cause anoxic deep layers, while rapid exchange between less productive surface waters and deep water masses favours an aerobic ocean situation.

Deep-ocean circulations are, however, not that rapid and the turnover of deep water masses requires hundreds or thousands of years, ample time to reduce the oxygen to zero without any chance of *in situ* supplementation. Within forty years the 5 mg oxygen per litre of a 4000 m deep ocean might theoretically be consumed by a surface primary production of 200 g of carbon per square metre per year.

Since this is not the case (Eppley and Peterson, 1979), the boundaries of the processes involving organic matter are in some way defined. Decomposition probably occurs predominantly in the productive zones where oxygen is transferred from and to the atmosphere. Organic matter in the deeper layers effectively decomposes very slowly and this matches the minor reductions in the oxygen concentration (Fig. 1).

Stability is a term frequently applied in connection with organic structures: "humic compounds" and "gelbstoff" are names used to describe ill-defined "stable" compounds. It is to the credit of the first oceanographers that with a lack of proper methods they tried to place the organic chemistry of the seas within the broad framework of oceanographic processes of physi-

2

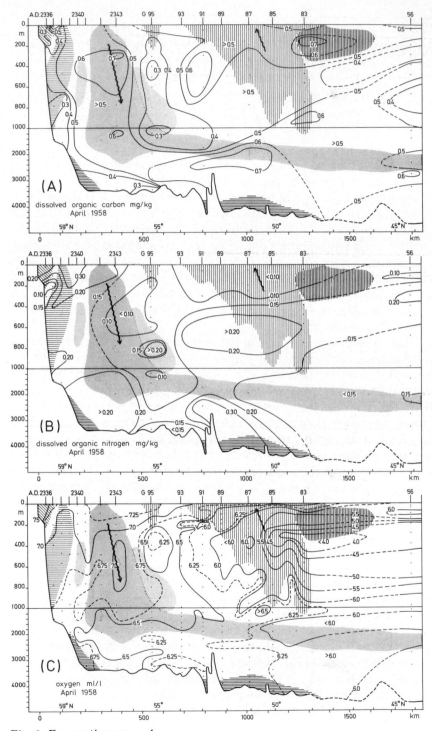

Fig. 1. For caption see p. 4.

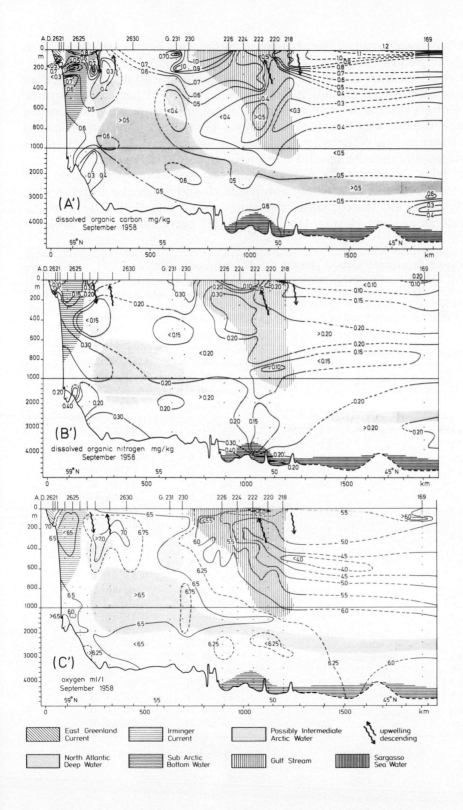

(A') dissolved organic carbon mg/kg
September 1958

(B') dissolved organic nitrogen mg/kg
September 1958

(C') oxygen ml/l
September 1958

East Greenland Current

Irminger Current

Possibly Intermediate Arctic Water

upwelling descending

North Atlantic Deep Water

Sub Arctic Bottom Water

Gulf Stream

Sargasso Sea Water

cal and biological nature. The chapter of Skopintsev is an example of such a struggle to resolve budgets and processes in the oceans.

Also the late Kurt Kalle (early personal communication), who worked extremely carefully, regarded, for example, a change of particulate organic matter along a vertical directly in terms of input and output. Minor increases over large depths would require enormous inputs, and any discussion on vertical distributions of organic substances is useless unless the origin of the water masses and their horizontal transports are considered as well.

Present-day oceanography has widened our knowledge of organic substances in the oceans. Elementary compositions and molecular structures are gradually being elucidated, while the functions of organic matter in biological systems become clearer. However, in this particular area of develop-

Fig. 1. Dissolved organic carbon (DOC) (A, A$'$), nitrogen (DON) (B, B$'$), and oxygen (O$_2$) (C, C$'$), in a north—south section off Greenland, as measured for April and September 1958. The watermasses are determined on basis of the temperature—salinity (T—S). Note the formation of Deep Atlantic Water in April and the southwards decrease of O$_2$ and DOC at 1000—2500 m depth. (Reproduced from Duursma, 1961.)

The bearings of the stations are:

Station	Date		Position	
			Latitude	Longitude
A.D. 2336	April	11, 1958	59°28′ N	43°54′ W
A.D. 2337		11	59°09′ N	43°50′ W
A.D. 2338		11	58°47′ N	43°50′ W
A.D. 2343		13	56°26′ N	43°25′ W
G 56		15	44°04′ N	39°26′ W
G 83		22	49°11′ N	40°30′ W
G 87		24	51°07′ N	40°54′ W
G 91		25	52°47′ N	42°18′ W
G 95		26	54°59′ N	42°25′ W
A.D. 2621	September	15, 1958	59°41′ N	43°56′ W
A.D. 2623		15	59°28′ N	43°54′ W
A.D. 2624		15	59°08′ N	43°57′ W
A.D. 2625		16	58°46′ N	43°50′ W
A.D. 2626		16	58°25′ N	43°47′ W
A.D. 2627		16	58°06′ N	43°42′ W
A.D. 2628		16	57°44′ N	43°38′ W
G 169	August	24, 1958	43°55′ N	38°57′ W
G 218	September	8	49°35′ N	40°05′ W
G 220		8	50°09′ N	40°28′ W
G 222		9	50°51′ N	40°48′ W
G 224		9	51°30′ N	41°14′ W
G 226		10	52°06′ N	41°45′ W
G 231		12	54°25′ N	42°49′ W

ment, many aspects were already anticipated by Krogh (1931, 1934a, b).

Ups and downs may be distinguished in certain historical aspects of the development of techniques and theories. The theories on complexation of trace metals are an example of this: when years ago copper and zinc were found to complex with dissolved organic matter, it was suggested that this was indicative for many other metals. Now we know that complexation is specific and competitive and that inorganic complexes may be dominant for many trace metals.

The study of biologically active substances is still a complicated field. The existence of such substances was suspected long ago but highly sensitive and selective techniques are required to detect the low concentrations of the substances which often have intricate molecular structures.

For these, but also for the less sophisticated techniques used to detect organic substances, the great problem has been the accuracy and precision of methods. In spite of the 2×10^{12} tons of dissolved organic matter in the world ocean, concentrations are at the low ppm level. Blanks for the DOC determination might easily contain 0.5 ppm C as many chemists have noted to their distress.

The battle for precise, accurately determined data, has been most severe in analysing of the DOC. Depending on the method, differences by a factor 2 to 3 have been observed for samples from the same ocean areas. As is gradually becoming clear, each method is determining something different with sometimes great accuracy. Some differences depend on the composition of volatile and easily or less easily oxidizable fractions. This may even differ from area to area as well as with depth.

Stress on the environment is caused not only by pollution but may also be a normal facet of oceanic conditions. In the course of evolution this has brought about the development of sometimes very specialised ecosystems or living communities which are adapted to such special conditions. Stress is equally known to influence the release of organic substances from phytoplankton. Thus, the excretion of organic matter from living algae or diatoms may differ from area to area or even seasonally. Data from one system are not necessarily applicable to another system. In particular this applies to experimental algal cultures (Sharp, 1977).

The authors of the chapters will present extensive information on these diverse aspects of marine organic chemistry. An epilogue is devoted to a discussion of pitfalls and the state of the art. Pollution aspects are not considered in spite of their increasing influence. For this the reader is referred to a review by Duursma and Marchand (1974) and to specialised papers in related journals.

The sequence of chapters has some logic as to the origin, processes and occurrence of organic substances. Although the approaches of the authors vary extensively, the book embraces the views of three generations of chemical oceanographers, all discussing their own material in their own way.

The organic chemistry of sea water is in a strong phase of development and elements of knowledge are constantly being added to the whole. This book will hopefully stimulate and strengthen this development.

REFERENCES

Duursma, E.K., 1961. Dissolved organic carbon, nitrogen and phosphorus in the sea. Neth. J. Sea Res., 1: 1—147.
Duursma, E.K. and Marchand, M., 1974. Aspects of organic marine pollution. Oceanogr. Mar. Biol. Annu. Rev., 12: 315—431.
Eppley, R.W. and Peterson, B.J., 1979. Particulate matter flux and planktonic new production in the deep ocean. Nature, 282: 677—680.
Krogh, A., 1931. Dissolved substances as food of aquatic organisms. Biol. Rev., 6: 412—442.
Krogh, A., 1934a. Conditions of life in the oceans. Ecol. Monogr., 4: 421—429.
Krogh, A., 1934b. Conditions of life at great depths in the ocean. Ecol. Monogr., 4: 430—439.
Sharp, J.H., 1977. Excretion of organic matter by marine phytoplankton: do healthy cells do it? Limnol. Oceanogr., 22: 381—399.

Chapter 2

CHEMICAL EVOLUTION — THE GENESIS OF THE FIRST ORGANIC COMPOUNDS

A.W. SCHWARTZ

1. INTRODUCTION

Organic molecules were synthesized before the Earth was formed. The solar nebula very probably already contained organic matter and more may have been formed during condensation of the planets, as evidenced by the presence of organic molecules in interstellar dust clouds, in comets and in meteorites. An important question is whether such primordial material could have contributed to chemical evolution on the Earth to any significant degree. The survival of complex molecules during formation and differentiation of the Earth appears extremely unlikely, although the possible contribution of the pyrolysis products of such material will be considered later. To begin with, I propose to discuss the probable range of conditions on the primitive Earth and to ask which processes may have led to the production of organic molecules and their further evolution to the first living organism. It is, however, beyond the scope of this chapter to review all published "prebiotic" syntheses. It will be necessary not only to select a few of the most significant areas of the problem for discussion, but to eliminate from consideration many reports which, within the boundaries of the approach to be described, no longer seem plausible. It will, necessarily, be a rather selective view of the possibilities. The reader interested in going deeper into the literature on chemical evolution is advised to consult the general sources listed in the bibliography.

2. GEOLOGICAL BACKGROUND

The oldest existing rocks on Earth, from the Isua supracrustal belt of West Greenland, are dated at 3.8×10^9 years (Moorbath et al., 1973) and consist of metamorphosed sediments. It has been suggested that the carbon in the Isua banded iron formation may be photosynthetic in origin (Schidlowski et al., 1979). Whether this remarkable conclusion is substantiated by further work remains to be seen. It is in any case abundantly clear from the available evidence that the conditions of deposition of the oldest known sediments were remarkably like those prevailing during the rest of the Archaean $(4.6 - 2.6 \times 10^9$ years). Furthermore, with the exception of the appearance of large quantities of free oxygen in the Earth's atmosphere during the middle Precambrian ($2.5 - 1.6 \times 10^9$ years), the geological record appears to be singularly free of large-scale changes in depositional environment (Holland, 1972). It is reasonable to begin with the assumption that the primitive oceans were probably not unique with regard to inorganic composition. What concerns us here, of course, is the problem of the source of organic compounds and their possible reactions in the primitive ocean. In the discussion following, I shall attempt to use geological constraints, where they are known, to develop a model of chemical evolution on the primitive Earth. Where such constraints are absent, however, it may be necessary to proceed in the opposite direction; that is, to ask whether any chemical pathways seem so strongly indicated that they may serve as constraints on physical models of the primitive Earth and aid us in constructing a scenario for at least some of the key stages in the initial production of organic compounds.

3. MODELS OF THE PRIMITIVE ATMOSPHERE

It is significant that one of the earliest recorded attempts to synthesize organic compounds from the components of a model primitive atmosphere was that of Groth and Suess (1938). By irradiating a mixture of CO_2 and H_2O with ultraviolet light, these authors were able to demonstrate the formation of formaldehyde and glyoxal (see discussion in Fox and Dose, 1977). Earlier attempts to generate organic compounds such as formaldehyde from CO_2 as models of photosynthesis are also known, although results were generally negative or unrepeatable. An example is the work of Moore and Webster (1913), who may have been the earliest workers to propose a gradual accumulation of reduced organic compounds as a necessary precursor to the origin of life. Moore (1913) may, in fact, have been the first to employ the term "chemical evolution". Other experiments demonstrated the formation of formaldehyde, formic acid and oxalic acid from CO_2 and H_2O (Garrison et al., 1951; Hasselstroom and Henry, 1956; Getoff et al., 1960). These experiments have generally been overshadowed by the work of Miller (1953) and subsequent "Miller-type" experiments. However, I shall have reason to return to them later.

In his book *The Origin of Life*, Oparin (1938) argued that the primary atmosphere of the Earth consisted of hydrocarbons and ammonia, produced by the hydrolysis of carbides and nitrides from within the Earth's crust. Spontaneous reactions would have produced a complex mixture of organic molecules which would have dissolved in the oceans and undergone an evolutionary process, producing more complicated organic systems and leading ultimately to the appearance of life. Similar ideas were expressed independently by Haldane (1929), who postulated an atmosphere consisting of carbon dioxide and ammonia, which would have reacted to produce organic compounds under the influence of ultraviolet light. A mélange of these proposals, together with the concept that the first form of life must have been not only anaerobic but heterotrophic, is usually referred to as the Oparin—Haldane Hypothesis. Bernal (1951) expressed similar views and a strong argument for a reducing atmosphere was developed by Urey (1952), based upon a consideration of the equilibria in a gas mixture of solar composition. These arguments formed a background for the experiment of Miller (1953), in which the synthesis of several amino acids and other compounds was demonstrated by passing a high-frequency electric discharge through a mixture of CH_4, NH_3 and H_2, and trapping organic products in a refluxing, ammoniacal solution. The concentration of ammonia which might have been available for organic syntheses on the primitive Earth has, in fact, become one of the more pertinent problems in this area of research. In recent years, the nature of the model atmosphere used by Miller in his original experiments has been criticized, both with regard to the presence of ammonia and the concentrations of methane and hydrogen employed. The concept of the "primitive soup" has also fallen somewhat into disrepute (see, for example, Sillèn, 1965 and Rutten, 1971). In order to understand the nature of the problem, it will first be necessary to review briefly some of the available information concerning the development of the Earth's atmosphere.

4. THE OUTGASSED ATMOSPHERE

Although there are many uncertainties concerning the primitive atmosphere of the Earth, there is one piece of information which provides a starting point for any discussion; the Earth has formed its own atmosphere by outgassing. Whatever may have been the gas composition within the solar nebula, it is clear that the Earth has retained no primordial gases, since it is depleted in all rare gases relative to the composition of the Sun (Mason, 1966). Rubey (1951) used the method of geochemical balances to demonstrate that the quantities of volatile substances now present on the surface of the Earth must have been produced by outgassing and presented data on the composition of volcanic gases to support his arguments. Such studies suffer from the criticism that gases from contemporary volcanoes may not be juvenile. Perhaps the best approach to a knowledge of the composition of an outgassed atmosphere is a study of chemical equilibria involving typical volcanic source rocks or meteoritic minerals. Holland (1962) has attempted to derive models of the primitive atmosphere by a consideration of data for equilibria in the systems $FeO—Fe_2O_3—SiO_2$ and $MgO—FeO—Fe_2O_3—SiO_2$. A key consideration in these studies is the presence or absence of metallic iron in the source material. Holland divided the early atmosphere into two stages, before and after formation of the core. In stage 1, the presence of free iron is assumed and the mixture of gases ejected at $1200°C$ consists primarily of H_2, H_2O, CO, N_2 and H_2S (in order of decreasing abundance).

The ratios of the partial pressures are:

$$\frac{H_2O}{H_2} = 0.44 \; ; \quad \frac{CO_2}{CO} = 0.17$$

In stage 2, the gases are N_2, H_2O, CO_2 and SO_2 with only very small amounts of H_2. The ratios are:

$$\frac{H_2O}{H_2} = 105 \; ; \quad \frac{CO_2}{CO} = 37$$

[In a more recent survey of this topic (Holland, 1978), these ratios have been reduced by about half.]

Upon cooling, the composition of these atmospheric models would, of course, be controlled by the equilibria prevailing at lower temperatures, and especially by the escape rate of H_2 from the Earth's atmosphere. Table I summarizes Holland's (1962) model for stages 1 and 2 after equilibrium at 25°C. Although methane is stable even at very low hydrogen pressures, the partial pressure of ammonia is more strongly dependent upon the availability of hydrogen. Table II summarizes some pertinent equilibrium constants. The problem becomes vastly more complicated, however, when the kinetics of the reactions are taken into consideration. Reduction of CO or CO_2 to yield CH_4, for example, although thermodynamically highly favored, are very slow processes at low temperature.

TABLE I

Summary of data on the probable chemical composition of the atmosphere (from Holland, 1962); for explanation see text.

	Stage 1	Stage 2
Major components $P > 10^{-2}$ atm	CH_4 H_2 (?)	N_2
Minor components $10^{-2} > P > 10^{-4}$ atm	H_2 (?) H_2O N_2 H_2S NH_3 Ar	H_2O CO_2 Ar
Trace components $10^{-4} > P > 10^{-6}$ atm	He	Ne He CH_4 NH_3 (?) SO_2 (?) H_2S (?)

TABLE II

Equilibrium constants at 25°C for selected atmospheric reactions

Reaction	K
$CO + 3H_2 \rightleftharpoons CH_4 + H_2O$	10^{25}
$CO_2 + 4H_2 \rightleftharpoons CH_4 + 2H_2O$	10^{20}
$N_2 + 3H_2 \rightleftharpoons 2NH_3$	$10^{5.8}$

At least two processes would have served to control the NH_3 concentration: photolysis by ultraviolet light to produce N_2 and solution of NH_3 in the oceans, with buffering of the NH_4^+ by clay minerals. Ferris and Nicodem (1972) have presented arguments that NH_3 would have been photolyzed to N_2 in geologically short periods of time (less than 10^6 years) and that buffering by adsorption of NH_4^+ would not have significantly extended the lifetime of NH_3. They conclude that the available form of nitrogen in the atmosphere has probably always been N_2, with no more than traces of NH_3. An atmosphere outgassed at 1200°C, corresponding to the precursor of Holland's stage 1, would probably be adequate to produce organic compounds whether or not low temperature equilibrium was attained. Abelson (1956) has shown that various mixtures of CO_2 or CO plus N_2 and H_2 will yield amino acids under the conditions of the Miller experiment, although higher levels of H_2 were used in his experiments than would be regarded as entirely realistic today. The stage 2 atmosphere, as already indicated, would be expected to produce formaldehyde and some simple carboxylic acids from the CO_2, although incorporation of nitrogen under these conditions is problematic. A possible mechanism is suggested by the report of Steinman and Lillevik (1964) of the synthesis of glycine by electric discharge through N_2 into a solution of acetic acid. Also of interest are the reported photochemical syntheses of amino acids from formaldehyde and nitrite (Bahadur, 1954), and in the presence of nitrogen and molybdenum oxide (Bahadur et al., 1958). These reactions require repetition and verification. It is unfortunate also that more quantitative data are not available on the production of organic compounds from atmospheric models intermediate between Holland's stages 1 and 2; that is, by inclusion of very low concentrations of H_2 or CH_4 in mixtures of CO_2 and N_2.

A serious consideration is whether iron could have remained in equilibrium with volcanic source rocks for a significant period of time. Holland (1962) suggests 500×10^6 years as an upper limit. However, more recent models for early core formation would reduce this period to perhaps 100×10^6 years (Fanale, 1971). There are also other objections to this model, among which is the hypothesis that core formation occurred prior to the Earth's acquisition of a volatile-rich "coat" (Turekian and Clark, 1969; Walker, 1976). If the formation of the first biologically interesting organic molecules must be restricted to such a geologically brief period, then it may be necessary to consider the composition of gases produced from much more primitive material than mantle rocks. The carbonaceous chondrites appear to be closer in composition to the source material from which the Earth was formed than any other known material (Anders, 1968; Mason, 1969; Wilkening, 1978). It is not inconceivable that pyrolysis products of the organic material known to be present in these meteorites formed an important part of the primitive atmosphere of the Earth during the first million years. Under such conditions, constraints from equilibrium studies are removed. Such an atmosphere may well have been rich in NH_3, CH_4 and even more reactive precursor molecules, including PH_3 (Schwartz, 1972). It must be pointed out that such an early start to chemical evolution, leading perhaps to a very early appearance of life on the Earth is, at least, not contradicted by the available fossil evidence (Schopf, 1972). However, leaving such speculation aside, a reasonable first approach is probably to choose an atmosphere intermediate between Holland's stages 1 and 2 as a working model, and explore its chemistry in some detail.

5. PREBIOTIC FORMATION OF ORGANIC MOLECULES

5.1. Initial gas-phase reactions

The most important forms of energy available for the activation of molecules in the primitive atmosphere probably were the various forms of electrical discharge and ultraviolet light (Miller and Urey, 1959; Miller et al., 1976). Electrical discharge is a relatively non-specific process, capable in principle of producing ions and radicals of all of the component gases. Ultraviolet irradiation can only cause dissociation in molecules by which it is absorbed, although secondary reactions may also be of importance. Since absorption by methane is only significant below 145 nm, while absorption by water tends to screen methane and extends to 200 nm, it has been suggested that the major photochemical step in mixtures of CH_4, N_2 and H_2O is the dissociation of H_2O, leading to the formation of aldehydes, ketones and alcohols (Ferris and Chen, 1975). However, in the upper atmosphere, where the vapor pressure of H_2O is extremely low, photodissociation of CH_4 would probably lead primarily to the production of hydrocarbons (Lasaga et al., 1971). Nitrogen does not participate in these photochemical reactions. The production of HCN from CH_4—N_2 mixtures, however, is favored by electrical discharge (Sanchez et al., 1966; Toupance et al., 1975). The pertinent transformations, therefore, can be schematically summarized as follows:

$$CH_4 + \tfrac{1}{2} N_2 \xrightarrow{\text{electrical discharge}} HCN + H_2 \qquad (1)$$

$$CH_4 + H_2O \xrightarrow{\text{discharge or uv}} CH_2O + 3 H_2 \qquad (2)$$

$$n \text{ CH}_4 \xrightarrow{\text{discharge or uv}} C_n H_{2n+2} + (n-1) \, H_2 \; ;$$

$$C_n H_{2n} + n \, H_2 \; ; \text{etc.} \tag{3}$$

It is clear that further reaction of the hydrocarbons produced in reaction (3) will produce a more complicated series of aldehydes and nitriles via reactions (1) and (2). In the presence of liquid water, a still more complicated reaction sequence becomes possible. Because of the limitations of space, I shall confine the discussion to the possible production of the components of proteins and nucleic acids. The principle pathways of interest here are the synthesis of carbohydrates from formaldehyde and the formation of amino acids, purines and pyrimidines by reactions involving aqueous solutions of HCN. [Although there have been suggestions that HCN can undergo direct polymerization in the gas phase (Matthews and Moser, 1966; Matthews et al., 1977), this reaction has never been directly observed.]

5.2. Aqueous reactions of formaldehyde — synthesis of carbohydrates

The Formose condensation, discovered by Butlerow (1861), remains in its various modifications the only candidate for the prebiotic synthesis of carbohydrates. The reaction is autocatalytic and involves the following series of reactions (Breslow, 1959):

$$2 \text{ CH}_2 O \rightarrow \text{CH}_2 OH-CHO \tag{1}$$

$$\text{CH}_2 O + \text{CH}_2 OH-CHO \rightarrow \text{CH}_2 OH-CHOII-CHO \tag{2}$$

$$\text{CH}_2 OH-CHOH-CHO \rightleftharpoons \text{CH}_2 OH-CO-CH_2 OH \tag{3}$$

$$\text{CH}_2 O + \text{CH}_2 OH-CO-CH_2 OH \rightarrow \text{CH}_2 OH-CHOH-CO-CH_2 OH \tag{4}$$

$$\text{CH}_2 OH-CHOH-CO-CH_2 OH \rightleftharpoons \text{CH}_2 OH-CHOH-CHOH-CHO \tag{5}$$

$$\text{CH}_2 OH-CHOH-CHOH-CHO \rightarrow 2 \text{ CH}_2 OH-CHO \tag{6}$$

Once the basic autocatalytic sequence is completed with the generation of two units of glycolaldehyde from a tetrose (6), the reaction accelerates since (1) is the rate-limiting step. The reaction is usually conducted with concentrated solutions of formaldehyde and a strong-base catalyst. However, the synthesis of ribose and other sugars has been reported to occur when a $0.01M$ solution of formaldehyde was refluxed in the presence of alumina or kaolinite (Gable and Ponnamperuma, 1967). A difficulty with this reaction is that unless arbitrarily stopped, the reaction continues to produce higher molecular-weight products of undefined structure, with destruction of the biologically interesting carbohydrates (Reid and Orgel, 1967). Synthesis of ribose and deoxyribose has been reported in solutions of 3×10^{-4} M formaldehyde, after irradiation with ultraviolet light or gamma rays (Ponnamperuma, 1965).

5.3. Aqueous reactions of HCN — amino acids and purines

Miller (1953) demonstrated that the most plausible mechanism for the production of amino acids in his experiment was a Strecker or cyanohydrin synthesis. In later work it has become clear that parallel reaction pathways are possible; an addition of HCN to an aldehyde to produce the hydroxy-nitrile:

$$
\overset{H}{RC}{=}O \; + \; HCN \; \rightleftharpoons \; R\overset{OH}{\underset{H}{C}}{-}CN
$$

$$
R\overset{OH}{\underset{H}{C}}{-}CN \; + \; NH_3 \; \rightleftharpoons \; R\overset{NH_2}{\underset{H}{C}}{-}CN \; + \; H_2O
$$

and addition of HCN to an imine to yield the aminonitrile:

$$
\overset{H}{RC}{=}O \; + \; NH_3 \; \rightleftharpoons \; \overset{H}{RC}{=}NH \; + \; H_2O
$$

$$
\overset{H}{RC}{=}NH \; + \; HCN \; \rightleftharpoons \; R\overset{NH_2}{\underset{H}{C}}{-}CN
$$

Since the imine will be in equilibrium with aldehyde and ammonia, and the aminonitrile will be in equilibrium with the hydroxynitrile and ammonia, it is clear that the production of amino acid will ultimately depend upon the available ammonia concentration. Because of the unlikelihood that the primitive atmosphere ever contained large concentrations of ammonia, Miller's original experiment has more recently been repeated using a mixture of CH_4 and N_2, and an aqueous phase consisting of a solution of $0.05M$ NH_4Cl titrated with NH_3 to a pH of 8.7 (Ring et al., 1972; Wolman et al., 1972). Under these conditions most of the protein amino acids as well as a large number of non-protein amino acids were synthesized. However, the question remains whether even these starting concentrations of NH_3 and NH_4^+ are reasonable (Bada and Miller, 1968). In addition, because of the closed system involved in the Miller apparatus, H_2 would have accumulated during the reaction.

An alternative route to certain of the amino acids as well as to the purines involves the oligomerization of HCN to produce the tetramer and further, as yet undefined oligomeric products:

$$
HCN \xrightarrow{HCN} NC{-}\overset{H}{C}{=}NH \xrightarrow{HCN} NC{-}\overset{H}{\underset{CN}{C}}{-}NH_2 \xrightarrow{HCN} \begin{matrix} NC \\ \; \\ NC \end{matrix}{>}{=}{<}\begin{matrix} NH_2 \\ \; \\ NH_2 \end{matrix} \xrightarrow{\;?\;} (HCN)x
$$

The aqueous solution chemistry of HCN leading to the production of biologically interesting products has been studied by Sanchez et al. (1967). In dilute solutions, where the concentration of HCN is less than about $0.01M$, the principle reaction is hydrolysis to formate and ammonia. Substantially higher concentrations, usually greater than $0.1M$, are necessary to permit a significant degree of oligomerization. In the presence of sufficient concentrations of ammonia, reactions leading to the synthesis of the purines either from the trimer or tetramer are known (Oro, 1961; Sanchez et al., 1967). However the necessity for ammonia can be by-passed through two alternative pathways. One involves a photochemical reaction of unknown mechanism which permits the formation of 4-aminoimidazole-5-carbonitrile from the HCN tetramer (Ferris et al., 1969):

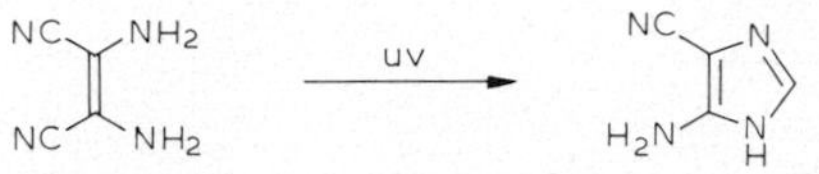

Subsequently the purines can be synthesized by further reactions with HCN, cyanogen or cyanate. Alternatively, hydrolysis of the oligomer under mild conditions yields adenine as well as a number of other products (Ferris et al., 1977 and 1978; see Table III). The purines guanine, hypoxanthine and xanthine have also been tentatively identified (Ferris et al., 1978; Schwartz and Goverde, unpublished). It is significant that no *additional* ammonia need be added for this reaction sequence (ammonia is, however, produced during the oligomerization and the hydrolysis).

5.4. Pyrimidines

None of the biologically important pyrimidines have been found to be formed in significant yield as products of the oligomerization of HCN. Small yields of orotic acid and of 5-hydroxyuracil (0.01% or less) have been reported, together with substantial yields of the unusual compound 4,5-dihydroxypyrimidine (Ferris et al., 1977). However, both hyproxypyrimidines are only isolated after strong acid hydrolysis of unknown precursors. The pyrimidines uracil, thymine and cytosine have been synthesized by reactions utilizing products of the gas-phase reactions of methane—nitrogen mixtures and of the oligomerization of HCN. The second most abundant product

TABLE III

Products identified after oligomerization of HCN and hydrolysis at pH 8.5 (from Ferris et al., 1977, 1978)

adenine	α-aminoisobutyric acid
orotic acid	guanidinoacetic acid
4-aminoimidazole-5-carboxamide	guanidine
glycine	diaminosuccinic acid
aspartic acid	urea
alanine	oxalic acid
β-alanine	

obtained in electrical discharge experiments with CH_4 plus N_2 is cyanoacetylene (Sanchez et al., 1966), which in aqueous solution hydrolyzes rapidly to cyanoacetyldehyde:

$$HC \equiv C-CN \ + \ H_2O \ \longrightarrow \ O = \overset{H}{C} - CH_2 - CN$$

Under quite dilute conditions, cyanoacetaldehyde cyclizes with guanidine, another product of HCN (see Table III), to produce 2,4-diaminopyrimidine (Ferris et al., 1974):

$$H_2N - \overset{\overset{\displaystyle NH}{\|}}{C} - NH_2 \ + \ O = \overset{H}{C} - CH_2 - C \equiv N \ \longrightarrow \ \text{[2,4-diaminopyrimidine]} \ + \ H_2O$$

Subsequently, 2,4-diaminopyrimidine hydrolyzes to produce cytosine (and isocytosine) and ultimately uracil:

$$\text{[2,4-diaminopyrimidine]} \ \xrightarrow[\text{NH}_3]{\text{H}_2\text{O}} \ \text{[cytosine]} \ \xrightarrow[\text{NH}_3]{\text{H}_2\text{O}} \ \text{[uracil]}$$

An alternate route to uracil which leads to the synthesis of thymine as well is formation and photolytic dehydrogenation of the dihydropyrimidines. Here we are dealing with one of many examples of prebiotic syntheses which require anhydrizing conditions. Urea and β-alanine are products of HCN (Table III). If a solution containing these compounds is evaporated to dryness and heated moderately, dihydrouracil is formed by the following reaction:

$$H_2N - \overset{\overset{\displaystyle O}{\|}}{C} - NH_2 \ + \ H_2N - CH_2 - CH_2 - COOH \ \longrightarrow \ \text{[dihydrouracil]}$$

It has been found that dihydropyrimidines are converted to the corresponding pyrimidines by irradiation with ultraviolet light in the presence of water vapor. The reaction is catalyzed by clay minerals, particularly montmorillonite, and an "evaporating pond" environment has been suggested as a reasonable prebiotic locale for such a reaction (Chittenden and Schwartz, 1976). If acetate salts are added to the system, thymine is formed in addition to uracil, presumably by addition of a methyl radical (Schwartz and Chittenden, 1977). An alternative route to thymine involves the addition of formaldehyde to uracil and subsequent reduction by formic acid (Choughuley et al., 1977). (Formic and acetic acids are major products of the Miller experiment and are probably also abundantly formed from CH_4—N_2 mixtures.)

The Fischer—Tropsch type of reaction has been claimed to produce all of the biologically important purines and pyrimidines (Hayatsu et al., 1968, 1972). This reaction is carried out by heating mixtures of CO and NH_3 to 600—700°C and quenching the products in water. The mechanism is unclear. For the reasons already discussed, however, this seems an unlikely set of circumstances on the primitive Earth, although it could conceivably have played a role during condensation of the solar nebula (Anders et al., 1974).

5.5. Nucleosides and nucleotides

Purine nucleosides are formed when a dilute solution of a purine, together with ribose or deoxyribose and $MgCl_2$ is evaporated to dryness and heated at 100°C (Fuller et al., 1972). Yields of several percent of the α- and β-isomers of adenosine, guanosine and xanthosine, and about 10% of α- and β-inosine are obtained. Interestingly, a mixture of the salts present in sea water produces even higher yields (15% of inosines). The simplicity of this reaction makes it attractive as a potential source of nucleosides. Unfortunately, it has not been possible to synthesize the pyrimidine nucleosides in this manner and there is as yet no satisfactory synthesis of these compounds.

Phosphorylation of nucleosides can be accomplished under similar, evaporating pond conditions. Mineral phosphate (fluoro- or hydroxyapatite) is solubilized and activated by the products of the oligomerization of HCN (Schwartz, 1974). The mechanism of this procedure probably involves two steps. The first involves complexing of calcium by oxalic acid to produce a precipitate of ammonium phosphate and calcium oxalate (ammonium oxalate and urea are major products of the oligomerization of HCN solutions — see Table III). The second stage of the reaction is the phosphorylation of the nucleoside by the mixture of ammonium phosphate, urea and possibly other products which serve to catalyze the phosphorylation reaction (Lohrmann and Orgel, 1971; Schwartz et al., 1973). Phosphorylations of nucleosides in the presence of urea have been studied extensively. It has also been claimed that ammonium chloride will solubilize apatite as well as ammonium oxalate (Lohrmann and Orgel, 1971; Orgel and Lohrmann, 1974). Among the products are inorganic polyphosphates, nucleotides, nucleoside polyphosphates, cyclic-2', 3'-phosphates and oligonucleotides. The oligonucleotides are invariably short, and consist of a mixture of linkage types, rather than solely the 3'—5' phosphodiester linkages of DNA and RNA. However, they could conceivably serve as the starting material for the template-directed synthesis of polynucleotides.

6. CHEMICAL EVOLUTION TO HIGHER ORGANIZATION LEVELS

6.1. Polynucleotides

Research on template-directed syntheses of polynucleotides has been reviewed by Orgel and Lohrman (1974). Briefly the technique involves using preformed polynucleotides as templates, by means of base-pairing, for the linking of suitably activated mono- or oligonucleotides in aqueous solution. For example, polyuridylic acid (poly U) will bind adenosine and adenosine cyclic-2′, 3′-phosphate and permit the formation of di- and trinucleotides. The reaction is catalyzed by di- and polyamines. More promising is the oligomerization of the preactivated nucleotide, adenosine-5′-phosphorimida-zolide on a poly U template to yield penta- and hexanucleotides. This type of process might have provided a mechanism for chain-lengthening and repli-cation of oligonucleotides in solution. If a supply of activated nucleotides and oligonucleotides were to have been formed under dehydrating condi-tions in evaporating ponds or tidal pools, then further evolution of such oligomers might have been possible by means of template-directed reactions when the initial products were redissolved. Provided, of course, that some of the initially formed products were of sufficient length to serve subsequently as templates. However, there are several difficulties with this model. In this and other template-directed syntheses, the primary mode of linkage in the product is 2′—5′ rather than 3′—5′. Furthermore, while stable complexes are formed between the polypyrimidines and derivatives of the purine nucleo-sides, no complexes are formed between the polypurines and monomeric pyrimidine nucleoside derivatives; presumably because of less stabilization by base-stacking effects between pyrimidines.

An interesting proposal which provides a possible solution for the first problem has been offered by Usher (1977). When incorporated into a helix, 2′—5′ phosphodiester linkages have a much greater rate of hydrolysis than 3′—5′ linkages. The rate is also greater than that of either bond type in a nonhelical configuration. Diurnal temperature cycling, such as has been postulated for the synthesis of nucleotides could therefore result in the synthesis of oligonucleotides by evaporation and heating during the day, while at night rehydration would take place, leading to partial hydrolysis. If the night-time temperature was low enough to favor helix formation, a preferential hydrolysis of the 2′—5′ linkages would result. Preferential hydro-lysis would, in any case, favor the 3′—5′ linkage above the 2′—5′, whether or not a helix was formed. In addition, recent results of Dhingra and Sarma (1978) indicate that the incorporation of 2′—5′ linkages into a helical poly-nucleotide tends to destabilize the helix, so that even in the absence of pre-ferential hydrolysis, further evolution of "non-natural" polynucleotides would not be possible.

The second problem mentioned above indirectly raises some interesting questions. As already mentioned, the purines are formed very readily by spontaneous reactions of HCN in aqueous solution. Under the same condi-tions no biologically significant pyrimidines are formed except for small amounts of orotic acid. The available syntheses of uracil and thymine involve more steps than the purines, since the reactants (β-alanine and urea) must first be formed, then undergo subsequent reaction on a clay surface. Cytosine presents a special problem. Although it can be formed directly in solution from cyanoacetaldehyde and guanidine, guanidine must also be produced by oligomerization and hydrolysis of HCN solutions. Furthermore, cytosine hydrolyzes sufficiently rapidly to uracil that there is doubt about its availability in primitive bodies of water (Miller and Orgel, 1974). One might speculate, therefore, that at a very early stage in chemical evolution purines might have been available but not pyrimidines. As already pointed out, the purines readily undergo dehydration—condensation reactions with ribose or deoxyribose to form nucleosides, while the pyrimidines do not. It has been suggested that a primitive, two-base genetic code might have been based on the purines adenine and hypoxanthine (Crick, 1968). As pyrimidine nucleosides became available, they could have been incorporated into polynucleotides by base-pair mutations, an adenine—uracil pair being formed rather than an adenine—hypoxanthine pair, for example. Guanine and cytosine could have been added considerably later in evolution, providing additional stabilization of the helix. (It is interesting that guanine and xanthine, although formed together with adenine and hypoxanthine from HCN, have extremely low solubilities in water.)

6.2. Polypeptides

Polymerization of amino acids in aqueous solution has, in general, only been partially successful. An exception is the remarkable reaction of aminoacyladenylates in the presence of a suspension of montmorillonite to produce chains of up to fifty amino acid residues in length (Paecht-Horowitz, 1974). By contrast, in the absence of the clay mineral, the aminoacyladenylates hydrolyze rapidly and produce, at most, short chains of four or five amino acids. The mechanism of this reaction is unclear and the possible origin of these highly unstable compounds on the primitive Earth has not been demonstrated directly. Possibly such activated amino acids could be formed under evaporating conditions from the free amino acids in the presence of adenosine cyclic-2', 3'-phosphate (Lohrmann and Orgel, 1973).

Direct polymerization of the unactivated amino acids under anhydrizing conditions seems at present to provide a more convincing model. Recently, the oligomerization of glycine has been demonstrated to occur by carrying out a series of evaporation—rehydration cycles in the presence of kalonite (Lahav et al., 1978). The maximum temperature used was 94°C and small yields of oligomers up to the pentamer were obtained. The interest of this reaction is that it demonstrates the feasibility of utilizing such natural cycles for the activation of amino acids on clay surfaces, although the degree of polymerization obtained is minimal (temperatures in this range are observable in desert areas today).

The direct polymerization of mixtures of all of the natural amino acids by heating has long been advocated as a model of the formation of primitive protein by Fox (see discussion in Fox and Dose, 1977). This technique has been criticized because of the use by Fox of temperatures between 160 and 200°C (Miller and Orgel, 1974). However, it has recently been demonstrated that the polymerization will also take place at lower temperatures, notably between 65 and 85°C (Rohlfing, 1976). Therefore, the thermal polymerization model continues to be of great interest. A question concerning the significance of these "proteinoids" is whether the mode of linkage is of the natural α-peptide type or a mixture of more complicated linkages. Until recently there have been few data available on the mode of linkage of these very complex products. However, some recent studies seem to indicate that while certain of the amino acids tend to react to form α-peptides, many unnatural linkages are undoubtedly present (Kokufuta et al., 1977, 1978). More structural studies on these interesting polymers are clearly desirable.

6.3. Protocells

One of Oparin's suggestions was that a key step in the emergence of primitive life on the Earth must have been the formation of a protocell (Oparin, 1938). It is extremely difficult to imagine chains of chemical reactions "evolving" to form self-perpetuating entities in a homogeneous environment, although theoretical attempts have been made to model such processes (Allen, 1972; Dekker and Speidel, 1972). Quite apart from the problem of dilution, is the unavoidable occurrence of undesirable reactions between components of different chemical systems (for example, the well-known "browning reaction" between amino acids and carbohydrates). As a possible solution to this problem Oparin suggested that the process of coacervation, discovered by the colloid chemist Bungenberg de Jong (1949) might provide an origin for such protocells. An interesting feature of the thermally synthesized amino-acid polymers described by Fox, is their propensity to spontaneously form spherical, micrometer-sized particles upon cooling of an aqueous solution of the material. A number of interesting properties of these "microspheres" have been described, and they have been proposed by Fox to be logical candidates for the role of a protocell (for a discussion of this and other protocellular models see Fox and Dose, 1977).

7. EVIDENCE FOR CHEMICAL EVOLUTION

The recent discoveries of complex organic molecules associated with interstellar dust clouds have, of course, attracted great interest among students of chemical evolution. Inspection of Table IV will reveal several compounds which have been referred to with regard to the syntheses of biologically significant compounds: HCN, formaldehyde and cyanoacetylene. It is perhaps unlikely that the processes which have produced these com-

TABLE IV

Molecules in interstellar dust clouds (as of August, 1978)

Number of atoms	Molecule		Number of atoms	Molecule	
2	H_2	hydrogen	5	HC_3N	cyanoacetylene
	OH	hydroxyl radical		HCOOH	formic acid
	SiO	silicon monoxide		CH_2NH	methanimine
	SO	sulphur monoxide		H_2CCO	ketene
	SiS	silicon monosulphide		NH_2CN	cyanamide
	NS	nitrogen sulphide			
	CH^+	methylidyne ion	6	CH_3OH	methyl alcohol
	CH	methylidyne		CH_3CN	methyl cyanide
	CN	cyanogen radical		NH_2CHO	formamide
	CO	carbon monoxide			
	CS	carbon monosulphide	7	CH_3C_2H	methylacetylene
				CH_3CHO	acetaldehyde
3	H_2O	water		NH_2CH_3	methylamine
	H_2S	hydrogen sulphide		CH_2CHCN	vinyl cyanide
	SO_2	sulphur dioxide		HC_5N	cyanodiacetylene
	HCN	hydrogen cyanide			
	HNC	hydrogen isocyanide	8	$HCOOCH_3$	methyl formate
	OCS	carbonyl sulphide			
	HCO^+	formyl ion	9	$(CH_3)_2O$	dimethyl ether
	HCO	formyl radical		CH_3CH_2OH	ethyl alcohol
	CCH	ethynyl radical		CH_3CH_2CN	ethyl cyanide
	N_2H^+	N_2H ion		HC_7N	cyanotriacetylene
4	NH_3	ammonia	11	HC_9N	cyanotetraacetylene
	H_2CO	formaldehyde			
	HNCO	isocyanic acid			
	H_2CS	thioformaldehyde			
	C_3N	cyanoethynyl radical			

pounds are directly related to those which seem reasonable for the primitive Earth. What is significant is the demonstrable fact that these and other, more complicated organic molecules are spontaneously produced in nature, and on a vast scale. This is a verification in principle of the concept of chemical evolution, and a demonstration of its universality.

The organic material in carbonaceous chondrites is of more direct interest with regard to the problem of the origin of life on the Earth, not only because of the possible participation of similar material in the formation of the Earth, but because of the possibility that the mechanisms of formation of terrestrial and meteoritic organic compounds may be related. It has been suggested by Anders et al. (1974) that many of the organic compounds in meteorites may have been formed by Fischer—Tropsch-type reactions. There is reasonable agreement, for example, between certain hydrocarbon isomers found in carbonaceous chondrites and those formed in synthetic reactions (Studier et al., 1972). On the other hand, agreement in composition has been shown between the amino acids extracted from the Murchison meteorite and those formed in electric discharge experiments. Data from several sources are summarized in Table V. Noteworthy is the absence of a number of amino acids in the meteorite and in the electric discharge experiments, although these are known to be formed in Fischer—Tropsch-type reactions. A difficulty in attempting to make such comparisons, however, is that not all of the amino acids known to be present in either the meteorite or in the synthetic experiments have yet been identified.

Fischer—Tropsch-type reactions have also been reported to synthesize

TABLE V

Partial list of amino acids in the Murchison meteorite and formed in electric discharge and Fischer—Tropsch type experiments

Amino acid	Murchison [1]	Electric discharge [2]	FTT [3]
Glycine	+	+	+
Alanine	+	+	+
Valine	+	+	+
Leucine	—	—	(+)
Isoleucine	—	—	(+)
Aspartic acid	+	+	+
Glutamic acid	+	+	+
Tyrosine	—	—	+
Proline	+	+	(+)
Histidine	—	—	+
Lysine	—	—	+
Ornithine	—	—	+
Arginine	—	—	+
N-methyl glycine	+	+	+
N-Ethyl glycine	+	+	NS
β-Alanine	+	+	+
N-Methyl alanine	+	+	NS
Isovaline	+	+	NS
Norvaline	+	+	NS
α-Aminoisobutyric acid	+	+	(+)
α-Amino-n-butyric acid	+	+	(+)
β-Aminoisobutyric acid	+	+	(+)
β-Amino-n-butyric acid	+	+	NS
γ-Aminobutyric acid	+	+	(+)
Pipecolic acid	+	+	NS

[1] Kvenvolden (1974).
[2] Miller et al. (1976).
[3] Anders et al. (1974). (+) = tentative identification; NS = not sought.

both purines and pyrimidines (Hayatsu et al., 1972). Although purines have been identified in meteorites, there has been some controversy concerning the possible presence of pyrimidines (Hayatsu, 1964; Hayatsu et al., 1968, 1975; Van der Velden and Schwartz, 1977). Earlier reports of unusual pyrimidine derivatives are now thought to have been artefactual (Folsome et al., 1971, 1973; Van der Velden and Schwartz, 1977). Very recently, however, uracil has been positively identified in water and acid extracts of the Orgueil, Murray and Murchison meteorites (Stoks and Schwartz, 1979).

It is noteworthy that a portion of the amino acids, and many of the purines are only identifiable after acid extraction or hydrolysis. It is therefore unlikely that they exist in the free form. A polymeric precursor has been suggested for the purines (Hayatsu et al., 1975). It is interesting that HCN oligomerization leads to the formation of all of the purines which have been identified in carbonaceous chondrites. Although HCN may have played a role in the formation of some of the organic compounds in meteorites, it is unlikely that any one mechanism can account for all of the products of these remarkably complex and interesting samples of primitive material.

8. ENVIRONMENTS FOR CHEMICAL EVOLUTION ON THE PRIMITIVE EARTH

The remarkable ease with which biologically important compounds can be formed spontaneously from precursors such as formaldehyde and HCN suggests strongly that such reactions did indeed take place on the primitive Earth. Yet there are many difficulties in conceptualizing such processes. Production of formaldehyde directly from bicarbonate in the oceans would undoubtedly have been an important process. However, as already mentioned, there is at present little evidence that nitrogen could have been incorporated solely from CO_2 and N_2. There is little doubt, however, that electrical discharge could have provided an enormous potential for *inorganic* nitrogen fixation on the primitive Earth. Although nitrate and nitrite are known to be the major products formed by electrical discharge through N_2 into water, ammonium ion is also produced (Yokohata and Tsuda, 1967). Furthermore, recent experiments on photolytic nitrogen reduction by natural minerals seem to provide an additional pathway to ammonia production on the primitive Earth (Schrauzer, 1978). However, HCN production from CH_4 and N_2 seems to be the most likely source of organic N.

It has been pointed out that the total amounts of carbon and nitrogen now present on the surface of the Earth, if dissolved in the present oceans, would produce a rather dilute solution (see for example Miller and Orgel, 1974). If all of the carbon were to be converted to formaldehyde, or all of the nitrogen to HCN, the maximum concentrations attainable would be about $1M$ and $0.2M$, respectively. Of course, it is extremely improbable that concentrations approaching these could ever have existed. It is clear that when the likely composition of the primitive atmosphere is taken into consideration, together with the expected rates of production of typical products and of the competing degradative reactions, these maximum concentrations would have to be lowered by several orders of magnitude at least.

Concentrations of HCN in the range of 0.01 to 0.1M appear to be necessary to produce products of biological interest. How could such concentrations have been attained? Adsorption of organic molecules on clay minerals, as suggested by Bernal (1951) may have functioned to concentrate certain secondary reactants (see for example Paecht-Horowitz, 1974; Schwartz and Chittenden, 1977). However no such adsorption has been observed with HCN. Solutions of HCN can be concentrated by evaporation under strongly basic conditions. However, since the oligomerization proceeds most rapidly when the pH is equal to the pK (9.2), the rate of the reaction would decrease to compensate for any increased accumulation of cyanide ion. For this mechanism to be useful, a pH change would have to occur subsequent to concentration; a somewhat unlikely sequence of events. Sanchez et al. (1967) have called attention to the tendency of HCN to form highly concentrated eutectic phases when dilute solutions are frozen. An ideal temperature for such reactions is about $-20°C$. The synthesis of amino acids and purines can thus be pictured as occurring during winter freezing of lakes and ponds, or in glaciers. This is an extremely attractive hypothesis and it is necessary to ask whether conditions permitting such climatic variation also existed on the primitive Earth.

It has recently become fairly well accepted that the Sun has been gradually increasing its luminosity during the entire course of the Earth's history. A consequence of this increase, in the absence of any other effect, would be that the temperature of the primitive Earth must have been substantially lower than at present. Such a conclusion, implying as it does the total absence of liquid water, is obviously unacceptable in view of the available geological evidence. Sagan and Mullen (1972) have argued that the presence of ammonia, in a mixing ratio of less than one part in 10^5, would have been sufficient, by virtue of the enhanced greenhouse effect, to keep the average surface temperature above the freezing point of water. Although substantial quantities of ammonia are very unlikely to have been present in the early atmosphere, a mixing ratio of less than 10^{-5} seems not unreasonable. Higher concentrations of ammonia, or the presence of other gases with absorptivities similar to ammonia in the infrared, would necessarily have produced elevated temperatures. Recently, Knauth and Epstein (1976), on the basis of hydrogen and oxygen isotope compositions for a small group of Precambrian cherts, have suggested the occurrence of temperatures as high as $52°C$ and $70°C$, 1.3 and 3×10^9 years ago, respectively. Such temperatures cannot, however, be considered to be representative of mean conditions, in view of the known occurrence of extensive Precambrian glaciations, a point acknowledged by these authors. At least until many more data are available which indicate the contrary, it would still seem that the most reasonable model of climatic conditions on the primitive Earth is one very much like the present. [Henderson-Sellers (1978) has recently argued that CO_2 and H_2O alone are adequate to keep the surface temperature of the primitive Earth above the freezing point of water.]

An atmosphere containing a preponderance of N_2 and a limited amount of CH_4 is a favorable mixture for the production of HCN by electric discharges (Sanchez et al., 1966; Toupance et al., 1975). In contrast, short-wave ultraviolet light produces primarily aldehydes from mixtures of methane, nitrogen and water vapor (Ferris and Chen, 1975). Taking these reports into consideration, as well as the direct photoreduction of CO_2, a picture emerges of a constant production of formaldehyde and other aldehydes over the entire surface of the Earth. Superimposed upon these processes would be a variable production of HCN, at high efficiency, in regions where electrical activity was common. There is evidence that electrical storms are much more common over land surfaces than over the oceans (Schonland, 1953). Furthermore, the largest part of the available electrical energy has been estimated to occur as corona discharges from pointed objects (Miller and Urey, 1959). HCN would then have been preferentially synthesized over precisely those areas of the Earth's surface where concentration mechanisms could operate, either in eutectics or by accumulation in alkaline pools. It has been pointed out by Schlesinger and Miller (1973) that formation of glyconitrile from HCN and formaldehyde in dilute solution is a highly favored process. This presents a problem, since the two compounds could not accumulate in the same environment. Schlesinger and Miller prefer the hypothesis that HCN oligomerization and formaldehyde oligomerization occurred at different times, to the possibility that they occurred in different places. However, the above-mentioned considerations with regard to production mechanisms suggest that production in different localities would have been probable. The most attractive scenario for further chemical evolution of HCN would be transportation of the products of oligomerization to locales where evaporation and heating were possible. A landscape dominated by glaciers and volcanism, not too dissimilar from locations on the contemporary Earth, would seem to be indicated. Mixing of the products of HCN oligomerization, which would be enriched in ammonium salts, with aldehydes or hydroxynitriles might also provide sufficient local concentrations of ammonia for the cyanohydrin pathway to operate, although this remains to be demonstrated.

Free formaldehyde could only accumulate over areas where relatively little HCN was being produced. This suggests a seasonal variation, or perhaps even an oceanic environment. Although the concentration problem for formaldehyde is less severe than for HCN, there would still be a formidable problem. Stratification of the primitive ocean would have reduced the effects of dilution by making only the upper 100—200 m readily accessible to the atmosphere (compare Weyl, 1968). A global ocean as suggested by Hargraves (1976) as a consequence of the early formation of a thin sialic crust, would be extensively stratified (LaBarbera, 1978; Chamberlain and Marland, 1977). Although this possibility hardly seems at first glance to be compatible with the terrestrial locales suggested above, these locales could have been pre-continental, volcanic island systems. The most ancient sediments do indeed all seem to have been produced in volcanic environments.

The concluding section of this chapter has necessarily been highly speculative, seeking as it does to paint a coherent picture from a very fragmentary set of data. During recent years, an increasing amount of attention has begun to be paid to the problem areas in chemical evolution. Hopefully, as more data accumulate in the near future, new and profitable areas of research will be indicated, and more coherence will emerge in our picture of the origins of living systems.

9. GENERAL BIBLIOGRAPHY

Fox, S.W. and Dose, K., 1977. Molecular Evolution and the Origin of Life. Dekker, New York, N.Y.

Kenyon, D.H. and Steinman, G., 1969. Biochemical Predestination. McGraw-Hill, New York, N.Y., 301 pp.

Miller, S.L. and Orgel, L.E., 1974. The Origins of Life on the Earth. Prentice-Hall, Englewood Cliffs, N.J., 229 pp.

Rutten, M.G., 1971. The Origin of Life by Natural Causes. Elsevier, Amsterdam, 420 pp.

10. REFERENCES

Abelson, P.H., 1956. Paleobiochemistry: Inorganic synthesis of amino acids. Carnegie Inst. Wash. Yearb., 1955—1956, No. 55: 171--180.

Allen, G., 1972. Chemical evolution under the bion hypothesis. In: A.W. Schwartz (Editor), Theory and Experiment in Exobiology, 2, Wolters-Noordhoff, Groningen, pp. 1--32.

Anders, E., 1968. Chemical processes in the early solar system as inferred from meteorites. Acc. Chem. Res., 1: 289—298.

Anders, E., Hayatsu, R. and Studier, M.H., 1974. Catalytic reactions in the solar nebula: implications for interstellar molecules and organic compounds in meteorites. Origins Life, 5: 57—67.

Bada, J.L. and Miller, S.L., 1968. Ammonium ion concentration in the primitive ocean. Science, 159: 423—425.

Bahadur, K., 1954. Photosynthesis of amino acids from paraformaldehyde and potassium nitrite. Nature, 173: 1141.

Bahadur, K., Ranganayaki, S. and Santamaria, L., 1958. Photosynthesis of amino acids from paraformaldehyde involving the fixation of nitrogen in the presence of colloidal molybdenum oxide as catalyst. Nature, 182: 1668.

Bernal, J.D., 1951. The Physical Basis of Life. Routledge and Kegan Paul, London, 80 pp.

Breslow, R., 1959. On the mechanism of the Formose reaction. Tetrahedron Lett., 21: 22—26.

Bungenberg de Jong, H.G., 1949. Morphology of coacervates. In: H.R. Kruyt (Editor), Colloid Science, II. Elsevier, Amsterdam, pp. 433—482.

Butlerow, A., 1861. Annalen, 120: 296.

Chamberlain, W.M. and Marland, G., 1977. Precambrian evolution in a stratified global sea. Nature, 265: 135—136.

Chittenden, G.J.F. and Schwartz, A.W., 1976. Possible pathway for prebiotic uracil synthesis by photodehydrogenation. Nature, 263: 350—351.

Choughuley, A.S.U., Subbaraman, A.S., Kazi, Z.A. and Chada, M.S., 1977. A possible synthesis of thymine: uracil—formaldehyde—formic acid reaction. BioSystems, 9: 73—80.

Crick, F.H.C., 1968. The origin of the genetic code. J. Mol. Biol., 38: 367—379.

Decker, P. and Speidel, A., 1972. Open systems which can mutate between several steady states ("Bioids") and a possible prebiological role of the autocatalytic condensation of formaldehyde. Z. Naturforsch., B27: 257—263.

Dhingra, M.M. and Sarma, R.H., 1978. Why do nucleic acids have $3'5'$phosphodiester bonds? Nature, 272: 798—801.

Fanale, F.P., 1971. A case for catastrophic early degassing of the Earth. Chem. Geol., 8: 79—105.

Ferris, J.P. and Chen, C.T., 1975. Photochemistry of methane, nitrogen and water mixtures as a model for the atmosphere of the primitive Earth. J. Am. Chem. Soc., 97: 2962—2967.

Ferris, J.P. and Nicodem, D.E., 1972. Ammonia photolysis and the role of ammonia in chemical evolution. Nature, 238: 268—269.

Ferris, J.P., Kuder, J.E. and Catalano, A.W., 1969. Photochemical reactions and the chemical evolution of purines and nicotinamide derivatives. Science, 166: 765—766.

Ferris, J.P., Zamek, O.S., Altbuch, A.M. and Freiman, H., 1974. Synthesis of pyrimidines from guanidine and cyanoacetaldehyde. J. Mol. Evol., 3: 301—309.

Ferris, J.P., Joshi, P.C. and Lawless, J.G., 1977. Pyrimidines from hydrogen cyanide. BioSystems, 9: 81—86.

Ferris, J.P., Joshi, P.C., Edelson, E.H. and Lawless, J.G., 1978. HCN: a plausible source of purines, pyrimidines and amino acids on the primitive Earth. J. Mol. Evol., 11: 293.

Folsome, C.E., Lawless, J.G., Romiez, M. and Ponnamperuma, C., 1971. Heterocyclic compounds indigenous to the Murchison meteorite. Nature, 232: 108—109.

Folsome, C.E., Lawless, J.G., Romiez, M. and Ponnamperuma, C., 1973. Heterocyclic compounds recovered from carbonaceous chondrites. Geochim. Cosmochim. Acta, 37: 455—465.

Fuller, W.D., Sanchez, R.A. and Orgel, L.E., 1972. Solid-state synthesis of purine nucleosides. J. Mol. Evol., 1: 249—257.

Gable, N.W. and Ponnamperuma, C., 1967. Model for origin of monosaccharides. Nature, 216: 453—455.

Garrison, W.M., Morrison, D.C., Hamilton, J.G., Benson, A. and Calvin, M., 1951. Reduction of carbon dioxide in aqueous solutions by ionizing radiation. Science, 114: 416—478.

Getoff, N.G., Scholes, G. and Weiss, J., 1960. Reduction of carbon dioxide in aqueous solutions under the influence of radiation. Tetrahedron Lett., 18: 17—23.

Groth, W. and Suess, H., 1938. Photochemie der Atmosphäre der Erde. Naturwissenschaften, 26: 77.

Haldane, J.B.S., 1929. The Origin of Life. Rational. Annu.

Hargraves, R.B., 1976. Precambrian geologic history. Science, 193: 363—371.

Hasselstroom, T. and Henry, M.C., 1956. New synthesis of oxalic acid. Science, 123: 1038—1039.

Hayatsu, R., 1964. Orgueil meteorite: organic nitrogen contents. Science, 146: 1291—1293.

Hayatsu, R., Studier, M.H., Oda, A., Fuse, K. and Anders, E., 1968. Origin of organic matter in early solar system — II. Nitrogen compounds. Geochim. Cosmochim. Acta, 32: 175—190.

Hayatsu, R., Studier, M.H., Matsuoka, S. and Anders, E., 1972. Origin of organic matter in the early solar system — VI. Catalytic synthesis of nitriles, nitrogen bases and porphyrin-like pigments. Geochim. Cosmochim. Acta, 36: 555—571.

Hayatsu, R., Studier, M.H., Moore, L.P. and Anders, E., 1975. Purines and triazines in the Murchison meteorite. Geochim. Cosmochim. Acta, 39: 471—488.

Henderson-Sellers, A., 1978. The evolution of the Earth's atmosphere. New Scientist, 26 October, pp. 287—289.

Holland, H.D., 1962. Model for the evolution of the Earth's atmosphere. In: A.E.J. Engel, H.L. James and B.F. Leonard (Editors), Petrologic Studies — A volume to Honor A.F. Buddington. Geological Society of America, Boulder, Colo., pp. 447—477.

Holland, H.D., 1972. The geologic history of sea water — an attempt to solve the problem. Geochim. Cosmochim. Acta, 36: 637—651.

Holland, H.D., 1978. The Chemistry of the Atmosphere and Oceans. Wiley—Interscience, New York, N.Y., 351 pp.

Knauth, L.P. and Epstein, S., 1976. Hydrogen and oxygen isotope ratios in nodular and bedded cherts. Geochim. Cosmochim. Acta, 40: 1095—1108.

Kokufuta, E., Suzuki, S. and Harada, K., 1977. Potentiometric titration behavior of poly-aspartic acid prepared by thermal polycondensation. BioSystems, 9: 211—214.

Kokufuta, E., Terada, T., Suzuki, S. and Harada, K., 1978. Potentiometric titration behavior of a copolymer of glutamic acid and alanine prepared by thermal polycondensation. BioSystems, 10: 299—306.

Kvenvolden, K.A., 1974. Natural evidence for chemical and early biological evolution. Origins Life, 5: 71—86.

LaBarbera, M., 1978. Precambrian geological history and the origin of the Metazoa. Nature, 273: 22—25.

Lahav, N., White, D. and Chang, S., 1978. Peptide formation in the prebiotic era: thermal condensation of glycine in fluctuating clay environments. Science, 201: 67—70.

Lasaga, A.C., Holland, H.D. and Dwyer, M.J., 1971. Primordial oilslick. Science, 174: 53—55.

Lohrmann, R. and Orgel, L.E., 1971. Urea—inorganic phosphate mixtures as prebiotic phosphorylating agents. Science, 171: 490—494.

Lohrmann, R. and Orgel, L.E., 1973. Prebiotic activation processes. Nature, 244: 418—420.

Mason, B., 1966. Principles of Geochemistry. Wiley, New York, N.Y., 329 pp.

Mason, B., 1969. Composition of stony meteorites. In: C.A. Randall, Jr. (Editor), Extra-terrestrial Matter. Northern Illinois University Press, DeKalb, Ill., pp. 3—22.

Matthews, C.N. and Moser, R.E., 1966. Prebiological protein synthesis. Proc. Acad. Sci. U.S., 56: 1087—1094.

Matthews, C.N., Nelson, J., Varma, P. and Minard, R., 1977. Deuterolysis of amino acid precursors: evidence for hydrogen cyanide polymers as protein ancestors. Science, 198: 622—625.

Miller, S.L., 1953. A production of amino acids under possible primitive Earth conditions. Science, 117: 528—529.

Miller, S.L. and Urey, H.C., 1959. Organic compound synthesis on the primitive Earth. Science, 130: 245—251.

Miller, S.L., Urey, H.C. and Oro, J., 1976. Origin of organic compounds on the primitive Earth and in meteorites. J. Mol. Evol., 9: 59—72.

Moorbath, S., O'Nions, R.K. and Pankhurst, R.J., 1973. Early Archaean age for the Isua iron formation, West Greenland, Nature, 245: 138—139.

Moore, B., 1913. The Origin and Nature of Life. (The Home University Library of Modern Knowledge. No. 63). Henry Holt and Co., New York, N.Y.

Moore, B. and Webster, T.A., 1913. Synthesis by sunlight in relationship to the origin of life. Synthesis of formaldehyde from carbon dioxide and water by inorganic colloids acting as transformers of light energy. Proc. R. Soc. (Lond.), B87: 163—176.

Oparin, A.I., 1938. The Origin of Life. (Translated by S. Morgulis). Macmillan, New York, N.Y., 270 pp.

Orgel, L.E. and Lohrmann, R., 1974. Prebiotic chemistry and nucleic acid replication. Acc. Chem. Res., 7: 368—377.

Oro, J., 1961. The mechanism of synthesis of adenine from hydrogen cyanide under possible primitive Earth conditions. Nature, 191: 1193 -1194.

Paecht-Horowitz, M., 1974. The possible role of clays in prebiotic peptide synthesis. Origins Life 5: 173—187.

Ponnamperuma, C., 1965. Abiological synthesis of some nucleic acid constituents. In: S.W. Fox (Editor), The Origins of Prebiological Systems and of their Molecular Matrices. Academic Press, New York, N.Y., pp. 221—241.

Reid, C. and Orgel, L.E., 1967. Synthesis of sugars in potentially prebiotic conditions. Nature, 216: 455.

Ring, D., Wolman, Y., Friedmann, N. and Miller, S.L., 1972. Prebiotic synthesis of hydrophobic and protein amino acids. Proc. Acad. Sci. U.S., 69: 765—768.

Rohlfing, D.L., 1976. Thermal polyamino acids: synthesis at less than 100°C. Science, 193: 68 -70.

Rubey, W.W., 1951. Geologic history of sea water — an attempt to state the problem. Bull. Geol. Soc. Am., 62: 1111—1148.

Rutten, M.G., 1971. The Origin of Life by Natural Causes. Elsevier, Amsterdam, 420 pp.

Sagan, C. and Mullen, G., 1972. Earth and Mars: evolution of atmosphere and surface temperatures. Science, 177: 52—56.

Sanchez, R.A., Ferris, J.P. and Orgel, L.E., 1966. Cyanoacetylene in prebiotic synthesis. Science, 154: 784—785.

Sanchez, R.A., Ferris, J.P. and Orgel, L.E., 1967. Synthesis of purine precursors and amino acids from aqueous hydrogen cyanide. J. Mol. Biol., 30: 223—253.

Schidlowski, N., Appel, P.W.U., Eichmann, R. and Junge, C.E., 1979. Carbon isotope geochemistry of the 3.7 × 10⁹ yr old Isua sediments, W-Greenland: Implications for the Archaean carbon and oxygen cycles. Geochim. Cosmochim. Acta, 43: 189—199.

Schlesinger, G. and Miller, S.L., 1973. Equilibrium and kinetics of glyconitrile formation in aqueous solution. J. Am. Chem. Soc., 95: 3729—3735.

Schonland, B.F.J., 1953. Atmospheric electricity. Methuen and Co., London, 46 pp.

Schopf, J.W., 1972. Precambrian paleobiology. In: C. Ponnamperuma (Editor), Exobiology. North-Holland, Amsterdam, pp. 16—61.

Schrauzer, G.N., 1978. As reported in Chem. Eng. News, Nov. 13, p. 7.

Schwartz, A.W., 1972. The sources of phosphorus on the primitive Earth — an inquiry. In: D.L. Rohlfing and A.I. Oparin (Editors), Molecular Evolution: Prebiological and Biological. Plenum, New York, N.Y., pp. 124—140.

Schwartz, A.W., 1974. An evolutionary model for prebiotic phosphorylation. In: K. Dose, S.W. Fox, G.A. Deborin and T.E. Pavlovskaya (Editors), The Origin of Life and Evolutionary Biochemistry. Plenum, New York, N.Y., pp. 435—443.

Schwartz, A.W. and Chittenden, G.J.F., 1977. Synthesis of uracil and thymine under simulated prebiotic conditions. BioSystems, 9: 87—92.

Schwartz, A.W., Van der Veen, M., Bisseling, T. and Chittenden, G.J.F., 1973. Nucleotide synthesis in the reaction system apatite-cyanogen—water. BioSystems, 5: 119—122.

Sillèn, L.G., 1965. Oxidation state of the Earth's ocean and atmosphere. I. A model calculation on earlier states. The myth of the "prebiotic soup". Ark. Kemi, 24: 431—456.

Steinman, G.D. and Lillevik, H.A., 1964. Abiotic synthesis of amino groups. Arch. Biochem. Biophys., 105: 303—307.

Stoks, P.G. and Schwartz, A.W., 1979. Uracil in carbonaceous meteorites. Nature, 282: 709—710.

Studier, M.H., Hayatsu, R. and Anders, E., 1972. Origin of organic matter in the early solar system — V. Further studies of meteoritic hydrocarbons and a discussion of their origin. Geochim. Cosmochim. Acta, 36: 189—215.

Toupance, G., Raulin, F. and Buvet, R., 1975. Formation of prebiochemical compounds in models of the primitive Earth's atmosphere. I: CH_4—NH_3 and CH_4—N_2 atmospheres. Origins Life 6: 83—90.

Turekian, K.K. and Clark, S.P., 1969. Inhomogeneous accumulation of the Earth from the primitive solar nebula. Earth Planet. Sci. Lett., 6: 346—348.

Urey, H.C., 1952. The Planets: Their Origin and Development. Yale University Press, New Haven, Conn.

Usher, D.A., 1977. Early chemical evolution of nucleic acids: a theoretical model. Science, 196: 311—313.

Van der Velden, W. and Schwartz, A.W., 1977. Search for purines and pyrimidines in the Murchison meteorite. Geochim. Cosmochim. Acta, 41: 961—968.

Walker, J.C.G., 1976. Implications for atmospheric evolution of the inhomogeneous accretion model of the origin of the Earth. In: B.F. Windley (Editor), The Early History of the Earth. Wiley, London, pp. 537—546.

Weyl, D.K., 1968. Precambrian marine environment and the development of life. Science, 161: 158—160.

Wilkening, L.L., 1978. Carbonaceous chondritic material in the solar system. Naturwissenschaften, 65: 73—79.

Wolman, Y., Haverland, W.J. and Miller, S.L., 1972. Nonprotein amino acids from spark discharges and their comparison with the Murchison meteorite amino acids. Proc. Acad. Sci. U.S., 69: 809—811.

Yokohata, A. and Tsuda, S., 1967. Silent discharge reactions in aqueous solutions. V. Nitrogen fixation in a heterogeneous system of nitrogen and water. Bull. Chem. Soc. Japan, 40: 1339—1344.

NOTE ADDED IN PROOF

Recent calculations on the possible effect of photolytic nitrogen reduction on the atmosphere of the early Earth suggest that this mechanism could have provided an important local source of NH_3 for prebiotic chemical synthesis without adversely influencing the "greenhouse effect" and consequently the surface temperature of the Earth (A. Henderson-Sellers, personal communication).

Chapter 3

DISTRIBUTION OF ORGANIC MATTER IN LIVING MARINE ORGANISMS

P.H. NIENHUIS

1. INTRODUCTION

Organic matter in the oceans may be broadly categorised as either dissolved or particulate organic matter. Particulate organic matter includes both living and non-living material in the sea and generally refers to particles larger than 0.5—1.0 μm diameter. This size group of particles is usually determined by the smallest pore size of a membrane filter commonly used to filter large volumes of sea water and so concentrate the particles for analysis. The definition of particulate organic matter is an arbitrary one and in reality the sea contains a continuous range of particle sizes from the smallest colloids up to larger organic aggregates, plankton and larger animals. From this diverse spectrum only the living organic matter, comprising the enormous wealth of organisms inhabiting the oceans will be discussed. The oceans have a total area of roughly 360×10^6 km^2 and account for more than 70% of the Earth's surface. As an environment for the support of life, the sea differs basically from the land; it has an average depth of approximately 3800 m, and only the superficial illuminated layers are capable of supporting plant growth.

In any discussion of organic carbon in living organisms, it should be borne in mind that this pool represents roughly only 2% of the total organic matter in the sea and that this amount is far outweighed by dissolved organic matter (89%) and dead particulate organic matter (9%) (cf. Parsons, 1975).

2. THE PROCESS OF PRODUCTION OF BIOMASS

Depending on their carbon source, living organisms can be classified as autotrophs, which can synthesise their organic components from CO_2, and heterotrophs, which depend on preformed organic material for energy and growth. A distinction may be made with respect to their energy source: photo-autotrophs (algae, blue-green algae and some bacteria) use light (photosynthesis) and chemo-autotrophs (restricted to bacteria) oxidise reduced or partially reduced inorganic compounds (chemosynthesis).

Carbon dioxide is the source of cell carbon during the photo-autotrophic growth of plants, where light energy is converted into the chemical energy of ATP, most of which is used to convert CO_2 into reduced carbon compounds. Photo-autotrophic growth is by far the dominant process among aquatic plants, but heterotrophy, defined as the utilisation of organic compounds for growth, also occurs. Species lacking photosynthetic pigments, occur in many plant groups. These are naturally heterotrophic, since none are known to be able to derive their energy from the transformation of inorganic compounds (Droop, 1974).

According to Droop (1974), species having photosynthetic pigments sometimes show also heterotrophy. However, there is some discussion as to the extent of this phenomenon in relation to phototrophy, because obligate phototrophy cannot always be conclusively demonstrated. Khoja and Whitton (1971) list 17 heterotrophic Cyanophyceae out of a collection of 24, and many littoral pennate diatoms have been shown to be facultative chemotrophs (Lewin and Lewin, 1960; Lewin and Hellebust, 1975). Droop (1974), however, found only 2 out of 24 diatom isolates from sand grains from a Scottish sea loch to be able to grow in the dark on organic carbon substrates.

Even when the main source of cell carbon is not CO_2 (heterotrophic growth), CO_2 fixation reactions are still important as the "dark" fixation reactions, essential for the operation of many biosynthetic pathways. In both autotrophic and heterotrophic growth, CO_2 can also play a catalytic role (Raven, 1974).

Chlorophylls are the essential pigments involved in light absorption and photochemistry in higher plants, algae and photosynthetic bacteria. Four major chlorophylls (Chl) have been described: Chl a, Chl b, Chl c and Chl d. Chl a is the primary photosynthetic pigment in all oxygen-evolving photosynthetic organisms investigated, but the other chlorophylls have a limited distribution and are considered as accessory or secondary photosynthetic pigments (Meeks, 1974).

Respiration is essentially the converse of the process of photosynthesis. If the two processes occur concurrently, then the rate of photosynthesis as actually observed (net photosynthesis) will be less than the total rate (gross photosynthesis) by an amount equal to the rate of consumption of the products of photosynthesis in respiration. At certain low light intensities or carbon dioxide concentrations the two processes balance so that there is no net gas exchange and the photosynthetic organism is then said to be at its compensation point. It has usually been assumed, according to Fogg (1975) without real justification, that the rate of respiration remains the same in light and dark, so that it is sufficient to correct observed rates of photosynthetic evolution of oxygen by adding an amount equal to the uptake of oxygen in otherwise similar samples in the dark, so as to obtain a value representing gross photosynthesis.

This view has been complicated by the knowledge that there is also a process of light-stimulated photorespiration. This may be defined as a light-dependent oxygen uptake and carbon dioxide release occurring in photosynthetic tissues (Fogg, 1975).

The production of a great variety of extracellular substances by aquatic vascular plants and algae is now well established. It is also clear that such substances often play important roles in algal growth and physiology, as well as in aquatic food chains and ecosystems in general (Hellebust, 1974). Carbohydrates — simple and complex polysaccharides — are liberated by a large number of taxonomically diverse algae. The amounts released may represent a considerable fraction of the photoassimilated carbon of some algae and higher plants during active growth. Other organic substances produced are: organic acids, of which glycollic acid is most commonly liberated, lipids, phenolic substances (from brown seaweeds), organic phosphates, volatile substances, enzymes, vitamins, sex factors, growth inhibitors and stimulators and toxins (Hellebust, 1974).

The release of simple substances such as sugars, and amino acids by healthy cells probably occurs chiefly by diffusion through the cell plasmalemma. The rate of such release will, therefore, depend on the concentration gradient of the substance across the membrane, and the permeability constant of the membrane for the substance. Large molecules, such as polysaccharides, proteins and polyphenolic substances, are probably excreted by more complex processes but these are quantitatively unimportant (Hellebust, 1974; Fogg, 1975).

There is good evidence that healthy, actively growing algae release a certain percentage of their photoassimilated carbon during their life cycle but there is some controversy about the reality and the extent of this phenomenon (cf. Fogg, 1977; Sharp, 1977; see also Section 3). In general, higher excretion rates have been observed with samples of natural phytoplankton than with laboratory cultures of such algae. One cannot always be certain, however, that this reflects with any accuracy how the algae behave *in situ*, because the process of incubating plankton samples in bottles for various lengths of time may itself be detrimental to natural populations and may result in high excretion rates (Hellebust, 1974).

Several factors influence extracellular excretion of organic substances. The highest percentage of extracellular production occurs in oligotrophic waters, and when cell densities are low. The lowest percentage occurs in eutrophic waters, but the total amount released is usually greater in eutrophic waters. Independent of the species under study, the algae show a strong increase of extracellular excretion near the surface where photosynthesis is light-inhibited (evidence summarised by Fogg, 1975). Exponentially growing cells release into the medium from less than 1% to over 50% of their photoassimilated carbon, depending on the species and growth conditions. Cells in lag and stationary growth phases, however, generally release more organic carbon than do exponentially growing cells (Hellebust, 1965; Nalewajko, 1966; Guillard and Hellebust, 1971; Hellebust, 1974).

Wangersky (1978), evaluating the meaning of the production of dissolved organic matter, stated that his own best estimate at present would be that between 10 and 20% of the carbon fixed by photosynthesis is released to the environment by actively growing plant populations. The amount released depends upon the species and the physiological state of the organisms involved. Higher rates of release may occur when the population is subjected to extreme stress, but in general the rates of release will parallel rates of photosynthesis. Some of the compounds released during normal growth will be used directly, either by the phytoplankters themselves or by associated marine bacteria, so that the standing crop of dissolved organic matter is likely to be considerably different from the material released, both in composition and quantity. Another large input of dissolved material will result from death and disintegration of the (planktonic) organisms. If the estimate of 10—20% release of organic material proves to be sound, then this source must be more significant in the long run than the addition due to death and disintegration (Wangersky, 1978).

3. METHODS OF MEASUREMENT OF PRIMARY PRODUCTION OF BIOMASS

Presenting a coherent general account of our present knowledge of marine production is not easy. It is further complicated by the lack of standardisation of terms and units. There are several different ways of measuring primary production and the accuracy of the measurements depends primarily on the sensitivity of the method. Secondary production is even more difficult to measure (Dunbar, 1975). Thus, to obtain a good estimate of primary production in seas and oceans the methods commonly used for its measurement must be treated first since their reliability determines the validity of the results.

Primary production can be measured by the amount of oxygen liberated by photosynthesis. The amount of oxygen used during respiration, measured in a dark bottle, is subtracted from the light bottle value. Oxygen concentration is usually measured by the Winkler titration which enables a reliable determination of changes in dissolved oxygen as small as 0.02 mg l^{-1}. The subtraction of the dark bottle value is based on the assumption that respiration in the dark is the same as in the light. Because of the rather low sensitivity this method is not recommended for primary production measurements in the seas and oceans, except in coastal regions (De Vooys, 1979).

An estimation of primary production is often attempted by determining the amount of functional chlorophyll present. Photosynthesis is dependent on chlorophyll so that it may be supposed that the chlorophyll content of a water sample may be taken as an index of its photosynthetic potential. Although methodological discussions continue (cf. Maerker and Szekielda, 1976) the chlorophyll content of sea water can readily be determined spectrophotometrically or fluorometrically. According to Fogg (1975) it is tempting to use the above assumption as a basis for the estimation of primary production in spite of its questionable theoretical validity. With a process so complex as photosynthesis, no constant relationship is to be expected between the photosynthetic rate (even under standard conditions of light intensity) and any single participant factor in the process. Nevertheless, Ryther and Yentsch (1957) have suggested the use of the following expression for the determination of the rate of photosynthesis, P in g carbon m^{-2} day^{-1}:

$$P = \frac{R}{k} \times C \times 3.7$$

R being the relative photosynthesis, determined empirically, for the appropriate value of surface radiation, k the extinction coefficient per metre of the water, C the amount of chlorophyll in g m^{-3}, and the factor 3.7 a mean value for the assimilation number. Values estimated by this method and those determined by the oxygen technique agreed reasonably well for a number of sea areas in which primary productivity varied by an order of magnitude (Fogg, 1975).

Chlorophyll concentration offers a static biomass parameter; productivity expresses a photosynthetic rate. The fact that ranges of chlorophyll concentrations occurring in the seas and oceans vary more than a thousand-fold and those of primary production less than fifty-fold (both per volume) illustrates according to De Vooys (1979) that the determination of primary production of phytoplankton by means of chlorophyll concentrations is not to be recommended.

Cadée and Hegeman (1977) found, nevertheless, that annual primary production of the benthic microflora in sediments of the Dutch Wadden Sea showed a good correlation with the annual average functional chlorophyll a content of the sediment.

The method of choice in routine productivity studies throughout the world nowadays is the radiocarbon method, introduced by Steemann Nielsen (1952). This method requires the addition of a small amount of $NaH^{14}CO_3$ to a sea-water sample with algae, and the measurement of the amount of ^{14}C taken up by the cells after a certain incubation time. The ^{14}C method is very sensitive and it can measure far smaller primary production rates than the O_2 method, and as such is preferable for measuring the relatively low primary production of the oceans. Measurements can be made *in situ*, incubating the samples at the same depths from where they were drawn, with the ship at anchor at a certain locality during the incubation period. Unfortunately there is no universally accepted procedure for measuring ^{14}C primary productivity, but the technique for analysis of plankton is described in detail by Strickland and Parsons (1972).

The techniques for measurement of photosynthesis in sediments and in macrophytes are less standardised. In a SCOR-workinggroup (UNESCO, 1973) an attempt was made to prepare guidelines for the measurement of marine primary production in microphytobenthos and in macrophytes communities. Many authors, however, do not give a description of their techniques in sufficient detail to allow any evaluation or a comparison with other work.

In spite of the wide acceptance of the radiocarbon method it should be realised that this method suffers from several drawbacks which detract from its use in making mondial estimates. De Vooys (1979) evaluated a number of methodological difficulties which may give rise to possible errors: (a) the absence of turbulence in the closed incubation glass bottle; (b) the duration of the incubation time; (c) the excretion of labelled organic compounds by algal cells during incubation; (d) the method of arresting photoassimilation after incubation; (e) the effect of filtration pressure on cell rupture; (f) the role of pH in removing the labelled bicarbonate by acidifying and aerating the filtrate; (g) drying of filters and storage of dried filters in a vacuum desiccator leads to losses in radioactivity; and (h) the share of dark fixation in the total carbon fixation.

To sum up, it should be realised that both methods, O_2 and ^{14}C method, have serious limitations. The ^{14}C method is certainly the more sensitive one, but from ^{14}C data no respiration can be measured and therefore no net production can be calculated whereas the oxygen method gives a good estimate of assimilation and dissimilation (Golterman, 1975). The question, at one time actively debated, on whether ^{14}C uptake gives a measure of net or gross photosynthesis or some intermediate value, has never been resolved (Bunt, 1975). Mostly the assumption is made that nearly all fixed carbon comes from directly assimilated CO_2 (and not partly from older respired carbon) and that the radiocarbon method more or less measures net primary production (De Vooys, 1979).

In a comparison of benthic microalgal production measured by ^{14}C and by the oxygen method Hunding and Hargrave (1973) came to the conclusion that both methods gave similar measures of the magnitude of production. Measures of ^{14}C uptake offer sensitivity when production is low, but when undisturbed sediment cores can be obtained, production is more easily measured by following changes in dissolved oxygen.

4. PRODUCTION AND ENVIRONMENTAL FACTORS

Primary productivity in the sea is controlled by a number of intricate environmental factors. Temperature in general does not seem to be a major factor in controlling productivity. Rates comparable in magnitude have been obtained in widely differing latitudes. Data collected by Bunt and Lee (1970) from samples from Antarctic Sea ice, however, demonstrate limitation at extreme low temperatures. Photosynthesis in the upper layers of the sea is not necessarily limited by light. It is generally assumed that the photic zone extends to the depth at which the light intensity is reduced to 1% of the value at the surface, which level should not be taken too literally (Bunt, 1975).

The availability of nutrients is probably the limiting factor in most marine systems. This may be illustrated by the example of the Arctic Ocean where winter photosynthesis is an impossibility, but where in the polar summer there is an abundance of light. It is not shortage of light that renders large areas of the Arctic Ocean so low in productivity then, but the scarcity of nutrients in the euphotic zone, a result of intense vertical density stratification (Dunbar, 1975). It is often considered that over much of the ocean's area the supply of available phosphorus is a critical factor for phytoplankton development; but both nitrogen and phosphorus may be limiting in some areas and in coastal waters nitrogen may exert primary control. Various trace elements are known to be necessary for algal growth, but it may be the nature of these substances in sea water, rather than their concentrations that control phytoplankton growth. Specific information on the inorganic nutritional requirements and nutrient uptake kinetics of the marine algae is quite limited (Bunt, 1975).

The supply of nutrients to the euphotic upper layer is entirely dependent upon instability of the water column, except where local direct outflows of nutrients from the land occur. Analysis of the world map of marine produc-

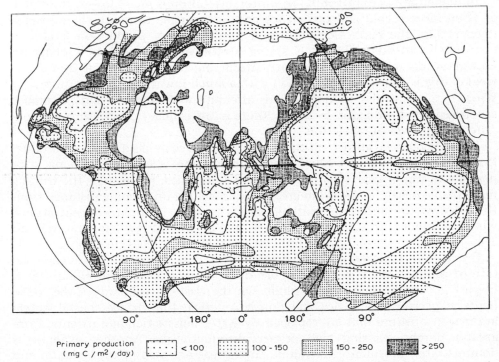

Fig. 1. Distribution of primary production in the oceans of the world (from Parsons et al., 1977, based on information of Koblentz-Mishke et al., 1970).

tion (Fig. 1) will show that it is this factor of vertical instability, and not temperature or latitude (light), which controls the pattern of productivity (Dunbar, 1975).

Thus the availability of nutrients to the phytoplankton is determined by hydrographical factors. An early evaluation of the importance of hydrography for primary production on a global scale was given by Sverdrup (1955) in a published world map based entirely on hydrographical data, where from theoretical considerations, polar and boreal regions were indicated as moderately productive, and large parts of the oceans in the lower latitudes (except near the equator) as having a low productivity. This global picture has largely been confirmed by subsequent investigations of primary productivity all over the world (see Section 5).

De Vooys (1979) summarised the influences of environmental factors on the primary production in the world ocean as follows: At lower latitudes, especially in subtropical regions, open ocean water contains little nitrogen and phosphorus and as a consequence these areas have a low primary production (about $30 \text{ g C m}^{-2} \text{ yr}^{-1}$). In anticyclones the water column has a great stability, an absence of silt and low amounts of algae and often no water transport, together determining primary production.

In higher latitudes (40—60°) in winter, insolation is much lower, the water is often more turbid and turbulence carries the algae to depths 5--10 times deeper than the euphotic zone. As a result primary production is almost completely arrested. When in spring stratification develops, a plankton bloom occurs, exhausting the nutrients in the euphotic zone within a few weeks; this is followed by a period with a much lower primary productivity. When in autumn new mixing periods alternate with stable periods, a smaller secondary bloom occurs. In winter there is low primary production due to a much lower level of insolation.

In polar regions, at high latitudes, the Arctic Ocean has a very low production (less than $10 \text{ g C m}^{-2} \text{ yr}^{-1}$) since this region is covered with pack-ice nearly the whole year. The ocean around Antarctica, however, is the world's most fertile ocean. Water from great depths welling up at some distance from the edge of the antarctic continent leads to high nutrient concentrations, which together with intense turbulence, and in spite of low temperatures and a short summer season, gives a production of about $100 \text{ g C m}^{-2} \text{ yr}^{-1}$.

On a more regional scale, hydrographic forces can greatly influence the rate of production. The most important of these is coastal upwelling, which in part is caused by the trade winds. On the western seaboards of the continents, especially off Africa and South America, the surface coastal waters are driven offshore and are replaced by water masses rich in nutrients, from greater depths, causing high primary productivities. Another example is demonstrated in equatorial waters. In the eastern part of the Atlantic and the Pacific Ocean a clearly higher primary productivity prevails in the region along the equator caused by the trade winds which also induce upwelling

(De Vooys, 1979; see also Fogg, 1975; Steemann Nielsen, 1975).

In turbid and eutrophicated coastal waters and estuaries solar radiation, and not nutrients, often limits production. Estuaries are continually replenished with nutrient enriched (river)water, and as a consequence (when light is not limiting) these waters often have a high primary productivity, which in low latitudes persists throughout the year (Fogg, 1975).

Ryther (1959) hypothesised why high production levels may occur in benthic populations, such as coral reefs, marine grassflats or beds of seaweed (Section 5 and 8), provided that seasonal temperature extremes and large amounts of suspended matter intercepting light do not impair growth. While the concentrations of nutrients in the surrounding waters may be very low, the fact that they are continually being replenished, as the water moves over the plants, probably prevents them from becoming limiting. Planktonic organisms on the other hand, suspended in the water mass, can probably never maintain high production rates in a given parcel of water, since their growth rapidly exhausts the nutrients from their surrounding environment and any mixing process which enriches the water must, according to Ryther (1959), at the same time dilute the numbers of organisms present.

Ryther's (1959) hypothesis may still hold today, but it is known now that conspicuous bacterial colonisation of benthic plants is also of importance in recycling nutrients (Section 7), explaining a higher production in benthic plant populations than in plankton communities.

5. ORGANIC MATTER IN AUTOTROPHS

5.1. Global primary production

From recent estimates of the world's net production and plant biomass (Whittaker and Likens, 1975; Woodwell et al., 1978) it appears that production and biomass on land predominate to a degree that was only recently recognised (Table I). Total production on land is somewhat more than twice that of the oceans on less than half as large an area (71% of Earth's surface is covered by water). A major cause of this contrast is the difference in nutrient function of plankton and land communities. The nutrients available to a plankton community are cycled relatively rapidly among its short-lived organisms. However, in stratified waters some distance from continents, the sinking of organisms and particles carries nutrients downward out of the euphotic zone, thereby impoverishing the plankton environment. In contrast, the characteristics of land ecosystems, which have evolved in relation to stable land surfaces, tend to hold a larger capital of nutrients in plant tissues and soil at the illuminated surface of the Earth where they may support primary productivity (Whittaker and Likens, 1975).

Photosynthesis in the sea is carried out by microscopic unicellular algae — the phytoplankton — freely suspended in the water, and also by

TABLE I

Global net primary production and standing crop (Woodwell et al., 1978) (t = 10^3 kg)

Ecosystem type	Area (10^6 km^2)	Total net primary production (10^9 t C yr^{-1})	Total plant mass of carbon (10^9 t C)
Continental ecosystems	149	52.8	826.5
Open ocean	332.0	18.7	0.45
Upwelling zones	0.4	0.1	0.004
Continental shelf	26.6	4.3	0.12
Algal bed and reef	0.6	0.7	0.54
Estuaries	1.4	1.0	0.63
Marine ecosystems (total)	361.0	24.8	1.74

attached plants, which may be microscopic algae, larger seaweeds or aquatic phanerogams.

The first systematic data series covering different marine ecosystem types which permitted an estimate of the primary productivity of the world oceans was collected during the Galathea Expedition (Steemann Nielsen and Jensen, 1957). On that cruise, 194 stations were sampled representing on the average 2×10^6 km^2 each. Since then both temporal and spatial data on primary productivity measurements have improved considerably: the most serious gaps in the present knowledge are situated in the tropical inshore waters and the polar waters in general (Platt and Subba Rao, 1975). Estimates of global primary production vary widely as shown in Table II. In most estimates benthic primary production is not included.

Riley (1944) obtained his estimates by measuring oxygen production of water samples. This method does not yield reliable results in oligotrophic waters (Section 3) and Riley's estimate is almost certainly much too high.

TABLE II

Estimates of global marine primary production

Production in 10^9 t C yr^{-1}	References
126	Riley (1944)
20—25	Steemann Nielsen and Jensen (1957)
20	Ryther (1969)
23	Koblentz-Mishke et al. (1970)
44	Bruyevich and Ivanenkov (1971)
31	Platt and Subba Rao (1975)
76	Sorokin (1978b)
25	Woodwell et al. (1978)
44	De Vooys (1979)

Ryther (1969) divided the oceans into three arbitrary regions: open ocean, coastal zone and upwelling areas, and considered mean production values of respectively 50, 100 and 300 g C m^{-2} yr^{-1}. From this he calculated a total production of 20×10^9 t C yr^{-1}.

The account of Koblentz-Mishke et al. (1970) contains a large amount of data of Russian workers. Based on data from over 7000 stations they divided the waters of the global ocean into five "types" with daily rates of productivity ranging from a mean of 70 mg C m^{-2} in oligotrophic waters in the central subtropics to a mean of 1000 mg C m^{-2} in open coastal waters.

From a recalculation of previous Russian estimates using radiocarbon, Bruyevich and Ivanenkov (1971) arrived at a figure of 44×10^9 t C yr^{-1}. Platt and Subba Rao (1975) taking new data into account, have given an estimate of 31×10^9 t C yr^{-1}. Neither of the last two reports take into account extracellular products and their estimates may therefore be too low.

Sorokin's (1978b) most recent estimate of world marine productivity is more than twice those of the other authors, because he assumes that the radiocarbon method on average underestimates primary production 2—3 times.

De Vooys (1979) is also of the opinion that the generally accepted estimates of primary productivity in the world ocean are too low, mainly as a result of shortcomings of the radiocarbon method. He proposes an overall correction of the existing values by an increase of 40% based on loss of radioactivity on storing plankton filters in a vacuum desiccator (20%), as is common practice in the standard method of Strickland and Parsons, (1972) and on account of loss of extracellular excretion products (20%). In his opinion the corrected estimate of Platt and Subba Rao (1975; 44×10^9 t C yr^{-1}) is the best approximation of primary production in oceans and seas.

5.2. Phytoplankton

Turning to Table I it is obvious that by far the largest share (75% according to Woodwell et al. 1978) in the global marine net primary production comes from the open oceans and hence from the phytoplankton in that system. It is at the same time remarkable that this production is out of proportion in relation to the phytoplankton biomass: 26% of the total aquatic marine plant mass could be estimated in the open ocean. It should be realised that the distribution of phytoplankton in the sea is rarely uniform and frequently extremely patchy. Horizontally, patches are usually elliptical and vary in size from a few metres to hundreds of kilometres across. Long narrow bands or streaks, a few metres in width, are common and may form a pattern superimposed on that of the patches. Under conditions of strong mixing, vertical distribution of phytoplankton may be uniform, but if the water column becomes stabilised, non-motile forms denser than water will tend to sink and become concentrated lower down while forms lighter than water rise to the surface and motile forms may congregate at particular intermediate depths. This happens particularly with relatively large fast-swimming species of dinoflagellates which may become concentrated in layers only a few centimetres thick (Fogg, 1975). It should be remembered that the usual oceanographic practice of taking discrete samples at successive depths can miss highly localised concentrations of phytoplankton and may sometimes fail to give an accurate estimate of abundance (Fogg, 1975). It is obvious that the patchy distribution of the phytoplankton in the world ocean presents difficulties in reliably estimating the production and biomass, especially in geographically confined areas.

On a global scale phytoplankton are the major producers. Phytoplankton have been divided into two size groups, net plankton and nannoplankton. Net plankton (larger than 60 μm), mainly consisting of large diatoms and dinoflagellates, have received considerable attention and form the basis of the established paradigm of the food web in the sea. Nannoplankton (smaller than 60 μm) however, are more difficult to study and until recently they have often been neglected in trophic considerations (Pomeroy, 1974). Quite a number of studies produced consistent results showing that the large diatoms and other net plankton account for a small fraction of total primary production. In studies estimating the relative importance of net phytoplankton and nannoplankton in photosynthesis, it appeared that nannoplankton may account for more than 90% of total photosynthesis (Pomeroy, 1974). This seems to be true not only in the central gyres of the ocean, but also in upwelling areas, coastal waters, and estuaries. Reynolds (1973), who used a fluorometric method for the determination of ultraplankton (defined as being less than 15 μm in diameter), concluded that in the Barents Sea and some other northern waters this algal group may sometimes contribute over 90% of the total chlorophyll a content in the water.

Watt (1971) measured the rates of photosynthesis of individual phytoplankters by ^{14}C autoradiography and found that in most cases the large diatoms and dinoflagellates showed little photosynthetic activity while nannoplankton showed considerable activity.

The relative importance of nannoplankton is also suggested by studies of numerical abundance. Conventional methods of studying phytoplankton populations from preserved samples often result in descriptions of dominance by a succession of large diatom and dinoflagellate species. In reality there is usually a constant numerical dominance of nannoplankton which are less well preserved in samples and often ignored (Bernard, 1967; Semina, 1972; Pomeroy, 1974).

5.3. Microphytobenthos

Benthic primary producers consist of a mixture of different groups of plants: microphytobenthos growing among and attached to sand and silt

TABLE III

Primary production of microphytobenthos (partly from Cadée and Hegeman, 1974)

Area	Production in g C m^{-2} yr^{-1}	References
Danish fjords	116	Grøntved (1960)
Danish Wadden Sea	115—178	Grøntved (1962)
False Bay (South Africa)	143—226	Pamatmat (1968)
Ythan Estuary (Scotland)	31	Leach (1970)
Southern New England	81	Marshall et al. (1971)
Western Dutch Wadden Sea	60—140	Cadée and Hegeman (1974)
Ems—Dollard Estuary	16—209	Colijn (1978)

grains, larger (macro) algae loose or attached to sediment, shells, stones, etc.; epiphytic macro- and microalgae growing attached to other plants; and, finally, aquatic phanerogams. Compared with the wealth of investigations carried out concerning planktonic primary production, only little has been reported on benthic primary production. However, interest in this research area is increasing: Parsons et al. (1977) recently devoted a chapter to benthic processes. Table III summarises a number of data for the microphytobenthos. It is difficult to separate the microphytobenthos from the other benthic plant populations. The areas where they grow are not known and there are no estimates of the surface area they occupy. The values in Table III may be compared only with caution. They were based on a variety of procedures, all with possible shortcomings, applied at different times of the year over nonuniform periods of incubation and depths, and with varying degrees of replication. For these reasons and because sufficiently detailed geographic information is lacking, it is not surprising that global or even regional estimates of production of microphytobenthos are not available. From Table III it appears that, locally, microphytobenthos can be at least as productive as the phytoplankton of coastal waters.

5.4. Macrophytes

The role of the macrophytes in global primary productivity has only recently been outlined. Within a comparatively narrow coastal zone, areas of extremely high primary productivity exist where in some instances the rate of carbon fixation is as high as for example in a luxurious tropical rain forest. The plant forms living in this zone are very varied, and include seaweeds, sea grasses, marsh grasses and mangroves. Semi-terrestrial and terrestric contributions to the carbon cycle (marsh grasses, mangroves) have been excluded from this discussion. The most spectacular rates of primary production are found in beds of the kelp and giant kelp, i.e. brown algae of the genera *Laminaria* and *Macrocystis*. In clear water, kelps flourish from the low water mark to a depth of 20—30 m. On gently sloping shores, they may extend 5—10 km from the coastline (Table IV). Wave action keeps the blades of the kelp in constant motion, providing maximum exposure to sunlight and contact with nutrients in an area where continuous renewal of these elements takes place. Kelp forests (*Macrocystis*, *Laminaria* and *Ecklonia*) are limited to cooler seas (with temperatures below 20°C) where the annual net production lies in the range 1000—2000 g C m^{-2}. Annual turnover in biomass (biomass increase divided by initial biomass) is 4—17 times. Intertidally, in the same general areas as the kelps, rockweeds are found such as *Fucus* and *Ascophyllum* which commonly produce 500—1000 g C m^{-2} yr^{-1}, with an increase of the biomass of a factor 5 per year (Mann, 1972b, 1973).

On sedimented shores in temperate climates, sea grasses such as *Zostera* live subtidally or intertidally and may fix annually up to 1500 g C m^{-2}. In tropical waters various marine grasses are important sublittoral producers. For example turtle grass (*Thalassia*) may have an annual productivity of 500—1500 g C m^{-2}. The data for these statements are summarised in Table IV.

During an International Biological Programme study of the productivity of the seaweed zone in a marine bay on the Atlantic coast of Canada (Mann, 1972a), it was discovered that *Laminaria longicruris*, *L. digitata* and *Agarum cribrosum* had particularly high rates of production, and that most of their growth in length occurred during winter and early spring. This has led to

TABLE IV

Net production of coastal macrophyte communities in g C m^{-2} yr^{-1} (Mann, 1972b)

Community	Location	Net production
Subtidal		
Laminaria	Atlantic Coast	1225—1900
Macrocystis	California	400— 820
	Indian Ocean	>2000
Thalassia	Caribbean area	590— 900
	Indian Ocean	500—1500
Zostera	Denmark and Washington State	58— 340
	Alaska	50—1500
Intertidal		
Fucoids	Atlantic Canada	640— 840 possibly >1000
Spartina	Atlantic Canada and U.S.A.	130— 897
Mangrove	Puerto Rico	ca. 400
	Florida	352
	Average net production	440—1100

some interest in the study of the mechanisms by which the high productivity is achieved in large marine algae (Mann and Chapman, 1975). Moreover, the results of the study suggest that the production/biomass ratios for attached algae, at least at high latitudes, may have been significantly underestimated (Conover, 1978).

Kelps are highly successful in achieving high levels of photosynthesis per unit area, and their success is apparently related to their habit of storing carbon reserves during summer and early autumn. However, in Canadian *Laminaria longicruris* and *L. digitata* the blade is completely renewed at least once over the course of the winter, by growth at the base and erosion at the tip. It is therefore difficult to envisage how this amount of growth can be supported by stored material from any part of the plant. Presumably these species are able to photosynthesise and produce an assimilatory surplus during the winter. This is a hypothesis of Mann and Chapman (1975), which in their own opinion requires verification.

Seaweeds have no root system and take their nutrients directly from the water. The sublittoral region which is subjected to strong wind-induced and tidal water movements is a good location for plants having to extract large amounts of nutrients from low concentrations in the water. In fact Ryther (1963) postulated that marine benthic plants are provided with a virtually inexhaustible supply of nutrients in the water moving along them. Marine angiosperms such as *Thalassia* and *Zostera*, however possess roots (although the xylem system tends to be somewhat reduced; Sculthorpe, 1967), and McRoy and Barsdate (1970), have demonstrated that ^{32}P may be taken up by the roots of *Zostera*.

The very productive macrophyte fringe of the ocean can make a major contribution to the input of primary production in bays and estuaries. In St. Margaret's Bay, Nova Scotia — east Atlantic Canadian coast — with a total area of about 130 km², the annual productivity of the seaweeds is estimated at three times that of the annual phytoplankton productivity (Mann, 1972a).

Based on world distribution maps and the average biomass per km coastline and assuming a P/B (production/biomass) ratio of 10, De Vooys (1979) calculated the yearly kelp production on a mondial basis. For rockweeds comparable calculations with a P/B ratio of 5 were made. De Vooys attempted also an alternative calculation of the production of benthic macrophytes on a world scale on the basis of estimates of standing crop of seaweed resources of the world. His results suggest a production of 0.02×10^9 t C yr^{-1} for kelps and 0.01×10^9 t C yr^{-1} for rockweeds (brown algae) and redweeds together. It should be realised that these figures are highly speculative and only indicate an order of magnitude, since insufficient data exist. Considering that the primary production values for the open ocean, the upwelling zones and the continental shelf together, as published by Woodwell et al. (1978), are based solely on phytoplankton data, this means that benthic macroalgae (excluding phanerogams) take approximately 1‰ of the phytoplankton share.

There are no global estimates of the production and biomass of aquatic phanerogams, but rough estimates of estuarine primary production (in which submerged phanerogams mainly share) do exist: 1×10^9 t C yr^{-1} (Woodwell et al., 1978), which is 4% of the total global marine primary production.

According to the calculations of Woodwell et al. (1978) the plant biomass in the narrow fringe along the continents where algal beds, reefs and estuaries occur (0.6% of the total water surface) account for an extremely high value of 67% of the total aquatic plant biomass (Table I). A striking feature is that, when comparing the estimates of Koblentz-Mishke et al. (1970), Bogorov (1969) and Rodin et al. (1975) with those of Woodwell et al. (1978), the aquatic plant biomass of the latter is far much higher (25 times) than that of the Soviet workers (0.07 $\times$ 10^9 t C). Rodin et al. (1975) estimate the phytoplankton biomass to be 0.06 $\times$ 10^9 t C (86%) and the biomass of the phytobenthos to be 0.01 $\times$ 10^9 t C (14%) of the total aquatic plant biomass. Although it is not possible on account of the compiled data of Woodwell et al. (1978) to estimate the phytoplankton/phytobenthos ratio, we may now assume that the phytobenthos makes the largest contribution to the total algal bed, reef and estuarine plant biomass, which might indicate that the Russian phytobenthos biomass estimate is more than 60 times lower than that of Woodwell et al. (1978).

It is generally stated that only a small fraction (kelps less than 10%; Mann, 1973) of macrophyte production normally enters grazing food chains and that the remainder enters detrital food chains, after being released as particulate or dissolved organic matter (Mann, 1972a, b, 1973). Most of the production should be exported away from the coastal fringe as detritus and doubtless contributes to the production of adjacent deeper waters. Menzies and Rowe (1969) and Rowe and Menzies (1969), have demonstrated that decaying turtle grass (*Thalassia testudinum*) contributes a significant amount of detritus to the sediments of the continental shelf, and may possibly even make a significant contribution to the particulate organic matter of abyssal regions, since Menzies et al. (1967) have observed debris of the phanerogam on the sea-floor off North Carolina at over 3000 m depth. This species does not grow off North Carolina but presumably was transported northwards by the Gulf Stream from a source in southern Florida or the Bahamas (Conover, 1978).

6. ORGANIC MATTER IN HETEROTROPHS

6.1. Position of heterotrophs

In the food chains the transfer of energy from the autotrophs through a series of organisms takes place by repeated consumption processes. At each transfer a large proportion (80–90%) of the potential energy is lost as heat. Food chains are of two basic types: the grazing food chain, which, starting from a green plant base, goes to grazing herbivores and on to carnivores; and the detritus food chain, which goes from dead organic matter into microorganisms and then to detritus-feeding organisms (detritivores) and their predators. The autotrophs occupy the first trophic level, herbivores the second level (primary consumer level), and carnivores the subsequent levels (secondary and tertiary consumer level) in complex natural communities.

We are far from an overall survey of biomass and production of marine animals, and each estimate contains many extrapolations. Relatively more information is available concerning the groups zooplankton, fishes and zoo-benthos. The gathering of quantitative information about these animal groups is strongly stimulated by their potential or actual economic value. In general the organisms constituting zooplankton are too small to be directly exploited by man at the moment, only species of sufficient size, such as fish, crustaceans, cephalopods and molluscs, which feed on plankton, small fish or other animals, which themselves are plankton eaters, are currently of economic importance (UNESCO, 1972).

6.2. Zooplankton

The zooplankton include a number of major taxonomic groups of animals, either for their entire life cycle or for short periods, as in the case of the larval stages of some fishes, polychaetes, crustaceans and molluscs. From among these types of zooplankton by far the most important group are the Crustacea, and of these, the copepods are the most predominant and have thus received most attention (Crisp, 1975).

Quantitative estimates of the global distribution and biomass of zooplankton are extremely difficult. The routine net sampling techniques probably fail to collect a considerable proportion of the microzooplankton which could itself equal the biomass of the net zooplankton (Pomeroy, 1974). Zooplankton biomass data were summarised for various oceanic and coastal waters by Raymont (1966). It is striking that Mullin (1969) in his survey of the production of zooplankton in the ocean does not himself venture any global estimate.

Productivity/biomass ratios are strongly dependent both on temperature and food supply. Mullin's (1969) review contains a summary table in which the season and location of the measurements are recorded: total production estimates in the literature vary between 4.9 and 70 (or 234) mg C m^{-2} day^{-1}. Moreover, this review draws attention to the many dubious approximations, assumptions, mistakes and ambiguities that are not uncommon in this literature (Crisp, 1975). Russian investigators (Bogorov, Moiseev, Vinogradov) estimated the zooplankton biomass in the world oceans at 1.6 to 1.9 × 10^9 t C (as cited in De Vooys, 1979). In the FAO atlas of the living resources of the sea (UNESCO, 1972) the zooplankton biomass is depicted on a map derived from Bogorov et al. (1968). The map shows that the areas rich in zooplankton (more than 200 mg m^{-3} wet weight averaged to a depth of 100 m) are the same as those with high phytoplankton productivity (cf. Fig. 1). Indeed biomass data of zooplankton herbivores seem to be linked up, both in space and in time with phytoplankton production. This does not hold, however, for extremely shallow coastal areas and estuaries where the euphotic zone may extend to the bottom and where the zoobenthos assumes an active grazing position.

The general vertical distribution of the zooplankton is described by an exponential curve decreasing with depth by about 1--1.5 orders of magnitude from the surface to 1000 m and about 1 order of magnitude from 1000 to 4000 m (Vinogradov, 1970; Menzel, 1974). Using this and the estimate for surface waters of Bogorov et al. (1968), the global zooplankton biomass integrated for the water column was estimated by Whittle (1977) at 1 to 2×10^9 t C; these calculations are in accordance with the Russian data as compiled by De Vooys (1979).

Wet weight values for net zooplankton biomass over large areas of the Pacific, Atlantic and Indian Oceans are 5--10 mg m^{-3} at 500 m and 0.1--1 mg m^{-3} at 4000 m (Vinogradov, 1970). In tropical and subtropical oceans, however, the population is comparatively sparse, becoming steadily more abundant towards higher latitudes. In mid- and higher latitudes a steady state may exist for most of the year in which phytoplankton primary production in the euphotic zone is balanced by zooplankton grazing in the entire water column. This implies a standing stock of zooplankton beneath unit area of sea surface several times larger than that of the phytoplankton (Strickland, 1965). Average zooplankton biomass is sometimes used as a positive parameter of the level of primary production in a watermass (Whittle, 1977). The almost linear relationship between the total zooplankton biomass and primary production in parts of the western Pacific Ocean, as found by Taniguchi (1972) lends evidence for this supposition.

6.3. Zoobenthos

The grouping of benthic organisms into size classes reflects the sampling and processing techniques. Thus it is generally accepted that the meiofauna comprises animals which are retained on screens of about 40--60 μm and pass through screens coarser than 300--500 μm. The macrofauna is retained on the latter screens, while the megafauna consists of larger animals which cannot be collected adequately with a grab or core sampler for quantitative evaluation (Wolff, 1977).

Most macrofauna data represent only a fraction of the total macrofauna. Grabs do not catch the deep-burrowing and the quick epibenthic species. Very large species which are not distributed densely enough are not collected in significant numbers by small numbers of grab samples. Macrofaunal figures, therefore, are minimum figures (Gerlach, 1978).

The data from quantitative investigations demonstrate a pronounced decrease in total biomass with increasing depth and distance from land reflecting the reduced food input. For example 80 km off California at about 200 m depth the total biomass is about 40 g m^{-2} wet weight, 80 km further out, at 4400 m the biomass drops to 3.5 g m^{-2}; a further decrease to 0.2—0.5 g m^{-2} occurs 960 km from the coast at about the same depth (Filatova, 1969).

Exceptions to the general rule of macrofaunal biomass decrease with depth and distance from land are the Arctic and Antarctic Oceans and deeper water off upwelling areas with a particularly high production in the euphotic zone. Most trenches close to the continents and larger islands are also relatively rich in fauna, probably because the trenches act as traps for organic material (Wolff, 1977). Biomass as high as 40 g m^{-2} wet weight has been reported from the Aleutian Trench (Filatova, 1969).

Seas with prolonged ice cover have low benthos biomass. Further, although there is no distinct relationship between salinity and benthic biomass, in regions where there are well-marked haloclines (the central and eastern Mediterranean) or stagnant bottom water layers (deeper parts of the Baltic, Black Sea and southern Caspian), the average biomass level is significantly lower, even though the shallower regions may be rich. Shallow seas receiving large rivers (e.g. Sea of Azov, northern Caspian) are as rich as shallow ocean basins (Crisp, 1975). Characteristic for shallow-water zoobenthos is the very high biomass and production of the populations of suspension feeders.

Menzies et al. (1973) have reviewed global estimates of macrozoobenthos biomass expressed in terms of wet weight, which are recalculated by Whittle (1977) as dry weight and organic carbon (Table V). The biomass value of

TABLE V

Marine benthic fauna biomass modified after Menzies et al. (1973) and Whittle (1977) (ratio wet weight/dry weight = 5; ratio dry weight/organic carbon = 2.5)

Range of depth (m)	Area (10^6 km^2)	Biomass		organic carbon total (10^9 t)
		dry weight		
		approximate mean (g m^{-2})	total (10^6 t)	
0— 200	27.5	40	1100	0.440
200—3000	55.2	4	220	0.088
>3000	278.3	0.04	10	0.004
Total	361.0	44.04	1330	0.532

Menzies et al. (1973) of 0.5×10^9 t C is about three times lower than the global estimate of Bogorov (in Moiseev, 1969): 1.67×10^9 t C. Table V supports the statement that there is an inverse relationship between benthic biomass and depth, but the often stated view that there is a direct relationship between surface productivity and benthic biomass holds primarily at depths above the thermocline (Menzies et al., 1973).

Both numerically and in biomass there is a significant difference between the macro- and meiofauna with increasing depth. The marked macrofaunal decrease is not typical for the meiofauna (Wolff, 1977). Meiofaunal biomass may be remarkably high at abyssal depths ($0.2-2.8$ g m^{-2} wet weight). Except for local abundance of foraminiferans, the overall dominating group in the deep-sea meiobenthos is that of the nematodes, in numbers as well as in biomass. Dominance of nematodes does not seem to vary with increasing depth and distance from land (Wolff, 1977). The meiofauna seems not only to play a significant role in bathyal and abyssal areas but there are also indications that the meiofauna may be important in shallow water areas, where the detrital energy component is particularly large (Crisp, 1975). Estimates of the overall contribution of the meiobenthos to the global carbon pool have so far not been supported by investigations.

6.4. Fishes

The estimations of fish biomass and production over large areas of the world ocean has been stimulated by the economic value of this group. Generally four biological groups are included under the term "fish", viz. bottom-living or demersal fish (plaice, cod, sole, etc.), pelagic fish (herring, sardine, tuna, etc.), large crustaceans (shrimp, crab, lobster, etc.) and cephalopods (squid, octopus).

Numerous attempts have been made to estimate the sea's production of fish and other organisms of existing or potential food value to man. These exercises, for the most part, are based on estimates of primary (photosynthetic) organic production rates in the ocean and various assumed trophic—dynamic relationships between the photosynthetic producers and the organisms of interest to man. Included in the latter are the number of links in the food chains and the efficiency of conversion of organic matter from each trophic level or link in the food chain to the next. Different estimates result from the choice of the number of trophic levels involved and in the efficiencies of conversion (Ryther, 1969). Ryther refined the existing estimates; the results of his calculations are shown in Table VI, a yearly world fish production of 0.24×10^9 t fresh weight (approx. 0.02×10^9 t C assuming a conversion factor for wet weight to carbon 12.5).

The open ocean is in fact a biological desert: it produces a negligible fraction of the worlds fish catch at present and has little or no potential for yielding more in the future. Upwelling regions, totalling no more than 0.1% of the ocean surface, produce about half the world's fish supply. The other

TABLE VI

Estimated fish production in three ocean provinces (Ryther, 1969)

Province	Primary production (t C)	Trophic levels	Efficiency (%)	Fish production (t fresh wt)
Oceanic	16.3×10^9	5	10	16×10^5
Coastal	3.6×10^9	3	15	12×10^7
Upwelling	0.1×10^9	1½	20	12×10^7
Total	20.0×10^9			24×10^7

half is produced in coastal waters and the few offshore regions of comparably high fertility (Ryther, 1969).

At abyssal depths fish (and amphipods) are the overall dominating scavengers. Their average size increases with increasing depths. It is remarkable that scavenging fish are particularly abundant in the central North Pacific Gyre where the benthic standing crop is extremely low. The ubiquitous scavengers must therefore rely on large "food parcels" descending from above, e.g. dead bodies of fish and marine mammals and food fragments from surface-feeding fish. Once on the bottom, the food is quickly located and consumed by opportunistic scavenging fish and amphipods (Wolff, 1977).

Knowledge about transport dynamics from the productive zone to the apparently impoverished regions in the deep sea has been reviewed by Fournier (1972). The constant rain of detritus is one of the explanations for the existence of deep-sea life. The quantity of organic matter reaching the deep sea via the rain of detritus must be a fraction of that produced in the sea above. Riley (1951) estimated that nine-tenths of the primary production is utilised in the upper 200 m. The remaining 10% is then subjected to further decomposition and, presumably, to constant grazing pressure as it settles through the remaining 4000 or 5000 m of water column to the bottom. While large dead diatoms may sink at more rapid rates, most actively growing cells sink at about 1 m day^{-1} or less. Faecal pellets of zooplankton may sink from 50 to nearly 900 m day^{-1} (Smayda, 1969, 1970, 1971). Riley (1972) estimated total primary production in the Sargasso Sea, including the portion excreted by living cells, to be 320 380 mg C m^{-2} day^{-1}. Of this source, 75% is utilised in the surface layer, about 20% in the mid-depths region, and the remaining 5% is divided equally between the rest of the pelagic zone and the bottom.

6.5. Total biomass and production

Whittaker and Likens (1973) ventured preliminary estimates of animal consumption, production and biomass, compiled perhaps from an amount

TABLE VII

Animal secondary production and biomass estimates for the biosphere (from Whittaker and Likens, 1973)

1 Ecosystem type	2 Area (10^6 km^2)	3 Animal consumption (%)	4 Herbivore consumption (10^9 t C yr^{-1})	5 Total animal production (10^9 t C yr^{-1})	6 Animal biomass (g C m^{-2})	7 Total animal biomass (10^9 t C)
Total continental	149	7	3.258	0.372	3.1	0.457
Open ocean	332.0	40	7.600	1.140	1.1	0.360
Upwelling zones	0.4	35	0.035	0.005	4.5	0.002
Continental shelf	26.6	30	1.300	0.195	2.7	0.072
Algal bed and reef	0.6	15	0.075	0.011	9	0.005
Estuaries	1.4	15	0.165	0.025	6.8	0.010
Total marine	361.0	37	9.175	1.376	1.24	0.449

TABLE VIII

Pools of organic carbon in secondary producers in the sea (in 10^9 t wet weight and 10^9 t C; ratio wet weight/organic carbon is 12.5; wet weight figures are from Conover, 1978).

Pool	Biomass wet wt.	C	Production wet wt.	C	P/B ratio
Zooplankton	21.5	1.72	53	4.24	2.5
Zoobenthos	10	0.8	3	0.24	0.3
Nekton	1	0.08	0.2	0.02	0.2
Total	32.5	2.60	56.2	4.50	1.7

of inadequate data (Table VII). The data are thought to give a relatively high estimate of "net" secondary productivity in the sea.

Column 4 of Table VII gives the net primary production consumed by herbivorous animals, based on estimates of percentage consumption in different ecosystem types (column 3 of Table VII) related to total net primary production. The values in column 5 in Table VII, animal production, are calculated by multiplying herbivore consumption (column 4) by gross efficiencies of 15%. Animal secondary production (1.376×10^9 t C yr^{-1}) represents 5—6% of net primary production in the sea (cf. Table I).

Rodin et al. (1975) arrive at a total living biomass of consumers and reducers in the sea (presumably including bacteria; cf. Section 7), based on the data of Bogorov (1969), of 3.3×10^9 t dry weight, or 1.32×10^9 t C (conversion dry weight/C = 2.5). Conover (1978) compiled data from Bogorov (1967) and Skopintsev (1971) and arrived at the survey given in Table VIII.

When comparing the Russian estimates (Rodin, et al., 1975; Conover, 1978) with those of Whittaker and Likens (1973) a striking feature is that

TABLE IX

Estimated production of organic matter measured in tonnes of carbon at different trophic levels in the food chain, if the annually synthesized amount of phytoplankton in the oceans is 1.9×10^{10} t of C (the calculations were based on ecological efficiencies of 10, 15 and 20%; Schaefer, 1965)

Trophic level	Tonnes of C synthesized		
	10%	15%	20%
Phytoplankton	1.9×10^{10}	1.9×10^{10}	1.9×10^{10}
Herbivores	1.9×10^9	2.8×10^9	3.8×10^9
1st Carnivores	1.9×10^8	4.2×10^8	7.6×10^8
2nd Carnivores	1.9×10^7	6.4×10^7	15.2×10^7
3rd Carnivores	1.9×10^6	9.6×10^6	30.4×10^6

animal production of the former is estimated approximately 3 times higher, and animal biomass 3—6 times higher than that of Whittaker and Likens (1973).

The values of Schaefer (1965) for rough estimates of animal production in the sea (Table IX), are comparable with the data of Whittaker and Likens (1973) if an ecological efficiency of 10% is sustained (herbivores and primary, secondary and tertiary carnivores production 2.11×10^9 t C yr^{-1}).

7. ORGANIC MATTER IN BACTERIA

7.1. Bacteria in the open water

With the exception of autotrophic bacteria which contribute a negligible amount of organic matter to the global primary productivity, the interest in the position of non-autotrophic bacteria as secondary producers of organic matter over recent years warrants the treatment of this subject in a separate section.

Since the work of Riley (1963) it has been recognised that a correlation exists between the productivity of a water mass and its content of non-living particulate matter. Oceanic particulates are partly considered to be products of the decomposition of phytoplankton with riverborne detritus being only a minor source of particulate organic matter in the coastal regions. According to Wangersky (1977), careful study of the nature of the particulate matter demonstrated that recognisable phytoplankton constitute a major portion of the particles only during bloom periods, and that during a considerable part of the year there is no close correlation between phytoplankton populations and particulate matter. This means that the correlation between productivity of the surface waters and content of particulate organic carbon exists in space, but not in time; at any given moment the more productive waters contain more particulate matter than do the marine deserts, but the yearly cycles of phytoplankton and particulate organic carbon at any given station are not necessarily correlated.

Possible reasons for the lack of temporal correlation became more apparent as mechanisms for the production of particulate matter were discovered (Pomeroy, 1974; Wangersky, 1977). All of these mechanisms involve the collection of surface-active material, either in solution or in the form of colloidal micelles, at an interface, followed by compression of the interface, both gas-liquid interfaces and liquid—solid interfaces (Wangersky, 1977). Nevertheless, the formation and ultimate fate of organic particles *in situ* is still poorly understood (Conover, 1978).

There is little doubt that marine bacteria are closely associated with the particulate matter. Microorganisms, however, are not concentrated on the particulate matter, almost every investigator, according to Wangersky (1977), has found that as much as 80% of the bacteria in the water column are to be found floating free (e.g. Sieburth, 1968; Sorokin, 1971a, b; Jannasch and Pritchard, 1972; Wiebe and Pomeroy, 1972). However, there are indications that the unattached bacteria may be in a resting stage, and may require 6 to 12 h of incubation at high nutrient levels before they begin to metabolise the added nutrients (Jannasch and Pritchard, 1972; Wiebe and Pomeroy, 1972), whereas the bacteria attached to particles are in an active physiological condition demonstrating growth and reproduction (Wangersky, 1977). Wangersky (1977) hypothesised that suspended bacteria when subjected to any turbulence in the water mass would move together with the parcel of water in which they were floating, and organic nutrients would reach them only by molecular diffusion through the surrounding envelope of water. In contrast, bacteria attached to particles metabolise at higher substrate concentrations. They are constantly exposed to new water; depending upon particle size and specific gravity it is highly likely that the particles will move either faster or slower than the surrounding water. The particle acts, moreover, as an absorbant surface for organic matter, and if the particle itself can be metabolised the advantage to the bacteria is increased.

According to Conover (1978) we have only recently become aware of the relatively large populations of microorganisms, apparently living heterotrophically and metabolising at all depths in the sea. Assuming that a portion of the large reservoir of dissolved organic matter in the sea is an immediate by-product of the photosynthetic process, the heterotrophic growth of bacteria probably represents a major pathway whereby this fraction of organic matter is converted into particulate matter and enters the food chain.

There is much discrepancy in literature concerning the extent of bacterial productivity in the sea. Whittaker and Likens (1973) demonstrated a theoretical approach in treating the biosphere as a system in steady state in which total respiration of all heterotrophic organisms essentially equals total net primary productivity. Given this equality, total reducer (bacterial and fungal) assimilation should approximately equal net primary production minus animal assimilation. With an assumed growth efficiency of 5—10% for marine reducers, then marine reducer production would be 0.7 to 1.4 $\times$ 10^9 t C yr^{-1}, or about half to about the same as the authors' estimated marine animal production (= 1.376 $\times$ 10^9 t C yr^{-1}).

In sharp contrast with these estimates Sorokin (1971a, b, 1978a, b) has suggested that the production of planktonic bacterial biomass may approach that of the phytoplankton, and estimated the total microbial production of bacteria in the world oceans to be 24 $\times$ 10^9 t C yr^{-1}. The total microbial biomass was determined at approximately 0.23 $\times$ 10^9 t C. He suggested that the bacterioplankton production in the tropical ocean equals or exceeds that of the phytoplankton. Moreover, the breakdown of organic matter during this process apparently exceeds primary production by 3.5 to nearly 40 times (Conover, 1978). Sorokin (1971a, b) could not explain the source of dissolved organic carbon for heterotrophic bacterial growth solely on the observed 25% phytoplankton excretion rate (Sieburth, 1977) and postulated that the excess organic matter must be produced elsewhere, presumably in colder regions, where formation exceeds utilisation, followed by transportation by water movements to the tropics. Regrettably, he does not provide any hydrographical evidence to support this hypothesis (Conover, 1978).

Banse (1974) heavily criticised some aspects of Sorokin's theories. He argued that bacterial production must be a fraction of phytoplankton production, and further suggested that Sorokin's values for bacterial production in the deep sea are an order of magnitude too high when compared with the reported values of oxygen consumption in the deep sea. Banse (1974) stated, moreover, that the accurate figure of annual production of particulate carbon by marine phytoplankton is likely to be higher than the currently accepted values, but is unlikely to be several times higher as believed by Sorokin (1971a).

Ferguson and Rublee (1976) estimated the amount of bacterial carbon in coastal waters as ranging between 4% and 25% of total plankton carbon biomass. Meyer-Reil (1977), using highly sensitive methods to determine the growth of bacteria under semi-natural conditions, arrived at an average bacterial biomass production in Kiel Fjord and Bight (Baltic) of 15- 29% of the phytoplankton primary production. Such values may indeed approach reality.

7.2. Bacteria in bottom sediments

The previous section dealt with the role of bacteria as producers of biomass in the open water. In this section the use of organic substrates in bottom sediments is also considered.

In coastal waters detritus is mainly derived from biodeposits (faeces and pseudofaeces) of animals and the decay of micro- and macrovegetation. The surfaces of organic particles constituting the detritus support large numbers of microorganisms (Tenore, 1977). Odum and De la Cruz (1967) and Fenchel (1969, 1970) showed an increase in the protein content of the particles derived from macrophytes with increasing age due to colonisation of microorganisms. According to Fenchel and Jørgensen (1977) it is now generally recognised that detritus feeders, invertebrates feeding on bottom sediment and deposited organic debris, digest only to a small extent the bulk of ingested organic material but mainly utilise the living microbial components. Moreover, the activities of the benthic detritus feeders accelerate the rate of detritus decomposition and hence enhance the microbial activity.

Fenchel and Jørgensen (1977) considered a hypothetical detritus feeder which ingests detrital particles as large as about 200 μm. This detritus feeder will ingest about 4 mg of bacteria, 5 mg of flagellates, 0.5 mg of ciliates and some other protozoans per gram of consumed detritus. This means that not only bacteria but a series of other microfaunal elements are ingested.

Bacteria only assimilate dissolved substrates; solid substrates are first hydrolysed by extracellular enzymes before being assimilated. Degradation of detritus starts with hydrolytic cleavage of the particulate material into small molecules which can be assimilated by the bacteria. The end-products of extracellular hydrolysis are most amino acids, mono- and disaccharides, and long-chain fatty acids. In aerobic environments these are taken up directly by heterotrophic bacteria, and further metabolism is intracellular. A variable fraction of the detritus in marine ecosystems is never completely remineralised, but accumulates mainly within the anoxic environment, and is gradually transformed into organic complexes refractory to microbial attack (Fenchel and Jørgensen, 1977).

Fenchel and Jørgensen (1977) argue that the synthesis of bacterial biomass available for higher trophic levels in the detritus food chain is a function of both the production of detritus and of the growth yield of the bacteria responsible for decomposition. As previously discussed, less than 10% of macrophyte biomass, such as kelps and seagrasses, are grazed directly. In tropical oceans where the zooplankton grazing reaches maximum efficiency, Jørgensen (1966) estimated that 40% of the net production would nevertheless ultimately end up as detritus. Thus, 40% of the total ecosystem production seems to be a minimum input to the detrital food chain. The maximum is close to 100% and the average is probably not far from 50%. As a rough estimate for aerobic environments, bacteria convert about 50% of the net primary production ending up as detritus; this means that, on average, for aquatic ecosystems the production of bacterial biomass is 25% of the photosynthetic production (including imported material) (Fenchel and Jørgensen, 1977).

Gerlach (1978) made some calculations for a hypothetical 30 m deep, marine silty sand station, and arrived at a figure of 2.5 g bacterial C m^{-2} (0.05 mg ml^{-1} over 5 cm sediment depth). Bacteria account for 27% of the living biomass in the sediment, whereas micro-, meio- and macrofauna contribute 6, 13 and 54% respectively. Gerlach (1978) assumed that bacteria are the main food of deposit-feeding macrofauna, meiofauna and microfauna and further that 21 bacterial generations per year would be sufficient to sustain their populations. Ankar (1977), using respiration data, reports a P/B ratio of 30 for bacteria in a benthic ecosystem in the Baltic, which is the same order of magnitude. Gerlach (1978) stated that the productivity of bacteria *in situ* in sediments is far below figures achieved in experimental cultures, where certain marine bacteria may duplicate within hours (Jannasch, 1969). Bacteria in sediments, therefore, according to Gerlach (1978), are supposed to exist under non-optimal environmental conditions and only a fraction of the total population exhibits high metabolic rates.

Recently the ATP method has been employed to determine the total biomass of bacteria, micro- and meiofauna and microphytes in sediments. In

1966 Holm-Hansen and Booth introduced the method to the field of aquatic ecology. The procedure is based on a quantitative assay for adenosine-5'-triphosphate (ATP), a molecule with a central function in the respiratory pathways in all living cells. Within the past decade a great deal of laboratory information has been compiled in support of the original theoretical basis for using ATP measurements as a biomass indicator (Holm-Hansen and Pearl, 1972; Holm-Hansen, 1973a, b). Although extensive laboratory investigations have indicated that the ATP content of various microorganisms (bacteria and algae) averages about 0.40% (range 0.10—0.70%) of the total cellular carbon (Hamilton and Holm-Hansen, 1967; Holm-Hansen, 1970a, b), a number of studies have arrived at significantly different ratios (Karl et al., 1978). These discrepancies may be attributed to actual metabolic and physiological transitions, differences between cells, or may, in fact be due to problems associated with the extraction and measurement of either ATP or total cell carbon (Karl et al., 1978).

In contrast to the results obtained from bacteria and small algae, Karl et al. (1978) found that the C/ATP ratios in multicellular organisms (copepods, isopods and worms) were less than 100, as compared with a ratio of 250 commonly found in microorganisms.

Gerlach (1978) compiled data concerning the living biomass in different types of marine sediments from all over the world to a water depth of 400 m, and found that roughly between 0.5 and 2 μg ml^{-1} ATP was present in those sediments, which approximately corresponds to 0.1—0.4 mg ml^{-1} C (assuming a living biomass/ATP ratio of 200 which is presumably too high; cf. Karl et al., 1978). These figures approach the calculations of Gerlach (1978) for the bacterial, micro- and meiofaunal biomass in an "ideal" 30 m deep marine sediment (in 5 cm sediment 0.08 mg ml^{-1} C). Whittle (1977) discussed the ATP method in relation to the biomass of small sediment inhabiting organisms, and recommended a ratio C/ATP of 250 : 1 which probably leads to an over-estimation of the true biomass.

8. THE DISTRIBUTION OF ORGANIC MATTER IN THE MARINE ENVIRONMENT

8.1. Open oceans, upwelling zones and continental shelves

Much of the above topic has been discussed already in previous sections in a different context. A short survey will be given here for practical reasons following the subdivision of Woodwell et al. (1978) and Whittaker and Likens (1973): open ocean, upwelling zones, continental shelves, algal beds and reefs and estuaries (excluding salt marshes and mangroves). Notwithstanding the fact that the open ocean, the continental shelves and the upwelling zones together account for 99% of the marine geographical area, for 92% of the marine primary productivity (Woodwell et al., 1978), and for 97% of the animal secondary production (Whittaker and Likens, 1973), these areas will only be briefly discussed here in view of the large number of review papers which have appeared on this subject within the past twenty years (e.g. Ryther, 1959, 1969; Raymont, 1963; Koblentz-Mishke et al., 1970; Petipa et al., 1970; Russell-Hunter, 1970; Riley, 1972; Conover, 1974; Walsh, 1974; Bunt, 1975; Fogg, 1975; Platt and Subba Rao, 1975; Steemann Nielsen, 1975; Lorenzen, 1976; Tranter, 1976; Whittle, 1977; Finenko, 1978; and Greze, 1978).

The majority of these papers deal with the "classical" marine food web, where photosynthetically produced organic matter in the form of phytoplankton is consumed by herbivorous zooplankters which are, in turn, consumed by carnivores. The tropho-dynamics of these processes have been postulated and documented for well over half a century (Strickland, 1970).

Ryther (1963) attempted a comparative analysis of global marine productivity, taking data at that time available from ^{14}C uptake or comparable techniques and with their shortcomings in mind. Between and within the major regions considered, there was both wide variation and considerable uncertainty. No global total was suggested. Later, Ryther (1969) accepted 15 to 18×10^9 t C yr^{-1} (rounded off to 20×10^9 t C yr^{-1} in Table II and Table X) as the most likely level of open ocean primary production. Dividing the oceans into three provinces, he suggested mean productivity values of 50, 100 and 300 g C m^{-2} yr^{-1} for the open ocean, coastal zones (including offshore areas of high productivity), and upwelling areas, respectively (Table X).

Fig. 1 shows the distribution of marine primary production as summarised by Koblentz-Mishke et al. (1970). Causal factors for this distribution pattern are discussed in Section 4. This map is only a first approximation of the true picture and requires updating in the future. Nevertheless, the map is highly representative, and it must be recognised that the efforts made by workers of different countries have produced quite substantial results (Finenko, 1978).

As shown in Fig. 1 photosynthesis is most intensive in regions where

TABLE X

Division of the ocean into provinces according to their level of primary organic production (after Ryther, 1969)

Province	Percentage of ocean	Area (km^2)	Mean productivity (g C m^{-2} yr^{-1})	Total production (10^9 t C yr^{-1})
Open ocean	90.0	326.0×10^6	50	16.3
Coastal zone	9.9	36.0×10^6	100	3.6
Upwelling areas	0.1	3.6×10^5	300	0.1
Total		362.4×10^6		20.0

upwelling water predominates over sinking water. In the Pacific Ocean such regions occur off the shores of Central and South America, along the Canadian—American coast and near the Kuril Ridge, to the east of Japan. In the Atlantic Ocean, such regions include those adjacent to the southwest coast of Africa. In the Indian Ocean, the most extensive upwelling of deep waters and the highest productivity levels are found in the northwest. In open-ocean waters, the most productive areas are those of equatorial and Antarctic waters, where photosynthetic rates are several times higher than those of anticyclonic oceanic regions characterised by the process of sinking water. Low-productivity waters are most extensive in Pacific areas, and occupy much smaller regions in the Atlantic Ocean and Indian Ocean. Pacific waters have the lowest mean productivity levels ($46 \text{ g C m}^{-2} \text{ yr}^{-1}$). Second in productivity ($69 \text{ g C m}^{-2} \text{ yr}^{-1}$), are the Atlantic waters (Koblentz-Mishke et al., 1970; Finenko, 1978).

It should be realised that both the attempts of Ryther (1969) and of Koblentz-Mishke et al. (1970) are primarily based on productivity of marine phytoplankton. Neither of these estimates takes into account benthic production, the magnitude of which can locally be significant (Bunt, 1975; Section 5).

Areas with high, moderate and low phytoplankton production roughly coincide with areas rich, moderate and poor in zooplankton biomass (Bogorov et al., 1968; Section 6). Zooplankton exhibits by far the largest biomass and production of all marine animal groups considered (viz. zooplankton, zoobenthos, nekton; Table VIII).

As regards the vertical distribution of biomass in the ocean, in general there is a sharp decrease from the amounts in the surface waters to the deeper water layers. Holm-Hansen (1970a) determined the ATP content in the particulate fraction ($0.45—150 \mu m$) in four profiles in the Pacific Ocean down to 3500 m. The ATP values were converted to cellular organic carbon values. The biomass of (mainly) bacteria and phytoplankton was very high ($15—200 \mu g \text{ C l}^{-1}$) in the euphotic zone and decreased rapidly to $1—2$ $\mu g \text{ C l}^{-1}$ at about 200 m. At the lower depths sampled, the calculated biomass contained about $0.1 \mu g \text{ C l}^{-1}$.

The biomass of macrobenthic invertebrates of the world has been estimated as 0.5 to $1.7 \times 10^9 \text{ t C}$ (Section 6), of which over 80% is to be found on the continental shelves ranging in depth from 0 to 200 m. The reason for this pattern of distribution is again that primary production is higher in coastal waters than in mid-ocean, and the fraction of fixed carbon reaching the bottom is inversely proportional to the depth of the water column.

8.2. Algal beds, coral reefs and estuaries

More recently attention has been focussed on the production of biomass in coastal areas with their algal beds, reefs and estuaries. These ecosystems cover only 0.6% of the world's sea area, but they produce 7% of the marine organic carbon in plants, and contain as much as 67% of the plant carbon standing stock in the sea (Woodwell et al., 1978). With regard to animal secondary production and biomass Whittaker and Likens (1973) stated that 3% of the yearly marine production and 3% of the biomass should be located in coastal reefs, algal beds and estuaries.

Mann and collaborators made explicit the significance of macrophytes (marine phanerogams and macroalgae) as producers of organic matter in the narrow fringe of sea along the continents (Mann, 1972a, b, 1973, 1976; Mann and Chapman, 1975; see also Section 5). The average net production of coastal macrophyte communities is estimated by Mann (1972b) at 440—1100 g C m^{-2} yr^{-1}. The most important groups of the large marine algae are the kelps (Laminariales) and rockweeds (Fucales). The kelps are found, attached to rocky substrates, below low-tide level on almost all temperate shorelines. They show extremely high rates of production, of the order of 1000—2000 g C m^{-2} yr^{-1} (Mann, 1973). Rockweeds have approximately the same geographical distribution as kelps, but occur chiefly between high- and low-tide marks. Their productivity is usually about half that of the kelps. Another important group of marine macrophytes are the sea grasses (marsh grasses and mangroves are excluded here). Flowering plants have a world-wide distribution, but colonise the more sheltered, sedimented areas of the coastline. Their productivity is of the order of 200—1000 g C m^{-2} yr^{-1} (Mann, 1972b). Since this production is concentrated at the land—water interface it is an important contributor to food chains leading to species of commercial importance in the coastal zone (Mann, 1976).

Recently Lewis (1977) presented a review of organic matter production in coral reefs. Reef-building corals (hermatypic corals) thrive only in oceanic zones between the latitudes 30°N and 30°S. Coral reefs predominate in the western regions of the oceans since eastern regions have lower water temperatures as a result of the trade wind westward transport of surface water, and induced upwelling. Coral reefs and atolls are most frequent in the Indo-Pacific and the Caribbean. De Vooys (1979) made a calculation of the total reef area in the oceans with the aid of coverage estimates given by Chave et al. (1972) and arrived at a figure of 0.1×10^6 km^2.

Sources of primary production within coral reefs include fleshy macrophytes, calcareous algae, filamentous algae on the coral skeletons or calcareous hard substrate, marine grasses and the zooxanthellae (autotrophic dinoflagellates) within coral tissue. Primary production on reefs has mainly been studied by flow respirometry, i.e. by measuring changes in oxygen or carbon dioxide concentrations in water flowing over reefs. Rates of gross primary production of reefs vary between 300 and 5000 g C m^{-2} yr^{-1} (Lewis, 1977), and as such these rates are as high or higher than those of the most productive marine macrophyte communities.

De Vooys (1979) calculated a gross primary production of coral reefs of 0.47×10^9 t C yr^{-1} and a net primary production of 0.30×10^9 t C yr^{-1}.

Woodwell et al. (1978) estimated the net production of algal beds and coral reefs together at 0.7×10^9 t C yr^{-1} (Table I). Assuming that the calculations of De Vooys (1979) are comparable with those of Woodwell et al. (1978), this means that algal beds produce on a global scale 0.4×10^9 t C yr^{-1}. De Vooys (1979) estimated along different lines (Section 5) a world production of 0.03×10^9 t C yr^{-1} for algal beds predominated by marine macro-algae.

There are large differences in the rates of primary production estimated on coral reefs. The reason for that lies apparently in the nature of the reef surface studied. For example, rates of primary production appear to be higher over lagoon areas covered by algae and marine grasses than over reefs dominated by coral growth (Lewis, 1977). The highest values reported are those from a Hawaiian fringing reef with 11 680 g C m^{-2} yr^{-1} gross primary production (Gordon and Kelly, 1962). In contrast, the estimates of gross production of the ambient water in the vicinity of the reefs are very low (21—50 g C m^{-2} yr^{-1} gross production, as summarised by Lewis, 1977).

The causes of the high productivity of coral-reef communities are still not clear. Concentrations of nitrogen and phosphorus in waters flowing over reefs are relatively low, but nevertheless there is a constant supply of nutrients. Lewis (1977) found some evidence to suggest that both these nutrients are recycled rapidly on the reef and that nitrogen is fixed by bacteria and primary producers.

In many cases the mass of detritus over coral reefs exceeds the biomass of zooplankton. While the quantitative significance of detritus as food for corals and other benthic organisms has not been evaluated, there is growing evidence to suggest that this may be the key to understanding secondary productivity (Lewis, 1977). Bacteria are a potential source of energy for secondary producers on reefs. Sorokin (1978a) reported that the stock of microbial biomass and the rate of its production in reef environments is extremely high. The microbial biomass in water much outweighs the biomass of phytoplankton, and becomes a fundamental source of food for the planktonic and benthic filter feeders and suspension feeders, including hermatypic corals. The rich microflora of sediments and detritus constitutes a food source for seston, and sediment-feeding bottom fauna. Sorokin's (1974, 1978a) calculations indicated that the bacterial biomass is an important intermediate link in the trophic chain through which energy produced by primary producers and from detritus is passed into the heterotrophic production of coral reefs (Lewis, 1977).

A crude estimate of the area of the world's estuaries indicates a total area of about 1.7×10^6 km^2 of which 3.8×10^5 km^2 is marsh and mangrove, and 1.4×10^6 km^2 is open water, including such areas as Chesapeake Bay (U.S.A.), St. Lawrence Gulf (Canada) and the Baltic Sea (Europe). Despite the numerous research efforts in recent decades, an evaluation of all relationships between net primary production, gross production and various aspects of total respiration in the estuarine environment cannot be given (Woodwell et al., 1973). Three different types of vegetation contribute to the productivity in estuarine areas: submerged angiosperms, epiphytic and benthic algae and phytoplankton. Assuming a mean net production for estuarine waters of 675 g C m^{-2} yr^{-1}, the world's total is estimated at 0.92×10^9 t C yr^{-1}, (Woodwell et al., 1973), which is 4% of total global marine primary production (cf. Table I; Woodwell et al., 1978).

Woodwell et al. (1973) summarised primary productivity data from a number of relevant papers and arrived at the following survey: submerged angiosperms in estuaries produce 880 to 1600 g C m^{-2} yr^{-1}, epiphytic and benthic algae have a production of 28 to 314 g C m^{-2} yr^{-1} (ratio dry weight to C is 2.5).

In contrast to coral reefs, estuaries are characterised as physically unstable areas between the open sea and the fresh waters. The allochthonous contribution to the pool of organic matter in the estuary may often be significant. For the Dutch Wadden Sea, an estuarine area bordering the North Sea, Essink and De Wolf (1978) calculated on the basis of an overall primary production of 120 g C m^{-2} yr^{-1} (which is low in comparison to the mean data of Woodwell et al., 1973), that 29×10^4 t C yr^{-1} is produced *in situ* by phytoplankton and microphytobenthos. Twice as much organic matter (58×10^4 t C yr^{-1}) is supplied by the North Sea, to which a discharge of organic waste from the mainland (6.5×10^4 t C yr^{-1}) should be added. This subsidy of organic matter to estuaries has, of course, far-reaching consequences for the secondary producers. The fish stocks of estuaries, for example, are commonly higher than in other coastal waters. Fish production in estuaries often ranges from 5 to 15 g dry weight m^{-2} yr^{-1} and is probably the highest of any water bodies (Woodwell et al., 1973).

9. REFERENCES

Ankar, S., 1977. The soft bottom ecosystem of the Northern Baltic proper with special reference to the macrofauna. Contrib. Askö Lab. 19: 1—62.

Banse, K., 1974. On the role of bacterioplankton in the tropical ocean. Mar. Biol., 24: 1—5.

Bernard, F., 1967. Contribution à l'étude du nannoplancton de 0 à 3000 m, dans les zones atlantiques lusitaniènne et mauritanienne. Pelagos 7: 4—81.

Bogorov, V.G., 1967. Productive regions of the ocean. Priroda Mosk., 1967 (10): 40—46 (Fish. Res. Board Can. Transl. 985).

Bogorov, V.G., 1969. Life of the oceans. Znaniye Biol. Ser., Moscow, 6: 3—5 (in Russian)

Bogorov, V.G., Vinogradov, M.E., Voronina, N.M., Kanaeva, J.N. and Suetova, J.A., 1968. Distribution of the biomass of zooplankton in the surface layers of oceans. Dokl. Akad. Nauk. S.S.S.R., Ser. Biol., 182 (5): 1205—1207.

Bruyevich, S. and Ivanenkov, V.N., 1971. Chemical balance of the world oceans. Okeanol. Mosk., 11: 694—699.

Bunt, J.S., 1975. Primary productivity of marine ecosystems. In: H. Lieth and R.H. Whittaker (Editors), Primary Productivity of the Biosphere. Ecological studies, 14. Springer Verlag, Berlin, pp. 169—183.

Bunt, J.S. and Lee, C.C., 1970. Seasonal primary production in Antarctic sea ice at McMurdo Sound in 1967. J. Mar. Res., 28: 304—320.

Cadée, G.C. and Hegeman, J., 1974. Primary production of the benthic microflora living on tidal flats in the Dutch Wadden Sea. Neth. J. Sea Res., 8: 260—291.

Cadée, G.C. and Hegeman, J., 1977. Distribution of primary production of the benthic microflora and accumulation of organic matter on a tidal flat area, Balgzand, Dutch Wadden Sea. Neth. J. Sea Res., 11: 24—41.

Chave, K.E., Smith, S.V. and Roy, K.J., 1972. Carbonate production by coral reefs. Mar. Geol., 12: 123—140.

Colijn, F., 1978. Primary production measurements in the Ems—Dollard estuary during 1975 and 1976. BOEDE Public. Verslagen No. 1 (1978): 1—21.

Conover, R.J., 1974. Production in marine planktonic communities. In: Proc. First Int. Congr. of Ecology, The Hague, The Netherlands, PUDOC, Wageningen, pp. 159—163.

Conover, R.J., 1978. Transformation of organic matter. In: O. Kinne (Editor), Marine Ecology, IV. Dynamics. Wiley, Chichester, pp. 221—499.

Crisp, D.J., 1975. Secondary productivity in the sea. In: D.E. Reichle, J.F. Franklin and D.W. Goodall (Editors), Productivity of World Ecosystems. Proc. Symp. U.S. Natl. Committee IBP, Natl. Acad. Sci. Washington, D.C., pp. 71—89.

De Vooys, C.G.N., 1979. Primary production in aquatic environments. In: B. Bolin, E.T. Degens, S. Kempe and P. Ketner (Editors), The Global Carbon Cycle. SCOPE 13, Wiley, Chichester, pp. 259—292.

Droop, M.R., 1974. Heterotrophy of carbon. In: W.D.P. Stewart (Editor), Algal Physiology and Biochemistry. Bot. Monogr., 10 (15): 530—559.

Dunbar, M.J., 1975. Productivity of marine ecosystems. In: D.E. Reichle, J.F. Franklin and D.W. Goodall (Editors), Productivity of World Ecosystems. Proc. Symp. U.S. Natl. Committee IBP, Natl. Acad. Sci. Washington, D.C., pp. 27—31.

Essink, K. and De Wolf, P., 1978. Pollution by organic waste in the Dutch Wadden Sea and the Ems—Dollard estuary. In: K. Essink and W.J. Wolff (Editors), Pollution of the Wadden Sea Area. Report 8 of the Wadden Sea working group, chapter 4.2. Stichting Veth Steun Waddenonderzoek, Leiden, pp. 39—45.

Fenchel, T., 1969. The ecology of marine microbenthos. IV. Structure and function of the benthic ecosystem, its chemical and physical factors and the microfauna communities with special reference to the ciliated protozoa. Ophelia, 6: 1—182.

Fenchel, T., 1970. Studies on the decomposition of organic detritus derived from the turtle grass, *Thalassia testudinum*. Limnol. Oceanogr. 15: 14—20.

Fenchel, T.M. and Jørgensen, B.B., 1977. Detritus food chains of aquatic ecosystems: the role of bacteria. In: M. Alexander (Editor), Advances in Microbial Ecology. Plenum Press, New York, N.Y., pp. 1—57.

Ferguson, R.L. and Rublee, P., 1976. Contribution of bacteria to standing crop of coastal plankton. Limnol. Oceanogr., 21: 141—145.

Filatova, Z.A., 1969. The deep-sea bottom fauna. In: The Pacific Ocean VII: Biology of the Pacific Ocean. Part 1: 202—216. (Trans. U.S. Naval Oceanogr. Office, 1970).

Finenko, Z.Z., 1978. Production of plant populations. In: O. Kinne (Editor), Marine Ecology, IV. Dynamics: Wiley, Chichester, pp. 13—87.

Fogg, G.E., 1975. Primary productivity. In: J.P. Riley and G. Skirrow (Editors), Chemical Oceanography. Academic Press, London, 2nd ed., pp. 385—453.

Fogg, G.E., 1977. Excretion of organic matter by phytoplankton. Limnol. Oceanogr., 22: 576—577.

Fournier, R.O., 1972. The transport of organic carbon to organisms living in the deep oceans. Proc. R. Soc. Edinb., B73: 203—212.

Gerlach, S.A., 1978. Food-chain relationships in subtidal silty sand marine sediments and the role of meiofauna in stimulating bacterial productivity. Oecologia (Berl.), 33: 55—69.

Golterman, H.L., 1975. Physiological Limnology. Developments in Water Science, 2. Elsevier, Amsterdam, 489 pp.

Gordon, M.S. and Kelly, H.M., 1962. Primary productivity of an Hawaiian coral reef: A critique of flow respirometry in turbulent waters. Ecology, 43: 473—480.

Greze, V.N., 1978. Production in animal populations. In: O. Kinne (Editor), Marine Ecology, IV. Dynamics. Wiley, Chichester, pp. 89—114.

Grøntved, J., 1960. On the productivity of microbenthos and phytoplankton in some Danish fjords. Medd. Danm. Fisk. Havunders. (N.S.), 3 (3): 55—92.

Grøntved, J., 1962. Preliminary report on the productivity of microbenthos and phytoplankton in the Danish Wadden Sea. Medd. Danm. Fisk. Havunders. (N.S.), 3 (12): 347—378.

Guillard, R.R.L. and Hellebust, J.A., 1971. Growth and the production of extracellular substances by two strains of *Phaeocystis poucheti*. J. Phycol., 7: 330—338.

Hamilton, R.D. and Holm-Hansen, O., 1967. Adenosine triphosphate content of marine bacteria. Limnol. Oceanogr., 12: 319—324.

Hellebust, J.A., 1965. Excretion of some organic compounds by marine phytoplankton. Limnol. Oceanogr., 10: 192—206.

Hellebust, J.A., 1974. Extracellular products. In: W.D.P. Steward (Editor), Algal Physiology and Biochemistry. Bot. Monogr., 10 (15): 838—863.

Holm-Hansen, O., 1970a. Determination of microbial biomass in deep ocean water. In: D.W. Hood (Editor), Organic Matter in Natural Waters. Occas. Publ., 1, Institute Mar. Sci. Univ. Alaska, pp. 287—300.

Holm-Hansen, O., 1970b. ATP levels in algal cells as influenced by environmental conditions. Plant Cell Physiol., 11: 689—700.

Holm-Hansen, O., 1973a. The use of ATP determinations in ecological studies. Bull. Ecol. Res. Comm. (Stockholm), 17: 215—222.

Holm-Hansen, O., 1973b. Determination of total microbial biomass by measurement of adenosine triphosphate. In: L.H. Stevenson and R.R. Colwell (Editors), Estuarine Microbial Ecology. University of South Carolina Press, Columbia, S.C., pp. 73—89.

Holm-Hansen, O. and Booth, C.R., 1966. The measurement of adenosine triphosphate in the ocean and its ecological significance. Limnol. Oceanogr., 11: 510—519.

Holm-Hansen, O. and Pearl, H.W., 1972. The applicability of ATP determination for estimation of microbial biomass and metabolic activity. Mem. Ist. Ital. Idrobiol., 29: 149—168.

Hunding, C. and Hargrave, B.T., 1973. A comparison of benthic microalgal production measured by C^{14} and oxygen methods. J. Fish. Res. Board Can., 30: 309—312.

Jannasch, H.W., 1969. Estimation of bacterial growth rates in natural waters. J. Bacteriol. 99: 156—160.

Jannasch, H.W. and Pritchard, P.H., 1972. The role of inert particulate matter in the activity of aquatic microorganisms. Mem. Ist. Ital. Idrobiol., (Suppl.), 29: 289—306.

Jørgensen, C.B., 1966. Biology of Suspension Feeding. Pergamon Press, Oxford, 357 pp.

Karl, D.M., Haugsness, J.A., Campbell, L. and Holm-Hansen, O., 1978. Adenine nucleotide extraction from multicellular organisms and beach sand: ATP recovery, energy charge ratios and determination of carbon/ATP ratios. J. Exp. Mar. Biol. Ecol., 34: 163—181.

Khoja, T. and Whitton, B.A., 1971. Heterotrophic growth of blue-green algae. Arch. Mikrobiol., 79: 280—282.

Koblentz-Mishke, O.J., Volkovinsky, V.V. and Kabanova, J.G., 1970. Plankton primary production of the world ocean. In: W.S. Wooster (Editor), Scientific Exploration of the South Pacific. Natl. Acad. Sci., Washington, D.C., pp. 183—193.

Leach, J.H., 1970. Epibenthic algal production in an intertidal mudflat. Limnol. Oceanogr., 15: 514—521.

Lewin, J.C. and Hellebust, J.A., 1975. Heterotrophic nutrition of the marine pennate diatom *Navicula pavillardi* Hustedt. Can. J. Microbiol., 21 (9): 1335—1342.

Lewin, J.C. and Lewin, R.A., 1960. Autotrophy and heterotrophy in marine littoral diatoms. Can. J. Microbiol., 6: 127—134.

Lewis, J.B., 1977. Processes of organic production on coral reefs. Biol. Rev., 52: 305—347.

Lorenzen, C.J., 1976. Primary production in the sea. In: D.H. Cushing and J.J. Walsh (Editors), The Ecology of the Seas. Blackwell, Oxford, pp. 173—185.

Maerker, M. and Szekielda, K.H., 1976. Chlorophyll determination of phytoplankton: a comparison of in vivo fluorescence with spectrophotometric absorption. J. Cons. Int. Explor. Mer., 36 (3): 217—219.

Mann, K.H., 1972a. Ecological energetics of the seaweed zone in a marine bay on the Atlantic coast of Canada. II. Productivity of the seaweeds. Mar. Biol. 14: 199—209.

Mann, K.H., 1972b. Macrophyte production and detritus food chains in coastal waters. Mem. Ist. Ital. Idrobiol., 29 (suppl.): 353—383.

Mann, K.H., 1973. Seaweeds: their productivity and strategy for growth. Science, 182 (4116): 975—980.

Mann, K.H., 1976. Production on the bottom of the sea. In: D.H. Cushing and J.J. Walsh (Editors), The Ecology of the Seas. Blackwell, Oxford, pp. 225—250.

Mann, K.H. and Chapman, A.R.O., 1975. Primary production of marine macrophytes. In: J.P. Cooper (Editor), Photosynthesis and Productivity in Different Environments. Cambridge University Press, Cambridge, pp. 207—223.

Marshall, N., Oviat, C.A. and Skauen, D.M., 1971. Productivity of the benthic microflora of shoal estuarine environments in southern New England. Int. Rev. Ges. Hydrobiol., 56: 947—995.

McRoy, P. and Barsdate, R.J., 1970. Phosphate absorption in eelgrass. Limnol. Oceanogr., 15: 6—13.

Meeks, J.C., 1974. Chlorophylls. In: W.D.P. Stewart (Editor), Algal Physiology and Biochemistry. Bot. Monogr., 10 (15): 161—175.

Menzel, D.W., 1974. Primary productivity and the oxidation of organic matter. In: E.D. Goldberg (Editor), The Sea, 5. Marine Chemistry. Wiley—Interscience, New York, N.Y., pp. 659—678.

Menzies, R.J. and Rowe, G.T., 1969. The distribution and significance of detrital turtle grass, *Thalassia testudinata*, on the deep-sea floor off North Carolina. Int. Rev. Ges. Hydrobiol., 54: 217—222.

Menzies, R.J., Zaneveld, J.S. and Pratt, R.M., 1967. Transported turtle grass as a source of organic enrichment of abyssal sediments off North Carolina. Deep-Sea Res., 14: 111—112.

Menzies, R.J., George, R.Y. and Rowe, G.T., 1973. Abyssal Environment and Ecology of the World Oceans. Wiley—Interscience, New York, N.Y.

Meyer-Reil, L.A., 1977. Bacterial growth rates and biomass production. In: G. Rheinheimer (Editor), Microbial ecology of a brackish water environment. Ecol. Stud., 25 (16): 223—236.

Moiseev, P.A., 1969. Living Resources of the World Ocean (Pishchevaia promyshlennost). Moscow, 338 pp. (English translation, 1971, Jerusalem, Israel Program for Scientific Translations, IPST Cat. No. 5954, 334 pp.)

Mullin, M.M., 1969. Production of zooplankton in the ocean: the present status and problems. In: H. Barnes (Editor), Oceanography and Marine Biology, 7. Allen and Unwin, London, pp. 293—314.

Nalewajko, C., 1966. Photosynthesis and excretion in various planktonic algae. Limnol. Oceanogr., 11: 1—10.

Odum, E.P. and De la Cruz, A.A., 1967. Particulate organic detritus in a Georgia salt-marsh—estuarine ecosystem. In: G.H. Lauff (Editor), Estuaries. Am. Assoc. Adv. Sci., Publ., No. 83: 383—388.

Pamatmat, M.M., 1968. Ecology and metabolism of a benthic community on an intertidal sandflat. Int. Rev. Ges. Hydrobiol., 53: 211—298.

Parsons, T.R., 1975. Particulate organic carbon in the sea. In: J.P. Riley and G. Skirrow (Editors), Chemical Oceanography. Academic Press, London, pp. 365—383.

Parsons, T.R., Takahashi M. and Hargrave, B., 1977. Biological Oceanographic Processes. Pergamon Press, Oxford, 332 pp.

Petipa, T.S., Pavlova, E.V. and Mironov, G.N., 1970. The food web structure, utilization and transport of energy by trophic levels in the planktonic communities. In: J.H. Steele (Editor), Marine Food Chains. Oliver and Boyd, Edinburgh, pp. 142—167.

Platt, T. and Subba Rao, D.V., 1975. Primary production of marine microphytes. In: J.P. Cooper (Editor), Photosynthesis and Productivity in Different Environments. IBP 3, Cambridge University Press, Cambridge, pp. 249—280.

Pomeroy, L.R., 1974. The ocean's food web, a changing paradigm. Bioscience, 24 (9): 499—504.

Raven, J.A., 1974. Carbon dioxide fixation. In: W.D.P. Stewart (Editor), Algal Physiology and Biochemistry. Bot. Monogr., 10 (15): 434—455.

Raymont, J.E.G., 1963 (reprint 1976). Plankton and Productivity in the Oceans. Pergamon Press, Oxford, 660 pp.

Raymont, J.E.G., 1966. The production of marine plankton. In: J.B. Cragg (Editor), Advances in Ecological Research. Academic Press, London, pp. 117—205.

Reynolds, N., 1973. The estimation of the abundance of ultraplankton. Br. Phycol. J., 8: 135—146.

Riley, G.A., 1944. The carbon metabolism and photosynthetic efficiency of the earth as a whole. Am. J. Sci., 32: 129—134.

Riley, G.A., 1951. Oxygen, phosphate and nitrate in the Atlantic Ocean, Bull. Bingham Oceanogr. Coll., 13 (1): 1—126.

Riley, G.A., 1963. Organic aggregates in seawater and the dynamics of their formation and utilization. Limnol. Oceanogr., 8: 372—381.

Riley, G.A., 1972. Patterns of production in marine ecosystems. In: J.A. Wiens (Editor), Ecosystem Structure and Function. Oregon State University Press, Corvallis, Ore., pp. 91—112.

Rodin, L.E., Bazilevich, N.I. and Rozov, N.N., 1975. Productivity of the world's main ecosystems. In: D.E. Reichle, J.F. Franklin and D.W. Goodall (Editors), Productivity of World Ecosystems. Proc. Symp. U.S. Natl. Committee IBP, Natl. Acad. Sci., Washington, D.C., pp. 13—26.

Rowe, G.T. and Menzies, R.J., 1969. Zonation of large benthic invertebrates in the deepsea off the Carolinas. Deep-Sea Res., 16: 531—537.

Russell-Hunter, W.D., 1970. Aquatic Productivity: An Introduction to Some Basic Aspects of Biological Oceanography and Limnology. Macmillan Company, Collier—Macmillan, London, 1, 306 pp.

Ryther, J.H., 1959. Potential productivity of the sea. Science, 130: 601—608.

Ryther, J.H., 1963. Geographic variations in productivity. In: M.N. Hill (Editor), The Sea, 2. Wiley—Interscience, New York, N.Y., pp. 347—380.

Ryther, J.H., 1969. Photosynthesis and fish production in the sea. Science: 166: 72—76.

Ryther, J.H. and Yentsch, C.S., 1957. The estimation of phytoplankton production in the ocean from chlorophyll and light data. Limnol. Oceanogr., 2: 281—286.

Schaefer, M.B., 1965. The potential harvest of the sea. Trans. Am. Fish. Soc., 94: 123—128.

Sculthorpe, C.D., 1967. The Biology of Aquatic Plants. Edward Arnold, London, 610 pp.

Semina, H.J., 1972. The size of phytoplankton cells in the Pacific Ocean. Int. Rev. Ges. Hydrobiol., 57: 177—205.

Sharp, J.H., 1977. Excretion of organic matter by phytoplankton: Do healthy cells do it? Limnol. Oceanogr., 22: 381—399.

Sieburth, J.Mc.N., 1968. Observations on planktonic bacteria in Narragansett Bay, Rhode Island; a resumé. Bull. Misaki Biol. Inst., Kyoto Univ., 12: 49—64.

Sieburth, J.Mc.N., 1977. International Helgoland Symposium: convener's report on the informal session on biomass and productivity of microorganisms in planktonic ecosystems. Helgol. Wiss. Meeresunters., 30: 697—704.

Skopintsev, B.A., 1971. Recent advances in the study of organic matter in the oceans. Oceanology, 11: 775—789.

Smayda, T.J., 1969. Some measurements of the sinking rate of fecal pellets. Limnol. Oceanogr., 14: 621—625.

Smayda, T.J., 1970. The suspension and sinking of phytoplankton in the sea. Oceanogr. Mar. Biol. Annu. Rev., 8: 353—414.

Smayda, T.J., 1971. Normal and accelerated sinking of phytoplankton in the sea. Mar. Geol., 11: 105—122.

Sorokin, Y.I., 1971a. On the role of bacteria in the productivity of tropical oceanic waters. Int. Rev. Ges. Hydrobiol., 56: 1—48.

Sorokin, Y.I., 1971b. Bacterial populations as components of oceanic ecosystems. Mar. Biol., 11: 101—105.

Sorokin, Y.I., 1974. Bacteria as a component of the coral reef community. In: A.M. Cameron et al. (Editors), Proc. Sec. Int. Symp. Coral Reefs, I. The Great Barrier Reef Committee, Brisbane, Qld., pp. 3—10.

Sorokin, Y.I., 1978a. Microbial production in the coral reef community. Arch. Hydrobiol., 83 (3): 281—323.

Sorokin, Y.I., 1978b. Decomposition of organic matter and nutrient regeneration. In: O. Kinne (Editor), Marine Ecology, IV. Dynamics. Wiley, Chichester, pp. 501—516.

Steemann Nielsen, E., 1952. The use of radio-active carbon (C^{14}) for measuring organic production in the sea. J. Cons. Int. Explor. Mer., 18: 117—140.

Steemann Nielsen, E., 1975. Marine Photosynthesis, With Special Emphasis on the Ecological Aspects. Elsevier Oceanography Series, 13. Elsevier, Amsterdam, 141 pp.

Steemann Nielsen, E. and Jensen, A., 1957. Primary oceanic production, the autotrophic production of organic matter in oceans. Galathea Rep., I: 47—135.

Strickland, J.D.H., 1965. Production of organic matter in the primary stages of the marine foodchain. In: J.P. Riley and G. Skirrow (Editors), Chemical Oceanography, Academic Press, New York, N.Y., pp. 477—610.

Strickland, J.D.H., 1970. Introduction. In: J.H. Steele (Editor), Marine Food Chains. Oliver and Boyd, Edinburgh, pp. 3—5.

Strickland, J.D.H. and Parsons, T.R., 1972. A practical handbook of seawater analysis. Fish. Res. Board Can. Bull., 167: 1—310 (2nd ed.).

Sverdrup, H.N., 1955. The place of physical oceanography in oceanographic research. J. Mar. Res., 14: 287—294.

Taniguchi, A., 1972. Geographical variation of primary production in the western Pacific Ocean and adjacent seas with reference to the interrelations between various parameters of primary production. Mem. Fac. Fish. Hokk. Univ., 19: 1—33.

Tenore, K.R., 1977. Food chain pathways in detrital feeding benthic communities: a review, with new observations on sediment resuspension and detrital recycling. In: B.C. Coull (Editor), Ecology of Marine Benthos. Belle W. Baruch Library Mar. Science No. 6. University of South Carolina Press, Columbia, S.C., pp. 37—54.

Tranter, D.J., 1976. Herbivore production. In: D.H. Cushing and J.J. Walsh (Editors), The Ecology of the Seas. Blackwell, Oxford, pp. 186—224.

UNESCO, 1972. Atlas of the Living Resources of the Seas. FAO Department of Fisheries, UNESCO, Rome.

UNESCO, 1973. A guide to the measurement of marine primary production under some special conditions. Monogr. Oceanogr. Methodol., 3: 1—73 (Unesco, Paris).

Vinogradov, M.E., 1970. Vertical distribution of oceanic zooplankton. Akad. Nauk, S.S.S.R., Inst. Okeanol. (English transl. by Israel Progr. for Sci. Transl., Jerusalem, for U.S. Dep. Interior and NSF, pp. 1—339).

Walsh, J.J., 1974. Primary production in the sea. In: Proc. First Int. Congr. Ecol., The Hague, The Netherlands. PUDOC, Wageningen, pp. 150—154.

Watt, W.D., 1971. Measuring the primary production rates of individual phytoplankton species in natural mixed populations. Deep-Sea Res., 18: 329—339.

Wangersky, P.J., 1977. The role of particulate matter in the productivity of surface waters. Helgol. Wiss. Meeresunters., 30: 546—564.

Wangersky, P.J., 1978. Production of dissolved organic matter. In: O. Kinne (Editor), Marine Ecology, IV. Dynamics. Wiley, Chichester, pp. 115—220.

Whittaker, R.H. and Likens, G.E., 1973. Carbon in the biota. In: G.M. Woodwell and E.V. Pecan (Editors), Carbon and the Biosphere. Proc. 24th Brookhaven Symp. Biol. New York, 1972. Technical Inform Center, Office Inform. Services U.S. Atomic Energy Comm. Symp. Ser., 30: 281—302.

Whittaker, R.H. and Likens, G.E., 1975. The biosphere and man. In: H. Lieth and R.H. Whittaker (Editors), Primary Productivity of the Biosphere. Ecological studies, 14. Springer Verlag, Berlin, pp. 305—328.

Whittle, K.J., 1977. Marine organisms and their contribution to organic matter in the ocean. Mar. Chem., 5: 381—411.

Wiebe, W.J. and Pomeroy, L.R., 1972. Microorganisms and their association with aggregates and detritus in the sea: a microscopic study. Mem. Ist. Ital. Idrobiol. (Suppl.) 29: 325—352.

Wolff, T., 1977. Diversity and faunal composition of the deep-sea benthos. Nature 267: 780—788.

Woodwell, G.M., Rich, P.H. and Hall, C.A.S., 1973. Carbon in estuaries. In: G.M. Woodwell and E.V. Pecan (Editors), Carbon and the Biosphere. Proc. 24th Brookhaven Symp. Biol. New York, 1972. Technical Inform Center, Office Inform. Services, U.S. Atomic Energy Comm. Symp. Ser., 30: 221—239.

Woodwell, G.M., Whittaker, R.H., Reiners, W.A., Likens, G.E., Delwiche, C.C. and Botkin, D.B., 1978. The biota and the world carbon budget. Science 199: 141—146.

Chapter 4

NON-LIVING PARTICULATE MATTER

G. CAUWET

1. INTRODUCTION

The carbon cycle in the oceans can be schematized by: (1) the fixation of carbon dioxide during primary production, (2) secondary production resulting from phytoplankton consumption, and (3) the destruction and mineralization of dead organisms restoring carbon dioxide and nutrients to the water. Nevertheless, this cycle is not perfect and a fraction of the organic matter escapes as soluble or particulate matter. This matter, also called detritus, supplements the organics added from atmospheric dusts, river inputs and land weathering, and together form the non-living organic matter which participates in the oceanic geochemical cycles and the sedimentation phenomena (Fig. 1).

Considering the volume of the whole oceans, these fractions are quantitatively very large. Estimates from the data of several reports (Cauwet, 1978) suggest that particulate organic carbon amounts to about 2×10^{16} g, which is five times more than the phytoplankton biomass and ten times less than dissolved organic carbon (Fig. 2). This enormous quantity of organic detri-

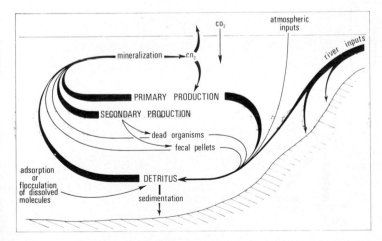

Fig. 1. Situation of detritus in the carbon cycle.

tus, disseminated in the oceans can be very complex and variable in nature, depending on its source, and we will see that the dividing line between dissolved and particulate matter is not well defined.

Most of the studies of particulate matter in the oceans concern themselves with suspensions recovered on filters and do not take into account the difference between detritus and living matter. Therefore, it is very difficult to estimate the respective impact of biological and chemical processes. Although there are many extensive studies of particles in the euphotic layer (Sutcliffe et al., 1970; Zeitschel, 1970; Flemer and Biggs, 1971; Wada and Hattori, 1976), only by examination of deep-sea suspensions can we understand the geochemistry of marine organic detritus.

Up until recently, only a few studies were concerned with the organic chemistry of deep-sea suspensions. Two major reasons explain the previous lack of interest: particulate organic carbon is about ten times more abundant in the euphotic layer than in deep water (Hobson and Menzel, 1969; Jacobs and Ewing, 1969; Menzel and Ryther, 1970; Williams, 1971; Copin-Montegut and Copin-Montegut, 1973) and the results of elemental analysis of deep-sea particles, showing nearly constant values of organic carbon (Copin-Montegut and Copin-Montegut, 1972; Wangersky, 1974, 1976), were in the past interpreted as being indicative of the stability of organic matter. As we know now, no organic matter can be considered as definitively non-degradable, but it is rather a question of degradation rate. Recent studies (Eadie and Jeffrey, 1973; Pustel'nikov, 1976) have shown that degradation occurs at any depth in the water column and at the sediment—water interface (Bianchi et al., 1977).

We can distinguish three zones where the phenomena are quite different:

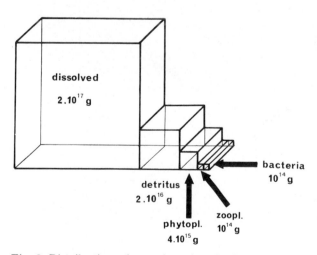

Fig. 2. Distribution of organic carbon in the oceans.

(1) The euphotic layer (0—500 m), where the living organisms produce the greatest part of organic matter, recycling more of the detritus.

The contribution of phytoplankton to the organic carbon of deep waters is estimated to be 4—6 g C m^{-2} yr^{-1}, which represents only 2—5% of the production (Skopintsev, 1972; Menzel, 1974).

(2) Deep waters, where the biota consists essentially of bacteria. During the settling of particles, with sinking rates depending upon their sizes and densities, chemical processes of sorption and desorption may occur. Slow mineralization of organic matter by microorganisms modifies the composition of the particles, returning some elements to the solution. In these waters the exchange processes between particulate and dissolved matter are of particular relevance.

(3) The sediment—water interface, where the organic matter slowly settles to undergo further transformation.

In this chapter particular attention is paid to the intermediate zone and the chemical reactions taking place.

2. RECOVERY TECHNIQUES

Two different techniques may be utilized to recover particulate matter: filtration or centrifugation. The latter is not widely used and is often reserved for chemical analysis of large volumes of water. Nevertheless, Breck (1978), has demonstrated that the method may produce a 30% better recovery.

Filtration is most widely employed, but the variety of filter types and porosities used makes any comparison between the results difficult. Three main types are commonly employed: membranes made of cellulose esters or polycarbonate, glass fiber filters and silver filters.

The membrane filters (numerous types and trade marks) are useful for weight determination, microscopic studies and inorganic analyses. The tendency seems to be now to change from thick, fibrous filters with a porosity defined as average, to thin perforated membranes. Glass fiber and silver filters are exclusively reserved for carbon analysis and chemical studies on organic matter. Their main disadvantage is to have relatively high and poorly defined porosities (Gordon, 1970a; Gordon and Sutcliffe, 1974).

Various pore sizes have been used, ranging from 0.1 to 5 μm; 0.45 μm is the theoretical limit discerning between particulate and dissolved material. Consequently, dissolved material includes organic matter in a true solution as well as colloidal material and particles up to about 0.5 μm. Although the porosity of filters is generally well defined, Sheldon and Sutcliffe (1969) have shown that they can retain a large number of smaller particles. Thus, studies on particulate matter can only be compared if material is recovered with identical filters. The size of particles retained can be very different from the assumed filter porosity. These filters and membranes can also adsorb significant amounts of dissolved organic matter; Saliot (1975) has shown that they can retain as much as 22% of a fatty acid solution and corroborated the previous results of Quinn and Meyers (1971), who found up to 80% retention of hepta decanoic acid on 0.45-μm filters.

The theoretical limit fixed at 0.45 μm is purely arbitrary, and the previous observations encourage us to use techniques to lower this limit, to allow the recovery of very small particles and colloids.

The volumes filtered are quite different, ranging from milliliters for microscopic studies to cubic meters for extensive chemical studies (Handa et al., 1970). Recovering sufficient material without undergoing any degradation is one of the main difficulties for precise organic chemistry studies. The utilization of immersed pumps allows the filtration of large volumes of sea water, even when the ship is underway (Krishnaswami and Sarin, 1976).

In all cases filtration must be performed soon after the sampling procedure in order to avoid any deposition onto the sampler or container surfaces or aggregation of small particles.

In some experiments, sediment traps have been used to collect particles (Spencer et al., 1978). They are generally a construction of plastic cylinders or funnels which are anchored on the bottom (Håkanson, 1976; Honjo, 1978) or attached to a wire at various depths. The majority of these experiments have been performed in lakes, due to the difficulty of handling and recovery in the open sea. Zeitschel et al. (1978) have recently employed a similar system in the Baltic Sea with some apparent success. Although such techniques are useful in evaluating the sedimentation rate and its seasonal variations, the wide range of results reported, from $0.2 \text{ g m}^{-2} \text{ day}^{-1}$ (Kimmel et al., 1977) to about $100 \text{ g m}^{-2} \text{ day}^{-1}$ (White and Wetzel, 1973), makes it difficult to assess whether these are true differences between studied areas or discrepancies in the techniques employed. Kirchner (1975) showed that size and shape of the collecting vessel are quite important in determining the quantity and type of material retained. Although the technique is generally used to recover the whole suspension, it has sometimes been utilized to evaluate the particulate organic-carbon flux in calm waters (Burns and Pashley, 1974).

Although there appear to be technical difficulties in employing the sediment traps, the methodology is a promising one which requires some perfecting. The utilization of floating traps, not fixed to anchor but carried in a water package will perhaps give more representative results of the true sedimentation patterns.

3. PARTICULATE ORGANIC CARBON (POC)

3.1. Methods

The introduction of wet combustion methods for dissolved organic carbon (Parsons and Strickland, 1959; Duursma, 1961) allowed a precise but time-consuming determination of carbon in sea water. Rapid determination of dissolved organic carbon (DOC) and POC became possible with the persulfate oxidation method developed by Menzel and Vaccaro (1964).

More recently, the development of dry combustion techniques led to rapid and reproducible methods (Sharp, 1974; Telek and Marshall, 1974). Although many laboratories are still utilizing a wet oxidation method, it seems that dry combustion will be the universal technique in the future. Glass fiber filters are generally folded in crucibles and burned in a furnace. The evolved CO_2 is then analysed in CHN analysers or specific carbon analysers (LECO Corp.). A recent intercalibration exercise (Cauwet, in prep.) compared the results obtained from different laboratories. No important variations were observed. A method derived from the one described previously for sediments (Cauwet, 1975) seems to give good reproducible results. According to the variability of the results from one author to another, it seems necessary to develop large intercalibration exercises in order to allow the comparison of the results.

Attempts to evaluate the organic matter content by microscopic observation seem to give results which are non-reproducible and quite different from the chemical analyses (Mel'nikov, 1976; Pustel'nikov, 1976).

3.2. Distribution of particulate organic carbon

There have been many studies of the distribution of POC in the oceans (Dalpoint and Newell, 1963; Bogdanov, 1965; Holm-Hansen et al., 1966; Hobson, 1967; Menzel, 1967; Riley, 1970; Loder, 1971; Wangersky, 1976; and many others), but many of these works were carried out by investigators using different sampling techniques and analysis methods. It is therefore difficult to compare the results. Nevertheless, it can be observed that there is a trend to find much smaller quantities of POC with the improvement of techniques.

Much of the data concerns samples taken at the surface or a few meters below, including both living and non-living particles. They are of little interest to the geochemist, except for evaluating the bulk of organic matter present in the euphotic zone. Of course, related to the productivity, POC is more abundant in the surface waters, but it seems to be particularly concentrated in the surface layer. The values reported by numerous investigators (Menzel and Ryther, 1964; Riley et al., 1965; Wangersky and Gordon, 1965; Szekielda, 1967; Gordon, 1970b; Melnikov, 1976) range from 20 to 200 μg C 1^{-1} in the surface layer compared with 10–60 μg C 1^{-1} in the underlying and deeper waters. According to recent data by Banoub and Williams (1972) and Wangersky (1976), these values seem slightly overestimated, particularly for the deep waters. We can probably consider that the true values are close to those reported by Wangersky (1976) for the Pacific and Atlantic Oceans: 20–100 μg C 1^{-1} for surface waters, 4–30 μg C 1^{-1} in the

76

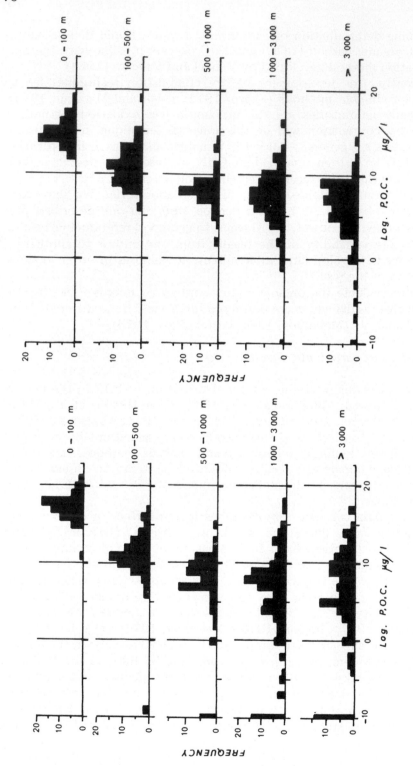

South Atlantic Ocean

Pacific Ocean

Fig. 3. Frequency distribution of log POC, partitioned by depth (from Wangersky, 1976, p. 453).

intermediate layer (down to 500 m) and about $0-10 \mu g \ C \ l^{-1}$ in deep waters.

The frequency distribution of POC with depth as reported by this author is of some interest (Fig. 3) and demonstrates a close agreement between the POC distribution in the Atlantic and Pacific Oceans, with the exception of surface values which are higher in South Atlantic waters. These differences in surface waters are however not significant because of the heterogeneity of the samples containing both living matter and detritus. The effects of seasonal or local hydrographic conditions are limited mainly to surface waters, which as a consequence show differences in productivity.

The slow decrease of the values with depth is more interesting, since it supports the idea that organic matter is utilized by organisms at any depth. Differences between old deep waters and deep layers of young cold waters, richer in organic carbon, seem to confirm this hypothesis.

In deep waters, when sampling and analyses are comparable, no marked difference exists between the oceans. The slight decrease with depth may be disrupted by higher values, corresponding to particular water masses or to nepheloid layers. The observation (Cauwet, unpublished) that particles from bottom water layers in the Mediterranean are richer (6% C) than surface samples (3—4% C) points to a resuspension of fine particles from the sediment, emphasizing the part played by bottom currents. This obliges us to consider that horizontal transport may be a deciding parameter with regard to sinking, and serve to explain the constant values obtained in deep waters (Menzel, 1967).

When studying the variations of POC with depth we have to consider that filtration is often made on filters of high porosity ($0.8-1.2 \mu m$). Any variation in size of particles with depth, occurring (as a product of chemical dissolution or biological processes) therefore, would lead to an apparent loss of suspended matter. The utilization of finer filters may limit this drawback.

4. SIZE AND MORPHOLOGY OF PARTICLES

Several investigators have been interested in the microscopic observation of particles (Gordon, 1970a; Mel'nikov, 1976, 1977). They agree as to the classes of particles greater than $1 \mu m$, in surface waters mainly comprising aggregates (60—80%), flakes (10—20%), fragments (3—5%) and non-identifiable particles.

Aggregates are composed of a variety of organic and inorganic subunits, bound through a gelatin-like matrix, consisting mainly of carbohydrates, ranging from 10 to 200 μm. These are very abundant in surface waters but disappear in deep waters, suggesting a possible digestion by organisms. Flakes are fine transparent plate-like particles, mainly proteinaceous, and their abundance in deep or surface waters is quite similar. Gordon (1970a) suggested that they probably include material from films produced by bubbling. These particles can be compared, according to their shape, appearance and

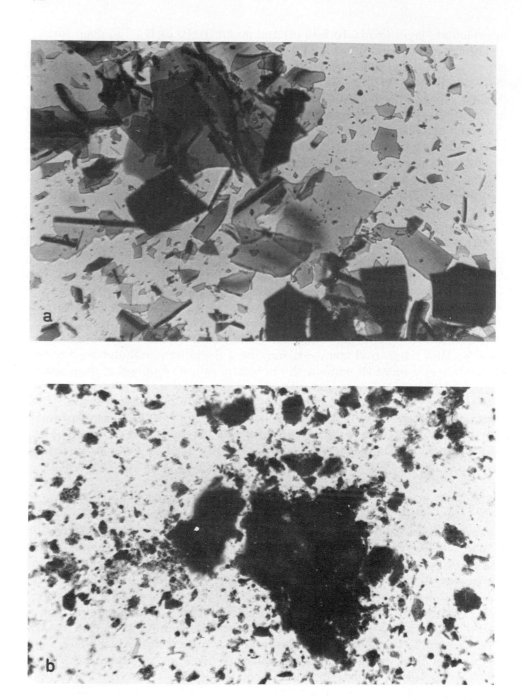

Fig. 4. Organic particles extracted from sediments. a. Humic acids. b. Kerogen.

colour to the flocculates of naturally occurring organic polymers, such as humic acids and kerogen issued from marine sediments (Fig. 4). It is probable that these particles are polymeric materials precipitated in sea water from organic colloids which may be composed predominantly of carbohydrates, proteins or lipids (Breger, 1970). Compared with aggregates they are more homogeneous and contain fewer mineral elements. On the other hand, they seem to be a suitable support for bacteria, being probably a reserve of substances, slowly assimilable, as is humic matter in soils. Consequently, their abundance is quite constant with depth and their slow consumption is in equilibrium with their formation from dissolved polymers. According to the work of Gordon (1970a), concentrations of aggregates are higher in surface samples and decrease sharply within the euphotic zone down to 1—2 particles ml^{-1} in deep waters. This may indicate that they are formed by bubbling in surface waters from dissolved molecules. These organic substances are easily assimilable and the aggregates are rapidly assimilated by organisms.

Smaller particles, described as unclassifiable particles (Gordon, 1970a) or fine fraction (Mel'nikov, 1977), are very abundant at any depth. Mel'nikov (1976) estimates their content to be about 60—90% of particulate organic carbon, even in deep waters. This seems to be slightly overestimated, probably because he measured POC by chemical analysis while the finely dispersed fraction was only estimated by microscopical analysis of the seston. These results are somewhat different from the observations of Gordon (1970a) who found that fine particles amount to 25—45% of POC in surface waters and 50—65% of POC in deep waters. This discrepancy emphasizes the need for further studies dealing with very fine particles down to the colloidal state, since these particles are able to form larger particles spontaneously by aggregation (Sheldon et al., 1967, 1972).

5. CHEMICAL COMPOSITION

Although the elementary composition of particulates has been widely reported (Parsons and Strickland, 1962; Menzel and Ryther, 1964; Holm-Hansen et al., 1966), this is not the case for the biochemical composition. The difficulty in collecting enough material for such studies is probably the cause of this discrepancy. Particles are more often than not composed of both mineral and organic substances and their elemental composition is not consistent. The separation of organic and mineral constituents reveals that organic particles represent 75—80% of the suspended matter (Lascaratos, 1974).

More widely used for descriptive purposes is the C/N ratio. From 5—8 in surface waters it increases to 10—12 with depth, indicating that proteins are more readily utilized than are carbohydrates. Handa (1970) and Handa et al.

(1972) confirmed this result, showing that the greater part of carbohydrates is present as polysaccharides which might be expected to be refractory. This is in opposition with Gordon's observation (1970a), that aggregates exhibiting intense coloration upon carbohydrate staining, quickly disappear within and below the euphotic zone, while flakes, mainly proteinaceous, are present in deep waters. Other works (Parsons and Strickland, 1962; Holm-Hansen et al., 1966; Hobson, 1967) report variable proportions of carbohydrates and ratios to protein-like material. Agatova and Bogdanov (1972) found a somewhat different distribution of compound classes with an increasing percentage of lipids with depth (10—20%) and few carbohydrates. The very complete works of Handa (1970), Handa and Tominaga (1969) and Handa et al. (1972), demonstrated the existence of a polysaccharide, which they found to be D-Glucan, and hydrolysable fractions yielding glucose, galactose, mannose, xylose and glucuronic acid. Their results suggest that carbohydrates may have been underestimated in the work of Agatova and Bogdanov (1972).

In some works (e.g. Parsons and Strickland, 1972), some discrepancies appear between the proportions of the biochemical constituents and the total organic carbon. This raises questions about the very nature of suspensions, especially in the deep sea.

With the exception of the works of Handa and co-workers, most of the determinations of the biochemical composition of suspensions have been made by direct colorimetric measurements. However, the molecular structure of particulate organic matter may be such that terminal functional

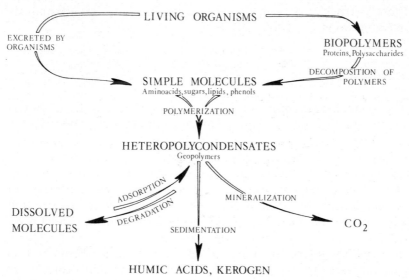

Fig. 5. Formation of condensed molecules from biogenic substances.

groups are hidden or blocked by the chemical environment, and do not react. On the other hand, some complex molecules may contain several functional groups and react to both reagents, being counted as two molecules. It is highly probable that most of the organic matter in the deep sea is composed of heteropolycondensed molecules, bearing numerous functional groups (e.g. humic-like material). These substances, resulting from the decomposition and copolymerization of biological material, are relatively stable towards assimilation, and slowly attain their definite composition during sinking, exchanging widely with dissolved organic molecules (Fig. 5).

As seen, we have very little information about the chemical composition of organic particles. It is obvious that we have much to learn about the chemical processes in sea water from the investigation of the chemical composition of dissolved and particulate organic matter. The following decades will probably give many answers to the numerous remaining questions.

6. ORIGIN AND REMOVAL OF ORGANIC PARTICLES

Particulate organic matter can be brought by rivers from land weathering or by atmospheric inputs to the oceans, but the major part seems to originate from *in situ* production of living organisms.

6.1. Land and atmospheric inputs

We have only a limited knowledge about inputs of organic matter to oceans via rivers, primarily due to the relatively few measurements of organic carbon in major river systems and to the uncertainty about the behaviour of organic substances upon entering the sea. A rough estimate of the particulate organic matter arriving from rivers may be about 10^{13} to 10^{14} g yr^{-1}, but this figure should be viewed with some reserve since the changes occurring in estuaries are complex. Many authors (Sieburth and Jensen, 1968; Nissenbaum and Kaplan, 1972; Hair and Basset, 1973) estimate that a major part of dissolved organic matter, mainly composed of humic substances, never reaches the open ocean, but is deposited in the estuarine and near-shore environment. However, the knowledge of some exchange mechanisms, and the comparison of the solubilities of calcium, magnesium and strontium humic complexes (Vile-Cambon, 1977), lead us to doubt the validity of these results. Sholkovitz (1976) estimates that only 3—11% of the dissolved organic matter is removed by flocculation upon mixing with seawater. The total particulate organic carbon input to the coastal seas from rivers should then be in the range of 0.5 to 1.1×10^{14} g yr^{-1}, where most of it never reaches the open ocean and is decomposed or deposited in coastal areas.

The atmospheric inputs are more widely dispersed and influence the whole ocean. The input of organic carbon by way of precipitation was estimated at 2.2×10^{14} g C yr^{-1} (Williams, 1971), but it can also be calculated

(Hunter and Liss, 1977) that bubble bursting injects into the atmosphere 1 to 2.5×10^{15} g C yr^{-1}, which is about one order of magnitude greater than the estimated inputs, but which may also be considered to be a recycling process.

Thus, the sea can apparently supply sufficient material to account for the carbon in rain, although no information exists concerning the efficiency of the transfer from the bubble bursting to the rainfall. It would appear therefore, that the atmospheric inputs enrich the sea with particles, while the surface layer increases the aerosol content of the atmosphere.

6.2. In situ formation of organic particles

The principal input of organic carbon to the ocean is constituted by the marine organisms, mainly through photosynthesis. Though the greater part is recycled, the primary production contributes strongly, by direct or indirect ways, to the formation of organic particulates. We can distinguish four main processes leading to particle production: (1) direct formation of detritus (fragments, faecal pellets); (2) agglomeration of bacteria; (3) aggregation of organic molecules by the bubbling in surface layers; and (4) flocculation or adsorption of dissolved substances onto mineral particles.

Fragments of organized tissues observed in the euphotic zone (Gordon, 1970a), are scarce in deep waters. In the previous chapter we saw that they represent only 3—5% of the particles present in surface waters. Faecal pellets are present in deep waters but due to their high settling rate it is difficult to recover this material on filters.

More complex is the *in situ* formation of organic particles. Baylor et al. (1962) and Sutcliffe et al. (1963) have shown that organic matter is adsorbed on bubbles in sea water. They found that phosphate was strongly adsorbed on bubbles and concentrated in the surface layer, and demonstrated that organic matter was associated with the adsorption process and may serve as an anion binder. Further experiments have shown that phosphates are bound to the organic molecules with molecular weights in excess of 300. They suggest that such large organic active molecules adsorb onto bubbles and produce monomolecular films which may be aggregated into insoluble organic particles. The agitation of foaming produces repeated collisions and results in semi-stable colloidal suspension of organic materials. Further aggregation allows the particles to settle out (Menzel, 1966; Kane, 1967). The same experiments applied to filtered sea water (Riley et al., 1965) revealed the formation of such particles, allowing the recovery of up to 70 μg l^{-1} of particulate carbon.

Experiments on water filtered through 0.22-μm filters showed a decreasing yield, pointing out the role of bacteria. This may explain that particle formation is rapidly inhibited unless a removable process is active in the water. Nevertheless, the phenomenon remains operative even in sterilized samples.

The formation of aggregates by bubbling of sea water seems to be a permanent but complex process, involving chemical and bacterial processes. Large molecules may play a major role in these phenomena either as source material or as support to the bacterial populations.

Besides the mainly organic particles, the interaction between mineral fragments and dissolved molecules leads to the formation of partly organic suspensions. This will be examined in a following section.

6.3. Removal of organic particles

As far as the organic content of sediments is concerned, it should be noted that the greater part of organic material undergoes recycling before it settles out. In the Baltic sea, Pustel'nikov (1976) estimated that the organic content in the last meter of the water column above the sediment represents about 12% of the surface water content and about 1.5% of the total primary production.

Some years ago, Menzel and Goering (1966) and Menzel (1967), while finding a homogeneous distribution of POC, concluded that particulate organic matter is not further oxidized below the euphotic layer. The sensitivities of analytical methods, however, were probably insufficient to detect small variations. Wangersky (1976) has recently demonstrated a slow decrease of POC with depth when verifying that the younger South Antarctic bottom water contains more POC (6.7 μg C 1^{-1}) than the older northern waters (4.9 μg C 1^{-1}). The dissolution of particulate trace metals associated with organic matter with depth suggests an oxidation during sinking (Krishnaswami and Sarin, 1976). The POC oxidation rate can be estimated to be 0.4×10^{-3} mg kg^{-1} yr^{-1} (Craig, 1971), based on an oxygen consumption of 1.5×10^{-3} ml kg^{-1} yr^{-1}.

Banoub and Williams (1972) measured the eutrophic activity in the Mediterranean Sea. Surprisingly they found no difference between this area and the English Channel, which is ten times more productive. This might be connected to the turnover rate of sugars and amino acids which is about 1—5% day^{-1}. While it is important in surface waters (20—40% day^{-1}), it decreases rapidly with depth (less than 1% day^{-1} below 300 m) with a much lower consumption of oxygen (1 μl yr^{-1}).

Apparently, the particulate organic matter is oxidized, even in deep waters, but at a very slow rate. This modification of organic matter tends to the formation of very inert substances, more aliphatic molecules, refractory to further decomposition. The organic matter first loses nitrogen, phosphorus and oxygen since the activation energies for cleavage of the C—C and C—H bonds are several kcal mole^{-1} higher than those of the C—N, C—P or C—O bonds (Toth and Lerman, 1977).

Only the large particles such as faecal pellets sink at a sufficient rate to reach the sediment in a more or less unmodified state. In fact, exceptionally we find these particles on the filters and only vertical sampling will allow us to recover enough material for studies. Except for these latter particles, all the organic matter seems to be recycled once or several times before it reaches the sediment. Riley (1970) has suggested that a dynamic balance must exist between the utilization by bacteria, filter feeders and other biological processes, and renewal as a result of absorption from the dissolved fraction. This may explain the differences observed between the chemical composition of organic matter in the sediment and in the overlying water. As far as mineral particles are concerned, it is possible to consider a residence time, but this is meaningless in the case of organic compounds which may be recycled more than once. We can only consider a mean residence time for carbon, in relation to the flux through a given area (MacCave, 1975).

7. RELATIONSHIPS WITH THE CHEMICAL ENVIRONMENT

Organic matter is often very reactive towards its environment, the nature of which can be mineral or organic.

This is especially true for the large molecules present as detritus or in the colloidal state (humic acids, polysaccharides). These compounds are relatively abundant in sea water, principally in the fraction between 0.5 and 0.05 μm. This polymeric material can be adsorbed onto mineral particles which are all to varying degrees coated with organic substances. Chave (1970) and Suess (1968) demonstrated the relationships between organic matter and carbonates. Experimental studies of the surface electrical charges of particles in sea water provide some information on the role of organic polymeric material in these processes. Neihof and Loeb (1972, 1974), measuring the surface charge of a variety of particles in sea water, UV-treated sea water and artificial sea water, demonstrated that all particles exhibit comparable negative charges in natural sea water, whereas they show different, positive or negative, charges in artificial or UV-treated sea water. The authors concluded that all types of particles are to some extent coated with substances of high-molecular weight; the nature of the suspended material determines the nature of the bond: hydrogen bonding, Van der Waals forces, electrostatic adsorption. The properties of the colloidal matter also affect the behaviour of dissolved or dispersed molecules. Thus, the adsorption of hydrocarbons (Boehm and Quinn, 1973), fatty acids (Meyers and Quinn, 1973) or pesticides (Pierce et al., 1974) by humic matter can seriously affect their solubility in sea water, a factor to be considered in studies dealing with pollution. The association of amino acids and sugars with humic acids have also been noted by several authors and discussed in detail by Degens and Reuter (1964).

8. DISCUSSION

The increasing interest in the study of organic matter over the last twenty years following the publication of Duursma (1961) has been stimulated by the awareness of the essential role of this material in biological and geochemical processes.

As far as the particulate matter is concerned, the numerous papers previously published gave us fundamental information, however we are far from understanding the whole concept. Most of the results give values for particulate organic carbon, even though each technical improvement reduces our confidence in previous data. With the exception of the works of Degens (1970) and Handa et al. (1972), we know nothing about the biochemical composition of the particulate organic matter. However, some sparse results are of some interest. For instance, the determination by Pocklington and MacGregor (1973) of lignin-like compounds which may be a good indicator for terrestrial material and thus may encourage further studies, allowing the characterization of molecular structures, related to the origin, the age or the geochemical pathways of the material. It is doubtless that particulate matter contains chemical species which may be employed as geochemical tracers, allowing the identification of water masses. Many classes of compounds, such as phenols, pigments, organosulfur compounds, etc., have not been investigated in detail, and could probably supply interesting information. It is obvious that such studies will develop in the next years and increase our knowledge of the chemical composition of the particles.

However, such studies are insufficient for our understanding of the mechanisms responsible for the formation of aggregates or the exchange processes between dissolved and particulate matter. The formation of aggregates has been studied by several authors (Baylor et al., 1962; Sutcliffe et al., 1963; Riley et al., 1965; Menzel, 1966; Kane, 1967); we know nothing, however, about the exchange processes. It is highly probable that polymeric material, disseminated in the colloids, plays a major role in the sorption—desorption phenomena. Breger (1970) was the first to call our attention to the role of the colloids. Few authors (Ogura, 1970, 1975; Stuermer and Payne, 1976) have undertaken a detailed study of this organic fraction, and though it is considered as dissolved, it is obvious that it has many similarities to the particulate matter and warrants extensive future study.

A number of questions remain: what is the origin of flakes as described by Gordon (1970a)? What is the meaning of the discrepancies observed between the age of particles, the settling rates measured and the organic content of sediments? What is the true residence time of organic carbon in the oceans and what is the meaning of such a concept for organic matter?

It is clear that to increase our understanding of marine particulates, our preliminary efforts should be focussed on standardizing the recovery techniques and on intercalibrating existing analytical methods.

9. REFERENCES

Agatova, A.T. and Bogdanov, Y.A., 1972. Biochemical composition of suspended organic matter from the Tropical Pacific. Okeanologyia, 12 (2): 267—276.

Banoub, M.W. and Williams, P.J. Le B., 1972. Measurement of microbial activity and organic material in the western Mediterranean sea. Deep-Sea Res., 19: 433—443.

Baylor, E.R., Sutcliffe, W.H. and Hirschfeld, D.S., 1962. Adsorption of phosphates onto bubbles. Deep-Sea Res., 9: 120—124.

Bianchi, A.J.M., Bianchi, M.A.G., Bensoussan, M.G., Boudabous, A., Lizarraga, M.L., Marty, D. and Roussos, S., 1977. Etude des potentialités cataboliques des populations bactériennes isolées des sédiments et des eaux proches du fond en Mer de Norvège. In: Géochimie organique des sédiments marins profonds, ORGON I, Mer de Norvège, 1974. C.N.R.S., Paris, pp. 15—32.

Boehm, P.D. and Quinn, J.G., 1973. Solubilization of hydrocarbons by the dissolved organic matter in sea water. Geochim. Cosmochim. Acta, 37: 2459—2477.

Bogdanov, Yu. A., 1965. Suspended organic matter in the Pacific. Oceanology, 5: 77—85.

Breck, W.G., 1978. Biomonitors of aquatic pollutants in time and space. Thalass. Yugosl., 14: 157—170.

Breger, I.A., 1970. Organic colloids and natural waters. In: D.W. Hood (Editor), Organic Matter in Natural Waters. Inst. Mar. Sci., Alaska, Occ. Publ., No. 1: 563—574.

Burns, N.M. and Pashley, A.E., 1974. In situ measurements of the settling velocity profile of particulate organic carbon in lake Ontario. J. Fish. Res. Board Can., 31: 291—297.

Cauwet, G., 1975. Optimisation d'une technique de dosage du carbone organique des sédiments. Chem. Geol., 16: 59—63.

Cauwet, G., 1978. Organic chemistry of sea water particulates. Concepts and developments. Oceanol. Acta, 1 (1): 99—105.

Chave, K.E., 1970. Carbonate-organic interactions in sea water, In: D.W. Hood (Editor), Organic Matter in Natural Waters. Inst. Mar. Sci., Alaska, Occ. Publ., No. 1: 373—386.

Copin-Montegut, C. and Copin-Montegut, G., 1972. Chemical analyses of suspended particulate matter collected in the northeast Atlantic. Deep-Sea Res., 19: 445—452.

Copin-Montegut, C. and Copin-Montegut, G., 1973. Comparison between two processes of determination of particulate organic carbon in sea water. Mar. Chem., 1: 151—156.

Craig, H., 1971. The deep metabolism: oxygen consumption in abyssal ocean water. J. Geophys. Res., 76: 5078—5082.

Dalpoint, C. and Newell, B., 1963. Suspended organic matter in the Tasman Sea. Austr. J. Mar. Freshwater Res., 14: 155—163.

Degens, E.T., 1970. Molecular nature of nitrogenous compounds in sea water and recent marine sediments. In: D.W. Hood (Editor), Organic Matter in Naturel Waters. Inst. Mar. Sci., Alaska, Occ. Publ., No. 1: 77—106.

Degens, E.T. and Reuter, J.H., 1964. Analytical techniques in the field of organic geochemistry. In: U. Colombo and G.D. Hobson (Editors), Advances in Organic Geochemistry. Pergamon, London, pp. 390—391.

Duursma, E.K., 1961. Dissolved organic carbon, nitrogen and phosphorus in the sea. Neth. J. Sea Res., 1: 1—148.

Eadie, B.J. and Jeffrey, L.M., 1973. δ^{13}C analyses of oceanic particulate organic matter. Mar. Chem., 1: 199—209.

Flemer, D.A. and Biggs, R.B., 1971. Particulate carbon nitrogen relations in northern Chesapeake Bay. J. Fish. Res. Board Can., 28 (6): 911—918.

Gordon, D.C., 1970a. A microscopic study of non living organic particles in the North Atlantic Ocean. Deep-Sea Res., 17: 175—185.

Gordon, D.C., 1970b. Some studies on the distribution and composition of organic particulates in the North Atlantic Ocean. Deep-Sea Res., 17: 233—243.

Gordon, D.C. Jr. and Sutcliffe, W.H. Jr., 1974. Filtration of sea water using silver filters for particulate nitrogen and carbon analysis. Limnol. Oceanogr., 19: 989—993.

Hair, M.E. and Bassett, C.R., 1973. Dissolved and particulate humic acids in an east coast estuary. Estuarine Coast. Mar. Sci., 1: 107—111.

Håkanson, L., 1976. A bottom sediment trap for recent sedimentary deposits. Limnol. Oceanogr., 21: 170—174.

Handa, N., 1970. Dissolved and particulate carbohydrates. In: D.W. Hood (Editor), Organic Matter in Natural Waters. Inst. Mar. Sci., Alaska, Occ. Publ., No. 1: 129—152.

Handa, N. and Tominaga, K., 1969. A detailed analysis of carbohydrates in marine particulate matter. Mar. Biol., 2: 228—235.

Handa, N., Yanagi, K, and Matsunaga, K., 1972. Distribution of detrital materials in the Western Pacific Ocean and their biochemical nature. IBP Unesco Symposium on Detritus and Its Role in Aquatic Ecosystems, Pallanza, Italy, pp. 248—273.

Hobson, L.A., 1967. The seasonal and vertical distribution of suspended particulate matter in an area of the Northeast Pacific Ocean. Limnol. Oceanogr., 12: 642—649.

Hobson, L.A. and Menzel, D.W., 1969. The distribution and chemical composition of organic particulate matter in the sea and sediments off the east coast of South America. Limnol. Oceanogr., 14: 159—163.

Holm-Hansen, O., Strickland, J.D.H. and Williams, P.M., 1966. A detailed analysis of biologically important substances in a profile off southern California. Limnol. Oceanogr., 11: 548—561.

Honjo, S., 1978. Sedimentation of materials in the Sargasso Sea at a 5,367 m deep station. J. Mar. Res. U.S.A., 36 (3): 469—492.

Hunter, K.A. and Liss, P.S., 1977. The input of organic material to the oceans: air—sea interactions and the chemical composition of the sea surface. Mar. Chem., 5: 361—379.

Jacobs, M.B. and Ewing, M., 1969. Suspended particulate matter: concentration in the major oceans. Science, 163: 380—383.

Kane, J.E., 1967. Organic aggregates in surface waters of the Ligurian Sea. Limnol. Oceanogr., 12: 287—294.

Kimmel, B.L., Axler, R.P. and Goldman, C.R., 1977 . A closing, replicate-sample sediment trap. Limnol. Oceanogr., 22: 768—772.

Kirchner, W.B., 1975. An evaluation of sediment trap methodology. Limnol. Oceanogr., 20 (4): 657—660.

Krishnaswami, S. and Sarin, M.M., 1976. Atlantic surface particulates: composition, settling rates and dissolution in the deep sea. Earth Planet. Sci. Lett., 32: 430—440.

Lascaratos, A., 1974. Contribution à l'étude granulométrique des particules en suspension dans l'eau de mer. Essai de différenciation de la granulométrie des particules minérales de celle des particules organiques. Thesis University of Paris VI, Paris, 90 pp.

Loder, T.C., 1971. Distribution of Dissolved and Particulate Organic Carbon in Alaskan Polar, Sub Polar and Estuarine Waters. Thesis, University of Alaska, College, Alaska, 236 pp.

MacCave, I.N., 1975. Vertical flux of particles in the ocean. Deep-Sea Res., 13: 707—730.

Mel'nikov, I.A., 1976. Finely dispersed fraction of suspended organic matter in eastern Pacific ocean waters. Oceanol. Acad. Sci. U.S.S.R., 15 (2): 182—186.

Mel'nikov, I.A., 1977. Morphological characteristics of organic detritus particles. Oceanol. Acad. Sci. U.S.S.R., 16 (4): 401—403.

Menzel, D.W., 1966. Bubbling of sea water and production of organic particles: a re-evaluation. Deep-Sea Res., 12: 963—966.

Menzel, D.W., 1967. Particulate organic carbon in the deep sea. Deep-Sea Res., 14: 229—238.

88

Menzel, D.W., 1974. Primary productivity, dissolved and particulate organic matter and the sites of oxidation of organic matter. In: E. Goldberg (Editor), The Sea, 5. Wiley, New York, N.Y., pp. 659—678.

Menzel, D.W. and Goering, J.J., 1966. The distribution of organic detritus in the ocean. Limnol. Oceanogr., 11: 333—337.

Menzel, D.W. and Ryther, J.H., 1964. The composition of particulate organic matter in the Western North Atlantic. Limnol. Oceanogr., 9: 179—186.

Menzel, D.W. and Ryther, J.H., 1970. Distribution and cycling of organic matter in the oceans. In: D.W. Hood (Editor), Organic Matter in Natural Waters. Inst. Mar. Sci., Alaska, Occ. Publ., No. 1: 31—54.

Menzel, D.W. and Vaccaro, R.F., 1964. The measurement of dissolved organic and particulate carbon in sea water. Limnol. Oceanogr., 9: 138—142.

Meyers, P.A. and Quinn, J.C., 1973. Factors affecting the association of fatty acids with mineral particles in sea water. Geochim. Cosmochim. Acta, 37: 1745—1759.

Neihof, R.A. and Loeb, G.J., 1972. Surface particulate matter in sea water. Limnol. Oceanogr., 17 (1): 7—16.

Neihof, R.A. and Loeb, G.J., 1974. Dissolved organic matter in sea water and the electric charge of immersed surfaces. J. Mar. Res. U.S.A., 32 (1): 5—12.

Nissenbaum, A. and Kaplan, I.R., 1972. Chemical and isotopic evidence for the *in situ* origin of marine humic substances. Limnol. Oceanogr., 17: 570—582.

Ogura, N., 1970. Dissolved organic matter in the sea, its production, utilisation and decomposition. Proc. 2nd CSK Symposium, Tokyo, pp. 201—205.

Ogura, N., 1975. Further studies on decomposition of dissolved organic matter in coastal sea water. Mar. Biol., 31: 101—111.

Parsons, T.R. and Strickland, J.D.H., 1959. Proximate analysis of marine standing crops. Nature, 184: 2038—2039.

Parsons, T.R. and Strickland, J.D.H., 1962. Oceanic detritus. Science, 136: 313—314.

Pierce, R.H., Olmey, C.E. and Felbeck, G.T. Jr., 1974. pp'DDT adsorption to suspended particulate matter in sea water. Geochim. Cosmochim. Acta., 38: 1061—1073.

Pocklington, R. and MacGregor, C.D., 1973. The determination of lignin in marine sediments and particulate form in sea water. Int. J. Environ. Anal. Chem., 3: 81—93.

Pustel'nikov, O.S., 1976. Organic matter in suspension and its supply to the bottom of the Baltic sea. Oceanol. Acad. Sci. U.S.S.R., 15 (6): 673—676.

Quinn, J.G. and Meyers, P.A., 1971. Retention of dissolved organic acids by various filters. Limnol. Oceanogr., 16: 129—131.

Riley, G.A., 1970. Particulate organic matter in sea water. Adv. Mar. Biol., 8: 1—118.

Riley, G.A., Van Hemert, D. and Wangersky, P.J., 1965. Oceanic aggregates in surface and deep waters of the Sargasso Sea. Limnol. Oceanogr., 10: 354—363.

Saliot, A., 1975. Acides gras, stérols et hydrocarbures en milieu marin: inventaire, applications géochimiques et biologiques. Thesis, University of Paris VI, Paris, 167 pp.

Sharp, J.H., 1974. Improved analysis for particulate organic carbon and nitrogen from sea water. Limnol. Oceanogr., 19 (6): 984—989.

Sheldon, R.W. and Sutcliffe, W.H., 1969. Retention of marine particles by screens and filters. Limnol. Oceanogr., 14: 441—444.

Sheldon, R.W., Evelyn, T.P.T. and Parsons, T.R., 1967. On the occurrence and formation of small particles in sea water. Limnol. Oceanogr., 12: 367—375.

Sheldon, R.W., Prakash, A. and Sutcliffe, W.H. Jr., 1972. The size distribution of particles in the ocean. Limnol. Oceanogr., 17: 327—340.

Sholkovitz, E.R., 1976. Floculation of dissolved organic and inorganic matter during the mixing of river water and sea water. Geochim. Cosmochim. Acta, 40: 831—845.

Sieburth, J. McN. and Jensen, A., 1968. Studies on algal substances in the sea. I. Gelbstoff (humic material) in terrestrial and marine waters. J. Exp. Mar. Biol. Ecol., 2: 174—189.

Skopintsev, B., 1972. On the age of stable organic matter: aquatic humus in oceanic waters. In: D. Dyrssen and D. Jagner (Editors), The Changing Chemistry of the Oceans. Wiley, New York, N.Y., pp. 205—207.

Spencer, D.W., Brewer, P.G., Fleer, A., Honjo, S., Krishnaswami, S. and Nozaki, Y., 1978. Chemical fluxes from a sediment trap experiment in the deep Sargasso Sea. J. Mar. Res. U.S.A., 36 (3): 493—523.

Stuermer, D.H. and Payne, J.R., 1976. Investigation of sea water and terrestrial humic substances with carbon 13 and proton nuclear magnetic resonance. Geochim. Cosmochim. Acta, 40 (9): 1109—1114.

Suess, E., 1968. Calcium Carbonate Interaction with Organic Compounds. Thesis, Marine Science Center, Lehigh University, Bethlehem, Pa., 153 pp.

Sutcliffe, W.H., Baylor, E.R. and Menzel, D.W., 1963. Sea surface chemistry and Langmuir circulation. Deep-Sea Res., 10: 233—243.

Sutcliffe, W.H., Sheldon, R.W. and Prakash, A., 1970. Certain aspects of production and standing stock of particulate matter in the surface waters of the Northwest Atlantic Ocean. J. Fish. Res. Board Can., 27 (II): 1917—1926.

Szekielda, K.H., 1967. Some remarks on the influence of hydrographic conditions on the concentration of particulate carbon in sea water. In: Chemical Environment in the Aquatic Habitat. Proc. IBP Symp., Amsterdam and Nieuwersluis, October 10—16, 1966, pp. 314—322.

Telek, G. and Marshall, N., 1974. Using a CHN analyser to reduce carbonate interference in particulate organic carbon analyses. Mar. Biol., 24: 219—221.

Toth, D.J. and Lerman, A., 1977. Organic matter reactivity and sedimentation rates in the ocean. Am. J. Sci., 227: 465—485.

Vile-Cambon, F., 1977. Relation entre les alcalino-terreux et les substances humiques dans le domaine marin. Thesis, University of Toulouse, Toulouse, 141 pp.

Wada, E. and Hattori, A., 1976. Natural abundance of ^{15}N in particulate organic matter in the North Pacific Ocean. Geochim. Cosmochim. Acta, 40: 249—251.

Wangersky, P.J., 1974. Particulate organic carbon: sampling variability. Limnol. Oceanogr., 19: 980—984.

Wangersky, P.J., 1976. Particulate organic carbon in the Atlantic and Pacific oceans. Deep-Sea Res., 23: 457—465.

Wangersky, P.J. and Gordon, D.C., 1965. Particulate carbonate, organic carbon and Mn^{++}, Limnol. Oceanogr., 10: 544—550.

White, W.S. and Wetzel, R.G., 1973. A modified sedimentation trap. Limnol. Oceanogr., 18: 986—988.

Williams, P.M., 1971. Distribution of organics in the ocean. In: S.J. Faust and J.V. Hunter (Editors), Organic Compounds in Aquatic Environment. Marcel Dekker, New York, N.Y., pp. 145—163.

Zeitschel, B., 1970. Quantity, composition and distribution of suspended particulate matter in the Gulf of California. Mar. Biol., 7 (4): 305—318.

Zeitschel, B., Diekman, P. and Uhlman, L., 1978. A new multisample sediment trap. Mar. Biol., 45: 285—288.

Chapter 5

PROCESSES CONTROLLING THE DISTRIBUTION OF BIOGENIC ORGANIC COMPOUNDS IN SEAWATER [1]

ROBERT B. GAGOSIAN and CINDY LEE

1. INTRODUCTION

When discussing factors controlling the distribution of organic compounds in seawater, several types of processes must be considered. These include biological production and consumption, geochemical and biological transport processes and chemically and biochemically controlled transformation reactions.

Probably the most well known organic compound produced by organisms in the sea is chlorophyll. Early interest in this molecule as an indicator of phytoplankton standing crop served as a predecessor for the study of other specific classes of organic compounds produced by phytoplankton, such as the normal and isoprenoid alkanes, alkenes, polyolefinic carotenoids and triterpenoid hydrocarbons. Marine organisms also biosynthesize heteroatomic compounds such as low molecular-weight aldehydes, ketones and halocarbons, amino acids and amino sugars, carbohydrates, urea, acyclic fatty alcohols, tetracyclic steroidal alcohols, xanthophylls, fatty acids and wax esters. *In situ* biological production is the major source for these compounds in seawater and sediments, while heterotrophic consumption is the major sink. The biota also mediate a large number of transformation reactions of specific organic compounds such as oxidation or reduction, bond formation or cleavage, or inorganic—organic binding reactions.

The transport of organic compounds can occur both horizontally and vertically through physical, geochemical or biological processes (see Fig. 1). Atmospheric transport of land-derived natural compounds to the open ocean can take place in a matter of days. Rivers can discharge large quantities of terrestrially derived organic matter into estuaries, either continuously or in pulse events such as storms. The surface layer of the ocean may serve as an interface for the exchange of organic matter between the atmosphere and the sea in either the vapor or particulate state. The vertical transport of organic matter in the water column on particulate material such as coccolithophore or diatom tests and in fecal pellets must be considered as a poten-

[1] Woods Hole Oceanographic Institution Contribution Number 4254.

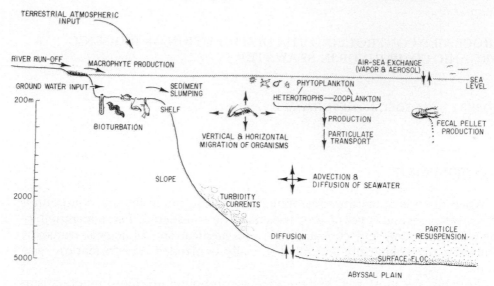

TERRESTRIAL ATMOSPHERIC
INPUT

RIVER RUN-OFF

GROUND WATER INPUT

MACROPHYTE PRODUCTION

AIR-SEA EXCHANGE
(VAPOR & AEROSOL)

SEA LEVEL

200m

SEDIMENT SLUMPING

PHYTOPLANKTON

HETEROTROPHS—ZOOPLANKTON

SHELF

BIOTURBATION

PRODUCTION

PARTICULATE TRANSPORT

FECAL PELLET PRODUCTION

VERTICAL & HORIZONTAL MIGRATION OF ORGANISMS

SLOPE

ADVECTION & DIFFUSION OF SEAWATER

2000

TURBIDITY CURRENTS

DIFFUSION

PARTICLE RESUSPENSION

5000

SURFACE FLOC

ABYSSAL PLAIN

Fig. 1. Scheme showing possible transport processes controlling the distribution of organic compounds in seawater.

tial source of carbon to the benthos. Advection and diffusion in water masses are important processes in transporting dissolved and small particulate material horizontally as well as vertically in the water column. Bottom currents are capable of moving resuspended sedimentary material both vertically and horizontally across the ocean floor, thus controlling grain-size distributions and the distribution of associated organic matter. Organic-rich sediments from highly productive areas or anoxic basins can serve as sources for gaseous organic compounds such as methane which diffuse into the upper water column and are transported advectively by water masses. All of these processes can interact with and affect the distribution of organic compounds in seawater.

Since several recent reviews summarize the classes and quantities of specific organic compounds found in seawater (Williams, 1975; Andersen, 1977) they will not be formally discussed in any detail in this chapter. Analytical methods will be addressed in Chapter 15 (Dawson and Liebeszeit). Rather, we will concentrate on the processes controlling the distribution of organic compounds and the mechanisms of transport and transformation in the water column. We have relied heavily on examples from our own laboratories and from the literature to explain the use of specific organic compounds to trace the effect of these processes on the total organic matter.

2. BIOLOGICAL PRODUCTION AND CONSUMPTION: MACROFAUNA, PHYTOPLANKTON, ZOOPLANKTON, AND BACTERIA

2.1. Production

Logically, we might expect to find in seawater any organic compound known to exist in marine macrofauna, phytoplankton, zooplankton, or microorganisms, such as yeasts or bacteria. These marine organisms excrete many organic compounds directly into seawater as a mechanism for removal of metabolic wastes, as chemical communicants, and for other reasons not yet clearly understood. Upon death of the organism, if it is not immediately consumed and passed along the food chain, the organic constituents of the body may be released into seawater, most likely in some state of physical, chemical, or biological decomposition. These exudates and decomposition products make up as diverse and complex a group of organic chemicals as found in any environment on earth. And, it is safe to assume that the list of known compounds and their functions is not yet complete.

In the sea, *in situ* photosynthetic production by phytoplankton is the major primary producer of organic carbon. Marine autotrophs (primary producers) can be classed as photo-autotrophs (algae, blue-green algae, and some bacteria) which use light as energy, and chemo-autotrophs (some bacteria) which use energy gained from the oxidation of reduced or partially reduced inorganic compounds. Marine heterotrophs gain their energy from preformed organic material. The zooplankton and most bacteria fall into this second category. The relative contribution of the various types of organisms to the carbon pool of the oceans is poorly understood. Whittle (1977) reviews what is known about the distribution of the biota and the relative amounts of living organic carbon produced by algae, zooplankton, bacteria, and benthic fauna. The location of these organisms both geographically and with depth in the water column will control the distribution of biogenic organic matter in the sea. Changes in productivity caused by seasonal fluctuations in water properties such as temperature can also be important in controlling organic matter distribution. The algal biomass is the most important source of primary organic carbon used by other organisms to synthesize new organic matter. Since the major algal biomass occurs in the euphotic zone, this is where concentrations of organic carbon will be highest. Distributions lower in the water column will be modified by the location of the chemo-autotrophic bacteria and heterotrophs.

What specific compounds are being produced by these organisms? Most of the classes of compounds present in seawater are also present in organisms and have important biological functions. Amino acids are the building blocks of proteins, the all-important structural component of living tissues. Proteins also serve as enzymes, antibodies, transport devices, and metabolic regulators (Mahler and Cordes, 1971). The excretion of proteins and free amino acids into seawater by phytoplankton, zooplankton, and bacteria has been a topic of study and controversy. Both phytoplankton and zooplankton apparently release dissolved free amino acids or protein into seawater (Hellebust, 1965; Webb and Johannes, 1967). The cell's behavior towards environmental conditions (e.g. light intensity, ambient nutrient levels) appear to play a role in these release processes (Williams, 1975). It is known that bacteria release extracellular proteins (Glenn, 1976) but this release process has not been measured in a marine system.

Catabolism of carbohydrates provides the major share of the energy requirement for activity and maintenance of life. In addition, carbohydrates act as energy reservoirs and serve architectural functions (Mahler and Cordes, 1971). Carbohydrates, like proteins, are polymers made up of smaller organic units, sugars in the case of the carbohydrates. When amino sugars are also present within the carbohydrate polymer, the compound is termed a mucopolysaccharide. Chitin is the most abundant marine mucopolysaccharide, and is probably second only to cellulose as the most abundant biogenic substance on earth. Carbohydrates also covalently bond with proteins to form glycoproteins. As the structural material of cell walls in plants, animals, and bacteria, glycoproteins, mucopolysaccharides, proteins, and carbohydrates make up the bulk of organic carbon in living organisms, marine and terrestrial. Excretion of carbohydrates into seawater by phytoplankton has been measured by Hellebust (1965) and Sieburth (1969) and these compounds appear to be major components of the phytoplankton excretion products.

Other compounds which have been studied in seawater are not produced in the same large quantities as protein and carbohydrate but have important biological activity. Fatty acids are usually incorporated into larger compounds such as triglycerides, wax esters, and phosphatides. Triglycerides, fatty acid esters of glycerol, are the most abundant lipids in phytoplankton (Lee et al., 1971) and can act as energy reservoirs, buoyancy controls, or as thermal or mechanical insulators. Wax esters, simple esters of long-chain fatty alcohols and long-chain fatty acids, are particularly abundant in zooplankton. Sargent et al. (1977) discuss the distribution of wax esters in calanoid copepods; they describe massive wax ester slicks which have been reported in the northwest Pacific Ocean and the North Sea presumably caused by a large mortality of copepods.

Sterols are another class of compounds also involved in metabolic processes. They act as hormonal regulators of growth, respiration, and reproduction in most marine organisms (Kanazawa and Teshima, 1971). Sterols, notably cholesterol, are also present in organisms as lipoproteins, probably a larger source of these compounds than as hormonal regulators. The presence of hydrocarbons in marine organisms is known, but their precise function is unclear. Farrington and Meyers (1975) review the present knowledge on biosynthesis of hydrocarbons by marine organisms. Alkenes appear to make up the major portion of hydrocarbons, especially in algae, while n-alkanes, branched alkanes, and cyclic alkanes and alkenes are present in smaller amounts. Although hydrocarbons and sterols are present only in small amounts in marine organisms, because of their species specificity, they have been used to show the possibility of terrestrial input of organic matter to the deep ocean. Farrington and Tripp (1977) showed that terrestrial n-alkanes (odd-chain 25—31 carbon atoms) as opposed to marine-derived n-alkanes (odd-chain 15—21 carbon atoms) were predominant in abyssal-plain surface sediments of the western North Atlantic Ocean. In a similar study, Lee et al. (1979) showed the predominance of terrestrial sterols (24-ethylcholesterol) over phytoplankton sterols (cholesterol) in the same western North Atlantic area.

Photosynthetic pigments such as chlorophyll, the carotenoids, and the phycobilins are used by plants to absorb and transfer light energy during photosynthesis. Chlorophyll is one of the few complex organic compounds which has historically been routinely measured in seawater and used as an indicator of photosynthetic activity (Yentsch and Menzel, 1963; Harvey, 1966). Since most phytoplankton contain chlorophyll as their primary photosynthetic pigment, the production of chlorophyll in the sea is quite large. Vitamins and co-enzymes are organic compounds which are required by enzymes for efficient performance of catalytic function. These compounds, e.g. ATP, co-enzyme A, thiamine, NAD^+, biotin, and vitamin B_{12}, have very diverse and complex structures and are usually produced in small amounts in living organisms. Using a bioassay technique, Carlucci and Bowes (1970) measured the release of biotin, thiamine, and vitamin B_{12} by algae. Holm-Hansen and Booth (1966) and later Hobbie et al. (1972) used an adaptation of the firefly luciferase assay to measure ATP in seawater as an estimate of microbial biomass and metabolic activity.

There are some classes of compounds present in seawater which have known biological activity but have received little attention. Compounds involved in chemical communication, especially halogenated compounds, will be discussed in other chapters. Other compounds of interest are those involved in the citric acid cycle, the major pathway of metabolism in almost all cells. Citric acid cycle intermediates and reactions are involved in the biosynthesis of most of the major metabolites, from amino acids and carbohydrates to long-chain fatty acids and porphyrins (Mahler and Cordes, 1971). The intermediates, such as succinic, fumaric, malic, oxaloacetic, citric, glyoxylic, and oxoglutaric acids are present in almost all organisms and are certainly being produced in the sea.

Other compounds serving essential biological functions are the products of purine and pyrimidine catabolism. Since these compounds are metabolic wastes, their production should be relatively large. The relatively high concentration of urea in seawater perhaps reflects this (McCarthy, 1970; Remsen et al., 1974). The release of relatively high amounts of glycolic acid (biochemically related to glyoxylic acid) by marine algae is well documented (Hellebust, 1965). Glyoxylic acid is an intermediate in the citric acid cycle as well as being an end-product in purine metabolism. Others included in this group are multi-nitrogen compounds such as hypoxanthine, allantoin, allantoic acid, and urea. Since the structural identity of so much of the nitrogen in the marine environment is unknown, perhaps more attention should be focused on these compounds. This reasoning would also apply to the nucleic acids, the purine or pyrimidine-containing macromolecules which function in the storage and transmission of genetic material in all living organisms.

2.2. Consumption

If all of the organisms existing in the oceans merely excreted their metabolic wastes and then died allowing their carcasses to pass unmolested through the water column, the oceans would contain far larger concentrations of the organic compounds described in the previous section. However, the relatively low concentration of organic compounds which have been identified in seawater shows that most biogenic organic carbon is removed from seawater at a rate comparable with its production. The distribution of biogenic compounds reflects this balance between production and consumption.

In general, phytoplankton are consumed almost as fast as they are produced (Steele, 1974). Ingestion of phytoplankton by zooplankton can introduce as well as remove organic material in seawater; feeding is not necessarily 100% efficient. Cells crushed during feeding can release particulate and soluble intercellular material (Whittle, 1977). The major mechanism for the removal of this and other dissolved and detrital material from the water column is thought to be heterotrophic consumption by bacteria. Williams (1975) discusses heterotrophic uptake of dissolved organic compounds from seawater. Most of this work has been carried out with free amino acids and sugars as ^{14}C-labelled substrates. As noted by Williams (1975), there is some controversy about which heterotrophs take up the organic compounds from natural systems. Size fractionation of organisms taking up dissolved glucose and amino acids has shown that bacteria are probably the major consumers (Williams, 1970). Less is known about the mechanism of biological consumption of particulate organic matter, partly due to the disputed suitability of radiotracer techniques to this area of research.

Algal uptake or consumption of organic matter from seawater has received considerable attention recently. Phytoplankton can apparently take up amino acids, other primary amines, and urea from natural seawater and synthetic seawater cultures (North and Stephens, 1971; McCarthy, 1972; Schell, 1974; North, 1975). Antia et al. (1975) compared phytoplankton growth on inorganic nutrients (nitrate, nitrite, and ammonia) with growth on organic species (glycine, urea, hypoxanthine, and glucosamine). From these results, it appears that phytoplankton can take up significant quantities of nitrogen-containing organic compounds from seawater under conditions of inorganic nutrient limitation. But, the quantitative importance of algal consumption of organic compounds (both N-containing and otherwise) for the world oceans is unclear.

Consumption of organic matter by phytoplankton or zooplankton does not necessarily lead to complete destruction. The organic compounds can be recycled intact or as a transformed version in the body of the consumer. However, the community of bacterial heterotrophs can complete the destruction of the organic matter to its inorganic decomposition products, primarily phosphate, nitrate, carbon dioxide and water. It is presently thought that regeneration of organic material in the marine environment takes place primarily in the euphotic zone. When organic matter escapes mineralization processes in the upper waters, regeneration can also take place deeper in the water column and on the sea floor. The recycling process in this case takes a little longer since the regenerated inorganic nutrients must be returned to a site of active primary production.

Historically, the precise location of water column regeneration of organic matter has been a subject of heated controversy. Traditional thought was that organic matter produced in the euphotic zone fell into deeper waters and gradually decomposed. The concomitant consumption of oxygen resulted in the "oxygen minimum zone" observed in ocean water columns (Redfield, 1942; Sverdrup et al., 1942). On the basis of dissolved organic carbon (DOC) measurements in the water column of several oceans, Menzel and Ryther (1968), however, felt that regeneration of organic matter was occurring only in the upper 500 m. They felt that the deeper oxygen minimum zone was an advective and not a biological feature. Later, Menzel (1974) reviewed his original position and discussed Craig's (1971a, b) arguments for *in situ* oxidation of organic matter based on a one-dimensional advection—diffusion model. More recently, Bishop (1977) found that in shallow areas (<1500 m) of higher productivity, a smaller proportion of the organic matter produced is remineralized in the upper 500 m, relative to areas of lower productivity. This indicates that the efficiency of remineralization is lower and increases the probability that labile (and not only resistant) organic material is being transported through the water column below 500 m, and is available for biological metabolism in these areas. More direct evidence for bacterial degradation in the deep sea is the measurement of organic-carbon consumption by a heterotrophic, low-nutrient bacterium obtained from deep waters of the northwestern Pacific by Williams and Carlucci (1976). Present thought seems to indicate that, although some organic matter may be decomposed at depth, the oxygen minimum is not being caused by this decomposition. However, the oxygen minimum could be a result of advected water low in oxygen due to decomposition and oxygen consumption at another location.

Because benthic regeneration of nutrients plays only a minor role in open ocean nutrient cycles, less attention has been focused on the biological oxidation of organic matter in sediments. Recently, however, the importance of coastal benthic nutrient cycling to the regulation of the concentration of organic matter in the overlying seawater has been shown. Rowe et al. (1975, 1977) found that ammonia-N remineralization from organic matter in the sediments of coastal northwest Africa, the New York Bight, and coastal Massachusetts supplied a large fraction of the nitrogen required for photosynthesis in the overlying waters. This means that the rate of benthic decomposition of organic matter which survived euphotic-zone regeneration processes acts as a control on the process of primary production of new organic matter at the surface in these areas. Benthic remineralization of deposited organic matter may also directly affect the concentration of dissolved organic compounds. Nixon et al. (1976) observed fluxes of dissolved organic nitrogen as well as inorganic nitrogen from the sediments of Narragansett Bay to the overlying seawater. They had predicted this flux on the basis of an unexplained nitrogen deficit in their nutrient regeneration model. It was assumed that incomplete metabolism of nitrogen-containing compounds was responsible for the flux.

Not all of the organic matter lost to the sediments from the overlying seawater is decomposed or diffused upward as incompletely metabolized dissolved organic matter. Some becomes part of the permanent deposit and is slowly transformed to a highly resistant and not easily characterized form of organic matter. These transformations may also occur during decomposition processes in the water column and are not necessarily slow. These and other reactions will be discussed in detail in the next section.

3. TRANSFORMATION REACTIONS

There are several types of chemical and biochemical transformation reactions that can occur in seawater. These reactions can be grouped into four major categories: oxidation--reduction (redox) reactions, carbon--carbon or carbon—heteroatom bond formation, carbon—carbon or carbon—heteroatom bond cleavage, and organic—inorganic interactions. Redox reactions would include the hydrogenation of olefins and acetylenic compounds, the loss of hydrogen through aromatization reactions, the oxidation or reduction of alcohols, aldehydes and ketones, and the oxidative cleavage of olefins. Oxidation reactions generally tend to be biochemical in nature although there is recent evidence to support photochemical, singlet-oxygen mediated reactions (Zepp et al., 1977; see also Chapter 10). Examples of bond formation transformations are polymerization or condensation reactions to form humic or fulvic type materials (see later discussion), esterification or amide formation and cyclization reactions. Several types of bond cleavage reactions which might affect compounds found in the marine environment are the decarboxylation of β-ketoacids, decarbonylation of aldehydes, deamination of amino acids, and the hydrolysis of esters and amides to form carboxylic acids. Organic—inorganic interactions include organometallic complexes and organomineral phase interactions. There are at least two types of organometallic complexes in the sea: (1) one containing covalently bound metals such as metalloenzymes, vitamin B_{12}, anthropogenic alkyl metal compounds, etc., and (2) the more abundant chelate-type complexes such as metal humates. Some algal products form complexes with metals possibly detoxifying the metals or making them otherwise available to the phytoplankton cells as micronutrients (see Chapter 7). Organomineral phase interactions involve the adsorption of highly surface active dissolved organic matter to ocean particulate matter (Pocklington, 1977). The mechanisms by which this takes place include ion-exchange (such as with $CaCO_3$), interlayering of organic compounds in clay minerals, clathration, Van der Waals interactions and hydrogen bonding (Meyers and Quinn, 1973; Hedges, 1977; Müller and Suess, 1977; Carter, 1978).

Although there are many potential types of transformation reactions in the sea as outlined above, very few studies have addressed this subject. Rather than list potential reaction schemes for known organic compounds in seawater, we wish to give a few examples of specific transformation studies which were chosen to show reactions involving compounds of varying reactivity and functionality. These include: (1) the formation of methyl chloride and subsequent hydrolysis to methanol; (2) steroidal alcohol oxidation, hydrogenation and dehydration reactions; (3) amino-acid transformations, and (4) humic substance formation in seawater via condensation of fatty acids, amino acids, sugars, amino sugars, and other organic compounds.

3.1. Nucleophilic reactions of methyl halides

Zafiriou (1975) recently reported that methyl iodide is thermodynamically unstable in seawater and its chemical fate is kinetically controlled. The equations showing the fate of methyl iodide are as follows:

$$CH_3I + Cl^- = CH_3Cl + I^- \tag{1}$$

$$CH_3I + Br^- = CH_3Br + I^- \tag{2}$$

$$CH_3Br + Cl^- = CH_3Cl + Br^- \tag{3}$$

$$CH_3X + H_2O = CH_3OH + X^-; X = Cl^-, Br^-, I^- \tag{4}$$

Chloride ion was theoretically predicted to be the most kinetically reactive species, with water second, and other anions of lesser importance. This suggested that methyl iodide in seawater would react predominantly via a nucleophilic substitution reaction with chloride ion (eq. 1) to yield methyl chloride. Methyl iodide and the methyl chloride produced by (1) would also

TABLE I

Rates of S_{N2} reactions of methyl halides in water (Zafiriou, 1975)

Halide	Nucleophile	T ($^{\circ}$C)	K' (sec^{-1}) [a]	Half-life (yr)
CH$_3$I	saturated NaCl	19.2	4.0×10^{-6}	0.0055
		10.8	7.3×10^{-7}	0.030
	19.0‰ chlorinity NaCl	19.2	3.5×10^{-7}	0.062
	19.8‰ chlorinity seawater	19.2	4.1×10^{-7}	0.054
		10.8 [b]	1.4×10^{-7}	0.16
	water (calc.)	0	9.4×10^{-10}	23
	water (calc.)	10	5.4×10^{-9}	4.0
	water (calc.)	20	3.2×10^{-8}	0.69
CH$_3$Cl	water (calc.)	0	2.5×10^{-10}	88
	water (calc.)	10	1.6×10^{-9}	14
	water (calc.)	20	8.9×10^{-9}	2.5

[a] Pseudo first-order rate constant.
[b] Over initial 42% of decay.

react with water, although more slowly, to yield methanol and halide ions according to eq. 4. Through methyl iodide/sodium chloride solution experiments, Zafiriou (1975) observed the loss of methyl iodide accompanied by the increase of methyl chloride concentration. Kinetic calculations were then performed to determine the half-lives of methyl iodide and methyl chloride during hydrolysis (Table I). According to these experiments, substantial amounts of methyl chloride should be formed in seawater. Methyl chloride has a long half-life for decomposition by known reactions in seawater (Table I). Hence, its presence could be a useful label for some surface-derived water masses. Methyl chloride is in fact found in the atmosphere, where compared to methyl iodide, it is less stable to photodegradation reactions (National Academy of Sciences, 1978).

3.2. Steroid degradation reactions

Steroids are a class of biogenic compounds which may serve as an indicator of certain processes transforming organic matter in seawater and sediments. The steroid hydrocarbon structure forms a relatively stable nucleus which may incorporate functional groups such as alcohols (stenols and stanols), ketones (stanones) and olefinic linkages (sterenes) either in the four ring system or on the side chain originating at C-17 (Fig. 2). These compounds are produced by a wide variety of marine and terrestrial organisms and often have specific species sources (Nes and McKean, 1977). Diagenetic alteration of steroids by geochemical and biochemical processes can lead to

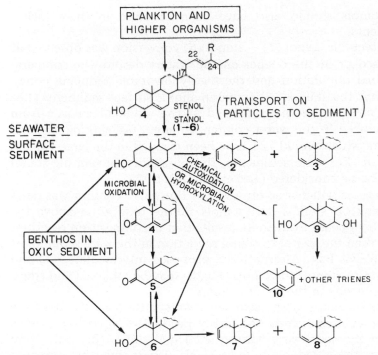

Fig. 2. A proposed mechanistic pathway for the microbiological or chemically mediated degradation of sterols in seawater and sediments. Compounds in parentheses are intermediates not isolated.

the accumulation of transformed products in seawater and sediments. The determination of the structures and concentrations of some of these transformation products has been accomplished recently in order to determine the pathways and rates of steroid degradation (Gagosian and Farrington, 1978; Gagosian and Heinzer, 1979; Gagosian and Smith, 1979). Fig. 2 is a composite reaction scheme summarizing the results of the above studies.

Seawater studies were done in the Black Sea (Gagosian and Heinzer, 1979) and in the anoxic waters of Walvis Bay (Gagosian and Nigrelli, unpublished results) in order to determine if reduction of stenols (*1*) to stanols (*6*) (Fig. 2) was occurring in the water column or at the sediment—water interface. The water below approximately 125 m is anoxic in the two central 2000 m deep basins of the Black Sea. The water column sampled near Walvis Bay is 40 m deep. At the time of sampling, the bottom 20 m were anoxic, although this condition changes periodically. The presence of high concentrations of hydrogen sulfide, as well as the absence of dissolved oxygen, restrict life in the anoxic zone to various species of microorganisms. Since stenol transformation reactions have been postulated to be microbially mediated, these regimes were ideal for examining these transformation reactions in the water column. If stenol reduction reactions were chemically mediated, then stanols should also have been observed in these highly reducing environments.

However, no detectable stenol $(1) \rightarrow$ stanol (6) conversion was observed at the O_2/H_2S interface or in the deeper anoxic waters despite the comparatively high microbial population and the highly favorable reducing conditions. It appears that the stanols (6) found in anoxic surface sediments (Lee et al., 1980) are either microbially formed at the sediment surface from stenols, or are produced in the water column on large, fast-sinking particles such as fecal pellets which would not have been trapped in the large-volume water samplers used. For oxic sediments, the possibility of benthic stanol production must also be considered (Lee et al., 1977).

However, the process which was observed in the water column was rapid metabolism of plankton stenols and stanols in the oxic zone above the O_2/H_2S interface. It appears from these results that decomposition reactions occur much faster than the stenol to stanol reduction in the water column of the Black Sea or Walvis Bay. Alternatively, proposed intermediates (4) and (5) in the conversion of stenols to stanols (Fig. 2) may be attacked and transformed by other processes in the water column.

The case is quite different at the sediment—water interface. Studies have been conducted in sediments from the highly productive, upwelling waters off Walvis Bay on the southwest African shelf and slope. Not only were stanols observed in these sediments (Lee et al., 1980), but the proposed stanone intermediates $(5;$ Fig. 2) were found in high concentrations (Gagosian and Smith, 1979). These stanones differed primarily in unsaturation at the C-22 position, methylation at C-4 and alkylation at C-24. The occurrence of 3-ketosteroids in biological systems is limited. Since planktonic or benthic faunal or floral sources for the stanones is unlikely and the stanones isolated correspond structurally to their stenol precursors detected in the sediments and overlying water column, the authors proposed stanones as microbially produced intermediates in the transformation of stenols to stanols in sediments.

In addition to the stanones and stanols, other stenol transformation products were observed. The stenol and stanol dehydration products, steradienes $(2$ and $3)$, and sterenes $(7$ and $8)$, respectively were isolated along with steratrienes presumably formed from dehydration of diol $(9;$ Fig. 2) (Gagosian and Farrington, 1978). Although high concentrations ($\mu g\ g^{-1}$ dry weight sediment) of these compounds were observed in the southwest African shelf and slope samples, lower or zero concentrations of the compounds were observed for samples from the continental shelf and slope of the western North Atlantic and the Black Sea.

Thus, it appears that there are highly unique environments in the upwelling area off southwest Africa providing the microbiological and chemical conditions for rapid diagenesis of steroids within a time scale of 100 years or less. Although these conditions exist in the sediments, they do not appear to exist in the water column of this region or in the waters of deep anoxic basins.

3.3. Amino acids

Transformation reactions involving amino acids have been well studied because of the interest in the formation and behavior of these compounds in the primitive ocean before the origin of life and in their later preservation in fossil materials. This interest led to the study of thermal decomposition kinetics of aqueous solutions of amino acids (Abelson, 1957; Vallentyne, 1964; Bada and Miller, 1968a, b). Bada (1971) later discussed the applicability of these and other amino-acid decomposition and transformation studies to natural waters. He suggested that oxidative deamination would be the major non-biological decomposition pathway under aerobic conditions but that the reaction would be slow even when catalyzed by trace metals. Decarboxylation would be even more insignificant, having a half-life on the order of a million years. Biologically mediated reactions would be much faster in the upper layers of the ocean and perhaps in deeper waters as well.

Studies of one of the transformation reactions mentioned by Bada (1971), racemization, particularly illustrates the control by living organisms of the distribution of amino acids in seawater. Protein amino acids, with the exception of glycine, are asymmetric at the α-carbon, and in some cases, also the β-carbon. This means that they can exist in several enantiomeric forms. However, most organisms, with the exception of bacteria, contain only the L-amino acid isomer. The fact that these enantiomers can interconvert, or racemize, with time to form an equilibrium mixture led to the idea that the extent of racemization of dead or fossil material would be related to its age. Hare and Abelson (1968) first observed this relationship in fossil shells. Bada (1971) measured amino-acid racemization rates in buffered aqueous solutions which led him to postulate that this reaction might be useful in determining the age of amino acids (and organic matter) in the deep sea. However, when Lee and Bada (1977) later measured enantiomeric ratios of amino acids in seawater from several water masses, they found the concentration of D-isomer too high to be explained by racemization over a time span consistent with ages estimated for these water masses by other methods. Three factors, (1) the distribution of amino acids both vertically and horizontally at the stations investigated, (2) the correlation of these distributions with the presence of marine bacteria, and (3) the behavior of bacteria with respect to standard seawater filtration procedures, led these investigators to postulate that production of D-amino acids by bacteria and not chemical racemization was the source of the high enantiomeric ratios observed. Because of the relative speed and facility of biologically mediated reactions, it would seem that the majority of transformation reactions undergone by amino acids and other labile organic compounds in the sea are controlled by living organisms.

Two issues presently cloud complete acceptance of the last statement. Are the metabolic processes of organisms active or even present in the deep sea? What distinguishes a compound as biologically labile or inert? The question of bacterial activity in the deep sea was discussed earlier in Section 2.2 dealing with nutrient recycling below the euphotic zone. The slowness of heterotrophic oxidation in the deep sea would allow time for other processes to occur, especially if the compounds were resistant to further biological decomposition after passing through the euphotic zone. Benthic macroorganisms may metabolize organic compounds at faster rates, but the relative scarcity of these organisms in the deep sea makes their overall contribution difficult to assess. Bada and Lee (1977) address this second question of lability of organic compounds with special reference to amino acids. Evidence from amino-acid uptake and excretion experiments and measured concentrations support the idea that the residence time of free amino acids in surface waters is short, on the order of days or less, due to rapid turnover by organisms. Concentrations of free amino acids in deeper waters are almost negligible, possibly due to fewer sources, such as zooplankton excretion, but similar consumption processes, such as microbial degradation or particle adsorption. Combined amino acids are present even in deep waters, suggesting a certain amount of resistance to biological attack. Combined amino acids are those which must be hydrolyzed in $6N$ HCl before free amino acids are released and can be measured. They are amino acids which are either part of a larger protein or polymeric molecule, or are absorbed onto submicron particles which pass the filtering procedure. Their presence suggests the possibility of another type of transformation reaction, i.e. condensation reactions which amino acids may undergo in seawater.

The condensation of amino acids with lignin oxidation products and sugars to form humic-type substances has been recognized for a long time. Hedges (1977, 1978) and Hedges and Parker (1976) have elegantly studied the probability of some of these condensation reactions occurring in seawater and the subsequent association of the products with clay minerals. Hedges and Parker (1976) found most of the lignin degradation products present in sediments from the Gulf of Mexico to directly result from plant fragments of terrestrial source. Less than 7% of these degradation products could be attributed to humic substances of terrestrial origin. This led Hedges (1978) to postulate that because carbohydrates are so much more abundant in seawater than are lignin polymers, a sugar—amino acid condensation reaction to form melanoidin polymers might be more likely. Laboratory simulation experiments in buffered aqueous solutions confirmed the formation of these polymers and their similarity to natural humic materials. Further, this study and Hedges (1977) showed that the melanoidins, especially those formed from basic amino acids, are adsorbed from seawater by clay minerals, effectively removing them from solution. Indeed, amino acids dissolved in seawater have a higher ratio of acidic and neutral to basic amino acids (Lee and Bada, 1977), but this probably reflects the source of these compounds more than the sink. Further studies of natural humic substances found in seawater might provide more insight into the role of melanoidin formation in humification processes.

3.4. The formation of seawater humic substances

Seawater humic substances are complex mixtures of molecules with several functional groups and a wide molecular weight range. They contain complex distributions of functional groups and possess some surface-active character. Marine sedimentary humic substances are chelators of metal ions (Rashid, 1971), growth-promoting agents of plants (Prakash et al., 1973) and solubilizing agents for hydrophobic compounds (Boehm and Quinn, 1973).

A study of the "bulk properties" of seawater humic substances was carried out by Kerr and Quinn (1975), while a detailed structural analysis was undertaken by Stuermer (1975) and Stuermer and Harvey (1978). Stuermer discussed the structural features in terms of origin, chemical and physical properties, interaction in the sea and eventual fate. As an example of the formation of a humic substance in seawater, we will discuss Stuermer's proposed structure of seawater fulvic material (Gagosian and Stuermer, 1977), the precursor compounds to its formation, and the condensation and polymerization reactions responsible for its synthesis. Although the material isolated by Stuermer represents only a small portion of the total humic material, it serves as an example of a possible condensation product.

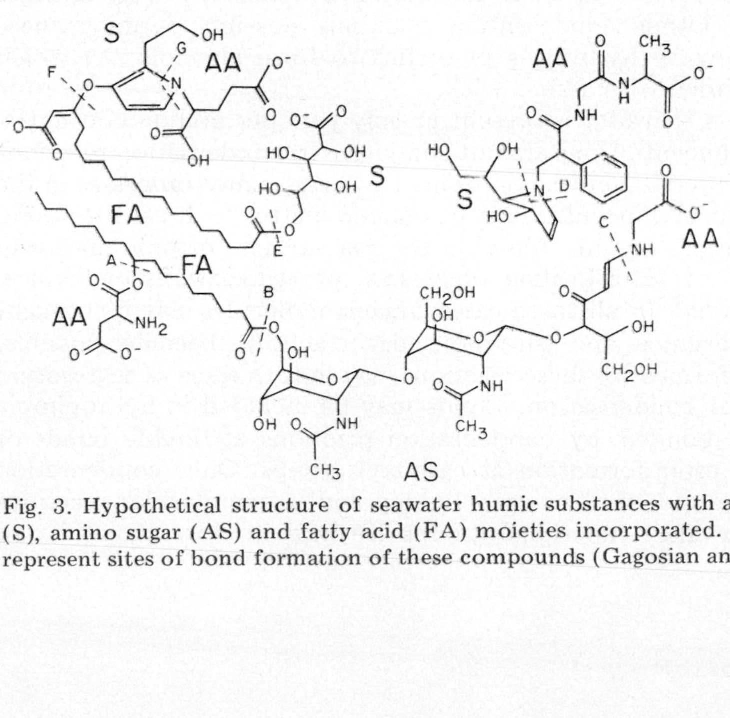

Fig. 3. Hypothetical structure of seawater humic substances with amino acid (AA), sugar (S), amino sugar (AS) and fatty acid (FA) moieties incorporated. The dashed lines A—G represent sites of bond formation of these compounds (Gagosian and Stuermer, 1977).

Fig. 3 presents the hypothetical structure for a typical marine humic substance. Its building blocks are many of the important biosynthetic molecules in the sea, such as amino acids, sugars, amino sugars, and fatty acids. The dashed lines show potential points of condensation with some rearrangement to form the structure. Other molecules such as carotenoid and chlorin pigments, hydrocarbons, and phenols may be incorporated into the structure but were not shown here. The structural features shown in Fig. 3 are based on elemental composition, molecular weight, titration, chemical reduction, and spectral data. Some of the major characteristics are the low abundance of aromatic relative to aliphatic protons, low phenol content, high nitrogen content, low average molecular weight (73% less than 700), low UV—VIS light extinction coefficients, and the 1560 cm^{-1} band in the IR spectra indicative of an amide linkage.

The first step in the condensation reaction of amino acids with sugars involves the formation of a Schiff base between the amino nitrogen of an amino acid and the aldehyde (or ketone) function of a sugar. Subsequent reactions include rearrangements, cyclizations, and decarboxylations to form complex, brown-colored mixtures (melanoidins) discussed earlier. The formation of marine humus or *gelbstoff* by a mechanism involving the condensation of sugars with amino acids has been proposed by Kalle (1966) and Nissenbaum (1974). However, the condensation of amino acids and sugars is not the only reaction occurring since it cannot account for the abundance of long-chain aliphatic structures in seawater humic substances. Components such as marine lipids must also be incorporated into the products via, for example, ester (*B* and *E* in Fig. 3), vinylogous ester (*F*), amide, or vinylogous amide (*C*) linkages. Attack at a site of unsaturation forming ether linkages may be involved. Other condensation reactions possibly forming these materials would involve hydrolysis of olefins to form alcohols (*A*) which could then be esterified with amino acids.

Organic matter in seawater is present in only part per million concentrations. These low concentrations are not conducive to condensation reactions which depend on intermolecular collisions. However, many processes in the sea result in high local concentrations of organic matter such as that found in decaying organisms, organic films on the sea surface, organic matter in particles, the guts of filter-feeding organisms, or aggregations of surface-active organic material. In all these cases, organic molecules may be brought within bonding distance and intermolecular reactions become possible. Humic substances formed by these reactions may in turn serve as aggregation centers and sites of condensation. Lipids may be included in hydrophobic sites and then be bonded by condensation reactions at double bonds or through amide or ester formation at carboxyl groups. Once condensation reactions have occurred, further cross-linking and intramolecular reactions are likely to occur since functional groups are held in close proximity to each other.

4. TRANSPORT OF ORGANIC SUBSTANCES BY PHYSICAL PROCESSES

In the previous sections we have described various biological and chemical production, consumption and transformation pathways of organic compounds in the sea. Yet another very important process is the physical transport of organic substances into, within, and out of the water column. Our discussion of physical transport will be divided into four subsections:

(1) The transport of organic compounds to the ocean via atmospheric processes and the exchange of this material across the air—sea interface.

(2) The input of organic materials to continental shelf areas from rivers.

(3) The interaction between physical processes of water masses with dissolved and particulate organic species.

(4) The contribution of large particle fluxes to the transport of organic matter through the water column to the benthos.

4.1. Atmospheric transport

Evaluating the role of the atmosphere in the transport of continentally derived materials to the oceans has been a challenging task. Our knowledge of atmospheric organic compounds, their structures, sources, fluxes and exchange mechanisms in the oceans is very limited. Organic matter in the atmosphere can exist either in the gas phase or on particles and can enter the sea in three major ways: precipitation (scavenging both gases and particles), dry fallout, and direct gas exchange. Calculations have been made on the flux of total organic carbon into the sea (Duce, 1978) by a combination of these mechanisms.

Hunter and Liss (1977) have reviewed the organic composition and physicochemical properties of the sea surface in order to assess the role of natural organic substances at the sea—air interface in modifying the transfer of organic matter from the atmosphere to the oceans. Liss and Slater (1974a, b) have also developed a gas-exchange model and applied it quite successfully to assess whether the sea is a source or a sink for a number of trace gases such as CH_4, CO, CCl_4, CCl_3F, CH_3I and $(CH_3)_2S$. With the exception of CCl_4 and CCl_3F, the ocean was calculated to be a net source for all these gases.

We clearly do not know the major sources of specific atmospheric organic compounds, be these sources continental, oceanic or atmospheric *de novo* synthesis of new structures. Ketseridis et al. (1976) undertook a fairly detailed study of the organic constituents of atmospheric particles in European and Atlantic air masses. They found a chemically complex mixture of fatty acids, phenols, amines, and aliphatic and aromatic hydrocarbons in their samples. They noted that ratios of the concentrations of some of the major groups were fairly constant for all the locations studied. They suggested either a common origin for all the substances or that the atmospheric processes responsible for producing, transporting, or transforming these compounds have similar effects on product distribution at all locations. On the other hand, Barger and Garrett (1976) found large concentrations of marine-derived lipid and surfactant material in marine atmospheric samples, possibly produced from bubble bursting (MacIntyre, 1974). From laboratory experiments, Hoffman and Duce (1976) observed that atmospheric particles produced by bubbling have a much higher organic-carbon content than would be expected on the basis of organic-carbon values in seawater. Thus, it is likely that a considerable amount of the organic carbon that enters the ocean from the marine atmosphere is simply recycled organic carbon from sea salt (Duce and Duursma, 1977).

These indications that the organic materials in the marine atmosphere are subject to intense chemical transformations and physical recycling processes imply that a total organic-carbon approach is not sufficient to resolve the numerous processes occurring. Thus, many investigators have studied specific compounds, choosing chemically unreactive materials as source markers for land and marine-derived compounds and more reactive compound classes to observe transformations occurring in the atmosphere. For example, Simoneit (1977) found land-derived plant waxes in aeolian dust off West Africa. Similar materials are widely distributed in abyssal plain sediments of the northwest Atlantic Ocean (Farrington and Tripp, 1977). Went (1960) and Rasmussen and Went (1965) point out that enormous fluxes of higher-plant derived volatile organic compounds are released to the atmosphere. Went et al. (1967) suggest that these are transformed into particulates that act as condensation nuclei.

A large literature describes the transport of anthropogenically produced or distributed organic compounds such as petroleum hydrocarbons and halogenated hydrocarbons, including the PCBs, the DDT family and the freons. Discussing this subject is beyond the scope of this chapter which deals with naturally produced organic compounds, and the reader is referred to the National Academy of Sciences Report (1978) on this subject.

4.2. River inputs

Very few studies have dealt with the temporal and spatial variations in the distribution of organic compounds in rivers. However, there have been some studies of the "bulk" properties of river organic material and the fate of dissolved and particulate riverine humic materials as they pass through the estuary into the coastal zone. Handa (1977) has reviewed the large number of studies attempting to separate and determine the molecular structures of complex colored materials extracted from rivers. He points out the many difficulties that still persist in unraveling this "organic soup". These materials were found to have a wide molecular-weight range (Gjessing, 1965; Ghassemi and Christman, 1968) and to contain polyhydroxy and methoxy carboxylic acids with aromatic and olefinic unsaturation (Christman and Ghassemi, 1966; Lamar and Goerlitz, 1966). Beck et al. (1974) undertook more comprehensive chemical and infrared and NMR (Nuclear Magnetic Resonance) spectroscopic studies. They found that riverine organic matter contained twice as many carboxyl groups as phenolic hydroxy functions. Comparing these data to soil humic substances, Handa (1977) concluded that river-water organic matter bears a closer resemblance to fulvic acid than humic acid.

The question of how much riverine organic matter is transported through the estuary into the open ocean has been a subject of great interest since Mackenzie and Garrels (1966) suggested that processes in estuaries were important in determining the chemical mass balance between rivers and oceans. Previous literature had suggested that humic substances occur in river water as negatively charged hydrophilic colloids in which the particle size is a function of pH. With this in mind, Beck et al. (1974) hypothesized that the particle structure was micellar in nature since humic molecules are thought to have defined polar and hydrophobic regions. In dilute solution, polymerization reactions would occur more readily if the reactants were in the form of a micellar aggregation. These reactions would result in the elimination of hydrophilic functional groups near micellar particles, thus decreasing the stability of the aggregate and eventually leading to the flocculation of organic matter. Handa (1977) summarizes laboratory and field evidence for this hypothesis.

Evidence from $\delta^{13}C$ studies (Sackett, 1964; Nissenbaum and Kaplan, 1972) and studies of the offshore decrease of terrestrially derived lignin material (Gardner and Menzel, 1974; Hedges and Parker, 1976) have led investigators to conclude that only a minor fraction of terrestrial humic substances are transported beyond the estuaries. The approach these investigators took was to compare sediment $^{13}C/^{12}C$ ratios with lignin oxidation products such as p-hydroxyl, vanillyl and syringyl phenolic aldehydes, ketones and carboxylic acids. A sharp offshore decrease of land-derived organic matter indicated that a major fraction of these terrestrial substances must be removed from seawater by precipitation of the flocculant in the estuary. On the other hand, Sholkovitz (1976) and Sholkovitz et al. (1978) have shown that only a small fraction of river dissolved organic matter is preferentially and rapidly flocculated during estuarine mixing. This fraction was found to be the high molecular-weight component, the dissolved humic acids. Although 60—80% of the dissolved humic acid flocculates during estuarine mixing, it represents only 3—6% of river dissolved organic matter. These investigators also found a similar salinity dependence for iron and humic acid flocculation, implying that both constituents may be removed from river water by a common mechanism of colloid flocculation.

There are other potential sources of organic matter from the coastal zone to the open ocean. However, we know little about the production of organic carbon from these sources. Coastal macrophyte production such as *Spartina* grass in marshes and eel grass and kelp in coastal waters are a possible source. Wheeler (1976) found that rivers transecting swamp areas had twice the dissolved organic carbon of other river waters. However, our knowledge in this area is also rudimentary and further research is needed.

4.3. Interaction of water masses with dissolved and particulate organic species

Dissolved compounds in seawater can be physically transported from one location to another by advection and can move from one parcel of water to another by eddy diffusion. But, the concentration of dissolved organic compounds is determined not only by these physical processes but also by the sinking of particulate material and vertical migration of organisms. Very little specific organic compound work has been accomplished in this area, but a few studies exist in the literature. We discuss four examples from these studies: (1) the "bulk" properties of particulate and dissolved organic carbon; (2) a dissolved gas, methane; (3) relatively labile organic compounds, amino acids; and (4) the slightly less labile steroids.

Differentiation between the effects of advective and sinking processes on the distribution of particulate organic carbon (POC) has been attempted by several investigators (see Gordon, 1971, and Menzel, 1974, for a review). Riley et al. (1965) and Gordon (1970) observed large temporal changes in particulate carbon in the Sargasso Sea which could not be associated with production processes occurring at the immediate surface. They found total accumulated carbon in the water column to be greater than that produced annually at the surface. To explain this observation they postulated advective processes displacing water from the entire water column seasonally. Higher carbon concentrations probably originated in water from more northern latitudes. Gordon (1971) also studied temporal changes of POC at a location near the Hawaiian Islands. Since seasonal variations in POC concentrations taken over a one-year period were detected for several vertical profiles above and below, but not at 1000 m, active sinking of particles from the surface to the deep water is not a reasonable process. Therefore, advective processes originating at the site of formation of Pacific Deep Water (60°S) were employed to explain the seasonal variations in the deep water (>1000 m). Menzel (1974) discusses the problems with this interpretation in terms of the age of the water and the variation of POC at depth both temporally and spatially.

Menzel (1974) has reviewed attempts made to relate the concentration of DOC to the movement of water masses. He feels that seasonal or geographical differences in DOC concentrations have failed to correlate with clearly identifiable water mass structure or movement. Transects in the southeast and southwest Atlantic (Menzel and Ryther, 1970) did not show correlations of DOC with the northerly transport of water originating at the subtropical convergence, although there was a slight correlation in the North Atlantic (Chapter 1; Duursma, 1965).

It will remain difficult to determine the relative importance of processes controlling the distribution of bulk organic matter in seawater until (1) the disagreement over whether variations of POC and DOC at depth are real or due to sample and analytical variability is resolved, (2) the methods for these analyses become more precise and sensitive, and (3) we can better assess what the terms POC and DOC really mean (Sharp, 1973, 1975). Thus, we will turn our attention to individual classes of organic compounds which are more easily definable in terms of their molecular structure.

Methane is one of a number of reduced gases present in the oceanic mixed layer in amounts greater than those suggested by solubility equilibrium calculations with the atmosphere, implying an active supply mechanism. Many investigators have dealt with the processes responsible for this excess (Lamontagne et al., 1973; Giger, 1977). Recent studies by Scranton and Brewer (1977) and Scranton and Farrington (1977) serve as an excellent example of the application of physical models to differentiate between the supply mechanisms. In the coastal upwelling area off Walvis Bay, Scranton and Farrington (1977) tried to determine the relative supply to the coastal mixed layer by *in situ* biological production of methane and by eddy diffusive and advective transport of methane originating from the sediments in contact

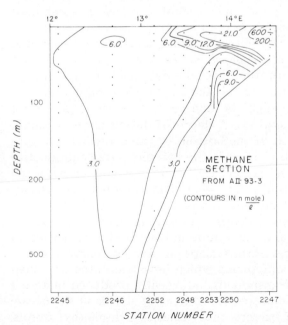

Fig. 4. Methane concentrations (n mole l^{-1}) as a function of depth on a section near Walvis Bay off the southwest African shelf (Scranton and Farrington, 1977).

with the bottom sediments of the coast. Methane concentrations from a westerly transect on the southwest African shelf show that mixed layer concentrations were high for shelf and slope stations (Fig. 4). Methane maxima were present in the top of the thermocline (10—20 m) for most stations. The methane distributions below the mixed layer on the shelf are consistent with some transport of low-methane offshore water onto the shelf. The high bottom water concentrations indicate that methane diffuses rapidly into the water from the sediments. The presence of a mid-depth minimum suggests that methane generation was not extensive in this part of the water column. Using Brewer and Spencer's (1975) coastal source model, Scranton and Farrington (1977) concluded that advective transport of coastal methane to offshore surface waters was sufficient to supply some of the excess methane. However, a significant amount of *in situ* methane production was also found to take place. Because the circulation patterns in this area are quite complex, the authors warn that caution should be exercised when trying to distinguish eddy diffusive and advective transport from *in situ* production in areas where complicated currents and an extremely time-dependent circulation pattern are present.

The interaction of water masses with amino acids has also been studied. Lee and Bada (1975) analyzed seawater samples from the eastern equatorial Pacific Ocean finding greater concentrations of combined than free amino acids. They noted that concentration variations of the combined amino-acid fraction in the water column suggests their potential use as a water mass indicator. This conclusion was based on the observation of a concentration minimum in the vertical profile of combined amino acids at 300—400 m at a station just west (101°W) of the Galapagos which was not present at a station east of the Galapagos. A profile of total amino acids from a station on the equator at 155°W also showed this same subsurface minimum. The authors felt that this minimum might be due to differences in the water masses present at the sampling stations. A depth of 300 m at the westerly stations corresponds to the depth of the bottom of the Cromwell Current. The source area of the Cromwell Current is the less biologically productive western Pacific equatorial region. Hence, this current would be expected to have lower amino-acid concentrations than Pacific equatorial surface water, causing a minimum where current waters are advected into the area. Since the Cromwell Current is blocked from proceeding eastward along the equator by the Galapagos Islands, no minimum in amino-acid concentration would be expected at a station east of the Galapagos; none was observed.

Sterols are another class of compounds which have been used to understand the processes controlling transport of biologically produced materials in seawater. There are many species-specific sources of sterols in the ocean, although few, if any, species of bacteria biosynthesize 4-desmethyl sterols. Thus, the interpretation of seawater sterol profiles is somewhat less complicated than similar studies with compounds important in bacterial metabolism, such as fatty acids or amino acids. Gagosian (1976) and Gagosian and Nigrelli (1979) have studied the distribution of sterols in the western North Atlantic. They have attempted to determine the relative roles which physical and biological processes play in controlling sterol distributions in seawater. They found large variations in individual sterol concentrations in the upper 100 m of the water column (sometimes as deep as 1000 m in the Sargasso Sea) with smaller variations from this depth to the sea floor.

Gagosian and Nigrelli (1979) hypothesized that the sterol concentration changes observed are controlled for the most part, by (a) processes which

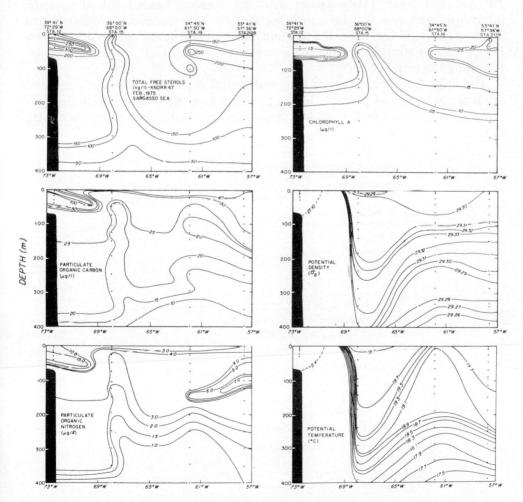

Fig. 5. Total free sterols, particulate organic carbon, particulate organic nitrogen, chlorophyll *a*, potential density (σ_θ) and potential temperature (θ) as a function of depth to 400 m for a section in the western North Atlantic Ocean (Gagosian and Nigrelli, 1979).

control surface primary productivity, and (b) particulate transport to the deeper water where grazing by bacteria and possibly zooplankton occurs. Sterols are hydrophobic compounds which tend to stay mostly in the particulate phase. The obvious correlation of total sterols with POC, PON, and chlorophyll (see Fig. 5) shows the importance of surface primary productivity to the distribution of sterols for a transect from the continental shelf across the Gulf Stream and into the Sargasso Sea east of Bermuda. Station 12 on the transect is in the cold continental shelf waters of the northeastern United States while Stations 19 and 2128 were taken in Sargasso Sea water. A water mass assignment for samples taken from Station 15 was not clear. Investigation into the reasons behind this deep-water intrusion in an area too far south for Gulf Stream water, and evidence from other studies in the area, led the authors to postulate that Station 15 was taken on the southwest edge of a cold core ring in Gulf Stream water surrounding the ring.

The chlorophyll *a* values in the continental shelf and slope surface waters are high, as expected (Fig. 5). In the Sargasso Sea, the subsurface chlorophyll *a* maximum is centered at approximately 50 m. Total free sterol concentra-

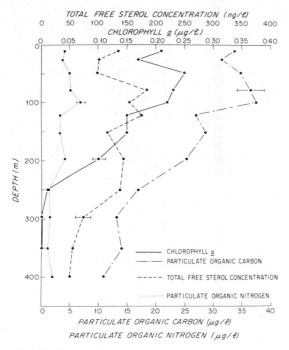

Fig. 6. Particulate organic nitrogen, total free sterol, particulate organic carbon, and chlorophyll *a* concentrations as a function of depth from 33°40.6′N, 57°36.8′W, February, 1975 (Gagosian and Nigrelli, 1979).

tions are quite high in the shelf waters and also exhibit a subsurface maximum in the Sargasso Sea at approximately 50 m. However, the chlorophyll *a* concentrations in the Station 15 data are lower than expected. POC and PON behave similarly. A distinct sterol minimum exists at Station 15 in water shallower than 350 m. The correlation between sterols POC, PON, and chlorophyll is seen more clearly by plotting the concentrations of the four chemical parameters as a function of depth for Station 2128 in the Sargasso Sea (Fig. 6). Previous occupation of this station by Gagosian (1976) showed a similar correlation.

From flux calculations at this Sargasso Sea station, Gagosian and Nigrelli (1979) found that a maximum of 0.05—0.3% of the sterols produced by phytoplankton in surface waters are deposited to the ocean floor. A similar calculation was done for hydrocarbons by Farrington and Tripp (1977) and found to be 0.01—1%. The sterol residence time (the average lifetime of a sterol molecule before it is metabolized) in the euphotic zone was calculated to be approximately one month, whereas the deep-water residence time value was found to be 20—150 years. This monthly turnover of surface water sterols is in contrast with that of more labile dissolved organic compounds such as amino acids whose turnover time has been estimated to be on the order of several days (Lee and Bada, 1977).

4.4. Particulate transport

Knowledge of the transport of organic compounds associated with particulate matter is presently in a considerable state of flux. Recent advances in sampling technology such as the Large Volume *in situ* Filtration System (Bishop and Edmond, 1976) and various moored (Honjo, 1978) and free-floating (Staresinic, 1978, and references therein; Staresinic et al., 1978) sediment traps have changed our ideas concerning the rate of transport and to some extent the composition and quantity of particulate matter in the sea. Previously, particulate matter analyses involved simple membrane-filtration procedures with small (~20 l) amounts of water. Riley (1970) and Menzel (1974) review some of the vast literature concerning studies of POC using conventional methods. These methods can produce reliable estimates of the quantity of particulate matter in the ocean. But, because of the flux of large particles not readily measured by conventional samplers, a new picture of transport is just beginning to emerge from the advanced technology.

Bishop et al. (1977) found that the large particles (mostly fecal material) they measured in the equatorial Atlantic were responsible for less than 4% of the total suspended mass concentration of particulate matter, but for 99% of the vertical flux at 400 m. Transit time for the fecal material through a 4 km deep water column was estimated to be on the order of 10—15 days, with a maximum lateral displacement due to advection of 30—40 km. These studies indicate that although the total amount of particulate organic carbon in a water column is not far from earlier estimates, the percentage of surface productivity rapidly reaching the sea floor directly below and potentially containing labile organic compounds, is much higher than previously thought. Bishop et al. (1977) calculated that 87% of the carbon, 91% of the nitrogen, and 97% of the phosphorus produced in the equatorial Atlantic surface waters are recycled in the top 400 m, leaving a flux to the bottom which is of sufficient quantity to meet the respiratory requirements of bottom-dwelling organisms.

Staresinic (1978) attempted to determine the quality (or nutritional value) of organic matter reaching the sea floor by measuring chlorophyll a and phaeopigments, POC, PON, and particulate protein in material collected from free-drifting sediment traps of Peru. The material was also analyzed by light microscopy to determine its gross qualitative composition and possible origin. The type of material at each location varied, from fecal rods dominating some stations and phytodetritus others. Distinct diurnal variations in the vertical flux of bulk organics were found within the euphotic zone, with an increased flux of phytodetritus at night. Analysis of euphausiid molts found in the traps showed molting frequencies to be much higher (44—70% of the population per day) than laboratory estimates, indicating an even larger input of this chitinous molt material through the euphotic zone than was previously thought. Sinking rates of 850 m day^{-1} for anchoveta fecal pellets, the principal recognizable component of trap material on the shelf, imply that these fecal casts rapidly reach the benthos in nearshore areas. However, particulate protein measurements of sediment trap material containing these casts indicated that the relative amount of protein decreased with depth in the water column and that this protein was depleted in nitrogen relative to a standard animal protein.

Further studies of the sediment-trap material collected by Staresinic (1978) are now being carried out by organic geochemists and will include analyses of lipids, carbohydrates, specific plant pigment components, PCBs, selected transuranic elements, and specific organic nitrogen compounds. The results of these and other similar investigations should tell us more about the quality of rapidly sinking large particles as a food source for benthic organisms and about transformation reactions involving organic compounds associated with these particles.

Sediment-trap measurements have sometimes allowed determination of the source as well as the rate of particulate organic transport. Spencer et al. (1978) deployed sediment traps 200 m off the Sargasso Sea floor within the bottom nepheloid layer. Resuspended bottom material, especially clays, made a major contribution to the total trap material collected. Interpretation of elemental and isotope data led them to postulate that 95% of the clay and 10% of the organic matter, but almost none of the calcium carbonate in the traps was contributed by resuspended bottom material instead of the rapidly settling large particles recently derived from the surface ocean.

The contribution of specific organic compounds from the sediment to the water column through diffusion and through resuspension processes, such as bioturbation and bottom current transport, has not yet been thoroughly investigated. The few measurements of dissolved organic carbon in interstitial waters show concentrations much higher than in seawater (Starikova, 1970; Krom and Sholkovitz, 1977). Henrichs and Farrington (1979) have shown that dissolved free amino-acid concentrations are about one hundred times greater in the interstitial waters of near-surface sediments than in the overlying seawater. Nissenbaum et al. (1972) and later Krom and Sholkovitz (1977) found that interstitial water DOC from estuarine sediments increased with depth in the core, which could indicate a possible flux of dissolved organic carbon upward from the sediment.

Particulate organic carbon is certainly being resuspended into the water column by the processes discussed above. These resuspension processes sometimes make a distinct delineation of the interface between the seawater and the sediment difficult. Lee et al. (1979) report measurements of sterols made on a surface floc material from the top of cores from the western North Atlantic. The sterol composition of this material, as well as the fatty acid and hydrocarbon content (J.W. Farrington, unpublished data) show that this floc material is of different composition than the underlying material. Whether this material is recently deposited and about to become part of the permanent deposit or whether it is material which is continually resuspended and moved along laterally to areas more conducive to permanent deposition is unknown. The measurement of specific organic compounds in this floc material and in samples from deep sediment traps or deep *in situ* pump filters should provide more insight into the question of net transport by resuspension in the deep sea.

Much has been accomplished over the past several years concerning the organic compound inventory of seawater, and it is clear from the previous discussions that we are beginning to understand more about the sources of organic compounds and the processes that control their transformation and transport in the sea. Several groups have recently discussed future approaches in this strongly emerging area of research, e.g. Symposium on Concepts in Marine Organic Chemistry (Andersen, 1977) and the National Science Foundation, International Decade of Ocean Exploration Meeting on Oceanographic Studies for the 1980's (NSF—IDOE Report, 1977; Gagosian et al., 1978). They have recommended that investigators in the field of marine organic chemistry now concentrate on the mechanisms and rates of processes controlling the distribution of organic compounds in the sea. The value of interdisciplinary studies including contributions from biologists, geologists, and physical oceanographers was especially recognized. We have presented several examples in this chapter of processes controlling the distribution of organic compounds which we hope will stimulate further investigations in this exciting field of research.

5. ACKNOWLEDGEMENTS

We wish to thank J.W. Farrington, S.M. Henrichs, R.F.C. Mantoura, and O.C. Zafiriou for their helpful comments on the manuscript. We also wish to thank the Oceanography Section of the National Science Foundation, Grants OCE 77-26180 and OCE 77-26084 for financial support.

6. REFERENCES

Abelson, P.H., 1957. Organic constituents of fossils. Treatise on Marine Ecology and Paleoecology, 2. Mem. Geol. Soc. Am., 67: 87—92.

Andersen, N.R., 1977. Concepts in marine organic chemistry. Mar. Chem., 5: 303—638.

Antia, N.J., Berland, B.R., Bonin, D.J. and Maestrini, S.Y., 1975. Comparative evaluation of certain organic and inorganic sources of nitrogen for phototropic growth of marine microalgae. J. Mar. Biol. Assoc. U.K., 55: 519—539.

Bada, J.L., 1971. Kinetics of the non-biological decomposition and racemization of amino acids in natural waters. Adv. Chem. Ser., 106: 309—331.

Bada, J.L. and Lee, C., 1977. Decomposition and alteration of organic compounds dissolved in seawater. Mar. Chem., 5: 523—534.

Bada, J.L. and Miller, S.L., 1968a. Ammonium ion concentration in the primitive ocean. Science, 159: 423—425.

Bada, J.L. and Miller, S.L., 1968b. Equilibrium constant for the reversible deamination of aspartic acid. Biochemistry, 7: 3403—3408.

Barger, W.R. and Garrett, W.D., 1976. Surface active organic material in air over the Mediterranean and over the eastern equatorial Pacific. J. Geophys. Res., 81: 3151—3157.

Beck, K.C., Reuter, J.H. and Perdue, E.M., 1974. Organic and inorganic geochemistry of some coastal plain rivers of the southeastern United States. Geochim. Cosmochim. Acta, 38: 341—364.

Bishop, J.K.B., 1977. The Chemistry, Biology and Vertical Flux of Oceanic Particulate Matter. Thesis, Joint Program in Oceanography, M.I.T./W.H.O.I., Woods Hole, Mass., 291 pp.

Bishop, J.K.B. and Edmond, J.M., 1976. A new Large Volume *in-situ* Filtration System for the sampling of oceanic particulate matter. J. Mar. Res., 34: 181—198.

Bishop, J.K.B., Edmond, J.M., Ketten, D.R., Bacon, M.P. and Silker, W.B., 1977. The chemistry, biology and vertical flux of particulate matter from the upper 400 m of the equatorial Atlantic Ocean. Deep-Sea Res., 24: 511—548.

Boehm, P.D. and Quinn, J.G., 1973. Solubilization of hydrocarbons by the dissolved organic matter in seawater. Geochim. Cosmochim. Acta, 37: 2459—2477.

Brewer, P.G. and Spencer, D.W., 1975. Minor element models in coastal waters. In: T.M. Church (Editor), Marine Chemistry in the Coastal Environment. Am. Chem. Soc., pp. 80—96.

Carlucci, A.F. and Bowes, P.M., 1970. Production of vitamin B_{12}, thiamine and biotin by phytoplankton. J. Phycol., 6: 351—357.

Carter, P.W., 1978. Adsorption of amino acid-containing organic matter by calcite and quartz. Geochim. Cosmochim. Acta, 42: 1239—1242.

Christman, R.F. and Ghassemi, M., 1966. Chemical nature of organic color in water. Am. Water Works Assoc. J., 58: 723—741.

Craig, H., 1971a. The deep metabolism: oxygen consumption in abyssal ocean water. J. Geophys. Res., 76: 5078—5086.

Craig, H., 1971b. Son of abyssal carbon. J. Geophys. Res., 76: 5133—5139.

Duce, R.A., 1978. Speculations on the budget of particulate and vapor phase non-methane organic carbon in the global troposphere. Pageoph., 116: 244—273.

Duce, R.A. and Duursma, E.K., 1977. Inputs of organic matter to the ocean. Mar. Chem., 5: 319—339.

Duursma, E.K., 1965. The dissolved organic constituents of seawater. In: J.P. Riley and G. Skirrow (Editors), Chemical Oceanography, I. Academic Press, London, pp. 433—475.

Farrington, J.W. and Meyers, P.A., 1975. Hydrocarbons in the marine environment. In: G. Eglinton (Editor), Environmental Chemistry, I. Specialists Periodical Report, The Chemical Society, London, pp. 109—136.

Farrington, J.W. and Tripp, B.W., 1977. Hydrocarbons in western North Atlantic surface sediments. Geochim. Cosmochim. Acta, 41: 1627—1641.

Gagosian, R.B., 1976. A detailed vertical profile of sterols in the Sargasso Sea. Limnol. Oceanogr., 21: 702—710.

Gagosian, R.B. and Farrington, J.W., 1978. Sterenes in surface sediments from the southwest African shelf and slope. Geochim. Cosmochim. Acta, 42: 1091—1101.

Gagosian, R.B. and Heinzer, F., 1979. Stenosis and stanols in the oxic and anoxic waters of the Black Sea. Geochim. Cosmochim. Acta, 43: 471—486.

Gagosian, R.B. and Nigrelli, G., 1979. The transport and budget of sterols in the western North Atlantic Ocean. Limnol. Oceanogr., 24: 838—849.

Gagosian, R.B. and Smith, S., 1979. Steroid ketones in surface sediments from the southwest African shelf. Nature, 277: 287—289.

Gagosian, R.B. and Stuermer, D.H., 1977. The cycling of biogenic compounds and their diagenetically transformed products in seawater. Mar. Chem., 5: 605—632.

Gagosian, R.B., Ahmed, S.I., Farrington, J.W., Lee, R.F., Mantoura, R.F.C., Nealson, K.H., Packard, T.T. and Reinhart, K.L., 1978. Future research problems in marine organic chemistry. Mar. Chem., 6: 375—382.

Gardner, W.S. and Menzel, D.W., 1974. Phenolic aldehydes as indicators of terrestrially-derived organic matter in the sea. Geochim. Cosmochim. Acta, 38: 813—822.

Ghassemi, M. and Christman, R.F., 1968. Properties of the yellow organic acids of natural waters. Limnol. Oceanogr., 13: 583—597.

Giger, W., 1977. Inventory of organic gases and volatiles in the marine environment. Mar. Chem., 5: 429—442.

Gjessing, E.T., 1965. Use of "Sephadex" gel for the estimation of molecular weight of humic substances in natural water. Nature, 208: 1091—1092.

Glenn, A.R., 1976. Production of extracellular proteins by bacteria. Annu. Rev. Microbiol., 30: 41—62.

Gordon, D.C., 1970. Some studies on the distribution and composition of particulate organic carbon in the North Atlantic Ocean. Deep-Sea Res., 17: 233—243.

Gordon, D.C., 1971. Distribution of particulate organic carbon and nitrogen at an oceanic station in the central Pacific. Deep-Sea Res., 18: 1127—1134.

Handa, N., 1977. Land sources of marine organic matter. Mar. Chem., 5: 341—359.

Hare, P.E. and Abelson, P.H., 1968. Racemization of amino acids in fossil shells. Carnegie Inst. Wash. Yearb., 66: 526—528.

Harvey, H.W., 1966. The Chemistry and Fertility of Sea Waters. Cambridge University Press, Cambridge, 240 pp.

Hedges, J.I., 1977. The association of organic molecules with clay minerals in aqueous solutions. Geochim. Cosmochim. Acta, 41: 1119—1123.

Hedges, J.I., 1978. The formation and clay mineral reactions of melanoidins. Geochim. Cosmochim. Acta, 42: 69—76.

Hedges, J.I. and Parker, P.L., 1976. Land-derived organic matter in surface sediments from the Gulf of Mexico. Geochim. Cosmochim. Acta, 40: 1019—1029.

Hellebust, J.A., 1965. Excretion of some organic compounds by marine phytoplankton. Limnol. Oceanogr., 10: 192—206.

Henrichs, S.M. and Farrington, J.W., 1979. Amino acids in interstitial waters of marine sediments. Nature, 272: 319—322.

Hobbie, J.E., Holm-Hansen, O., Packard, T.T., Pomeroy, L.R., Sheldon, R.W., Thomas, J.P. and Wiebe, W.J., 1972. A study of the distribution and activity of microorganisms in ocean water. Limnol. Oceangr., 17: 544—554.

Hoffman, E.J. and Duce, R.A., 1976. Factors influencing the organic carbon content of atmospheric sea salt particles: a laboratory study. J. Geophys. Res., 81: 3667—3670.

Holm-Hansen, O. and Booth, C.R., 1966. The measurement of adenosine triphosphate in the ocean and its ecological significance. Limnol. Oceanogr., 11: 510—519.

Honjo, S., 1978. Sedimentation of materials in the Sargasso Sea at a 5367 m deep station. J. Mar. Res. U.S.A., 36: 469—492.

Hunter, K.A. and Liss, P.G., 1977. The input of organic material to the oceans: air—sea interactions and the organic chemical composition of the sea surface. Mar. Chem., 5: 361—379.

Kalle, K., 1966. The problem of gelbstoff in the sea. Oceanogr. Mar. Biol. Annu. Rev., 4: 91—104.

Kanazawa, A. and Teshima, S., 1971. *In vivo* conversion of cholesterol to steroid hormones in the spiny lobster, *Panulirus japonica*. Bull. Jap. Soc. Sci. Fish., 27: 207—212.

Kerr, R.A. and Quinn, J.G., 1975. Chemical studies on the dissolved organic matter in seawater. I. Isolation and fractionation. Deep-Sea Res., 22: 107—116.

Ketseridis, G., Hahn, J., Jaenicke, R. and Junge, C., 1976. The organic constituents of atmospheric particulate matter. Atmos. Environ., 10: 603—610.

Krom, M.D. and Sholkovitz, E.R., 1977. Nature and reactions of dissolved organic matter in the interstitial waters of marine sediments. Geochim. Cosmochim. Acta, 41: 1565—1573.

Lamar, W.L. and Goerlitz, D.F., 1966. Organic acids in naturally colored surface waters. U.S. Geol. Surv. Water Supply Pap., 1817-A, 17 pp.

Lamontagne, R.A., Swinnerton, J.W., Linnenbom, V.J. and Smith, W.D., 1973. Methane concentrations in various marine environments. J. Geophys. Res., 78: 5317—5324.

Lee, C. and Bada, J.L., 1975. Amino acids in equatorial Pacific Ocean water. Earth Planet. Sci. Lett., 26: 61—68.

Lee, C. and Bada, J.L., 1977. Dissolved amino acids in the equatorial Pacific, Sargasso Sea and Biscayne Bay. Limnol. Oceanogr., 22: 502—510.

Lee, C., Gagosian, R.B. and Farrington, J.W., 1977. Sterol diagenesis in Recent sediments from Buzzards Bay, Massachusetts. Geochim. Cosmochim. Acta, 41: 985—992.

Lee, C., Farrington, J.W. and Gagosian, R.B., 1979. Sterol geochemistry of sediments from the western North Atlantic Ocean and adjacent coastal areas. Geochim. Cosmochim. Acta, 43: 35—46.

Lee, C., Gagosian, R.B. and Farrington, J.W., 1980. Geochemistry of sterols in sediments from the Black Sea and the Southwest African shelf and slope. Org. Geochem., 2: 103—113.

Lee, R.F., Nevenzel, J.C. and Paffenhofer, G.-A., 1971. Importance of wax esters and other lipids in the marine food chain: phytoplankton and copepods. Mar. Biol., 9: 99—108.

Liss, P.S. and Slater, P.G., 1974a. Fluxes of gases across the air—sea interface. Nature, 247: 181—184.

Liss, P.S. and Slater, P.G., 1974b. Use of a two layer model to estimate the flux of various gases across the air—sea interface. Nature, 247: 818—827.

MacIntyre, F., 1974. Chemical fractionation and sea surface microlayer processes In: E.D. Goldberg (Editor), The Sea, 5. Wiley—Interscience, New York, N.Y., pp. 245—300.

Mackenzie, F.T. and Garrels, R.M., 1966. Chemical mass balance between rivers and oceans. Am. J. Sci., 264: 507—525.

Mahler, H.R. and Cordes, E.H., 1971. Biological Chemistry. Harper and Row, New York, N.Y., 1009 pp.

McCarthy, J.J., 1970. A urease method for urea in seawater. Limnol. Oceanogr., 15: 309—313.

McCarthy, J.J., 1972. The uptake of urea by natural populations of marine phytoplankton. Limnol. Oceanogr., 17: 738—748.

Menzel, D.W., 1974. Primary productivity, dissolved and particulate organic matter, and the sites of oxidation of organic matter. In: E.D. Goldberg (Editor), The Sea, 5. Wiley—Interscience, New York, N.Y., pp. 659—678.

Menzel, D.W. and Ryther, J.H., 1968. Organic carbon and the oxygen minimum in the South Atlantic Ocean. Deep-Sea Res., 15: 327—337.

Menzel, D.W. and Ryther, J.H., 1970. Distribution and cycling of organic matter in the oceans. In: D.W. Hood (Editor), Organic Matter in Natural Waters. Inst. Mar. Sci., Univ. Alaska, Occas. Publ., No. 1: 31—54.

Meyers, P.A. and Quinn, J.G., 1973. Factors affecting the association of fatty acids with mineral particles in seawater. Geochim. Cosmochim. Acta, 37: 1745—1759.

Müller, P.J. and Suess, E., 1977. Interaction of organic compounds with calcium carbonate — III. Amino acid composition of sorbed layers. Geochim. Cosmochim. Acta, 41: 941—949.

Nes, W.R. and McKean, M.L., 1977. Biochemistry of Steroids and Other Isoprenoids. University Press, Baltimore, Md., 690 pp.

National Academy of Sciences, 1978. The Tropospheric Transport of Pollutants and Other Substances to the Oceans. National Research Council, Washington, D.C., 243 pp.

National Science Foundation—IDOE, 1977. Ocean research in the 1980's: recommendations from a series of workshops on promising opportunities in large scale oceanographic research. Center for Ocean Management Studies, URI, Kingston, R.I.

Nissenbaum, A., 1974. The organic geochemistry of marine and terrestrial humic acids: implications of carbon and hydrogen isotope studies. In: B. Tissot and F. Bienner (Editors), Advances in Organic Geochemistry. Editions Technip, Paris, pp. 39—52.

Nissenbaum, A. and Kaplan, I.R., 1972. Chemical and isotopic evidence for the in situ origin of marine humic substances. Limnol. Oceanogr., 17: 570—582.

Nissenbaum, A., Baedecker, M.J. and Kaplan, I.R., 1972. Studies on dissolved organic matter from interstitial water of a reducing marine fjord. In: H. von Gaertner and H. Wehner (Editors), Advances in Organic Geochemistry, 1971. Pergamon Press, London, pp. 427—446.

Nixon, S.W., Oviatt, C.A. and Hale, S.S., 1976. Nitrogen regeneration and the metabolism of coastal marine bottom communities. In: J.M. Anderson and A. Macfadyen (Editors), The Role of Terrestrial and Aquatic Organisms in Decomposition Processes. Blackwell, Oxford, pp. 269—283.

North, B.B., 1975. Primary amines in California coastal waters: utilization by phytoplankton. Limnol. Oceanogr., 20: 20—27.

North, B.B. and Stephens, G., 1971. Uptake and assimilation of amino acids by Platymonas. 2. Increased uptake in nitrogen-deficient cells. Biol. Bull., 140: 242—254.

Pocklington, R., 1977. Chemical processes and interactions involving marine organic matter. Mar. Chem., 5: 479—496.

Prakash, A., Rashid, M.A., Jensen, A. and Subba Rao, D.V., 1973. Influence of humic substances on the growth of marine phytoplankton: diatoms. Limnol. Oceanogr., 18: 516—524.

Rashid, M.A., 1971. Role of humic acids of marine origin and their different molecular weight fractions in complexing di- and trivalent metals. Soil Sci., 111: 298—305.

Rasmussen, R.A. and Went, F.W., 1965. Volatile organic material of plant origin in the atmosphere. Proc. Natl. Acad. Sci. U.S.A., 53: 215—220.

Redfield, A.C., 1942. The processes determining the concentration of oxygen, phosphate and other organic derivatives within the depths of the Atlantic Ocean. Pap. Phys. Oceanogr., Meteor., 9: 1—22.

Remsen, C.C., Carpenter, E.J. and Schroeder, B.W., 1974. The role of urea in marine microbial ecology. In: R.R. Colwell and R.Y. Morita (Editors), Effects of the Ocean Environment on Microbial Activities. University Park Press, Baltimore, Md., pp. 286—304.

Riley, G.A., 1970. Particulate organic matter in seawater. Adv. Mar. Biol., 8: 1—118.

Riley, G.A., Van Hemert, D. and Wangersky, P.J., 1965. Organic aggregates in surface and deep waters of the Sargasso Sea. Limnol. Oceanogr., 10: 354—363.

Rowe, G.T., Clifford, C.H., Smith, K.L., Jr. and Hamilton, P.L., 1975. Benthic nutrient regeneration and its coupling to primary productivity in coastal waters. Nature, 255: 215—217.

Rowe, G.T., Clifford, C.H. and Smith, K.L., Jr., 1977. Nutrient regeneration in sediments off Cap Blanc, Spanish Sahara. Deep-Sea Res., 24: 57—63.

Sackett, W.M., 1964. The depositional history and isotopic organic carbon compositions of marine sediments. Mar. Geol., 2: 173—185.

Sargent, J.R., Gatten, R.R. and McIntosh, R., 1977. Wax esters in the marine environment — their occurrence, formation, transformation, and ultimate fates. Mar. Chem., 5: 573—584.

Schell, D.M., 1974. Uptake and regeneration of free amino acids in marine waters of southeast Alaska. Limnol. Oceanogr., 19: 260—270.

Scranton, M.I. and Brewer, P.G., 1977. Occurrence of methane in the near-surface waters of the western subtropical North Atlantic. Deep-Sea Res., 24: 127—138.

Scranton, M.I. and Farrington, J.W., 1977. Methane production in the waters off Walvis Bay. J. Geophys. Res., 82: 4947—4953.

Sharp, J.H., 1973. Size classes of organic carbon in seawater. Limnol. Oceanogr., 18: 441—447.

Sharp, J.H., 1975. Gross analysis of organic matter in seawater: why, how, and where? In: T.M. Church (Editor), Marine Chemistry in the Coastal Environment. Am. Chem. Soc., pp. 682—696.

Sholkovitz, E.R., 1976. Flocculation of dissolved organic and inorganic matter during the mixing of river water and seawater. Geochim. Cosmochim. Acta, 40: 831—845.

Sholkovitz, E.R., Boyle, E.A. and Price, N.B., 1978. The removal of dissolved humic acids and iron during estuarine mixing. Earth Planet. Sci. Lett., 40: 130—136.

Sieburth, J. McN., 1969. Studies on algal substances in the sea. III. Production of extracellular organic matter by littoral marine algae. J. Exp. Mar. biol. Ecol., 3: 290—309.

Simoneit, B.R.T., 1977. Organic matter in eolian dust over the Atlantic Ocean. Mar. Chem., 5: 443—464.

Spencer, D.W., Brewer, P.G., Fleer, A., Honjo, S., Krishnaswami, S. and Nozaki, Y., 1978. Chemical fluxes from a sediment trap experiment in the deep Sargasso Sea. J. Mar. Res., 36: 493—523.

Staresinic, N., 1978. The Vertical Flux of Particulate Organic Matter in the Peru Coastal Upwelling as Measured with a Free-Drifting Sediment Trap. Thesis, Joint Program in Biological Oceanography, W.H.O.I./M.I.T., Woods Hole, Mass., 255 pp.

Staresinic, N., Rowe, G.T., Shaughnessy, D. and Williams, A.J., III, 1978. Measurement of the vertical flux of particulate organic matter with a free-drifting sediment trap. Limnol. Oceanogr., 23: 559—563.

Starikova, N.D., 1970. Vertical distribution patterns of dissolved organic carbon in sea-water and interstitial solutions. Oceanology, 10: 796—807.

Steele, J.H., 1974. The Structure of Marine Ecosystems. Harvard University Press, Cambridge, Mass., 128 pp.

Stuermer, D.H., 1975. The Characterization of Humic Substances in Seawater. Thesis, W.H.O.I./M.I.T., Woods Hole, Mass., 188 pp.

Stuermer, D.H. and Harvey, G.R., 1978. Structural studies on marine humus: a new reduction sequence for carbon skeleton determination. Mar. Chem., 6: 55—70.

Sverdrup, H.U., Johnson, M. and Fleming, R., 1942. The Oceans. Prentice-Hall, Englewood Cliffs, N.J., 1087 pp.

Vallentyne, J.R., 1964. Biogeochemistry of organic matter. II. Thermal reaction kinetics and transformation products of amino compounds. Geochim. Cosmochim. Acta, 28: 157—188.

Webb, K.L. and Johannes, R.E., 1967. Studies of the release of dissolved free amino acids by marine zooplankton. Limnol. Oceanogr., 12: 376—382.

Went, F.W., 1960. Organic matter in the atmosphere and its possible relations to petroleum formation. Proc. Natl. Acad. Sci. U.S.A., 46: 212—224.

Went, F.W., Slemmons, D.B. and Mozingo, H.N., 1967. The organic nature of atmospheric condensation nuclei. Proc. Natl. Acad. Sci. U.S.A., 55: 69—74.

Wheeler, J.R., 1976. Fractionation by molecular weight of organic substances in Georgia coastal waters. Limnol. Oceanogr., 21: 846—852.

Whittle, K.J., 1977. Marine organisms and their contribution to organic matter in the ocean. Mar. Chem., 5. 381—411.

Williams, P.J. LeB., 1970. Heterotrophic utilization of dissolved organic compounds in the sea. I. Size distribution of population and relationship between respiration and incorporation of growth substrates. J. Mar. biol. Assoc. U.K., 50: 859—870.

Williams, P.J. LeB., 1975. Biological and chemical aspects of dissolved organic matter in seawater. In: J.P. Riley and G. Skirrow (Editors), Chemical Oceanography, 2. Academic Press, London, pp. 301—363.

Williams, P.M. and Carlucci, A.F., 1976. Bacterial utilization of organic matter in the deep sea. Nature, 262: 810—811.

Yentsch, C.S. and Menzel, D.W., 1963. A method for the determination of phytoplankton carbon and phaeophytin by fluorescence. Deep-Sea Res., 10: 221—231.

Zafiriou, O.C., 1975. Reaction of methyl halides with seawater and marine aerosols. J. Mar. Res., 33: 75—81.

Zepp, R.G., Wolf, N.L., Baughman, G.L. and Hollis, R.C., 1977. Singlet oxygen in natural waters. Nature, 267: 421—423.

Chapter 6

DECOMPOSITION OF ORGANIC MATTER OF PLANKTON, HUMIFICATION AND HYDROLYSIS

B.A. SKOPINTSEV

1. DECOMPOSITION OF ORGANIC MATTER OF PHYTOPLANKTON

Organic matter (OM) is represented on this planet by living (autotrophic and heterotrophic) organisms and their excretory products and after-death remains, but may also be "inert" or non-living (refractory) (Vernadsky, 1934). The latter is found in large accumulations (fuel deposits, soils), as well as in the dispersed state in most mountain rocks and ocean waters. Many authors believe that "inert" OM is of biogenic origin and that in reservoirs it dominates "living" OM.

In seas and oceans refractory OM is mainly of autochthonous origin, the intake from land being comparatively small. Its primary source is phytoplankton. Heterotrophic organisms use phytoplankton as food for growth and for replacing expended energy. Excretory products of organisms and their remains are consumed by bacteria. As a result, the primary produced OM undergoes various conversions caused mainly by the activity of enzymes (Kreps, 1934). But despite all transformations, the OM of phytoplankton is not completely converted into the mineral components which initially served as nutrients. Experiments testify that the remaining part is not large, and the rate of its biochemical conversion is slow. Krogh (1934) considered the dissolved OM of oceans to be the result of the total OM turnover. One way of studying the processes of OM decomposition is to perform a series of long-term experiments on decomposition of OM from dead algae.

1.1. Decomposition experiments carried out between 1935 and 1940

The earliest studies of this kind seem to be those of Cooper (1935) and Von Brand et al. (1937). The amount of phytoplankton, zooplankton or mixed plankton they incubated in experimental bottles filled with water almost corresponded to that of organic P and N in the sea. The experiments were carried out under dark, aerobic conditions for periods of 3—5 weeks, the temperature being 15—19°C and 20—25°C, respectively. The amounts of phosphates and mineral nitrogenous compounds accumulated exceeded the organic P and N supplied by the zooplankton. The authors attributed this to partial mineralisation of dissolved OM of sea water. The fact is of great significance in studying the OM turnover in a water basin. In experiments with phytoplankton, 60% of organic phosphorus and 78% of organic nitrogen were mineralised during 26 and 31 days, respectively. In all the experiments ammonification was duly followed by nitrification. When analysed, the results of these studies (Skopintsev, 1938) indicated that accumulation of the products of plankton decomposition proceeded at a progressively lower rate. In this case, ammonification and phosphatization rates decreased as the quantity of decomposing plankton decreased. Initially, a relatively rapid decomposition of labile organic constituents prevailed over slow decomposition of refractory constituents, but then the latter prevailed. A number of processes are usually described with equations of first-order reactions. The oxidation of OM in domestic waste waters may prove a good example. According to Felps (Theriault, 1927), the rate of the process can be measured by the decrease in dissolved oxygen. For practical estimations the method of Biochemical Oxygen Demand (BOD) is applied. In waters receiving waste effluents, the decrease of the oxygen concentration with time is described by the equation:

$$-\frac{\mathrm{d}D_t}{\mathrm{d}t} = k'D_t \tag{1}$$

which after transformation yields:

$$\log \frac{D}{D_t} \text{ or } \log \frac{D}{D - x} = k_{\mathrm{BOD}}t \tag{2}$$

where D is the total oxygen (mg l^{-1}) consumed for oxidation of OM; D_t is the oxygen requirement of the sample at the time t; x is the oxygen consumed in t days; and k_{BOD} is a constant of BOD day^{-1}. For a given waste flow the mean value of k at 20°C is 0.1 mg l^{-1} day^{-1}: such value of k gives 68% oxidation of the initial OM after five days and 90% after ten days.

Using eqs. 1 and 2, and having exchanged D for N and P, in the experiment where decomposition of planktonic OM was controlled according to the accumulation of NH$_4^+$ and PO$_4^{3-}$, rate constants for ammonification and phosphatization were determined (Skopintsev, 1938). The following equations may be of practical value in such calculations:

$$k_{\mathrm{N}} = \frac{1}{t} \log \frac{\mathrm{N}_t}{\mathrm{N}_{2t} - \mathrm{N}_t} \tag{3}$$

and:

$$k_{\mathrm{P}} = \frac{1}{t} \log \frac{\mathrm{P}_t}{\mathrm{P}_{2t} - \mathrm{P}_t} \tag{4}$$

where N$_t$ and P$_t$ are N—NH$_4^+$ and P—PO$_4^{3-}$ concentrations formed in t days and N$_{2t}$ and P$_{2t}$ for double the number of days. The necessary information on

TABLE I

The rate constants of ammonification (k_N) and phosphatization (k_P) of phytoplankton according to Cooper (1935) and Von Brand et al. (1937)

No. exper.		Mean	Extremes
Values given by Von Brand et al. (1937)		k_N day^{-1}	
1	Mixed plankton	0.056	0.062—0.054
4	Plankton mixed and filtered through sieve No. 8	0.092	0.112—0.073
6	*Nitszchia* culture	0.038	0.045—0.031
Values given by Cooper (1935)		k_P day^{-1}	
1	Mixed plankton with a preponderance of zoo-plankton	0.064	0.090—0.054
2	Zooplankton	0.063	0.068—0.055
4	Phytoplankton—diatoms	0.074	0.092—0.052

accumulation of ammonia and phosphate in water was obtained by inter-polation of data presented by Cooper (1935) and Von Brand et al. (1937), throughout the experimental period. Extreme values of these rate constants calculated for various periods of time and their mean values are summarised in Table I. Usually, the rate constant values tend to decrease, which can be attributed to the gradual rise of the refractory fraction of organic biomass. The labile constituents degrade readily in the first phase.

The estimated mean rate constants are usually lower than 0.1 day^{-1}. The k_N values are lowest for experiment no. 6. Possibly in this case, consider-able detritus was formed together with the cultivation of algae. The same was observed in later experiments, where the algal cultures were used. From the mean values of k_N and k_P using the formulas:

$$N_t = 100(1 - 10^{-k_N t}) \tag{5}$$

and:

$$P_t = 100(1 - 10^{-k_P t}) \tag{6}$$

the percentage of these biogenic elements was estimated on successive days of various experiments, upon which the corresponding curves were based (Fig. 1A and 1B). The analytically determined values on accumulation of NH_4^+ or PO_4^{3-} fitted the curves or were close to them.

Thus, decomposition of OM in phytoplankton can be approximately described by a first-order reaction equation. Proceeding from these observa-tions, the approximate amount of regenerated biogenic elements can be estimated at a certain time, and also the time required for mineralisation of a given amount of organic N and P in the phytoplanktonic OM. The values

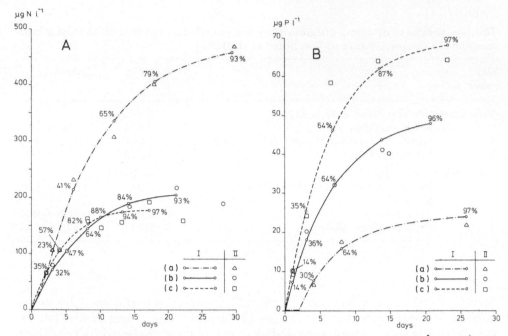

Fig. 1. A. The formation of NH_4-N in water from decomposing planktonic OM: I, according to the experimental data of Von Brand et al. (1937), and II, according to calculations of Skopintsev (1938); a, b, c refer to the experiments 6, 1 and 4, respectively. B. The formation of $PO_4^{3-}-P$ in water from decomposing planktonic OM: I, according to the experimental data of Cooper (1935), and II, according to calculations of Skopintsev (1938); a, b, c, refer to the experiments 4, 1 and 2, respectively. (The numbers on the curves are percentages of the final NH_4-N and $PO_4^{3-}-P$ concentrations, respectively.)

are useful for calculation of the turnover of biogenic elements in reservoirs. The temperature of water should be taken into account by using the corresponding temperature coefficient. The works of Maksimova (1972, 1974) and Arjhanova (1976) for seas and oceans follow this scheme.

As it has been already stated, the process of decomposition of OM in phytoplankton is similar to that in domestic waste waters. In both cases it proceeds at a decreasing rate, approximately exponentially, though it has been estimated by different parameters of OM (by BOD for waste waters and by the rate of accumulation of NH_4^+ and PO_4^{3-} as the main products of mineralisation of phytoplankton). In order to prove this, identical criteria of OM had to be used in each experiment. To find those, two studies were conducted. The first (Skopintsev and Ovtshinnikova, 1934), carried out under aerobic conditions on natural water mixed with domestic waste flows, revealed that (in two experiments) ammonification proceeds at a decreasing rate before nitrification, and the curve of NH_4^+ accumulation nearly repeats that of BOD. The rates of both processes were similar, the mean k_{BOD} at 20°C equalled 0.081 day⁻¹, 0.096 day⁻¹, while k_N equalled 0.11 day⁻¹ and 0.13 day⁻¹, respectively. Oxygen consumption ranged from 8 to 10 mg per 1 mg of the $N-NH_4^+$ formed. Since BOD was not estimated in the plankton experiment, this was done in a further experiment with fresh water and fresh-water plankton (Skopintsev and Brook, 1940). The phytoplankton in the bottles exceeded by a factor of four its own concentration in a reservoir, and mainly consisted of blue-green algae. The experiments were conducted in darkness for periods of 1—2 months, the temperature ranging from 22 to 26°C in the first experiment, from 15 to 18°C in the second and at 6°C (A) and 16°C (B) in the third experiment. Simultaneously, control experiments with water without plankton were carried out. Incubation took place in large bottles filled with water and stoppered with glass stoppers. The bottles were not filled to the top and were stirred regularly to maintain aerobic conditions. Tap water, stored for a long time in darkness, served as reference blank.

The first two experiments yielded the following results: the NH_4^+ concentration began to increase from the first to the second day, it reached its maximum at the seventh, then nitrification followed. The PO_4^{3-} concentration increased from the second to the fifth day. The concentration of the dissolved OM in the first experiment determined by permanganate oxidisability, the albuminoid nitrogen method and BOD exceeded the concentrations observed in the control experiment without plankton by the sixtieth day (Fig. 2).

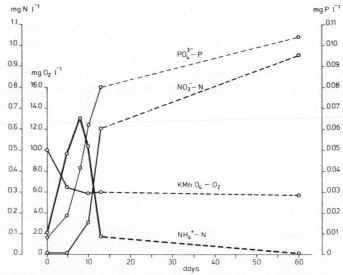

Fig. 2. Organic matter and nutrient changes according to the results of experiment 1 (Skopintsev and Brook, 1940).

In the second experiment, after the sixtieth day oxygen of permanganate oxidisability in acid medium equalled 120% in comparison with the control. According to numerous studies of rivers, lakes and reservoirs the oxygen of permanganate oxidisability to organic carbon ratio varied from 0.6 to 1.2 with an average of 1.0 (Krylova and Skopintsev, 1959). The lowest values are typical of less coloured waters, especially in the period of phytoplankton mass growth. The organic C was determined using a variant of wet combustion using dichromate oxidation (Krylova, 1953). The bacteria growth on meat-pepton agar reached its peak at the first and second days and then began to decrease. The third experiment gave the following measurements: k_{BOD} at $6°C$ equalled 0.035 day^{-1} and 0.082 day^{-1} at $16°C$, the constant values of ammonification k_N being 0.061 day^{-1} and 0.108 day^{-1}, respectively. The maximum abundance of NH_4^+ calculated was 0.24 mg N l^{-1} at $16°C$ before nitrification in this experiment, while the concentration found by analysis equalled 0.22 mg N l^{-1}. The obtained values appear similar. The calculation was carried out according to the formula:

$$N_s = \frac{N_t}{1 - 10^{-k_N t}} \tag{7}$$

where N_s is the maximum concentration of $N-NH_4^+$ and N_t the amount of $N-HN_4^+$ over the period of t days. As can be seen, a temperature fall will considerably reduce decomposition of OM. Thus, results of the estimated BOD in these experiments gave evidence that the process of decomposition of OM in phytoplankton and in waste domestic water goes along the same lines. The possible reason for this similarity is that both domestic waste waters and phytoplankton are rich in proteins, carbohydrates and fats. The evaluation of amounts of organic P at the start and at the end of experiments 1 and 2 and in the control experiment also made it possible to calculate the degree of decomposition of OM in phytoplankton (see Table II).

In both experiments, the initial OM in the water did not mineralise, which is in agreement with the results observed in the English and American studies of phytoplankton described above.

TABLE II

Mineralisation of phytoplankton OM, calculated from a decrease in organic phosphorus

Number of experiment	Initial mg P l^{-1}		Final mg P l^{-1}		Degree of decomposition
	total P [1]	planktonic P	total P [1]	planktonic P	
1	0.0299	0.0174	0.0195	0.0068	61%
2	0.0302	0.0177	0.0190	0.0063	64%

[1] Total P is inorganic P of the initial water + organic P of phytoplankton.

Conclusions. The processes and the rates of decomposition of OM of marine and fresh phytoplankton in the sea and in fresh water are virtually identical. Temperature changes considerably influence the rate of decomposition. The amount of the dissolved OM in water tends to increase towards the end of the experiment. This is, probably, a refractory OM or water humus of planktonic origin. The remaining non-degradable particulate OM (detritus) also shows a relatively high stability.

1.2. Decomposition experiments carried out between 1948 and 1964

The study of regeneration of biogenic elements in the decomposition of killed Baikal plankton (Votintsev, 1948, 1953) at 16°C indicated that the degree of decomposition of zooplankton OM determined by the gain of mineral N and P was 94 and 87% (k_N = 0.079 to 0.076 and k_P = 0.085, all per day) respectively. In experiments with phytoplankton, the degree of OM decomposition equalled 95 and 82%; k_N = 0.085, k_P = 0.075 and k_{BOD} = 0.088, all per day. The estimated degree of decomposition of OM of the Baikal plankton corresponds to the values of rate constants for N and P regeneration and to the rate of BOD. These rate constant values are similar.

Skopintsev and Krylova (1955) studied the accumulation of dissolved OM in fresh water of 16—18°C after 160 days of plankton decomposition in the dark. The phytoplankton was sampled from a pond (experiment no. 1) and from a lake (experiment no. 2). Its amount from the first was about 270 mg l^{-1}, and from the second ~130 mg l^{-1} dry weight, mainly consisting of blue-green algae. The water for the experiment was previously stored in the dark for a long period. After filtration through a glass filter no. 4 (mean pore size 10 μm), the initial colour (according to the platinum—cobalt scale) was 5°. Carbon was estimated using the wet combustion method by dichromate oxidation (Krylova, 1953). At the same time the initial water without plankton was exposed under the same conditions. The results are shown in Table III. The monitoring of mineralisation of OM was by following the formation of mineral nitrogen compounds. Their successive conversions from NH_4^+ to NO_3^- provided evidence for aerobic conditions of the experiment. Glass filters of 5 μm pore size were used for separating particulate OM from DOM.

It thus appears that the content of dissolved OM in the experiments with phytoplankton was, after 160 days, twice that in the control experiments. The accumulation of this refractory OM — the water humus — constituted 2—3% of the organic mass of plankton. In the fresh-water control the content of organic C showed no evidence of changing. The decomposition degree of OM thus evaluated by C equalled 93 and 82%.

Corresponding results were obtained using Black Sea plankton in dark experiments with surface Black Sea water (salinity 18‰). The experiments

TABLE III

The dynamics of organic carbon in the process of decomposition of freshwater phytoplankton for 160 days

Organic C (mg l⁻¹)	Without plankton		With pond plankton (experiment 1)				With lake plankton (experiment 2)			
	A¹	B	C	D	E	F	C	D	E	F
Total	3.03	3.09	165.4	16.9	162.4	13.5	92.2	21.2	89.2	18.1
Dissolved	2.94	2.94	2.94	5.76	0	2.82 (2)²	2.94	5.64	0	2.70 (3)
Particulate	0.09	0.15	162.5	11.1	162.4	11.0 (7)	89.3	15.6	89.2	15.4 (17)

¹ A = at the beginning of the experiment; B = at the end of the experiment; C and D = at the beginning and and at the end of the experiment without organic C correction in initial water; E and F = with organic C correction in initial water.
² Figures in parentheses give the percentages of carbon at the end of the experiment compared to its inital concentration in the plankton (in particulate fraction).

lasted from July 1955 to October 1960 (Skopintsev et al., 1964) and were maintained under aerobic and anaerobic conditions. These studies were conducted with freshly collected Black Sea diatoms (*Coscinodiscus* prevailed) and copepods. The Black Sea water was stored in the dark and filtered before use. The aerobic experiments were carried out in two large bottles (volume 20 l). The wet plankton (20 g) was introduced into one of the bottles, the other served as a control; the bottles were periodically stirred. The oxygen level in the bottles remained constant throughout the experiment, the value of redox potential being about +0.5 V. The plankton decomposition resulted in a change in the chemical composition of the water, typical for aerobic conditions.

The anaerobic experiments were carried out in sixteen dark bottles (4 l in volume each); 4 g of wet plankton were added to each of eight bottles; the other bottles served as controls. Oxygen was purged from the water with nitrogen and the bottles were closed with ground stoppers covered with paraffin and placed in the dark. Samples were taken periodically. The anaerobic conditions were characterised by hydrogen sulphide, the redox potential was about −0.2 V, the sulphate content decreased, the alkalinity increased, nitrification did not take place, i.e. the conditions were similar to those typical for Black Sea deep waters. During the first 220 days average water temperature in the experiment remained at 14°C. From then until the 1935th day the bottles were kept in the refrigerator at 7--9°C, i.e. at the temperature of Black Sea deep waters.

Organic carbon was estimated using the dry combustion method (Skopintsev and Timofeeva, 1961); organic nitrogen and phosphorus were determined according to Kjeldahl and by hydrolysis in an autoclave, respectively. For separation of dissolved and particulate OM the water was filtered through glass filter no. 4 with a 1-mm layer of barium sulphate. Using the data obtained by Black and Christman (1963), and our own evaluation of colour in coloured fresh water before and after filtration, the mean pore size (diameter) of filters was less than 0.5 μm. The results are summarised in Table IV. Since organic carbon in the control water contained a small amount of particulate matter (see Table III), which showed no signs of changing, estimations of POM in the described experiment were neglected.

These studies again revealed that the content of dissolved OM in phytoplankton experiments at the 220th day and 1935th day was respectively 1.5 and 2 times more than its total in the control experiment at the 220th day. This fact shows that the refractory dissolved OM accumulates in the process of decomposition. In ratio to the initial content of C in plankton it was estimated to be 1% in both experiments in spite of the prolonged period of time. It should be stressed that the amount of dissolved carbon was evidently lowered as a result of filtration through the very thick filters. The degrees of mineralisation of planktonic OM at the 220th day and 1935th day were equal (79%), being only slightly lower in the anaerobic experiment

TABLE IV

The dynamics of carbon content in the decomposition of the Black Sea plankton

Organic C (mg l⁻¹)	Aerobic conditions					Anaerobic conditions				
	without plankton		with plankton			without plankton		with plankton		
	A^1	B	A	B	C	A	B	A	B	C
Total	3.4	2.5	38.4	11.8	10.6	3.5	1.9	38.6	14.2	12.3
Dissolved	—	—	—	4.3 (1)²	3.3 (1)	—	—	—	4.7 (1)	3.7 (1)
Particulate	—	—	35.0	7.5 (21)	7.3 (21)	—	—	35.1	9.5 (27)	8.6 (26)

[1] A = at the beginning of the experiment; B = at the 220th day; C = at the 1935th day.
[2] Figures in parentheses give the percentages of organic carbon compared to its initial concentration in the phytoplankton (in particulate fraction).

TABLE V

The degree of plankton mineralisation as estimated by decrease of organic N and P in particulate matter in % of their initial content

The time of storage (days)	Aerobic conditions		Anaerobic conditions	
	N	P	N	P
60	71	66	55	69
220	74	74	77	76
1935	84	84	74	72

(73%). It thus appears that decomposition of OM is practically complete by the seventh month. The data describing the intensity of plankton mineralisation, estimated by other elements, are presented in Table V, which shows that the values for the extent of decomposition of the OM of phytoplankton obtained by the decrease of N, P and C are similar. The remaining 25% (or even less in some experiments) of the initial OM represents a particulate fraction of high biochemical stability.

The extent of decomposition in aerobic and anaerobic conditions hardly varies. If the results of these experiments are compared with those described earlier, there is a strong similarity between the regularity and rate of the decomposition of plankton in fresh and sea water. This is important since a number of analytical determinations are more easily carried out using fresh water.

The analysed data on carbon content and also on content of organic N and P in dissolved and particulate state at the end of all the experiments illustrate the stability of OM. The experiments simulate accumulation of resistant insoluble and soluble organic compounds in reservoirs and soils. These compounds appear as the result of decomposition of dead "living" matter and its "living" excretions. The processes are not only responsible for transformation of initial OM (oxidation, polymerisation, condensation), but also for synthetic processes, caused by bacterial activity (Kononova, 1963), and this is the "water humus" of planktonic origin.

Later investigations dealt with biochemical constituents both in insoluble remains of decomposing plankton and in its soluble products. Krause (1959) studied the decomposition of OM of fresh-water zooplankton at $18-20°C$ in aerobic and anaerobic conditions. The organisms were killed by short-term exposure to heat at $40°C$ before the experiment and this obviously violated natural conditions. Therefore, immediately after such processing the mean loss of the initial mass of his experiments constituted 26%. The high degree of decomposition of protein and carbohydrate was typical for the first day; it reached 50% including the initial loss. During the same period the decrease of the inorganic fraction was estimated to be more than 70%. From the second to the sixth day the lipid (ether) fraction diminished rapidly, especially under aerobic conditions; under anaerobic conditions the proteins showed marked signs of decrease. By the 13th day the total loss of initial zooplankton mass under aerobic conditions equalled 71%, and under anaerobic conditions 79% (i.e. both values were similar but high). On the basis of such studies, Krause (1959) assumed that complete mineralisation of planktonic OM will occur only after several months. But this suggestion can not be regarded as convincing since it fails to take into account the formation of humus (soluble and insoluble forms) in the decomposition of the remaining "living" matter.

Grill and Richards (1964) studied the regeneration of biogenic elements during OM decomposition of diatoms. The latter were grown from plankton samples collected at Puget Sound off Seattle (Wash., U.S.A.). Phosphates and nitrates were introduced into filtered sea water and kept at 11°C. The initial concentrations of phosphates, nitrates and silicates in water were 3.6 μg-at PO_4^{3-}—P l^{-1}, 51 μg-at NO_3^-—N l^{-1} and 54 μg-at Si l^{-1}, respectively. The pH was adjusted to 8.2. The algal culture was exposed to fluorescent light. After a month the phosphate concentration decreased to 0.08 mg at PO_4^{3-}—P l^{-1}. Two days later illumination ceased and the process of regeneration of biogenic elements in darkness was observed.

The experiment lasted for more than one year. The maximum content of particulate organic N and P, being estimated at the eighth and tenth day, equalled 54.7 and 3.49 μg-at l^{-1}, respectively (Fig. 3A and 3B). The N : P ratio was about 16, which coincides with mean ratio values of these elements in phytoplankton (Fleming, 1940). According to Fleming, mean atomic ratio C : N in phytoplankton equals 106 : 16 = 6.6, thus maximum initial content of particulate organic C in the experiment may be considered to be about 360 μg-at l^{-1} or 4.3 mg C l^{-1}. The authors note that there was no anaerobiosis

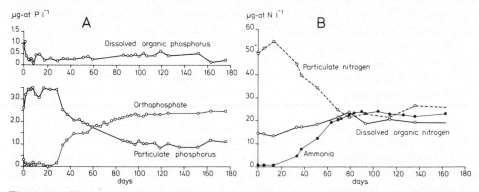

Fig. 3. A. Phosphorus concentrations of a plankton suspension and its remains after storage in the dark. B. Idem for nitrogen concentrations. (After Grill and Richards, 1964.)

in the experimental bottle throughout the whole experiment due to the regular stirring of water during the sampling period. But nevertheless, the regenerated NH_4^+ was not subject to nitrification; the given amount of planktonic OM (in mg C l^{-1}) could not prevent the process. From the final concentrations of particulate organic N and P and their maximum concentrations, we can assume that the mineralisation degree of plankton for the given period of time equalled 54% and 66%, respectively. Such comparatively low values may be due to an amount of formed detritus of the grown algae culture. The authors believe that the most considerable changes in the content of the components under study occurred during the first five months. The phosphatization process was described by an equation of the first-order reaction; two labile fractions and one refractory were determined. The corresponding rate constants of decomposition were estimated to be: $k_1 = 0.0380$ day^{-1}, $k_2 = 0.0381$ day^{-1}, $k_3 = 0.0174$ day^{-1} at $11°C$. The given equation was at first also readily ascribed to the process of ammonification with application to two labile and one refractory fractions of organic nitrogen.

1.3. Decomposition experiments carried out between 1970 and 1978

Studies on a larger scale were carried out by the American scientists Forée and McCarty (1970), and Jewell and McCarty (1971), made necessary by the growing eutrophication of water basins as a result of man's impingement on the environment. According to data presented by these authors, the degree of decomposition of OM in algae in aerobic conditions varied between 24 and 67% of the original, the remaining 33—76% representing the refractory fraction. Other investigators considered the algae to be subject to complete degradation, due to the oxygen content of the water. The American scientists used cultures of pure and mixed algae, and on some occasions plankton from natural water bodies. The cultures were grown under artificial illumination and constant air + CO_2 flow at $20°C$. Aerobic experiments on decomposition of OM (lasting from 23 to 295 days) were often made on algae of various ages (from 20 to 211 days) in fresh or sea water in the dark, also using constant aeration with a mixture of air and CO_2 (Jewell and McCarty, 1971). The studies were carried out at different temperatures. The estimation of OM was achieved by: (a) wet oxidation with dichromate — Chemical Oxygen Demand (COD) method using silver sulphate as catalyst and mercuric sulphate for binding the chlorides; and (b) by the total content of particulate matter. The initial value of the total COD in different experiments varied from 32 to 1260 mg O_2 l^{-1}. The resistant fraction of OM, which remained non-degraded, was estimated by COD, and ranged from 19 to 86%, the mean value being 44%.

It thus appears that the degree of OM decomposition varied from 14 to 81% (with 56% as a mean value). The lowest degree of decomposition was typical for both old and young cultures. Decomposition of the labile fraction of phytoplanktonic OM was described by a first-order reaction equation with k' constants from 0.010 to 0.15 day^{-1}, with a mean constant of 0.040 day^{-1}. Rate constants of decomposition were higher for young cultures ($k' = 0.01-$0.06 day^{-1}) compared with the old ($k' = 0.01-0.03$ day^{-1}). Decomposition of OM in pure cultures initially follows a second-order reaction but later follows a first-order reaction. The experiments carried out at other temperatures did not permit an overall view of the influence of temperature on the rate of OM decomposition, though in some cultures the amount of resistant OM decreased with the rise of temperature.

In this respect, the result of determination of some OM indices at the first day of the experiment on decomposition of algal cultures of different ages taken from the eutrophic Lake Searsville are of interest. The amount of dissolved OM in the youngest culture (12 days growth) constituted 35% of the total COD. Later (at the 59th and 211th day) this fraction decreased to 15—20%. The COD values of particles compared to their total weight, and to the particulate N and P (from 12 to 211 days), as a rule, increased. It appears reasonable to assume that the relative participation of the labile OM fraction in particulate matter grew with time. The COD : OM ratio in suspended particles, even in the oldest culture, equalled 1.32. Assuming the organic C content in these particulate matters to be about 50%, the value of the given COD : C ratio will constitute 2.64. Thus, in the particulates the resistant OM made up a considerable portion of the total contents.

During the vegetation period the ratio of COD to organic C in particulate matter equalled 3.34 in Lake Nero (Yaroslavl, U.S.S.R.). In autumn and winter months it decreased to 2.94 (Larionov and Skopintsev, 1977). It may be seen that the degree of oxidation of OM particles in Lake Nero during the vegetation period was considerably below that of the 211 days algal culture from Lake Searsville. It follows that the amount of labile OM in particulate matters of cultures sampled from Lake Searsville was considerably smaller.

A question arises concerning possible reasons for the low degree of decomposition of planktonic OM estimated by Jewell and McCarty (1971) in a number of experiments and the low values of decomposition rate constants. Perhaps such values are due to accumulation of large quantities of resistant particulate and dissolved OM — the water humus — in the water in which the algal cultures were grown. Under constant aeration of the experimental bottles, particles might be created from the dissolved OM as a result of a bubbling effect. This may account for the amount of refractory OM which already existed in the experimental bottles from the beginning of the incubation period and obscured the real values of rate of OM decomposition. In the experiment described above this value was calculated from the total COD at the start and at the end of the experiment. In such a case, the estimation of the degree of decomposition of particulate OM introduced into the experimental bottle would be misleading.

The works of Forée and McCarty (1970) present a detailed analysis of the degradation of planktonic OM under the anaerobic conditions according to the pattern of methane fermentation and sulphate reduction, provided there is a large excess of sulphates in the water. It appeared that anaerobic degradation of planktonic OM proceeds in two phases, as is typical for decomposition of OM in sludges of the domestic waste waters in methane tanks (Korolykov, 1926; McCarty, 1964).

Two stages of methane formation of complex organic compounds are shown in Fig. 4 (from McCarty, 1964). The first stage of acid formation consists of two processes, hydrolysis and fermentation with little or no change in BOD or COD of the OM in suspension and solution. A portion of the initial OM is converted into bacterial cells. In this phase the labile organic constituents of the sludges and plankton decompose, forming considerable amounts of soluble organic compounds, fatty acids (acetic and propionic acid in the case of methane fermentation) and also CO_2 and NH_3. In the second phase the enzymatic decomposition of organic acids occurs with formation of CH_4 and CO_2 which proceeds only at pH values between 7.0 and 8.0. Under sulphate reduction in the first phase, only acetic acid is formed, in the second H_2S and CO_2. When sulphate is present in water under anaerobic conditions, this substance acts as hydrogen acceptor. Forée and McCarty (1970) believe that the sulphides, formed concurrently, have a toxic effect on the methane-forming bacteria.

The study of decomposition of planktonic OM with methane fermentation showed for experiments with various algal cultures at 20°C, that the amount of refractory OM on the 200th day ranged from 18 to 64% (average 41%) of the initial amount. The estimation of the OM content in these experiments was also made using the COD method. Thus, for the given period, the amount of OM decomposed ranged from 36 to 82% (average 59%). The decomposition of the labile fraction of OM in all the experiments except three (their particulate matter contained a large amount of lipids), could be described as a first-order reaction. The rate constants k of this process varied from 0.011 to 0.032 day^{-1}, the mean value being 0.022 day^{-1}.

In the experiments with three cultures at 15, 20 and 25°C the temperature factor was proved to influence the decomposition of OM. The highest degree of decomposition was found in five experiments with algae sampled from reservoirs; it ranged from 67 to 82% (with a mean value of 76%). Over a further 413 days the degree of decomposition increased only by 2%. In other experiments with algal cultures, the degree of decomposition of OM changed from 36 to 79% and the average values constituted 47% from the

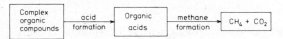

Fig. 4. Stages of methane formation (after McCarty, 1964).

initial. In all the aerobic experiments the extreme values of degree of decomposition were five times higher. As was already stated, the main reason for such discrepancies in the values is the accumulation of refractory dissolved and particulate OM in the algal cultures during their growth.

On the basis of the estimation of organic N, COD and the loss of OM of plankton at the beginning of the experiment and its particulate remnants at the end, the authors estimated the initial and final content of protein, lipids and carbohydrates. These substances constituted 26.6, 11, 62.4% at the start and 37.3, 14.0, 48.7% respectively at the end of the experiments. Unexpected was the increase of the relative amount of protein at the end of the experiment.

Discussing the similarity of the degree of decomposition of planktonic OM under aerobic and anaerobic conditions (which contradicts the general view), the authors come to a conclusion that anaerobic conditions exist in nature as a result of accumulation of OM and not vice versa. Such a view may hold true for the initial phase of the formation of a hydrogen sulphide zone. In the next phase the living organisms die because of the water stratification; this is the reason of the decrease of the degree of OM decomposition and its accumulation.

In Japan, the experimental study of decomposition of planktonic OM was carried out on the grown green alga *Scenedesmus*, which was killed by freeze-drying before the experiment (Otsuki and Hanya, 1972). A quantity of 15 g algae (dry weight) was incubated in the experimental 3-l bottles with distilled water, containing a phosphate buffer. The medium was inoculated with a water extract from a wet lake mud. The suspension contained in the bottle was constantly bubbled with O_2 at a rate of 200 ml min^{-1}. After 30 and 220 days of incubation in darkness at 20°C the following compounds were mineralised (as percentage of their initial content): C-containing compounds (in C) about 55 and 71%, N-containing compounds (in N) about 60 and 70%. Some 6 and 7% of C and 7 and 12% of N was present in dissolved

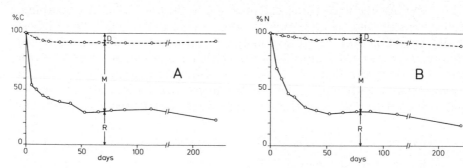

Fig. 5. A. Decomposition pattern of algal cell carbon under aerobic conditions: D = dissolved organic, M = mineralised and R = residual as particulate. B. Idem for algal cell nitrogen. (After Otsuki and Hanya, 1972.)

states. The particulate matter contained about 39 and 23% of C and about 33 and 18% of N (Fig. 5). The authors note that as a result of freeze-drying, part of the cell components immediately dissolved. Nitrification practically did not occur, probably due to the large amount of OM (more than 100 mg C l^{-1} at the 200th day). It follows that the degree of mineralisation of phyto-planktonic OM was higher than that in the American experiments and approached the results of the earlier studies. The decomposition of phyto-planktonic OM was calculated to proceed according to a first-order reaction equation between the fifth and the thirtieth day. The rate constants of decomposition for C averaged 0.0156 day^{-1} and for N 0.0296 day^{-1}.

The Japanese authors consider the OM of the dead cells to consist of labile and refractory constituents. The major attention is given to the formation of the soluble fraction of OM during the process of decomposition. When the aqueous phase was bubbled, organic aggregates were observed. The dissolved OM was fractionated at the 40th day between butanol and acidified water. One fraction, insoluble in butanol, after vacuum-drying turned into a white amorphous solid. Its infrared spectrum revealed its proteinaceous origin; after acid hydrolysis fourteen different amino acids were found. This fraction may possibly be called the biochemically resistant proteinaceous material of the water humus. The fraction soluble in butanol after drying produced a vis-cous substance. From its infrared spectrum carbonyl groups typical for lipids, were detected.

Similar studies were carried out with the same plankton but under an-aerobic conditions (Otsuki and Hanya, 1972). At the beginning of the exper-iment and during the sampling period the water was bubbled with H$_2$. After the 30th and 200th day the following compounds were found to be mineral-ised (given in percent of their initial content): C-containing compounds (in C) about 14 and 31%, N-containing compounds (in N) about 40 and 64%; 29 and 31% C and 8 and 5% N were present in dissolved states, 57 and 38% C and 52 and 31% N in particulate states (Fig. 6).

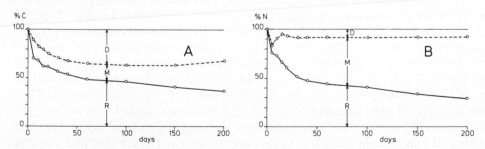

Fig. 6. A. Proportional change of algal cell carbon at 20°C in % under anaerobic con-ditions: D = dissolved organic, M = mineralised and R = residual as particulate. B. Idem for algal nitrogen. (After Otsuki and Hanya, 1972.)

Compared with the aerobic experiment, under the anaerobic conditions after the 200th day, the particulate matter contained 1.6 times more of C and N, i.e. the rate of decomposition was considerably lower. Larger amounts of dissolved organic compounds were also found under anaerobic conditions. Acetic acid prevailed in the medium together with smaller quantities of propionic and formic acids. This is in good agreement with observations made by Forée and McCarty (1970). Since the Japanese experiments showed no evidence of a large decrease of dissolved organic C at the 200th day (in the American studies such decrease was observed already at the 40—60th day), there being no methane in the water, it can be assumed that methane fermentation did not occur. The main reason for this may be the low value of pH. The rate constants of OM decomposition at 20°C calculated for periods from 0 to 35 days by organic carbon equal 0.0088 day^{-1} and by organic nitrogen 0.0156 day^{-1}, i.e. twice as low as in the aerobic experiment. At 30°C the decomposition rate increased.

In conclusion, it is of interest to analyse the results of the similar experiments, carried out in the Institute of Biology of Inland Waters of the Academy of Sciences of the U.S.S.R. (Bikbulatova et al., 1977). The studies were conducted on freshly collected blue-green algae of the Rybinsk reservoir during the period of their mass growth (*Microcystis* prevailed). They were filtered through a plankton net for separation from mechanical admixture and major zooplankton. Under the aerobic conditions the green algae (700 mg l^{-1} of dry weight) were incubated in a large bottle filled with natural yellow-coloured water, stored for a long time in darkness prior to the experiment. The water with plankton was kept in the dark at 20°C and aerated with clean air at a rate of 200 ml min^{-1}. Samples were taken periodically after stirring the bottle thoroughly. For analysis of the particles, the water was filtered through a membrane filter covered with a layer of SiO_2 (mean pore size 0.5 μm) (Larionov and Skopintsev, 1974).

For the anaerobic conditions, sodium sulphate was added to the natural water in an amount calculated to be sufficient for complete oxidation of planktonic OM during sulphate reduction. This was calculated from an average value of the oxygen equivalent of plankton OM being 3.0 mg O_2/mg C. Usually a value of 2.67 is used, but besides carbohydrates plankton contains also proteins and fats (Skopintsev, 1947a). The atomic ratio of the elements with the formation of mineral products from plankton OM is: under aerobic conditions O : C : N : P = 240 : 106 : 16 : 1, and under anaerobic conditions by sulphate reduction S : C : N : P = 60 : 106 : 16 : 1 (Skopintsev, 1975a). The amount of algal suspension mentioned above was added to a large bottle with water. The contents were thoroughly mixed and poured into eight dark bottles (2 l in volume each). Oxygen was purged with nitrogen and the bottles sealed with ground glass stoppers, covered with paraffin and placed in the dark. The contents of each bottle was sampled at specified times.

For chemical analysis the following methods were applied: the organic

C in particles was determined by the method of dry combustion (Krylova, 1957) and in the water by the persulphate method (Bikbulatov and Skopintsev, 1977); organic N by Kjeldahl's oxidation with estimation of NH_3 in the distillate after the Nessler's method, organic P by the oxidation with persulphate and measurements of the colour of "Molybdenum blue" (Bikbulatov, 1974). Particulate proteins were determined with Folin's reagent (Bogdanov et al., 1968); carbohydrates with phenol-sulphuric reagent (Bikbulatov and Skopintsev, 1974); lipids (from particulate matter) by extraction with petroleum ether according to the method of Jewell and McCarty (1971), amino acids by the ninhydrin reaction (Semenov et al., 1960); and biogenic elements (NH_4^+, NO_2^-, NO_3^-, PO_4^{3-}) by standard methods.

It is reasonable to assume that there may be a loss of a volatile fraction of OM because of the bubbling of the water in the aerobic experiment. But this loss was considered negligible since all forms of N and P at all periods of the experiment were balanced. The organic acids in the anaerobic experiments were determined by a variant of potentiometric titration suggested by E.S. Bikbulatov. The results of these investigations are summarised in Table VI.

At the beginning of the experiment the plankton content was 700 mg l^{-1} (in dry weight), and the organic C, N and P concentrations were 385, 52 and 3.6 mg l^{-1}, respectively; the carbohydrates and protein concentrations were 243 and 284 mg l^{-1}, respectively. On the final day of the experiment the contents of the above-mentioned components can be calculated from Table VI. The relative content of carbohydrates and protein in dry mass of particles at the start was 35 and 28%, at the end 40 and 22%. By the 22nd day the decrease of the majority of components in particulate matter was about 50% of the amounts initially present; the protein content diminished by 65%

TABLE VI

The relative content of organic components in the particulate matter in the successive periods of the aerobic experiment (in %)

Periods of analysis (days)	Dry weight of particles	Organic			Carbohydrates	Proteins
		C	N	P		
1	100	100	100	100	100	100
2	92	96	96	114	96	105
4	72	75	85	100	76	84
7	66	63	81	94	67	81
11	60	54	67	88	61	70
22	52	51	52	50	51	35
56	40	39	33	36	33	17
196	31	30	—	31	25	14
391	18	18	—	23	15	10

and lipid content by 22%. At the end of the experiment (which lasted more than one year) the degree of decomposition of OM (by all parameters) varied from 77 to 85%, with an exception of protein (90%). The determination of lipids was stopped on the 75th day, when their decrease reached 53%. On the second day there was an increase of protein and organic P content in the particulate matter, while the decrease of other components slowed down. This tendency was observed also on the fourth day. These effects were probably caused by the growth of phytoplankton on the first and second day of the experiment (at the expense of NO_3^- and PO_4^{3-} present in the water at the start of the experiment) and the development of large numbers of flagellates which afterwards rapidly decreased.

Of all the dissolved organic components present at the beginning of the experiment, organic C and carbohydrates constituted the major part (39 and 16 mg l^{-1}). This is due to the considerable amount of coloured humus compounds, found in the initial water. The experimental evidence (Bikbulatov and Skopintsev, 1974) strongly suggested that the addition of phenol-sulphuric reagent to such water caused a considerable increase in light absorption in the 480—490 nm sector. This is evidently caused by: (a) the conversion of organic components of the high sulphuric acid medium into more coloured forms, and (b) the hydrolysis of carbohydrate compounds of the coloured water humus.

During the next days, the content of dissolved organic C and carbohydrates varied considerably, probably due to additions from the products of particle decomposition. On the 56th day their content was 27 and 24 mg l^{-1}, respectively (i.e. an increase of carbohydrate C with respect to total organic C). At the end of the experiment the content of dissolved organic C decreased to 12 mg l^{-1} — this could be the remaining water humus of the fresh water and the produced water humus of planktonic origin. Organic N and P (which were measured by the difference between total and mineral N and P) were almost below the limits of detection on the seventh day.

The amino acids could still be detected for nearly a month. The analysis of the absorption spectrum of light in the filtered water in the 250—320 nm sector showed that the extinction values were the highest on the first day and changes in extinction were not exponential as was typical during later periods for the control water. The deviation (due to decomposition of some labile organic compounds) was hard to detect on the 7th day. On the 210th day the extinction values in the given sector of spectrum were lower. The changes of mineral nitrogen compounds in the water corresponded to regular conversion of these into more oxidised forms (NO_3^-). In all periods a balance of all forms of N and P under study was observed.

In the anaerobic experiment (at $20°C$) the initial and final contents of total organic mass in dry weight, organic C, N and P, carbohydrates, protein, lipids in the particulate matter were respectively 720 and 252, 380 and 129,

75 and 23, 9 and 1.7, 240 and 68, 310 and 57, 63 and 30 mg l^{-1}. The relative changes in their concentration is shown in Table VII.

The decrease of dry weight and organic carbon on the 12th day was practically equal to that found in the aerobic experiment on the 11th day. At the same time the decrease of organic N and P, carbohydrates and protein by this time in the anaerobic experiment was higher: this can be attributed to the termination of photosynthetic activity of phytoplankton on the first day and the absence of flagellates. On the 25th day the decrease of all organic compounds in the anaerobic experiment was still greater than that in the aerobic on the 21st day. By the 196th day the degree of decomposition of OM in both experiments was nearly the same. It was for all components (with an exception of the lipids) 69—86% and 65—82% under the aerobic and anaerobic conditions, respectively. Usually, the degree of decomposition of protein was higher, and that of lipids lower; the other components degraded at similar rates. The relative content of carbohydrates and protein in planktonic OM at the beginning of the anaerobic experiment and in particulate matter at the end, constituted 33 and 27%, 43 and 22% (in dry weight), respectively.

Practically, the same values were obtained in the aerobic experiment; in both experiments the decrease of protein exceeded that of carbohydrates. The lipid content in the anaerobic experiment increased from 9 to 12%. The concentration of dissolved organic C during the 26 day period in this experiment increased from 65 to 206 mg l^{-1}, which is in good correlation with the increase of organic acids (from 30 to 300 mg l^{-1}). Later their concentration decreased and on the 196th day amounted to 62.5 mg C l^{-1} and 30 mg l^{-1} of acid. The decrease of organic acids was caused by the activity of sulphate-reducing bacteria; the amount of H_2S in the water increased. Thus, the decomposition process of planktonic OM in this experiment is in good agreement with the results obtained by Forée and McCarty (1970). The content of carbohydrates changed irregularly throughout the experi-

TABLE VII

The relative content of organic components in the particulate matter in the successive periods of the anaerobic experiment (in %)

Periods of analysis (days)	Dry weight of particles	Organic			Carbo-hydrates	Proteins	Lipids
		C	N	P			
1	100	100	100	100	100	100	100
5	82	79	79	67	48	85	93
12	61	57	59	30	45	50	85
26	42	45	45	23	32	37	79
77	41	42	33	20	29	20	63
196	35	34	31	19	28	18	47

146

ment with a tendency to decrease on the 196th day. The dissolved organic P decreased regularly from the beginning of the experiment up to the 26th day (0.95 and 0.0 mg l^{-1}). The amount of dissolved organic N increased at the 5th day (from 10.0 to 12.7 mg l^{-1}), and then decreased, the concentrations of amino acids followed the same pattern.

From measurements of the spectral changes in the filtered water (Fig. 7), two maxima of light absorption at the 610 and 275 nm were apparent on the 5th and 26th day of the anaerobic experiment, respectively. Later these maxima of absorption disappeared. This may be attributed to the change of content and composition of the dissolved organic compounds. Hence, this method seems to be promising for studying the dynamics of small concentrations of dissolved OM. The light absorption in the long wave part of the spectrum is due to the input of phycocyanin (the pigment of blue-green algae). The concentration of this pigment, which is of proteinaceous origin (Strickland, 1965), correlated with the concentrations of the free amino acids (which may be considered to be natural). The long-term preservation in the anaerobic conditions is in good agreement with the similar observation of chlorophyll in the bottom deposits of the Black Sea (Drozdova and Gursky, 1972).

Mineralisation of organic nitrogen compounds in this experiment ended with NH_3 formation. In all periods of the experiment the nitrogen forms, particulate and dissolved, were in complete equilibrium.

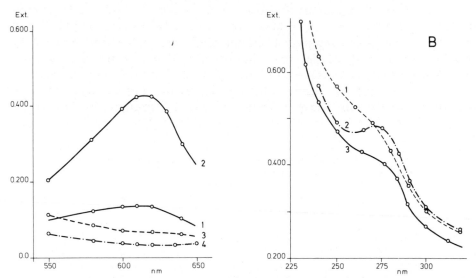

Fig. 7. A. Light absorption in the visible part of the spectrum: 1 = at the start of the experiment; 2 = at the 5th day; 3 = at the 26th day; and 4 = at the 196th day. B. Light absorption in the ultraviolet part of the spectrum: 1 = at the start of the experiment; 2 = at the 26th day; and 3 = at the 196th day. (After Bikbulatova et al., 1977.)

The formation of phosphates proceeded more intensively, but the P forms failed to make a closed budget. A phosphate decrease was observed at the end of anaerobic experiment on decomposition of the Black Sea plankton (Skopintsev et al., 1964). Deep waters of reservoirs containing hydrogen sulphide, such as Lake Nitinat (Richards et al., 1965), the Black Sea (Skopintsev, 1975a) are also depleted in phosphates. The pH value at the beginning of the experiment (7.40) decreased until the 12th—26th day (5.86—5.88) but increased by the end of the experiment again to 6.55. It was evidently connected with the formation of organic acids in the first phase of OM decomposition. Their presence influenced the alkalinity, which increased at the beginning of the experiment (from 2.45 to 6.64 mg-eq. l^{-1}). Being initially controlled by carbonates it later became influenced by the acetic acid produced; decomposition of the organic acids produced again carbonate together with sulphide. This forms a good example of how plankton in the anaerobic conditions may influence considerably the inorganic composition of water.

In both experiments, during the first 6—10 days, when the labile organic compounds were mostly subject to degradation, the process was described by a first-order reaction equation, the rate constants of which were less than 0.1 day^{-1}. The rate constants according to organic carbon were: 0.041 day^{-1} at the fourth day and 0.039 day^{-1} at the seventh in the aerobic experiment, and 0.024 day^{-1} at the fifth and 0.022 day^{-1} at the twelfth in the anaerobic experiment. The decrease in their value in later periods is attributed to the increasing influence from the decomposition of more stable organic constituents.

In conclusion, it can be said that the degree of decomposition of phytoplanktonic OM estimated by the decrease in dry weight, organic C, N and P, carbohydrates and protein in the last two experiments on the 196th day at approximately 20°C amounted to about 70% of its initial content; on the 391st day of the aerobic experiment it constituted about 80%. These data are in good agreement with those found in the long-term experiments, described at the beginning of this chapter. The lower degree of decomposition estimated in the previous experiments was evidently caused by more considerable difference between the experimental and natural conditions.

The study of the influence of lower temperature in both the aerobic and anaerobic environment showed that the decomposition rate of planktonic OM even at a low temperature of 8°C (estimated by the dry weight, organic C, N and P, carbohydrates and protein) decreased with time (Bikbulatov et al., 1978). The main difference in the decomposition at 8°C and 20°C was observed in the initial stages, where mainly the labile OM was subjected to degradation. At the same time a drop in temperature of 10°C decreased the rate of the process by 2.5 times, according to the Van't Hoff rule. For lipids this coefficient rose to 7.6. With the gradual increase of the resistant fraction of OM in particles, the rate of the process became slower and slower.

The accumulation of dissolved organic compounds in water at 8°C is not as rapid as it is at 20°C. The same holds true for their ensuing degradation.

1.4. Evaluation of the experimental results

The experimentally found time-related changes in the biochemical content of particles are similar to those observed in the Pacific Ocean from the surface to deep waters. The amount of "stable" OM in the particles increases with depth (Agatova and Bogdanov, 1972).

The similarity between the degree of decomposition of OM in the aerobic and anaerobic experiments appeared unusual to all investigators. It is generally known that the decomposition of any OM under anaerobic conditions, especially in the presence of hydrogen sulphide, proceeds at a lower rate. The Black Sea may serve as a good example: its waters contain hydrogen sulphide below 200 m depth. As it was shown by Kriss et al. (1951), these deep waters contain partly decomposed organic remains. The authors attribute this to the sinking remains of animals and plants which have not become food for the deep-water fauna.

It is also known that the bottom deposits of the Black Sea contain higher amounts of organic carbon (Skopintsev, 1975a). According to Odum (1968), small animals and also bacteria and fungi "work" together: small animals split the large pieces of food and thus make it more available for microorganisms. Both in aerobic and anaerobic experiments, the OM of plankton and its remains is almost exclusively used by bacteria and not by zooplankton and other organisms.

Thus the degree of OM decomposition, amounting to 70—80% of the initial amount (at 20°C) in the long-term experiment, is close to that in water bodies containing hydrogen sulphides; it is evidently lower for the oxygen-containing ones.

As it was estimated by Kuznetsov (1970), for a water layer of 12 m thickness in Lake Beloye (Kosino, U.S.S.R.), 90% of the OM from the annually produced phytoplankton is degraded. The same value was suggested for the Rybinsk reservoir, U.S.S.R. (Kuzin, 1972). Studies carried out by Ohle (1962), for three Holstein lakes, showed that the annually produced OM was mineralised to extents of 86, 89 and 96%.

The dissolved organic components studied in the aerobic experiment decreased more rapidly than the particulate ones (Bikbulatova et al., 1977). Only the carbohydrates (their content usually correlated with that of dissolved organic C) were also present at the end of the experiment, evidently being part of "stable" OM complexes. It is clear that in the natural environment, especially during vegetation time, the dissolved labile organic compounds are constantly replenished as a result of the activity of hydrobionts.

A number of well-known organic compounds (labile among them) have been shown to be present in ocean waters. They belong to various classes

(Duursma, 1965), but their relative content in the water is small. According to Williams (1971), they constitute about 10% of the total organic C in the northwest part of the Pacific Ocean. The average concentration of the total organic C is shown in Table VIII. The same relative value for the identified organic fraction was adopted for the oceanic water in 1975 (Williams, 1975). In this case not only free carbohydrates, but also the carbon from all the carbohydrates was taken into account and estimated to be about 0.2 mg l^{-1}. It appears reasonable that the "bound" carbohydrates partly constitute the refractory OM.

As the water humus is the dominating component of OM in ocean waters, the formulations below concern the humus in the first place:

(a) "It is the OM dissolved and suspended "non-living" and "living" which determines the difference in all the properties which exists between natural waters and solutions of the same salts and gasses in distilled water. These distinctions affect the physical, chemical and physico-chemical properties in water." (Skopintsev, 1950)

(b) "Dissolved and suspended OM are the major determinants of the biological difference between natural sea water and saline solutions of the same mineral composition. These differences can be demonstrated physically, chemically and biologically." (Wagner, 1969)

(c) "The same has been noticed for the land humus. One of the main indicators, which distinguishes the soil from crust material is the presence of OM, the quantity and nature of which define to a great extent the direction of the transformation process, the biochemical, physical, chemical properties and the fertility of the soil." (Kononova, 1968)

Here, it is worthwhile discussing the results of C estimations in the ocean. According to the method of dry combustion, adopted in the U.S.S.R. some time ago (Skopintsev, 1960; Skopintsev and Timofeeva, 1961), the content of organic C is 2—3 times higher than that estimated by the persulphate method (Menzel and Vaccaro, 1964). The corresponding data can partly be found in William's paper (1975) and in Chapter 14 of this book. Some opponents considered that the high values, obtained by the dry combustion method are caused by contamination during preparation of the water samples for the analysis. This argument does not hold since all the investigators who used this method must have contaminated their samples equally, which is unlikely. In addition, the initiator of this method (Krylova, 1953) and her followers (Skopintsev and Timofeeva, 1961) tested the possibility of such factors. The application of the persulphate method to determine organic C in fresh waters showed that the bubbling of the acidic water with nitrogen in the presence of persulphate led to the decrease of organic C (Bikbulatov and Skopintsev, 1972).

It is feasible that such a procedure could cause a partial oxidation of OM. This fact was confirmed by Sharp (1973a), who modified the persulphate method, but the results obtained were still lower than those obtained by the application of the dry combustion method. The fact was attributed to the presence of organic compounds resistant to oxidation with persulphate, which disagrees with the results of the estimation of organic C in the river Volga (Skopintsev et al., 1972). According to this paper, the results of carbon estimation by persulphate and photochemical methods practically coincided; the dry combustion method yielded somewhat (6%) lower values (caused by the loss of organic fraction during evaporation of the water).

The amount of purified NaCl, added to the fresh water, was the cause of low values of organic carbon; most probably, the persulphate was partly used for Cl⁻ oxidation (Skopintsev et al., 1976).

According to Gordon and Sutcliffe (1973), who made use of another variant of dry combustion method, the contents of organic C in the waters of the west Atlantic were in close agreement with those obtained by myself and co-workers in the same region, but they were higher than those found by Sharp (Wangersky, 1975). Later MacKinnon (1978) showed that his data were on the average 50% lower than ours, so there is still some confusion about the actual concentration of organic C in ocean waters.

Since the average concentration of organic C, estimated by the dry combustion method, in the water layer from 0 to 100 m equals about 1.8 mg l⁻¹ and at depths of more than 2000 m from 1.2 to 1.4 mg l⁻¹ for three oceans (Skopintsev et al., 1976), the fraction of carbon from identified labile organic compounds in the deep oceanic waters will be less than 10%. The remaining 90% are probably made up of refractory (stable) organic complexes — the water humus.

2. WATER HUMUS

The analysis of the results of the experiments on decomposition of OM in dead plankton presents a view of the time-related changes of particulate and dissolved organic C, N and P, and some components of OM. The processes occurring are responsible not only for oxidation of the initial OM, but also in soils (Kononova, 1963) for the polymerisation (condensation of the more biochemically resistant dissolved and particulate fractions of OM.

Synchronically, new forms of OM are synthesised by bacteria. The combined processes lead to the formation of the refractory organic substance — water humus — in particulate and dissolved state. This surely occurred initially in the Precambrian period with blue-green algae and bacteria in water reservoirs under anoxic conditions (Rutten, 1971). As shown by experiments, the degree of decomposition of the OM of dead hydrobionts under such conditions was less. That is why more organic residues settled on the bottom of the reservoir. Naturally this influenced the further transformation of OM in sediments and the accumulation of oil precursors: their concentration in the anaerobic conditions of the Precambrian must have been larger.

Berzelius (1833) appears to have pioneered a serious study of OM in natural waters. He discovered the dissolved organic compounds with acidic properties in the mineral spring Porla (Sweden). These compounds were called crenic and apocrenic acids. Berzelius assumed that their salts were washed out from soil humus "intact". Oden (1919) found the common term for these acids — fulvic acids (fulvus = yellow). Aschan (1932) attributed them to the water humus group, being typical for peat, river and lake waters.

Thus, fulvic acids may originate from continental plants, and when carried by rivers and streams to the reservoirs, seas and oceans, they constitute the major part of the allochthonous water humus. Kalle (1966) suggested they were only partly responsible for the "yellow substance" (*Gelbstoffe*) present in oceanic waters, the concentration of which in the water decreases with the increase of salinity. He considered that a part of the "yellow substance" was of autochthonous origin. It seems probable, that its existence in water is due to transformed products of excretion of dissolved and thinly dispersed OM which was discovered in littoral algae during studies conducted by Khailov (1962), Sieburth and Jensen (1969) and Sieburth (1969).

According to Birge and Juday (1934), the refractory OM appears in the process of decomposition of dead plankton in lakes. This is the water humus of autochthonous origin. Krogh (1934) considered that in the deep waters of the ocean OM consists partly of "humus", which is resistant even for bacteria. Waksman (1936), studying the question of the formation of OM in reservoirs, distinguished three types of humus: river, lake and sea humus. Besides this natural water humus, there may enter and be formed a humus from sewage of domestic origin. It was Odum (1975) who considered humic substance (humus), the most stable product of OM decomposition in nature, to be a necessary component of the ecosystem.

Thus, it appears that water humus is the indispensable resistant product resulting from transformation, decomposition and synthesis of OM of excretions and dead remains of plankton in the ocean: its final structure is largely determined by the activity of micro-organisms (Skopintsev, 1950).

2.1. The annual input and formation of water humus in the waters of the world ocean and its chemical characteristics

Table VIII summarises data to allow an approximate budget of water humus in the ocean. It is assumed that the annual decrease as a result of oxidation, aggregation and adsorption on the surface of particles with their subsequent sedimentation, equals its input.

The main differences observed between the calculated elements in a budget are typical in the amounts of organic C from the annual input reaching the ocean bottom. The bottom deposits also contain the aggregation products of dissolved OM. Trask (1939), Bogdanov et al. (1971) and other scientists estimated it to be $n \times 10^{15}$ g C, where $n < 5$. River inputs are relatively low in comparison to the total organic production. The input of OM from atmospheric precipitation has been neglected, since a similar amount of OM seems to be carried out from the ocean surface by strong winds. In a review paper, Duce and Duursma (1977) draw special attention to the works devoted to studies of the release of marine OM into the atmosphere and the input of river OM into the oceans. Recent data obtained give some estimation of OM content in the atmosphere. Blanchard (1964), and Barger and Garret (1970, 1976) investigated the surface-active OM fraction collected on platinum wires or glass fibre filters. As a result of analyses of the chloroform extracts, carried out with a gas chromatograph, fatty acids were found (C_{14}—C_{18}). The concentration ratios of single fatty acids in the sea aerosols were the same as those in the surface layer of oceanic waters. The contents of methylesters formed from lipids in aerosols collected off the coast of Hawaii constituted 5%, and in the samples taken from the equatorial Pacific Ocean were 4% of the total OM isolated from sea aerosols. According to Hoffman

TABLE VIII

Elements of the annual balance of OM in the water of the world ocean (in carbon)

Content, input and loss of OM	Williams (1971)	Skopintsev (1971) [1]	Williams (1975)
Average concentration of C (mg l^{-1}) (dissolved and particulate)	*From 0 to 300 m* 1.0 *From 300 to 3000 m* 0.5	≤1.5	0.70
Total C ($\times 10^{18}$ g)	0.76	2.0	1.03
Phytoplankton production ($\times 10^{16}$ g) [2]	3.6 (100)	3.8 (120)	3.6 (100)
Phytoplankton excretions of OM (10% of the annual production of OM) ($\times 10^{15}$ g)	—	—	3.6
Input from atmospheric precipitations ($\times 10^{14}$ g) [3]	2.2 (1.0)	—	2.2 (1.0)
Input by rivers ($\times 10^{14}$ g) [3]	0.31 (2.0)	1.8 (5.0)	1.8 (5.0)
OM of phytoplankton production used by organisms ($\times 10^{16}$ g) [4]	—	3.5 (92)	—
Input in water of the resistant soluble OM of planktonic origin ($\times 10^{15}$ g) [4]	—	1.1 (3)	1.8 (5)
Sedimentation of particulate organic remains ($\times 10^{15}$ g) [4]	~0.1 (0.3)	1.9 (5)	~0.1 (0.3)

[1] For world oceans without adjacent seas.
[2] Figures in parentheses: calculated as g m^{-2} yr^{-1}.
[3] Figures in parentheses: in mg l^{-1}.
[4] Figures in parentheses: in percentage of the annual production of phytoplankton.

and Duce (1974) and Duce and Duursma (1977), the total content of organic C in the investigated sea aerosols ranged from 0.2 to 1.0 μg m^{-3}. In the aerosols, collected on the coast of the Bermuda Islands, the ratio of organic C to the total content of included salts ranged from 0.010 to 0.19, with an average of 0.051.

Thus in aerosols, the value of this ratio is hundreds, or even thousands times greater than the values of the corresponding ratio in sea water. As a result of laboratory studies, Hoffman and Duce (1976) found that the ratio of C : Na depends on the quantity and on the nature of the OM (dissolved and colloidal). According to these authors there may probably also be other sources of OM assimilation in sea aerosols; the latter may be associated with particles of sea salts.

The annual amount of the terrigenous water humus entering rivers might have been overestimated as a result of, (a) its partial coagulation by the salts of the sea water (Skopintsev, 1946, 1947a), and (b) its extensive oxidation in the presence of decomposing remains of plankton. The latter phenomenon has been reported by many authors (see Skopintsev and Bikbulatova, 1978, and section 2.2). According to Sholkovitz (1976), only 3—11% of dissolved OM from waters of four Scottish rivers when mixed with sea water flocculated within half an hour. The humic-acid content (colorimetrically determined) of river waters was 4—20% of the total OM. Fulvic acid and possibly non-humic compounds amounted to 80—90%. Fulvic acids form water-soluble compounds with Ca^{2+} and Mg^{2+}; a large proportion of the humic acid (HA) would have aggregated in the mixing zones between river and sea water. According to calculations, approximately 60% of the total HA content coagulated in these experiments.

Colloidal properties are characteristic of humic compounds of river waters. Owing to their aggregation, the colour of the filtered water decreased only by 10—25% of its initial value, which indicates only a partial removal of OM of river water on mixing with sea water. At the same time, the concentrations of the elements Fe, Mn, Al and P in the water were markedly decreased (relative to their initial content). Similar results were obtained by Sholkovitz et al. (1978) during experiments carried out in the waters of two rivers, the Luce (Scotland) and the Amazon (Brazil). Rapid flocculation was characteristic for the high molecular-weight fractions of HA. This resulted in an aggregation of 60—80% of HA within half an hour (this amount represented 3—6% of the initial dissolved OM).

The results of these works are principally in agreement with earlier published data (Skopintsev, 1947b). In these experiments filtered marsh water was mixed with sea water in various proportions. Whilst carrying out these experiments, the water was kept in the dark at room temperature. After 30 days the colour of the filtered (size = 1.5 μm) water with a salinity of 0.2‰ was 98% of the initial value. At a salinity of 28.9‰ the colour was equal to 70% of the initial value of 62° (colour grade). In a similar experiment after 200 days the colour of the filtered water with a salinity of 0.2‰ and 32.5‰ was 50% of the initial value (44°); in such a continuous experiment a partial oxidation of the water humus might have taken place. The greatest rate of coagulation was observed on the first day. In some experiments with peat water, the coagulation effect was practically zero when a number of salts ($CaCl_2$, $MgCl_2$, $MgSO_4$ and HCl) were introduced. Equal effects might be found by mixing peat water with sea water. Another effect is caused by the load of suspended particles. As has been shown by experiments, coagulation (or aggregation) of suspended particles in sea water (Skopintsev, 1946) is less effective at a low particulate load even when humic compounds are present.

Thus, the sharp decrease in colour, which is observed in estuaries with increasing distance from shore, is obviously due to the increasing dilution with sea water and not necessarily due to coagulation processes; the results of fluorescence measurements of oceanic water support this fact (Kalle, 1962, 1966; Duursma, 1974). The intensity of the fluorescence of water is inversely proportional to its salinity. In this case the fluorescence is mainly due to the humic compounds of terrestrial origin. Its effect becomes weaker as the dilution of river water with sea water becomes greater. According to Table VIII the annual input of the dissolved terrigenic OM in the ocean is approximately 0.01% of the total OM content in the ocean water. Based on the existing dynamic equilibrium in ocean waters the quantity of terrigenous OM entering the ocean annually is balanced by its removal. This occurs by means of biochemical oxidation and aggregation followed by the settling of suspended particles.

Bordovsky (1974) is right in his belief that HA of marine deposits is of autochthonous and allochthonous origin, the latter playing a minor role. Based on chemical investigations and isotope determinations of C in HA isolated from ocean sediments, Nissenbaum and Kaplan (1972) do not agree with this point of view. The allochthonous HA in the bottom sediments of the ocean is evidently at the limits of detection of the analytical methods employed. According to Table IX the total annual input of water humus to the ocean constitutes 1.5×10^{15} g C, provided the distribution is proportional, this amounts to 1 μg C l^{-1}.

Convincing evidence for a genetic relationship between the water humus of the ocean and the resistant fraction of planktonic OM is supplied by data describing the ratio of stable isotopes ^{13}C and ^{12}C, i.e. the values of $\delta^{13}C$. Table X lists some typical values together with corresponding data of the bottom deposits. Their chemical composition and physicochemical properties yield valuable information for revealing the nature of particulate matter in water (Strakhov, 1954). Thus it appears that in most biochemically resistant components of planktonic OM, specifically in "cellulose" and "lignin", the organic carbon is depleted of the heavier isotope in comparison to amino acids, hemicelluloses and sugars. The same is true for dissolved and particu-

TABLE IX

Approximate annual budget of water humus in the world ocean (in carbon)

Reservoir of water humus in the ocean $= 1.8 \times 10^{18}$ g C ($\sim$90% of total content of organic carbon in the ocean water, which is equal to 2×10^{18} g)

Input	Output
1. From atmospheric precipitates: 2.2×10^{14} g (1.0)	1. From surface water to the atmosphere?
2. From rivers: 1.8×10^{14} g (5.0)	2. Biochemical oxidation (mainly by bacteria), sorption onto particles, aggregation (followed by partial use of aggregates by deep-water organisms) and sedimentation at bottom; (a) humus of terrigenous origin: 1.8×10^{14} g, (b) humus of planktonic origin: 1.1×10^{15} g,
3. The relative resistant organic matter of planktonic origin: 1.1×10^{15} g ($\sim$3% of annual productivity equals 3.8×10^{16} g)	
$\Sigma = 1.5 \times 10^{15}$ g	$\Sigma = 1.3 \times 10^{15}$ g

Number in brackets: in mg l^{-1}; annual input of water humus in the oceans equals $< 0.1\%$ of its total content in ocean water.

late organic carbon in ocean water and deep-sea bottom deposits. Williams and Gordon (1970) support the view that OM of the deep-ocean waters is biochemically resistant. In fact, it is subject to decay at a slower rate and is capable of aggregating. The values of $\delta^{13}C$ in FA (fulvic acid) and HA (humic acid) extracted from sea waters and bottom deposits are similar. It should be noted that the carbon of these humic acids often equals <10% of its total content, estimated by persulphate method, in the Sargasso Sea the ratios of humic acid to fulvic acid fluctuate from 0 to 100 (surface layer) and 22 to 78 (deep layers) (Stuermer and Harvey, 1977). The values of $\delta^{13}C$ in the organic carbon of land plants and soil humus are even lower. The same is true for OM of fresh water.

The study of hydrogen isotope contents (Nissenbaum, 1974) also testifies to the similarity between δD (D = deuterium) of the humus of marine sediments and OM of marine plankton: $\delta D = -105\%_0$. In humic acids of soils δD varies in the range of -57 to $-97\%_0$. Other studies revealed the difference between the composition of planktonic and terrigenous humus. This is due to the nature of precursors of humus: while lignin and proteins are essential for the formation of soil humus, plankton-derived carbohydrates and amino acids have the same function for water humus (Skopintsev, 1950). The spectral characteristics in the ultraviolet region of FA concentrates of the dissolved OM from estuaries differ considerably from those observed in the water concentrates of the Sargasso Sea (in all cases there was an equivalent concentration of organic C). Still greater differences were found between the

TABLE X

The values of $\delta^{13}C$ in OM for some marine and land sources (in ‰)

Organic objects	$\delta^{13}C$	References
Total OM of mixed plankton	from —17.4 to —21.2 from —17 to —24 by 15 to 27°C and from —24 to —31 by 2 to 10°C	Degens et al., 1968; Williams, 1971
The labile fraction of planktonic OM	from —16.7 to —19.2	Degens et al., 1968
The resistant fraction of planktonic OM	from —22.4 to —23.1	Degens et al., 1968
Dissolved and particulate OM in Pacific Ocean water	from —21.2 to —24.4	Williams, 1968; Williams and Gordon, 1970
OM in deep-water bottom deposits	average value —22,3	Williams, 1968 Williams and Gordon, 1970
"Fulvic acids" (FA) from the waters of the Sargasso Sea	—22.8 and —23.7	Stuermer and Harvey, 1974
"Humic acids" (HA) from the waters of the Sargasso Sea	—22.8	Stuermer and Harvey, 1974
"Humates" of marine bottom deposits	from —20 to —22	Nissenbaum and Kaplan, 1972
OM of land plants	from —20 to —32	Williams, 1971
FA of soils	from —25 to —27	Stuermer and Harvey, 1974
HA of soils	from —25 to —26	Nissenbaum and Kaplan, 1972
OM of water and particulate matter of the Amazon river	—28.5 and —29.4	Williams and Gordon, 1970

Notes: The so-called FA and HA, isolated from the water and bottom deposits are different in nature from the soil humates; common for them is only the method of extraction.

spectral characteristics of FA concentrates of the OM sampled from the Sargasso Sea and soils (Kerr and Quinn, 1975).

The study of humic substances by nuclear magnetic resonance of the ^{13}C isotope and the proton suggests that aliphatic structures prevail over aromatic structures for the fulvic acids extracted from marine water, contrary to the case for soil fulvic acids. The high values of the H : C ratio in marine fulvic acids also point to an aliphatic nature (Stuermer and Payne, 1976).

The different nature of fulvic acids (FA) of, on the one hand, oceanic, coastal and open waters (Sargasso Sea), and, on the other hand, soil FA is confirmed by the results of their reductive degradation (Stuermer and Harvey, 1978). The main products of FA decomposition of the above-mentioned sea waters were hydrocarbons (n alkane). These have a carbon range from C_{14} to C_{31} with the maximum at C_{18}. These values are characteristic for the fatty acids of marine organisms. These compounds may be the source for alkane formation after high-pressure hydrogenation of sea water FA. The alkanes of soil FA have a carbon range from C_{16} to C_{35} with the maximum at C_{24}. Their source in soil humus may be the fatty acids of land plants and animal organisms. The authors believe that marine lipids may be incorporated into the structure of sea water FA; the influence of soil lipids even in coastal ocean waters is negligible. Among the products of high-pressure hydrogenation of marine FA are aromatic hydrocarbons. Their source is unknown although a source from micro-organisms is suspected. The ratio of saturated to aromatic hydrocarbons is approximately 20 : 1. The total amount of these hydrocarbons (in carbon) constitutes approximately 3% of the FA carbon. HA of marine bottom deposits (amounting to 70% of the total OM in some of the samples) contained more organic N and S than the soil HA; the aliphatic structures are less condensed and not rich in phenol groups (Nissenbaum and Kaplan, 1972).

Destructive oxidation of the previously methylated HA of deep oceanic sediments with alkaline permanganate revealed that aromatic systems with fewer aliphatic structures are the major constituent features of HA from these sediments, in contrast to HA of lake sapropels (Vasilevskaya et al., 1977). Analogous chemical processing of near-shore deposits yields decomposition products typical for lignin. However, the relative participation of phenol groups in the above-mentioned aromatic structures is considerably lower than in the HA of continental sediments. The above authors believe that the probable reason for the appearance of aromatic structures in the deep-water oceanic deposits (in spite of their low abundance in plankton) might be: (a) dehydrogenation of the cyclic polymers of unsaturated fatty acids of lipids in marine organisms; (b) the accumulation of aromatic metabolites of micro-organisms; and (c) the formation of melanine-like products.

The second assumption finds support in a number of works. These are reviewed in an article by Mishustin et al. (1956) and in the monograph by Kononova (1963). It was reported that when growing a number of bacteria, moulds and actinomycetes in mineral media with sugar, aromatic compounds appeared in the culture media during the process of resynthesis. Thus, the assimilable carbohydrates are used not only as an energy source, but may also be a precursor of the formed humus. This phenomenon may possibly occur in the euphotic zone of the ocean.

The possibility of formation of humic substances from a melanoidin (melanin-like) type in the ocean water was underlined by Kalle (1962), Duursma

(1965) and for the soils by Kononova (1963), and Manskaya and Drozdova (1964). The latter authors and also Williams (1975) consider also the part OM plays in metal migration as a result of formation of organometallic complexes.

The organic acids extracted from the lake sapropels with alkali are of interest in the light of investigations described by Karavayev and Budyak (1960) and Karavayev et al. (1964). After permanganate oxidation in alkaline media and chemical analysis and infrared spectroscopy of the resulting products, it was determined that aliphatic, and probably, heterocyclic nitrogen-containing structures lie at the basis of the structure of these acids. There was an observed absence of condensed aromatic structures. It is generally known that plankton is a primary source of sapropels. Otsuki and Hanya (1967) made a special study of HA extracted from bottom deposits of Lake Haruna (Japan); their infrared spectra turned out to be different from that of soil HA and similar to that of dried lake organisms.

Finally let us deal with a more primitive but still effective method of differentiating between planktonic water humus and terrigenous humus. The values of the ratio of oxygen of permanganate oxidisability (in acid medium) in mg l^{-1} to organic C (method of dry combustion) in mg l^{-1} in river waters during vegetation time varied from 0.73 to 1.2 (average = 1.06), and in the lake water from 0.65 to 1.1 (average = 0.87). The lowest ratios are typical for the lake water in the period of mass phytoplankton production and low amounts of coloured humic compounds (Krylova and Skopintsev, 1959). Special investigations testified that the highest degree of OM oxidation by permanganate is characteristic for phenol-containing compounds (Skopintsev, 1950). Karavayev and Budyak (1960) found the "humolits" constructed from the cyclic aromatic nuclei to be more resistant to oxidation, except for phenol derivatives.

In Atlantic waters where alkaline permanganate oxidation was used, the values of the mentioned ratio were always less than 1.0: they ranged from 0.35 up to 0.75 with an average of about 0.5 (Skopintsev et al., 1966). In the northeast Atlantic and in the Norwegian Sea (Skopintsev et al., 1968) the values of the given ratio varied from 0.32 to 0.71 with an average of 0.55. The alkaline permanganate method of estimation of OM in marine water has also been used during other investigations: Gillbricht (1957), Szekielda (1968a), and Lundstedt (1970). Studies made by Carlberg (1972) in the Baltic Sea testify to the effectiveness of this method for the detection of the humic material. Permanganate oxidisability (oxidation) under alkaline conditions of Baltic water in May—June 1972 was 2—3 times higher than in the Kattegat, even though the organic carbon content in the Baltic Sea was only slightly higher. There was a tendency towards an inverse relationship between permanganate oxidation and the salinity of water, and at the same time a proportional relationship existed between the content of "yellow substance" and the salinity. The correlation coefficient in this case was 0.71. The calculated ratios of permanganate oxidation to organic carbon in the surface water were from 0.47 to 0.56 in the Kattegat and from 0.87 to 1.10 in the Baltic Sea. The difference is caused by considerably higher amounts of terrigenous OM entering the Baltic Sea in comparison to the North Sea. In the latter the water humus of planktonic nature is dominant.

The results of ultra filtration of sea water are significant in describing water humus. Sharp (1973b) gave the following measurements: the particulate C (caught by the 0.8-μm filter) in the North Atlantic at various depths was on average equal to 2%; the colloid fraction (from 0.003 to 0.8 μm) made up only 16% of the total organic C. Thus, the content of dissolved organic carbon averaged around 80%. Baturina and Mishustina (1975), working with membrane ultra filters ("Diaflo"), made a more detailed fractionation of OM of the equatorial Pacific Ocean. The filters (the mean pore size from 0.002 to 0.4 μm) caught 6—15% of the initial content of organic C in the upper layer, and 1—5% in deeper layers. The filters (mean pore size from 0.0005 to 0.002 μm) caught from 19 to 37% and from 43 to 48%, respectively. Thus, the amount of the dissolved organic C in the final filtrate was from 54 to 70% and from 51 to 52%, respectively. The study of the enzyme activity showed the high molecular fraction of OM in deep waters to be of little proteolytic activity in contrast to the subsurface waters of the open ocean. The same goes for the amylolytic activity. A biochemical reaction revealed protein components to be present in the above fraction; those were apparently the protein-like compounds of water humus (Skopintsev, 1950). After long storage in darkness, the isolated and sterilized OM concentrates formed particles in the absence of air bubbling (Mishustina et al., 1976). Thus, the OM of the ocean waters is, to a great extent, colloidal and tends to aggregate, which supports the earlier suppositions (Skopintsev, 1947b). The formation of organic aggregates from dissolved OM is evidently of great ecological importance. These particles as well as particulate products of the organic life activity and their dead remains may serve as additional food sources for the biota inhabiting lower depths and ocean sediments. In the aggregation process of OM in the upper layers the so-called "bubbling effect" seems to play a major role (Batoosingh et al., 1969).

It is highly remarkable that the total content of organic C in the waters of the world ocean and that in the layer of the earth soil from 0 to 30 cm (Waksman, 1936) are quite similar: 2.0×10^{18} g and 0.4×10^{18} g, although their origin is different (2.4×10^{18} g in 0—100 cm soil according to Kovda and Yakushevskaya, 1971). The water humus as well as the soil type is a significant reserve of biogenic elements, biologically active substances and micro-elements for metal binding. From the data presented by Williams (1975) we can approximately assume that the mean concentration of organic N and P in oceanic water (below the photic layer) equals 50 μg N l^{-1} and 4 μg P l^{-1}. For the world ocean (without adjacent seas) it will constitute 7×10^{16} g of organic N and 0.5×10^{16} g of organic P. According to Vaccaro

(1965) the total content of mineral N in ocean water equals 58×10^{16} g N. From the mineral N : mineral P ratio, which is 7.2 (by weight: Fleming, 1940), we find that the total content of mineral P in ocean water equals 8×10^{16} g. Thus, the total content of organic N and P is approximately one order of magnitude lower than that of mineral N and P. Thus, there exists an enormous reserve of the given nutrients in organic form.

Water humus of the ocean, as well as other natural organic substances, accumulate energy. To what extent is energy accumulated in OM of ocean water? A direct determination of the heat of combustion has yet to be performed. The same can be said about complete set of data on the elemental (C, H, N, O) content of water humus, which is helpful for the calculation of the characteristics mentioned. Tyurin (1937) suggested using the results of estimating bichromate oxidisability (COD) and organic carbon for measuring the energy reserve in the soil humus. He based this idea on the fact that when different natural organics combust (oxidise), the amount of energy released per oxygen unit measured is similar: 1 mg-at. of oxygen equals 52 to 56 cal., which is an average of 3.4 cal. per 1 mg of oxygen. This is the so-called oxycaloric coefficient which is, according to Ivlev (1934), for the OM of different water organisms 3.38 cal. per mg of oxygen. The amount of energy bound by humus in the 0—20-cm layer of different types of soils was estimated from COD with the measurements by Aliev (1973) and agreed favourably — 91 to 103% (average 97%) — with the values obtained by the calorimetric method.

The method of estimation used by Tyurin permits an approximate measure of the energy reserve in OM of natural waters. It is known that OM of inland waters is oxidised completely under acidic conditions by potassium bichromate in the presence of silver sulphate. However, this method cannot be used for ocean waters because of the high concentration of Cl^-. Nevertheless, it was successfully employed for estimation of organic C in 123 samples of particulates from the Gulf of Guinea. The samples were taken from vertical profiles from 0 to 6000 m (Copin-Montegut and Copin-Montegut, 1973). The mean ratio between the organic C as determined in these particles by dry combustion and COD was 1.09 ± 0.03. The authors attributed the low values of the wet combustion to the incomplete oxidation of OM in the particles. A reliable criterion of the oxidation degree of OM is the value of oxygen equivalent (OE), i.e. oxygen of COD : organic C ratio. As it was mentioned earlier, the OE value of the organic mass of plankton (dry weight) averages around 1.50. Assuming that the C content in it is about 50%, the value of OE calculated from C will constitute about 3.0 mg O_2 per mg C (the same method of calculation will be used below).

If the organic carbon content in OM, measured by COD, equals the amount measured by the dry combustion method, then the OE values of the given OM average 2.67 mg O_2, which is typical for glucose. The mean OE value of OM particles of the Gulf of Guinea will be lower: $2.67 \times 0.92 =$

2.44. This is explained with analytical data which are obtained for OM of suspended particles in the mesotrophic Lake Plestcheevo (Yaroslavl region, U.S.S.R.). During the year the OE value varies here from 3.51 (vegetation time) to 2.42 in winter (under the ice cover), when suspended particles contained practically no labile OM (Larionov and Skopintsev, 1977). The same is typical for suspended particles of the Gulf of Guinea where the OE value of C in these particles is evidently more than 2.67 in the photic layer, but lower in layers deeper than 200 m; this agrees with the results of calculations made above. From the assumption given above we can assume that the mean OE value of water humus in the ocean is close to 2.5 mg O_2. Multiplying OE values by the number of calories corresponding to 1 mg of oxygen (this product is called energetic conversion factor by Hallegraeff, 1978) and by the total content of OM in the ocean water (in C) we obtain a figure for the amount of potential energy: $2.5 \times 3.4 \times 1.8 \times 10^{18}$ g $\simeq 15 \times 10^{18}$ kcal. This gives an idea of the size of the reserves of accumulated energy in the humus of the oceans (of planktonic and terrigenous origin).

To evaluate how correct this calculated figure of potential energy is in water humus, other corresponding estimations will be reviewed below. According to Bogorov (1967), the amount of energy found in the biomass and in annual production of phytoplankton of the world ocean is 5.2×10^{14} kcal and 19×10^{16} kcal, respectively. This difference is caused by the fact that the annual production of phytoplankton exceeds its biomass by more than 300 times. Odum (1975) gives the value of potential energy of the total primary annual production in the open ocean as 32.6×10^{16} kcal. Taking into consideration the total primary production in coastal waters, estuaries and reefs, this constitutes 43.6×10^{16} kcal per year. Thus it follows that the calculated energy, contained in the annual production of phytoplankton, given by both authors is similar.

Assuming that the average annual production of phytoplankton in the world ocean is 120 g C m^{-2} (see Table VIII), the value of potential energy of phytoplankton will be: 3.8×10^{16} g $\times 3.0 \times 3.4 \simeq 39 \times 10^{16}$ kcal, which is similar to values given by Odum and supports the assumptions made in the calculations above. It is worthwhile to note that the potential energy of water humus in the ocean is about 100 times higher than that of the annual production of phytoplankton. The energy reserve in the 0—100-cm layer of soil humus of the earth equals 12×10^{18} kcal (Kovda and Yakushevskaya, 1971) which practically is equal to the energy of water humus of the world ocean. The annual production of higher plants on land (according to the cited authors) is 11.5×10^{16} g of dry OM, while energy found in it is 57×10^{16} kcal. Odum (1975) gives the same figure. Thus, the energy reserve in non-living OM exceeds the potential energy of the annual production of photosynthesising plants on the land. We should note that the total annual production of OM in the ocean water and on the land is similar, the same can be said about the humus content in both spheres.

The pathways of transformation of dissolved and particulate OM in the ocean water (considered above) and also the data on the residence time (from 1500 to 3400 years) allows some insight into mode and scale of the conversion of energy bound in the water humus. Its reserve in the dissolved (prevailing) fraction is used mainly by micro-organisms. As a result of sorption on particles, the potential energy of the latter rises. The particulate fraction of water humus appears to be a source of OM and energy for the more highly organised hydrobionts. One should stress again the importance of the aggregation processes of the colloid—dissolved fraction of water humus in forming particulate OM in natural waters. This is an essential form of transformation of dissolved organic matter and the energy accumulated in it. Bottom-depositing particles with organic matter and energy reserve in them have a considerable influence on the diagenetic processes in bottom sediments. If we take the organic residues depositing to the bottom (given in Table VIII) and assume their oxygen equivalent to be 2.4 mg O_2, then the amount of energy deposited together with those residues will be:

from $2.4 \times 3.4 \times 0.1 \times 10^{15}$ g $\simeq 0.8 \times 10^{15}$ kcal
to $2.4 \times 3.4 \times 1.9 \times 10^{15}$ g $\simeq 15 \times 10^{15}$ kcal, annually.

2.2. Hydrolysis and stability of water humus

Hydrolysis, i.e. decomposition of humus is usually determined by bacterial activity (Waksman, 1936). The use of humus by other organisms has been studied by numerous authors. The study by Khailov et al. (1973) may serve as an example. Their work supplies a number of references to the literature on the subject.

Sepers (1977) reviews a number of papers devoted to the study of the use of OM by bacteria, phytoplankton and invertebrates. Recent investigations have revealed that the main consumers of the above-mentioned OM are bacteria, which agrees with the view expressed earlier.

There are, however, other factors which determine the water humus decrease: partial loss of colour in the surface layers, sorption onto particles, aggregation followed by partial consumption by deep-sea inhabitants and sedimentation; possibly also by chemical oxidation.

The intensity of these processes cannot be estimated quantitatively yet, but, apparently, their role is comparatively small.

A dynamic equilibrium can be said to exist between the annual input and decrease of water humus in the world ocean (Skopintsev, 1950). Its annual average input equals 1 μg C l^{-1} (see Table IX). In reality, according to the vertical distribution of phytoplankton, zooplankton (Yashnov, 1962) and bacteria (Sorokin, 1962), its input should be highest in the euphotic zone and zones adjacent to it. The trophic factors determine the regularity of the vertical distribution of zooplankton (Vinogradov, 1968) and bacteria. Thus, in the above-mentioned zones the concentration of excretions of organisms and their after-death remains (the initial sources of water humus) are the highest. Experimental evidence, obtained for sea water (Cooper, 1935; Von Brand et al., 1937) and fresh water (Skopintsev and Bikbulatova, 1978) suggests that the decomposition rate of initial dissolved OM in the water follows the decay of the dead planktonic OM. The addition of biologically assimilated OM to the soil also stimulates the decrease of the soil humus (Kononova, 1963).

The upper layers of ocean water have the most intensive biochemical decay of organic remains and have also the most intensive decomposition of water humus and nutrient formation, especially in upwelling areas. This phenomenon is usually ascribed to microbial metabolism (co-metabolism), which causes the decomposition of stable organic compounds (Horwath, 1972).

This phenomenon has been carefully dealt with in the work of De Haan (1972a, b, 1975) for Lake Tjeukemeer (The Netherlands). Humic compounds (mainly fulvic acids) of allochthonous origin dominate in winter. They enter from adjacent polders. On the basis of quality estimations it was determined that these fulvic acids contained polioxycarbon acids, sugars and amino acids and also phenol structures, all of these being characteristic of soil fulvic acids (Kononova, 1963).

During vegetation time, when the influxes from the polders cease, Lake Tjeukemeer appears to be a eutrophic reservoir; its primary production reaches 1.5 g C m^{-2} per day. At the same time the organic C content and the optical density decreases. The Cl$^-$ concentration increases. Over the year the variation in colour in this lake was approximately from 15 to 100° on the colour scale, the fulvic acid concentration varied from 3 to 14 mg l^{-1}. It was assumed that such variations of OM are determined not only by changing hydrography, but also by microbial activity. The fading of the colour of fulvic acids as a result of short-wave solar radiation and their aggregation or coagulation will also lead to a decrease in concentration in the water.

Experiments testified to the fact that the addition of labile organic compounds to bacterial media containing fulvic acid caused intensive degradation (De Haan, 1972c, 1974). In this case it was named co-metabolism. De Haan (1975, 1977) isolated from water of Lake Tjeukemeer the corresponding specific micro-organisms. It is evident that under natural conditions the rate of bacterial decomposition of fulvic acids will depend on the intensity of primary production in the reservoir. This fact is of a great ecological significance for the energy resources of water humus and probably applicable in the case of marine humus. Together with a decrease of the total content of dissolved organic carbon in the water in summer, the relative proportion of higher molecular-weight fulvic acids decreases also, while the proportion of lower molecular-weight compounds increases. The ratio values of optical density E_{250}/E_{365} increase during vegetation time. This also suggests a preponderance of comparatively small molecules in the dissolved autochthonous OM.

The percentage of sugars and amino acids in higher molecular-weight fulvic acids is higher in winter than in summer. This was ascribed to their lower rate of degradation in the given period as compared with the summer. In summer the contents of sugars and amino acids of autochthonous origin in fulvic acids (evidently, less resistant) increase considerably. During the vegetation time the autochthonous OM plays the dominant role in formation of water humus. The values of the organic C : N ratios in the comparatively large molecules of fulvic acid are higher in winter than in summer which points to a predominance of autochthonous OM. When bacteria were grown on fulvic acid-containing media, the relative proportion of lower molecular-weight compounds increased; most probably this is a first step in the degradation of polymeric fulvic acid compounds.

As a result of a reduced input of labile OM, processes leading to formation of water humus decreased with depth, and hence the concentration of humus also decreased, although at the lower water temperature the rate of decomposition is lower. The observed decrease of organic C (Skopintsev et al., 1966; Starikova, 1970; Menzel and Ryther, 1970; Williams, 1975) content in the ocean with depth, is in agreement with the above-mentioned opinion.

The highest discrepancy of organic C content in oceanic waters, estimated by the method of dry combustion, ranged in values from 1.5 to 2.1 times in the 0—300-m layer, and from 1.4 to 1.7 times in deeper layers. The variations of values of primary production in the areas studied are much higher with a good correlation to the content of particulate organic C (Bogdanov and Lisitsyn, 1968). It is noteworthy that the relative decrease of the average total organic C content down the vertical lies well below the decrease of zooplankton and bacteria populations in the water and highlights the role of water humus in the OM of ocean waters.

The mechanism of its input to the ocean (specifically, to the deeper layers) is the sinking process of highly productive cold water masses in the areas of the North and South Polar fronts (Skopintsev, 1950; Szekielda, 1968b). This follows the conception of Redfield (1942) that these waters supply the deep ocean with dissolved oxygen and biogenic elements. It seems therefore logical that water humus should also reach the deeper waters.

2.3. Water humus and oxygen demand

The question arises, as to how the decomposition of water humus in the ocean proceeds. The method of estimation of oxygen decrease in water, incubated in the dark at specific temperature, is commonly applied as a qualitative index of OM decomposition. The biochemical oxygen demand (BOD) method was suggested to estimate the degree of purification of waste waters and polluted rivers and proved to be very effective for this application. However, the storage of non-polluted natural and particularly marine (oceanic) waters in bottles revealed an increase of bacteria after only a few days (Waksman and Carey, 1935). The fact was ascribed to the sorption of OM and bacteria to the walls of the bottles. As a result, the measurements of values of BOD obtained for the natural waters were too high and are only valid for the "potential OM decomposition". This fact should be borne in mind especially for BOD measurements of deep-ocean waters. Rakestraw (1947) carried out a series of experiments on the estimation of BOD in water from the North Atlantic surface waters, the oxygen minimum layer, and from 1240 m and 1725 m depths (the temperatures *in situ* were 25°, 8—9° and 4—5°C, respectively). The volume of the bottles was 250 ml and the experimental time from 1 to 2 years. The oxygen content in the surface water decreased more than in the other samples, though irregular variations were observed.

In the same water samples, stored at 8—9° and 4—5°C, the results proved to be more constant: after 100 days the oxygen content remained at the same level. In samples taken from the two deeper layers, the content remained practically unaltered after the first 50 days. The high oxygen consumption in the surface layers was attributable to planktonic material. BOD increased slightly in the water from the oxygen minimum layer when after 42 days the bottles were incubated at 25°C.

During 305 days the oxygen decrease at the *in situ* temperature in those successive layers constituted 0.61, 0.08 and 0.14 ml O_2 l^{-1} for the waters from surface, the oxygen minimum layer and the deep layers, respectively. Thus, OM is biochemically more resistant in the oxygen minimum layer than in the deeper layers. It is noteworthy that BOD in the oxygen minimum layer did not exceed that in the deeper layer. Plunkett and Rakestraw (1955) indicated that in this layer of the Pacific Ocean there was evidence neither for a decrease of dissolved C nor for an accumulation of particulate OM (Rakestraw, 1958). The same results were obtained for the Atlantic Ocean (Menzel and Ryther, 1968).

From these results the conclusion can be reached that there is no direct evidence for the existence of processes which cause a higher oxygen demand in the oxygen minimum layer. The oxygen minimum layer is rather the consequence of the mixing of upwelling waters from the eastern part of the ocean poor in oxygen and then brought to the west by the passat (Skopintsev, 1961; Menzel and Ryther, 1968).

Experimental values of BOD, obtained by Rakestraw (1947) are considerably below those found by Seiwell (1937) in the North Atlantic. The annual BOD, estimated from the surface layer, from 2000 and 4600 m depth (at the temperature *in situ*) amounted to 4.3 to 0.6 and 0.3 ml O_2 l^{-1}, respectively. They were calculated using the results of the BOD estimation after ten days, assuming the annual rate of BOD to be constant. In reality, as was already mentioned, the rate of oxygen consumption in the water decreases with time. The BOD curves in the water for two North Atlantic stations, presented in Novoselov's work (1962), also support the above. The initial phase can be characterised as "exponential" with prevailing oxidation of the labile fraction of OM. In the next period, during the oxidation of the more resistant OM, the rate of the process decreases and the concentrations remain practically constant. Such results were already obtained for domestic sewage waters (Theriault, 1927). According to Novoselov, BOD after 200 days (BOD_{200}) was 1.48 ml O_2 l^{-1} in the 10-m water layer (18°C) at the station near Iceland and 0.83 ml O_2 l^{-1} in the open ocean; at 300, 900 and 1700 m depth (8°C) BOD_{200} equalled 0.04, 0.08 and 0.20 ml O_2 l^{-1}, respectively. These values vary little from those obtained by Rakestraw (1947) for the deep waters at the temperature *in situ*. Novoselov's (1962) BOD_{180} values in the waters of the open ocean at depths of 600 and 1150 m at 18°C were higher: 0.63 and 0.69 ml O_2 l^{-1}. If we take the value of the oxygen equivalent of the refractory OM to be 2.10 ml O_2 per mg C (Skopintsev, 1964), then, according to the organic C content in those depths (estimated by the dry combustion method) about 25% of the initial OM is oxidised during a period of 180 days at 18°C.

Barber (1968) showed that about 50% of the dissolved OM of the surface Atlantic waters is oxidised during 1—2 months at 20°C, but there was no change of organic C content during the same period for deep waters. These values agree with the above-mentioned results of the BOD estimation in the surface and deep waters. Ogura (1972a, b, 1975), studying the decrease of organic C and O_2 in the surface oceanic and sea waters incubated in the darkness at 25°C, also found that the process of OM decomposition proceeds in two phases.

In the East China Sea (Ogura, 1972a) the rate constant decomposition was measured at 0.005 day^{-1}, supposing a first-order reaction for an average period of 50 days. The rate constant for the north-equatorial waters of the Pacific (Ogura, 1972b) was, however, for the first period higher: 0.033 day^{-1}, a difference of one order of magnitude. The absolute value of the determined oxygen decrease after 50 days was near the values obtained in Novoselov's experiments. In the coastal waters (Ogura, 1975) the rate constant ranged from 0.01 to 0.09 day^{-1} for the first phase and from 0.001 to 0.009 day^{-1} for the second period. These values support the assumption that resistant OM may also prevail in surface ocean waters.

The vertical profiles of BOD in the North Atlantic waters studied by Aisatullin and Leonov (1972) show, however, a relationship between the profiles of the OM input to the water and its oxidation. The ability to decompose either particulate or dissolved OM in the ocean water seems to be potentially similar. According to Menzel and Goering (1966), the decrease of particulate organic C is for surface water from 16 to 57% of the initial for a period of 90 days at 20°C. At 200 m and 1000 m depth the changes are within the accuracy of the method. Gordon (1970) reported that less than 20% of particulate OM in the deep Atlantic waters are subject to biochemical hydrolysis.

As was mentioned already, the results for oxygen demand (OD) by the BOD method in natural waters are too high. To what extent can be judged by calculation. The annual oxygen demand evaluated from hydrophysical parameters in the North Atlantic, according to Wyrtki (1962), is in the 1000 m to bottom layer equal to 0.001 ml O_2 l^{-1} (average). According to Arons and Stommel (1967) the value ranges from 0.002 to 0.0025 ml O_2 l^{-1}. In the Pacific Ocean at greater depth than 1000 m, according to Munk (1966), it varies in the range from 0.0027 to 0.0053 ml O_2 l^{-1}; further values are from Craig (1971): 0.004 ml O_2 l^{-1} (average), and Tsunogai (1972) who estimated the OD at 3000 m depth to be 0.0015 ml O_2 l^{-1}.

The estimation made by Riley (1951) revealed that the curve of vertical distribution of OD values in the North Atlantic waters from the surface to 3000 m depth follows the curve of planktonic respiration. The maximum value of OD at more than 1000 m equals 0.002 ml O_2 l^{-1} per year, which corresponds to the oxidation of about 1 μg organic C l^{-1} yr^{-1}.

Since the vertical distribution of zooplankton and bacteria in the ocean is determined by the available total OM, the calculation of OD for separate depth intervals was carried out by Skopintsev (1966, 1975b). Based on an average annual value of phytoplankton production of 120 g C m^{-2} with about 90% available as OM, it turned out that in 0—100 m, 100—1000 m and 1000—4100 m layers the annual OD at the *in situ* temperature constitutes 2.15, 0.07 and 0.003 ml O_2 l^{-1} or 75, 22 and 3% of the total OD of the 4000-m water column, respectively. The last value is at the limits of determination. From the evaluation of the electron transport system in the tropical regions of the Pacific Ocean it was calculated that the annual value of OD at 3000 m depths equals 0.003 ml O_2 l^{-1} (Packard et al., 1971). According to Riley (1951), about 90% of the OM annually produced by phytoplankton, is consumed in the upper layer (up to 200 m depth).

The observed similarity, as well as the proximity of the calculated OD values in the deep oceanic waters to the above-mentioned values, suggests some confidence in the employed methods. Naturally, the estimated values will vary in accordance with the actual values of primary production in various oceanic regions (Ivanenkov, 1977). In their turn, the low values of OD point to the considerable biochemical resistance of OM of deep ocean waters, especially at low temperatures. This factor apparently determines the almost constant values of ^{13}C in dissolved OM (Williams, 1968; Williams and Gordon, 1970), and its apparent conservative nature (Menzel, 1974).

The residence time of OM in the ocean water is also evidence of its resistance. This view is shared by Wangersky in his excellent review (1978). Estimated from the ^{14}C content of ocean water (Williams et al., 1969), the mean residence time of OM is around 3400 years. Based on the total content of organic C in the ocean water and the annual input of water humus, a residence time of about 1500 years (Skopintsev, 1971, 1972) is calculated.

Sorokin in his recent study (1977) as well as in his past report (1971) (which was criticised by Banse, 1974), states that in deep ocean waters water humus does not prevail. From his point of view this is supported by the facts that: (a) a relatively small change of the absolute OM content exists over great depths, (b) there is an almost constancy in its composition, and (c) some constancy of the composition of the labile OM fraction. The labile OM was determined by the BOD and by the potential production of bacteria incubated at 28—30°C. As a matter of fact, as it was already shown, organic C and OD decrease with depth. The considerable variations between the calculated OD values in the deep water and those, obtained by the BOD method allow us to assume that the latter can give evidence only of potential bacterial ability to utilise OM, but does not show the degree of its lability. For this purpose a determination of the potential bacterial production after incubation of water over 4—5 days at 28—30°C (especially in deep-water layers) is perhaps an inapplicable method. The conditions of this method differ too much from the natural conditions. Sorokin (1977) ascribes the low OD in the deep oceanic waters not to the resistance of OM but to the negative influence of low temperature and high pressure on bacteria. The physical factors mentioned should apparently affect the functioning of the deep-water microflora.

In this respect some of the recent investigations of this problem are of obvious interest. The organotrophic activity of bacteria in some samples, taken from the 50-cm water layer above the Pacific Ocean bottom enriched with OM and measured *in situ*, approached that in the surface samples (Seki et al., 1974). On inoculation of sterile labile OM with deep-water bacteria after one year of incubation *in situ*, only little consumption (a few percent) was observed (Jannash and Wirsen, 1973). The authors note that such a low rate of OM conversion as well as the nature and amount of OM reaching the ocean bottom determines the activity of bacteria. They also assume that elevated hydrostatic pressure and low temperature retards the rate of metabolism (biosynthesis rather than respiration) in all isolates tested, but these effects are considerably less in those of psychrophylic character (Wirsen and Jannash, 1975).

This opinion has found confirmation in the investigations of Stupakova (1975): the strains of barotolerant bacteria of *Pseudomonas* had a larger variety of catabolic enzymes than in the strain of the same bacteria non-reproducible at 500 atm. A pressure of 265 atm did not affect mixed epipelagic and bathypelagic zooplankton (King and Packard, 1975). It should also be noted that the corpses of the oceanic macro-inhabitants were not found on the ocean bottom.

Finally, as a result of increasing anthropogenic influence more and more humus of sewage origin enters the seas and the oceans. In a number of cases this humus is characterised not only by higher toxicity to aquatic organisms, but also by a higher biochemical resistance (Rogers and Landreth, 1975) than the natural water humus. The formation of water humus will increase together with the discharge of industrial sewages, specifically organic synthetic materials, into the water. At the same time the rate constants of the decomposition of this sewage water "humus" will be lower than that of the natural products. So it may be assumed that in reservoirs receiving such effluents not only the nature of water humus will change, but also its concentration will increase.

3. REFERENCES

Agatova, A.I. and Bogdanov, Ju.A., 1972. Biochemical composition of organic matter in the tropical Pacific Ocean. Okeanologiya, 12: 267—276 (in Russian).

Aisatullin, T.A. and Leonov, A.B., 1972. On kinetics and vertical profile of oxygen consumption rate. In: S.V. Bruevitch (Editor), Studies on Theoretical and Applied Chemistry of the Sea. Nauka, Moscow, pp. 11—17 (in Russian).

Aliev, S.A., 1973. Bioenergy of Organic Matter of Soils. ELM, Baku, 65 pp. (in Russian).

Arjhanova, N.V., 1976. The effective use of biogenic elements (nitrogen and phosphorus) by phytoplankton in the Skotii Sea. Tr. VNIRO, 112: 87—96 (in Russian).

Arons, A.B. and Stommel, H., 1967. On the abyssal circulation of the world ocean. Deep-Sea Res., 14: 441—457.

Aschan, O., 1932. Om Wattenhumus och dess medverkan vid sjomalms bildingen. Arch. Kemi, Mineral. Geol., 10A: 1—143.

Bakulina, A.G. and Skopintsev, B.A., 1969. Determination of the total organic carbon content in natural waters by the method of dry combustion. Gidrokhim. Mater., 52: 133—141 (in Russian).

Banse, K., 1974. On the role of bacterioplankton in the tropical ocean. Mar. Biol., 24: 1—5.

Barber, R.T., 1968. Dissolved organic carbon from deep water resists microbial oxydation. Nature, 220: 274—275.

Barger, W.R. and Garret, W.D., 1970. Surface active organic material in the marine atmosphere. J. Geophys. Res., 75: 4561—4566.

Barger, W.R. and Garret, W.D., 1976. Surface active organic material in air over the Mediterranean and over the eastern equatorial Pacific. J. Geophys. Res., 81: 3151—3157.

Batoosingh, E., Riley, G. and Keshwar, B., 1969. Analysis of experimental methods for producing organic matter in sea water by bubbling. Deep Sea Res., 16: 213—219.

Baturina, M.P. and Mishustina, I.E., 1975. Electron-microscopy study of the fraction of organic matter from sea water. Izv. Akad. Nauk S.S.S.R., Ser. Biol., No. 2: 281—288.

Berzelius, J.J., 1833. Sur deux acides organiques qu'on trouve dans les eaux minerales. Ann. Chim. Phys., 54: 219—231.

Bikbulatov, E.S., 1974. On estimation of total phosphorus in natural waters Gidrokhim. Mater., 60: 167—173 (in Russian).

Bikbulatov, E.S. and Skopintsev, B.A., 1972. On the possible use of the variated Menzel—Vaccaro method to estimate organic carbon in fresh water. Inf. Byull. Inst. Biol. Vnutr. Vod. Akad. Nauk S.S.S.R., No. 14: 65—69 (in Russian).

Bikbulatov, E.S. and Skopintsev, B.A., 1974. Estimation of the total content of dissolved carbohydrates in natural waters in the presence of humic substances. Gidrokhim. Mater., 60: 179—185 (in Russian).

Bikbulatov, E.S. and Skopintsev, B.A., 1977. Estimation of organic carbon in natural

170

waters and particles. In: M.M. Senyavin (Editor), Methods of Analysis of Natural and Waste Waters. Nauka, Moscow, pp. 171—176 (in Russian).

Bikbulatov, E.S., Skopintsev, B.A., Bikbulatova, E.M. and Melnikova, N.I., 1978. Decomposition of organic matter of phytoplankton at ~8°C under aerobic and anaerobic conditions. Vodn. Resur., No. 1: 135—142 (in Russian).

Bikbulatova, E.M., Skopintsev, B.A. and Bikbulatov, E.S., 1977. Decomposition of organic matter of blue-green algae under aerobic and anaerobic conditions at room temperature (~20°C). Vodn. Resur., No. 6: 132—147 (in Russian).

Birge, B.A. and Juday, C., 1934. Particulate and dissolved organic matter in inland waters. Ecol. Monogr., 4: 440—474.

Black, A.P. and Christman, R.F., 1963. Characteristics of colored surface waters. J. Am. Water Works Assoc., 35: 753—770.

Blanchard, D.C., 1964. Sea-to-air transport of surface active material. Science, 146: 396—397.

Bogdanov, Ju.A. and Lisitsyn, A.P., 1968. Distribution and content of particulate organic matter in the Pacific Ocean. Okeanol. Issled., No. 18: 75—155 (in Russian).

Bogdanov, Ju.A., Lisitsyn, A.P. and Romankevitsh, E.A., 1971. Organic matter of particles and bottom deposits of seas and oceans. In: N.V. Vassoevitsh (Editor), Organic Matter of Recent and Fossil Deposits. Nauka, Moscow, pp. 35—103 (in Russian).

Bogdanov, Ju.A., Grigorovitsh, Ju.A. and Shaposhnikova, M.G., 1968. Estimation of protein in the aqueous particulate matter. Okeanologiya, 8: 1087—1090.

Bogorov, V.G., 1967. Biological transformation, energy and material metabolism in the ocean. Okeanologiya, 7: 839—859 (in Russian).

Bordovsky, O.K., 1974. Organic Matter of Sea and Ocean Sediments in the Stage of Early Diagenesis. Nauka, Moscow, 104 pp. (in Russian).

Carlberg, S.R., 1972. A study of the distribution of organic carbon and oxidability in Baltic waters. Medd. Havsfiskelab. Lysekil, N141: 1—22.

Cooper, L.H.N., 1935. The rate of liberation of phosphate in sea water by the breakdown of plankton organisms. J. Mar. Biol. Assoc. U.K., 20: 197—200.

Copin-Montegut, Cl. and Copin-Montegut, G., 1973. Comparison between two processes of determination of particulate organic carbon in sea water. Mar. Chem., 1: 152—156.

Craig, H., 1971. The deep metabolism. Oxygen consumption in abyssal ocean water. J. Geophys. Res., 76: 5078—5086.

Datsko, V.G., 1939. Organic matter in some sea waters. Dokl. Akad. Nauk S.S.S.R., 24: 294—297 (in Russian).

Degens, E.T., Behrendt, M., Gotthardt, B. and Reppman, E., 1968. Metabolic fractionation of carbon isotopes in marine plankton. Deep-Sea Res., 15: 11—20.

De Haan, H., 1972a. Some structural and ecological studies on soluble humic compounds from Tjeukemeer. Verh. Int. Ver. Theor. Angew. Limnol., 18: 685—695.

De Haan, H., 1972b. Molecular-size distribution of soluble humic compounds from different natural waters. Freshwater Biol., 2: 235—241.

De Haan, H., 1972c. The biological transformation of soluble humic substances in Tjeukemeer, the Netherlands: A preliminary report. In: H.L. Golterman and D. Povoledo (Editors), Humic Substances, Their Structure and Function in the Biosphere. Pudoc, Wageningen, pp. 63—69.

De Haan, H., 1974. Effect of a fulvic acid fraction on the growth of a *Pseudomonas* from Tjeukemeer (The Netherlands). Freshwater Biol., 4: 301—310.

De Haan, H., 1975. Limnologische Aspecten van Humusverbindingen in Het Tjeukemeer. Thesis, Lemmer, 94 pp.

De Haan, H., 1977. Effect of benzoate on microbial decomposition of fulvic acids in Tjeukemeer (the Netherlands). Limnol. Oceanogr., 22: 38—44.

Drozdova, T.V. and Gursky, Ju.I., 1972. Conditions of preservation of chlorophyll, pheophitin and humic substances in the Black Sea sediments. Geokhimia, No. 3: 323–334 (in Russian).

Duce, R.A. and Duursma, E.K., 1977. Inputs of organic matter to the Ocean. Mar. Chem., 5: 319–339.

Duursma, E.K., 1965. The dissolved organic constituents of sea water. In: J.P. Riley and G. Skirrow (Editors), Chemical Oceanography, 1. Academic Press, London, pp. 433–475.

Duursma, E.K., 1974. The fluorescence of dissolved organic matter in the sea. In: N.G. Jerlov and E. Steeman Nielsen (Editors), Optical Aspects of Oceanography. Academic Press, New York, N.Y., pp. 237–256.

Fleming, R.H., 1940. The composition of plankton and units for reporting population and production. Proc. Sixth Pacif. Sci. Congr. Calif., 3: 535–540.

Forée, E.G. and McCarty, P.L., 1970. Anaerobic decomposition of algae environment. Sci. Technol., 4: 842–849.

Gillbricht, M., 1957. Ein Verfahren zum oxydativen Nachweiss von organischer Substanz im Seewasser. Helgol. Wiss. Meeresunters., 6: 76–83.

Gordon, D.C., 1970. Some studies on the distribution and composition of particulate organic carbon in the North Atlantic Ocean. Deep-Sea Res., 17: 233–243.

Gordon, D.C. and Sutcliffe, W.H., 1973. A new dry combustion method for the simultaneous determination of total organic carbon and nitrogen in sea waters. Mar. Chem., 1: 231–244.

Grill, E.V. and Richards, F.A., 1964. Nutrient regeneration from phytoplankton decomposing in seawater. J. Mar. Res., 22: 51–69.

Hallegraeff, C.M., 1978. Caloric content and elementary composition of seston of three Dutch freshwater lakes. Arch. Hydrobiol., 1978(83): 80–98.

Hoffman, E.J. and Duce, R.A., 1974. The organic carbon content of marine aerosols collected on Bermuda. J. Geophys. Res., 79: 4474–4477.

Hoffman, E.J. and Duce, R.A., 1976. Factors influencing the organic carbon content of marine aerosols: A laboratory study. J. Geophys. Res., 81: 3667–3670.

Horwath, R., 1972. Microbial co-metabolism and the degradation of organic compounds in nature. Bacteriol. Rev., 36: 146–155.

Ivanenkov, V.N., 1977. Processes controlling oxygen distribution in the ocean. In: B.A. Skopintsev and V.N. Ivanenkov (Editors), Chemical Oceanologic Studies. Nauka, Moscow, 224: 206–209 (in Russian).

Ivlev, V.S., 1934. Eine Mikromethode zur Bestimmung des Kalorien Gehalt von Nährstoffen. Biochem. Z., 275: 49–55.

Jannash, H.W. and Wirsen, C.C., 1973. Deep-sea microorganisms: In situ response to nutrient enrichment. Science, 180: 103–105.

Jewell, W.I. and McCarty, P.L., 1971. Aerobic decomposition of algae environment. Sci. Technol., 5: 1023–1031.

Kalle, K., 1962. Über die gelösten organischen Komponenten in Meerwasser. Kiel. Meeresforsch., 18: 128–131.

Kalle, K., 1966. The problem of the Gelbstoff in the sea. Oceanogr. Mar. Biol. Annu. Rev., 4: 91–104.

Karavayev, N.M. and Budyak, N.F., 1960. Study of the so-called humic acids of freshwater sapropels. Dokl. Akad. Nauk S.S.S.R., 132: 192–194 (In Russian).

Karavayev, N.M., Vener, R.A. and Korolyova, K.I., 1964. On composition and chemical properties of sapropel acids. Dokl. Akad. Nauk S.S.S.R., 156: 877–879 (in Russian).

Kerr, R.A. and Quinn, J.G., 1975. Chemical studies on the dissolved organic matter in sea water. Isolation and fractionation. Deep-Sea Res., 22: 107–116.

Khailov, K.M., 1962. Some unknown organic substances of marine water. Dokl. Akad. Nauk S.S.S.R., 147: 1200–1203 (in Russian).

Khailov, K.M., Finenko, G.A., Burlakova, Z.P. and Smirnov, V.A., 1973. On the relation between stationary concentrations of the basic forms of organic matter in marine water and specific rate of trophic usage of coastal communities by organisms. Dokl. Akad. Nauk S.S.S.R., 209: 1210—1214 (in Russian).

King, F.D. and Packard, T.T., 1975. The effect of hydrostatic pressure on respiratory electron transport system activity in marine zooplankton. Deep-Sea Res., 22: 99—105.

Kononova, M.M., 1963. Organic Matter of Soil. Izd. Akad. Nauk S.S.S.R., Moscow, 314 pp. (in Russian).

Kononova, M.M., 1968. Transformation of organic matter and its relation to soil fertility. Pochvovedenie, No. 8: 17—26 (in Russian).

Korolykov, K.P., 1926. Decay of sewage sediment in anaerobic environment. Tr. Sov. Otshist. Stotsh. Vod Uprav. Kanal. Mosk. Komm. Khoz., No. 8: 58—67. (in Russian).

Kovda, V.A. and Yakushevskaya, I.V., 1971. Biomass and humic cover of land. In: V.A. Kovda (Editor), Biosphere and Its Resources. Nauka, Moscow, pp. 132—141 (in Russian).

Krause, H.R., 1959. Biochemische Untersuchungen über den postmortalen Abbau von totem Plankton unter aeroben und anaeroben Bedingungen. Arch. Hydrobiol., Suppl., 24: 297—337.

Kreps, E., 1934. Organic catalysts or enzymes in sea water. In: A.E. Daniel (Editor), James Johnston Memorial. Liverpool University Press, Liverpool, pp. 193—202.

Kriss, A.E., Rukina, E.A. and Biryuzova, V.N., 1951. Fate of dead organic matter in the Black Sea. Microbiologya, 20: 90—102 (in Russian).

Krogh, A., 1934. Conditions of life at great depth in the ocean. Ecol. Monogr., 4: 430—439.

Krylova, L.P., 1953. Micromethod to estimate carbon in organic matter. Gidrokhim. Mater., 20: 63—69 (in Russian).

Krylova, L.P., 1957. Estimation of carbon from organic matter of natural waters by dry combustion method. Gidrokhim. Mater., 26: 237—242 (in Russian).

Krylova, L.P. and Skopintsev, B.A., 1959. Content of organic carbon in lakes and rivers of Podmoskovye and large rivers of the Soviet Union. Gidrokhim. Mater., 28: 28—44 (in Russian).

Kuzin, B.S. (Editor), 1972. The Rybinsk Reservoir and Its Life. Nauka, Leningrad, 364 pp. (in Russian).

Kuznetsov, S.I., 1970. Microflora of Lakes and Its Geochemical Activity. Nauka, Leningrad, 440 pp. (in Russian).

Larionov, Ju.V. and Skopintsev, B.A., 1974. Sampling of particles from natural waters for studies of organic fraction. Gidrokhim. Mater., 60: 192—194 (in Russian).

Larionov, Ju.V. and Skopintsev, B.A., 1977. Indexis of lability of organic particles. Gidrobiol. Zh., 13: 95—101 (in Russian).

Lundstedt, K., 1970. Jämförelse mellan permanganattal och halt av organiskt kol i några olika yt watten (Comparison of permanganate number and content of organic carbon in some surface waters). Vatten, 26: 126—134.

MacKinnon, M.D., 1978. A dry oxidation method for the analysis of the TOC in sea water. Mar. Chem., 7: 13—37.

McCarty, P.L., 1964. The methane fermentation. In: H. Heukelekian and H.C. Dondero (Editors), Principle Applications in Aquatic Microbiology. Wiley, New York, N.Y., pp. 314—343.

Maksimova, M.P., 1972. Estimation of regeneration rates of nitrogen and phosphorus in the Indian Ocean. Okeanologya, 12: 1003—1009 (in Russian).

Maksimova, M.P., 1974. N/P and Si/P ratios in the Indian Ocean. Okeanologiya, 14: 830—839 (in Russian).

Manskaya, S.M. and Drozdova, T.V., 1964. Geochemistry of Organic Matter. Nauka, Moscow, 315 pp. (in Russian).

Menzel, D.W., 1974. Primary productivity and particulate organic matter and the sites of oxidation of organic matter. In: E. Goldberg (Editor), The Sea, 5. Wiley—Interscience, New York, N.Y., pp. 659—678.

Menzel, D.W. and Goering, J.J., 1966. The distribution of organic detritus in the ocean. Limnol. Oceanogr., 11: 333—337.

Menzel, D.W. and Ryther, J.H., 1968. Organic carbon and the oxygen minimum in the South Atlantic Ocean. Deep-Sea Res., 15: 327—337.

Menzel, D.W. and Ryther, J.H., 1970. Distribution and cycling of organic matter in the oceans. In: D.W. Hood (Editor), Symposium on Organic Matter in Natural Waters. Inst. Mar. Sci., Alaska, Occ. Publ., No. 1: 31—54.

Menzel, D.W. and Vaccaro, R., 1964. The measurement of dissolved organic and particulate carbon in seawater. Limnol. Oceanogr., 9: 138—142.

Mishustin, E.N., Dragunov, S.S. and Pushkinskaya, O.I., 1956. The role of microorganisms in the synthesis of soil humus compounds. Izv. Akad. Nauk S.S.S.R. Ser. Biol., No. 6: 83—94 (in Russian).

Mishustina, I.E., Rossova, E.Ja. and Baturina, M.B., 1976. Electronic microscopy in studies of bacterial population and ultramicroscopic forms in sea water. Izv. Akad. Nauk S.S.S.R., Ser. Biol., No. 6: 840—848 (in Russian).

Munk, W.H., 1966. Abyssal recipes. Deep-Sea Res., 13: 707—730.

Nissenbaum, A. and Kaplan, J., 1972. Chemical and isotopic evidence of in situ origin of marine humic substances. Limnol. Oceanogr., 17: 570—582.

Nissenbaum, A., 1974. Deuterium content in humi acids from marine and nonmarine environments. Mar. Chem., 2: 59—64.

Novoselov, A.A., 1962. Study of biochemical oxygen consumption in the North Atlantic waters. Okeanologiya, 2: 84—91 (in Russian).

Oden, Sv., 1919. Die Huminsäuren. Kolloidchem. Beih., 11: 75—260.

Odum, E., 1968. Ecology. Prosvestshenie, Moscow, 168 pp. (in Russian).

Odum, E., 1975. The Essentials of Ecology. Nauka, Moscow, 740 pp. (in Russian).

Ogura, N., 1972a. Dissolved organic matter in the sea, its production, utilization and decomposition. In: K. Sugawara (Editor), The Kuroshio, II. Saigon Publ. Co., Tokyo, pp. 201—205.

Ogura, N., 1972b. Rate and extent of decomposition of dissolved organic matter in surface water. Mar. Biol., 13: 89—93.

Ogura, N., 1975. Further studies on decomposition of dissolved organic matter in coastal seawater. Mar. Biol., 31: 101—111.

Ohle, W., 1962. Der Stoffhaushalt der Seen als Grundlage einer allgemeine Stoffwechseldynamik der Gewässer. Kiel. Meeresfor., 18: 107—120.

Otsuki, A. and Hanya, T., 1967. Some precursors of humic acid in recent lake sediment suggested by infrared spectra. Geochim. Cosmochim. Acta, 31: 1503—1515.

Otsuki, A. and Hanya, T., 1972. Production of dissolved organic matter from dead green algae cells. I. Aerobic microbial decomposition. Limnol. Oceanogr., 17: 248—257; II. Anaerobic microbial decomposition. Limnol. Oceanogr., 17: 258—264.

Packard, T.T., Healy, M.L. and Richards, F.A., 1971. Vertical distribution of the activity of the respiratory electron transport system in marine plankton. Limnol. Oceanogr., 16: 60—70.

Plunkett, M.A. and Rakestraw, N.V., 1955. Dissolved organic matter in the sea. Deep-Sea Res., 3 (Suppl.): 12—18.

Rakestraw, N.V., 1947. Oxygen consumption in sea water over long periods. J. Mar. Res., 6: 259—267.

Rakestraw, N.V., 1958. Particulate matter in the oxygen minimum layer. J. Mar. Res., 27: 429—431.

Redfield, A.C., 1942. The process determining the concentration of oxygen, phosphorus and other organic substances within the depths of the Atlantic Ocean. Papers Phys. Ocean. Meteorol., 9: 22 pp.

Richards, F.A., Cline, I.D., Broenkow, W.W. and Atkinson, L.P., 1965. Some consequences of the decomposition of organic matter of lake Nitinat, an anoxic fjord. Limnol. Oceanogr., 10 (Suppl.): 185—201.

Riley, G.A., 1951. Oxygen, phosphate and nitrate in the Atlantic Ocean. Bull. Bingham Oceanogr. Coll., 13: 1—126.

Rogers, Ch.J. and Landreth, R.R., 1975. Degradation mechanisms: controlling the bioaccumulation of hazardous materials. Natl. Environ. Res. Centre, Cincinnati, Ohio, pp. 1—15.

Rutten, M.G., 1971. The Origin of the Life by Natural Causes. Elsevier, Amsterdam, 420 pp.

Seiwell, H.R., 1937. Consumption of oxygen in sea water under controlled laboratory conditions. Nature, 140: 504—507.

Seki, H., Wada, E., Koike, J. and Hattori, A., 1974. Evidence of high organotrophic potentiality of bacteria in the deep ocean. Mar. Biol., 26: 1—4.

Semenov, A.D., Ivleva, I.N. and Datsko, V.G., 1960. Determination of amino acids in natural waters. Gidrokhim. Mater., 33: 172—179 (in Russian).

Sepers, A.B.J., 1977. The utilization of dissolved organic compounds in aquatic environments. Hydrobiologia, 52: 39—54.

Sharp, I.H., 1973a. Total organic carbon in sea water — comparison of measurements using persulphate oxidation and high temperature combustion. Mar. Chem., 1: 211—229.

Sharp, I.H., 1973b. Size classes of organic carbon in sea water. Limnol. Oceanogr., 18: 441—447.

Sholkovitz, E.R., 1976. Flocculation of dissolved organic and inorganic matter during the mixing of river water and sea water. Geochim. Cosmochim. Acta, 40: 831—845.

Sholkovitz, E.R., Boyle, E.A. and Price, N.B., 1978. The removal of dissolved humic acids and iron during estuarine mixing. Earth Planet. Sci. Lett., 40: 130—136.

Sieburth, J.M., 1969. Studies on algal substances in the sea. III. The production of extracellular organic matter by littoral marine algae. J. Exp. Mar. Biol. Ecol., 3: 290—309.

Sieburth, J.M. and Jensen, A., 1969. Studies on algal substances in the sea. II. The formation of gelbstoff (humic material) by oxidates of Phaeophyta. J. Exp. Mar. Biol. Ecol., 3: 275—289.

Skopintsev, B.A., 1938. On the regeneration rate of biogenic elements (N and P) under the bacterial decomposition of planktonic organisms. Mikrobiologiya, 7: 755—765 (in Russian).

Skopintsev, B.A., 1946. On the coagulation of terrigenous suspended particles in the river flow in sea water. Izv. Akad. Nauk S.S.S.R., Ser. Geogr. Geofiz., 10: 357—371 (in Russian).

Skopintsev, B.A., 1947a. On the oxygen equivalent of organic substances in natural waters. Dokl. Akad. Nauk S.S.S.R., 58: 2089—2092 (in Russian).

Skopintsev, B.A., 1947b. On coagulation of humic substances in the river flow in sea water. Izv. Akad. Nauk S.S.S.R., Ser. Geogr. Geofiz., 11: 21—36 (in Russian).

Skopintsev, B.A., 1950. Organic matter in natural waters (water humus). Tr. Gos. Okeanogr. Inst., No. 17(29): 1—290 (in Russian).

Skopintsev, B.A., 1960. Organic matter in sea water. Tr. Norsk. Gidrofis. Inst. Akad. Nauk S.S.S.R., 19: 3—30 (in Russian).

Skopintsev, B.A., 1961. Study of oxygen minimum layer in the North Atlantic Ocean, autumn, 1959. Okeanol. Issled., No. 13: 108—114 (in Russian).

Skopintsev, B.A., 1964. Estimation of formation and oxidation of organic matter in sea water. Okeanol. Issled., No. 13: 96—107 (in Russian).

Skopintsev, B.A., 1966. Some considerations on the distribution and condition of organic matter in oceanic water. Okeanologiya, 6: 441—450 (in Russian).

Skopintsev, B.A., 1971. Recent progress in study of organic matter in ocean water. Okeanologiya, 11: 939—956 (in Russian).

Skopintsev, B.A., 1972. On the age of stable organic matter aquatic humus in the oceanic waters. In: D. Dyrsen and D. Jahner (Editors), The Changing Chemistry of the Oceans. Nobel Symposium 20. Wiley—Interscience, New York, N.Y., pp. 205—207.

Skopintsev, B.A., 1975a. Formation of the Present Chemical Composition in the Black Sea Waters. Gidrometizdat, Moscow, 336 pp. (in Russian).

Skopintsev, B.A., 1975b. Oxygen consumption in the deep waters of oceans. Okeanologiya, 15: 830—838 (in Russian).

Skopintsev, B.A. and Bikbulatova, E.M., 1978. Behaviour of dissolved coloured humus substances in the presence of decomposing blue-green algae. Gidrobiol. Zh., 14: 87—91 (in Russian).

Skopintsev, B.A. and Brook, E.C., 1940. Investigation of oxidation processes in the water under decomposition of phytoplankton in aerobic conditions. Mikrobiologiya, 9: 596—607 (in Russian).

Skopintsev, B.A. and Krylova, L.P., 1955. The results of some aspects of organic matter dynamics study in the natural waters. Tr. Vses. Gidrobiol. O.va., 6: 35—45 (in Russian).

Skopintsev, B.A. and Ovtshinnikova, Ju.S., 1934. Study of oxidation processes in the polluted waters under aerobic conditions. Mikrobiologiya, 3: 138—147 (in Russian).

Skopintsev, B.A. and Timofeeva, S.N., 1961. Application of a dry combustion method suggested by L.P. Krylova to estimation of organic carbon in marine waters. Gidrokhim. Mater., 32: 153—176 (in Russian).

Skopintsev, B.A., Lyubimova, E.M. and Timofeeva, S.N., 1964. Study of mineralization of organic matter from dead plankton in anaerobic conditions. Sb. Khim. Tekhnol. Inst. Prague, 8: 209—224 (in Russian).

Skopintsev, B.A., Timofeeva, S.N. and Vershinina, O.A., 1966. Organic carbon in waters of preequatorial and southern Atlantic and Mediterranean. Okeanologiya, 6: 251—260 (in Russian).

Skopintsev, B.A., Romenskaya, N.N. and Sokolova, M.V., 1968. Organic carbon in the Norwegian Sea and north-east Atlantic. Okeanologiya, 8: 225—234 (in Russian).

Skopintsev, B.A., Bakulina, A.G., Bikbulatova, E.M., Kudryavtseva, N.A. and Melnikova, N.I., 1972. Organic matter in the water of Volga and its reservoirs. Tr. Inst. Biol. Vnutr. Vod., Akad. Nauk S.S.S.R., No. 23(26): 39—53 (in Russian).

Skopintsev, B.A., Bikbulatov, E.S. and Melnikova, N.I., 1976. On the estimation of organic carbon by the persulphate method in waters rich in chlorids. Okeanologiya, 16: 1109—1114 (in Russian).

Sorokin, Ju.I., 1962. Microflora of the water mass of the Central Pacific Ocean. Okeanologiya, 2: 922—931 (in Russian).

Sorokin, Ju.I., 1971. On the role of bacteria in the productivity of tropical oceanic waters. Int. Rev. Ges. Hydrobiol., 56: 1—48.

Sorokin, Ju.I., 1977. Microflora production. In: M.E. Vinogradov (Editor), Oceanology. Biology of the Ocean, 2. Nauka, Moscow, pp. 209—232 (in Russian).

Starikova, N.D., 1970. Regularities of vertical distribution of dissolved organic carbon in marine waters and ground solutions of marine sediments. Okeanologiya, 10: 988—1000 (in Russian).

Strakhov, N.M., 1954. Sedimentation in the Black Sea. In: D.S. Belyankin (Editor), Sedimentation in Contemporary Reservoirs. Izd. Akad. Nauk S.S.S.R., Moscow, pp. 81—136 (in Russian).

Strickland, J.D.H., 1965. Production of organic matter in the primary stages of the marine food chain. In: J.P. Riley and G. Skirrow (Editors), Chemical Oceanography, 1. Academic Press, London, New York, N.Y., pp. 477—610.

Stuermer, D.H. and Harvey, G.R., 1974. Humic substances from sea water. Nature, 250: 480—481.

Stuermer, D.H. and Harvey, G.H., 1977. The isolation of humic substances and alcohol-soluble organic matter from sea water. Deep-Sea Res., 24: 303—309.

Stuermer, D.H. and Harvey, G.R., 1978. Structural studies of marine humus: a new reduction sequence for carbon skeleton determination. Mar. Chem., 6: 55—70.

Stuermer, D.H. and Payne, J.R., 1976. Investigation of seawater and terrestrial humic substances with carbon-13 and proton nuclear magnetic resonance. Geochim. Cosmochim. Acta, 40: 1109—1114.

Stupakova, G.P. 1975. Effect of elevated hydrostatic pressure on the activity of metabolic enzymes in barotolerant bacteria. Mikrobiologiya, 44: 174—175 (in Russian).

Szekielda, K.H., 1968a. Vergleichende Untersuchungen über den Gehalt an organischen Kohlenstoff in Meerwasser und den Kalium-permanganatverbrauch. J. Cons. Perman. Int. Explor. Mer, 32: 17—24.

Szekielda, K.H., 1968b. The transport of organic matter by the bottom water of the oceans. Sarsia, 34: 243—252.

Theriault, E.I., 1927. The oxygen demand of polluted waters. Publ. Health Bull., No. 173: 1—185.

Trask, P.D., 1939. Organic content of recent marine sediments. In: P.D. Trask (Editor), Recent Marine Sediments. American Association of Petroleum Geologists, Tulsa, Okla., pp. 428—453.

Tsunogai, Sh., 1972. An estimate of the rate of decomposition of organic matter in the deep water in Pacific Ocean. In: A.Y. Takenouti (Chief Editor), Biological Oceanography of the North Pacific Ocean, pp. 517—533.

Tyurin, I.V., 1937. Organic Matter of Soils and Its Role in the Soil Formation and Fertility. Selchozgiz, Moscow, 275 pp. (in Russian).

Vaccaro, H.F., 1965. Inorganic nitrogen in sea water. In: J.P. Riley and G. Skirrow (Editors), Chemical Oceanography I. Academic Press, London, pp. 365—408.

Vasilevskaya, N.A., Galyashin, B.H., Denisenko, N.M. and Maksimov, O.B., 1977. Chemical study of humic acids from bottom deposits in the west regions of the Pacific Ocean. Okeanologiya, 17: 459—469 (in Russian).

Vernadsky, V.I., 1934. Essays on Geochemistry. Gorgeoneftizdat, 4th ed., Moscow, 380 pp. (in Russian).

Vinogradov, M.E., 1968. Vertical Distribution of Organic Zooplankton. Nauka, Moscow, 320 pp. (in Russian).

Von Brand, Th., Rakestraw, H.W. and Renn, C.E., 1937. The experimental decomposition and regeneration of nitrogenous organic matter in sea water. Biol. Bull., 72: 165—175.

Votintsev, K.K., 1948. Observations of regeneration of biogenic elements under decomposition of *Episura b.* Dokl. Akad. Nauk S.S.S.R., 63: 741—744 (in Russian).

Votintsev, K.K., 1953. On regeneration rates of biogenic elements at decomposition of dead *Melosira b.* Dokl. Akad. Nauk S.S.S.R., 92: 687—690 (in Russian).

Wagner, F.S., 1969. Composition of the dissolved organic compounds in sea water; a review. Contrib. Mar. Sci., Univ. Texas, 14: 115—153.

Waksman, S.A., 1936. Humus. Bailliere, Tindall and Cox, London, 494 pp.

Waksman, S.A. and Carey, C., 1935. Decomposition of organic matter in sea water by bacteria. 1. Bacterial multiplication in stored sea water. J. Bacteriol., 29: 531—543.

Wangersky, P.J., 1975. Measurement of organic carbon in seawater. In: Th.R. Gibb (Editor), Analytical Methods in Oceanography. Am. Chem. Soc., Washington, D.C., pp. 148—162.

Wangersky, P.J., 1978. Production of dissolved organic matter. In: O. Kinne (Editor), Marine Ecology, 4. John Wiley, New York, N.Y., pp. 115—200.

Williams, P.J. le B., 1975. Biological and chemical aspects of dissolved organic material in sea water. In: J.P. Riley and G. Skirrow (Editors), Chemical Oceanography, 2. Academic Press, London, 2nd ed., pp. 301—363.

Williams, P.M., 1968. Stable carbon isotopes in the dissolved organic matter of the sea. Nature, 219: 152—153.

Williams, P.M., 1971. The distribution and cycling of organic matter in the ocean. In: S.D. Faust and I.P. Hunter (Editors), Organic Compounds in Aquatic Environment. Marcel Dekker, New York, N.Y., pp. 145—163.

Williams, P.M. and Gordon, L.J., 1970. Carbon-13: Carbon-12 in dissolved and particulate organic matter in the sea. Deep-Sea Res., 17: 19—27.

Williams, P.M., Oeschger, H. and Kinney, P., 1969. Natural radiocarbon activity of the dissolved organic carbon in the northeast Pacific Ocean. Nature, 224: 256—258.

Wirsen, C.O. and Jannash, H.W., 1975. Activity of marine psychrophilic bacteria at elevated hydrostatic pressures and low temperatures. Mar. Biol., 31: 201—208.

Wyrtki, K., 1962. The oxygen minimum in relation to ocean circulation. Deep-Sea Res., 9: 11—23.

Yashnov, V.A., 1962. Plankton of the tropical Atlantic Ocean. Tr. Morsk. Gidrofiz. Inst. Akad. Nauk S.S.S.R., 25: 195—207 (in Russian).

Chapter 7

ORGANO-METALLIC INTERACTIONS IN NATURAL WATERS [1]

R.F.C. MANTOURA

1. INTRODUCTION

Organic matter dissolved in the sea may interact with metal ions to form organo-metallic complexes. This phenomenon is not new, but had been postulated fifty years ago in order to explain several observations linked with the apparent supersaturation of ferric ions in sea water (Harvey, 1928, p. 51; Cooper, 1935, 1937, 1948a, b) and the need to include metal sequestering agents as a means of rendering iron biologically available for the growth of algae in sea water (Gran, 1931, 1933) and in artificial media (Provasoli et al., 1957; Johnston, 1964). Since then, organo-metallic interactions have been invoked in a wide range of investigations such as the speciation of trace metals (Lerman and Childs, 1973; Stumm and Brauner, 1975), heterogeneous reactions with authigenic minerals (Chave, 1970; Parks, 1975; Hedges, 1977), the outbreak of red tides (Martin and Martin, 1973), and the inhibition of metal toxicity to marine algae (Steemann-Nielsen and Wium-Anderson, 1970; Sunda and Guillard, 1976) and copepods (Whitfield and Lewis, 1976). However, only in recent years with the advent of new analytical techniques has it been possible to demonstrate the presence and characterise the nature of these complexes in natural waters. The purpose of this chapter is to review our knowledge of the chemistry of organo-metallic complexes in natural waters.

2. CLASSIFICATION OF COMPLEXES

Of the wide range of organo-metallic complexes likely to exist in natural waters only a small number have been structurally identified. Two examples of identified complexes found in sea water are cobalt(II)-containing cyanocobalamine or vitamin B_{12} (Carlucci, 1970) and magnesium in chlorophyll a (Riley and Chester, 1971). Although a great deal is known about the properties of the unidentified complexes, it is not possible to classify them according to the systematic molecular nomenclature used in classical chemistry (IUPAC, 1973; Cahn, 1974). This undoubtedly has contributed

---

[1] Institute for Marine Environmental Research Manuscript No. 272.

to the inconsistent usage of nomenclature found in the literature (see e.g. Siegel, 1971; Williams, 1975; Whitfield, 1975; Reuter and Perdue, 1977). The necessarily less rigorous classification adopted in this review is based on two criteria: (1) chemical classification according to the type of bond linking the metal ion to the organic moiety; and (2) operational classification based on the procedures used to isolate, fractionate and characterise the complexes.

2.1. Chemical schemes

The term "organo-metallic" will be used to describe molecules in which the metal is bonded to organic matter via (a) covalent bonds with carbon atoms giving rise to *organo-metallic compounds*; (b) carboxyl groups leading to *organic salts*; (c) electron donating atoms (O, N, S, P, As, etc.) forming *coordination complexes*; or (d) overlap of π electron orbitals to form π complexes. In many stable naturally occurring complexes, two or more types of bonds may be involved in a cooperative binding of the metal atom to the organic component. Chelation is a particularly important type of cooperative interaction occurring between a central metal ion and two (or more) electron-donating sites on a bidentate (or polydentate) organic ligand resulting in a heterocyclic structure (Martell, 1971). These chelate structures are thermodynamically very stable, and have been known to survive diagenetic transformations over several millenia (Baker and Smith, 1977). Chelation is also involved in the binding of metals with humic compounds — probably the most abundant ligands in natural waters (Rashid, 1972; Schnitzer and Khan, 1972) as well as with biological ligands such as metallo-enzymes, nucleic acids and peptides (Eichhorn, 1975). Although the metals of Groups IA and IIA of the Periodic Table will be mentioned, we shall be concerned mainly with the first-row transition metals and Groups IB and IIB.

2.2. Operational schemes

Organo-metallic complexes in natural waters have mainly been characterised using five operational classifications: size, solubility, stability, degradability and lability of the complexes with respect to a wide variety of analytical conditions (Stumm and Brauner, 1975; Whitfield, 1975).

a. Size. The division of the chemical form of an element into "dissolved" and "particulate" fractions by filtration is the most common operational classification in marine chemistry (Riley, 1975). This analytically convenient classification nevertheless represents an arbitrary cut-off point within a polydisperse size spectrum. Further fractionation of "dissolved" organo-metallic complexes into size classes by molecular filtration through gels (Gjessing, 1965, 1973; Ghassemi and Christman, 1968; Means et al., 1977; Sugai and

Healy, 1978), ultramembranes (Gjessing, 1970; Sharp, 1973; Smith, 1976) and dialysis bags (Dawson and Duursma, 1974) are a particularly useful way of differentiating macromolecular and colloidal constituents from the truly dissolved species.

b. *Solubility*. Variable fractions of copper and other metals dissolved in sea water have been recovered by solvent extraction (Slowey et al., 1967) or by adsorption onto hydrophobic resins (Montgomery and Santiago, 1978). These have been categorised as "organically bound" metals.

c. *Stability*. The degree of metal dissociated from complexes resulting from variations in pH, ionic strength, concentration of competing metals, has been used by Rashid (1974) to classify these complexes in order of increasing stabilities. These stabilities include thermodynamic and kinetic contributions and are thus empirical.

d. *Degradability*. The chemical degradation of the organic matter by ultraviolet photo-oxidation (Foster and Morris, 1971), oxidising agents (Alexander and Corcoran, 1967) or ozonation (Clem and Hodgson, 1978) may give rise to an increase in the analytical concentration of some trace metals. This increase has been explained in terms of the presence of organo-metallic complexes.

e. *Lability*. There are a host of analytical techniques available for the determination of trace metals in natural waters. However, the ability of those techniques to measure organically associated metals is linked with the kinetic lability of the complexes under the experimental conditions used in the analysis. For example, Piro et al. (1973) and Bernhard et al. (1975) have employed polarographic and radiochemical techniques to investigate the organic speciation of zinc in sea water. Marchand (1974) and Batley and Florence (1976b) used cation exchange and chelating resins to assess the lability and extent of organic speciation of Cu, Zn, Cd and Pb.

In the final analysis, the selection of criteria for classifying organo-metallic complexes must be based on a physically and chemically meaningful characterisation of the interactions and a more detailed structural elucidation of the complexes.

3. COMPLEXING CAPACITY OF NATURAL WATERS

Many natural waters possess the property of partially masking the presence of metal ions by rendering them unavailable for measurement by conventional chemical and biological metal-sensing techniques. Although it is as yet not possible to distinguish which of the many possible mechanisms (e.g. precipitation, colloidal adsorption, complexation, etc.) gives rise to this

TABLE I

A compilation of metal-complexing capacities attributed to organic matter in fresh and sea water

Metal complexed	Sample location (number of samples)	Complexing capacity [1,2]		Method	Reference
		$\mu M \, l^{-1}$	$\mu M \, (mg \, DOC{-}C)^{-1}$		
Cu^{2+}	Canadian lakes (12)	0.0—1.59 (0.61)	0.0—0.35 (0.15)	amperometric titration using DPASV [3]	Chau and Wong (1976)
	Sudbury lakes (9), Canada	0.0—0.63 (0.31)	—	amperometric titration using DPASV [3]	Chau et al. (1974)
	Ottawa River (1), Canada	25.4	4.23	potentiometric titration using Cu—ISE [4]	Ramamoorthy and Kushner (1975a,b)
	Montana stream (2), U.S.A.	15.6	—	chelate solubilisation of $Cu(OH)_2$ precipitate	Kunkel and Manaham (1973)
	Maine lakes (12), U.S.A.	0.0—8.6 (2.43)	0.0—0.62 (0.19)	potentiometric titration using Cu^{2+}—ISE [4]	J.P. Giesy et al. (pers. comm.)
	Ogeechee Estuary (25), Ga., U.S.A.	0.0—0.5 (0.2) [5]	0.0—0.38 (0.09)	amperometric titration using DPASV [3]	Smith (1976)
	Boston harbour (2) seawater, Mass., U.S.A.	0.46—0.60 (0.53)	—	chromatography — PES [6]	Stolzberg and Rosin (1977)
	Pacific Ocean (1), off Australia	0.01	0.014 [7]	amperometric titration using DPASV [3]	Florence and Batley (1977)
	North Sea (1), Dutch coast	0.36	0.30 [8]	amperometric titration using DPASV [3]	Duinker and Kramer (1977)
	Coastal fjords (15), Vancouver, Canada	0.13—0.60 (0.42)	—	copepod nauplii bioassay	Whitfield and Lewis (1976)
	Narragansett Bay (2), U.S.A.	0.04—0.27	0.02—0.09	phytoplankton bioassay	Davey et al. (1973)
	Sargasso Sea (1)	0.05	0.094	bacterial bioassay	Gillespie and Vaccaro (1978)
	Vineyard Sound (1), Mass., U.S.A.	0.11	0.055	bacterial bioassay	
	Saanich Inlet (1), B.C., Canada	0.19	0.047	bacterial bioassay	

Cu²⁺	eutrophic marine marsh	0.50	—	bacterial bioassay	Gillespie and Vaccaro (1978)
Pb²⁺	sea water	0.32	—	amperometric titration using DPASV[3]	Clem (1973, cited by Whitfield, 1975)
Hg²⁺	Ottawa River (2)	20—55	3.3—9.2	potentiometric titration using Hg^{2+}—ISE[4]	Ramamoorthy and Kushner (1975a,b)
Cd²⁺	S. Carolina Lake (1), U.S.A.	0.01—0.86[5]	0.035—0.129	potentiometric titration using Cd^{2+}—ISE[4]	J.P. Giesey et al., pers. comm.
Zn²⁺	Coastal fjord (2), Vancouver, Canada	0.00—0.19	—	copepod nauplii bioassay	Whitfield and Lewis (1976)
Co²⁺	N. Carolina streams (32) U.S.A.	0.5—2.4 (1.13)	—	DPP[9] of Co^{2+} in ethylenediamine medium	Hanck and Dillard (1977)
	= Sewage effluent	4.1	—		
In³⁺	activated sewage, Ohio River, U.S.A.	5.16[10]	—	DPP[9] of uncomplexed In^{3+} after preconcentration of ligand	Haberman (1971)

[1] Results expressed as μM of complexed metal per litre, assuming a 1:1 metal-binding site stoichiometry; the range and average (bracketed) values are reported.
[2] Complexation capacity normalised in terms of the dissolved organic carbon (DOC) concentration, the absence of DOC data is indicated by —.
[3] Differential pulse anodic stripping voltammetry.
[4] Ion-selective electrodes.
[5] Range corresponds to molecular size fractions of organic ligands obtained by ultra-filtration.
[6] Plasma emission spectrometry.
[7] Calculation based on mean DOC of 0.7 mg C l⁻¹ reported by Holm-Hansen et al. (1966) for surface waters of the northeastern Pacific Ocean.
[8] Calculation based on a mean DOC of 1.2 mg C l⁻¹ reported by Duursma (1961) for the North Sea, off the Dutch coast.
[9] Differential pulse polarography.
[10] Calculation based on the excess In³⁺ spikes required to produce detectable $In^{3+} \rightarrow In^0$ polarographic wave at a potential of —0.61 V vs. SCE.

masking effect, there is mounting evidence (Chau et al., 1974; Perdue et al., 1976; Duinker and Kramer, 1977; Gillespie and Vaccaro, 1978) favouring organic complexation of metals. The complexing capacity of a sample, defined as the moles of added metal (usually Cu^{2+}) which are complexed per litre of sample, is a measure of the *metal-buffering capacity* and is of fundamental importance for a quantitative assessment of the fate of polluting metals in natural waters. Gächter et al. (1973) have suggested that temporal variability in complexing capacity may influence the succession of phytoplankton species in natural waters.

The complexing capacity may be determined using chemical (Chau et al., 1974; Ramamoorthy and Kushner, 1975a; Duinker and Kramer, 1977; Hanck and Dilland, 1977; Stolzberg and Rosin, 1977) and bioassay techniques (Davey et al., 1973; Whitfield and Lewis, 1976; Gillespie and Vaccaro, 1978). The two approaches are fundamentally similar since they are based on the microtitrimetry of complexing ligands by the successive additions of reacting metal ions (usually Cu^{2+}). However, they differ from one another in that the determination of the end-point relies on detecting a particular form(s) of the metal to which the technique is sensitive. For example, the two most popular chemical techniques, differential pulse anodic stripping voltammetry (DPASV) and ion-selective potentiometry (ISP) do not measure the same end-point in the titration of complexing ligands with ionic copper, since DPASV senses the electroactive forms of copper ($[Cu(H_2O_6)]^{2+}$, $Cu(OH)^+$, $Cu(OH)_2^0$, $CuCO_3^0$, $CuHCO_3^+$ and *labile* Cu-organic complexes; Chau et al., 1974), whereas ISP responds essentially to the activity of copper ions in solution (Whitfield, 1975). This may explain the elevated values of complexing capacities obtained by Ramamoorthy and Kushner (1975b) using ISP. On the other hand, the end-point as determined by the bioassay procedures corresponds to the inflection point of the sigmoid growth depression curves in response to increasing concentration of *toxic* forms of copper (Davey et al., 1973). Thus, the complexation capacity as measured by differing techniques will not necessarily be the same although they may all respond quantitatively to the addition of synthetic but functionally analogous chelators, such as EDTA (see e.g. Chau et al., 1974; and Gillespie and Vaccaro, 1978).

The values of complexing capacity of different waters reported by various workers show considerable variation (Table I). Although some of this variation may be attributable to the different experimental conditions and methods employed, there is, nevertheless, a trend in the data. The mean complexing capacities (weighted for the number of samples analysed, see Table II) for sewage, inland waters and sea water are 4.63, 1.35 and 0.27 μmole l^{-1} respectively, indicating the greater metal-buffering capacities of organic-rich inland waters, relative to marine waters. Additional information on the organic complexing characteristics of these waters may be gained by examining the data normalised in terms of their DOC concentrations (see column

TABLE II

A comparative summary of the apparent metal-complexing properties of natural waters and waters reconstituted with natural and synthetic organic ligands

Sample	Complexing capacity	
	$\mu M\ l^{-1}$	$\mu M\ (mg\ DOC-C)^{-1}$
Estuarine and marine waters [1]	0.27	0.093
Inland waters [1]	1.35	0.168
Sewage [1]	4.63	—
Marine humics [2]	0.28	0.208
Swampy river humics [3]	5.67	25.57
Glycine [4]	8.53	41.67
NTA [4]	3.77	13.89
EDTA [4]	2.81	8.33

[1] Calculations for mean complexing capacity $\overline{x}$ is weighted for the number n_i of samples analyzed by the i different authors whose individual mean complexing capacities is x_i (shown in Table I) using the equation: $\overline{X} = \Sigma\ n_i \overline{x}_i / \Sigma_i\ n_i$; Ramamoorthy and Kushner's (1975a,b) and Kunkel and Manahan's (1973) results have been omitted.
[2] Calculation based on humic cation-exchange capacity data obtained by Stuermer and Harvey (1974, 1977): fulvic acid (FA) concentration in Sargasso Sea = 0.134 mg l^{-1} (49.98% C); equivalent weight = 473 g; the FA carbon represents 5.1% of the DOC: it is assumed that recovery of FA is 100%, and that FA is the only metal-binding ligand. Lower recoveries and/or the presence of other non-humic ligands would tend to increase values shown.
[3] These high values represent organically rich rivers containing an average of 22.2 mg C l^{-1} (Beck et al., 1974).
[4] Calculations based on model system containing 2 mg l^{-1} of synthetic ligand characterized by a 1:1 metal-binding stoichiometry.

4, Table I). It is clear from Table II that the mean complexing capacities based on direct measurements of marine and fresh waters appear to represent only a small proportion ($\sim 20\%$) of the potential metal-complexing capacities expected from the known cation exchange properties of typical organic ligands found in marine and fresh waters, e.g. humic compounds (Beck et al., 1974; Stuermer and Harvey, 1974). One explanation is that complexation capacities carried out directly on water samples measure the concentration of complexing sites which are either unoccupied or occupied by metals which are easily displaced by copper (e.g. Ca^{2+}, Mg^{2+}, Ni^{2+}, Cd^{2+}; Reuter and Perdue, 1977). Organic ligands whose sites are occupied by thermodynamically more stable Fe^{3+} ions or kinetically inert Pb^{2+} ions (Whitfield and Turner, 1978) may not be displaced by copper ions (Chau et al., 1974; Reuter and Perdue, 1977) and will therefore not be measured by these techniques. Another explanation is that only a small proportion of the binding sites on the ligands possess favourable stereochemical conformations necessary to compete with inorganic ligands for the complexation of copper. This is supported by the works of Ernst et al. (1975), Mantoura and Riley (1975) and others who have demonstrated that Cu(II) is bound to fulvic acid through two different binding sites possessing quite different stability constants, and by the experimental results of Chau et al. (1974), which indicate that the DPASV technique only measures ligands forming complexes with copper with a log conditional constant greater than 10 (see also Davison, 1978). Similar considerations apply to the other methods for the measurement of complexation capacity such as the chromatographic technique of Stolzberg and Rosin (1977).

On further inspection of the papers listed in Table I, it appears that the chemical methods for the determination of complexing capacity produce higher results than the bioassay techniques, indicating that there are "complexed" forms of copper which are toxic to the bioassay organisms. However, preliminary attempts by Gächter et al. (1973) to correlate the DPASV method with an algal bioassay technique using a *Chlorella* sp. were not conclusive. Additional information on the molecular structures and sizes of the ligands contributing to the complexing capacities of natural water have been obtained by ultra-filtration through molecular membranes (Ramamoorthy and Kushner, 1975b; Smith, 1976; Gillespie and Vaccaro, 1978). Using the DPASV technique, Smith (1976) found that 90% of the copper-complexing capacity of estuarine waters (4—13‰) was in the <1000 molecular weight fraction, which is in accordance with Gillespie and Vaccaro's (1978) bioassay experiments on Saanich Inlet waters which showed that 92% of the copper-complexing capacity was in the <10,000 molecular weight ultrafiltrate. In contrast using ISP, Ramamoorthy and Kushner (1975a, b) reported a loss of 82.6% of the copper-binding capacity in the <45,000 molecular weight range.

Further progress in this expanding field of aquatic chemistry will be made along with recent advances in (a) the thermodynamics and kinetics of speciation reactions in sea water (Mantoura et al., 1978; Whitfield and Turner, 1978) and (b) in correlating the electrochemical with the biological availability and transport mechanisms of the various organic complexes and their synthetic analogues (O'Shea and Mancy, 1976; Raspor et al., 1977; Turner and Whitfield, 1979).

4. MOLECULAR POLYDISPERSITY AND STRUCTURE OF ORGANIC LIGANDS

4.1. Low molecular weight ligands

It is convenient to classify organic ligands according to their molecular weights. Many of the characterised organic compounds in sea water such as the "free" amino acids (Siegel and Degens, 1966; Bohling, 1972; Bada and Lee, 1977) and fatty acids (Williams, 1961; Zsolnay, 1977) are typical of the low molecular weight (LMW; <200 amu [1]) organic ligands which are potentially capable of binding Cu^{2+}, Ni^{2+}, Co^{2+}, Fe^{3+}, Zn^{2+} (Sillén and Martell, 1971). However, the concentration of these compounds is very low ($<10^{-8}$ M) since they tend to be rapidly removed from the productive zone by microheterotrophic activity (Hobbie et al., 1968) making them unlikely candidates for the chelation of transition metals in sea water (Duursma, 1970; Stumm and Morgan, 1970). Instead, these LMW ligands are either protonated (depending on their pK_a values and pH) or associated with the very abundant Ca^{2+} and Mg^{2+} ions in sea water (Dyrrsen and Wedborg, 1974). In situations where the concentration of LMW ligands exceeds 10^{-5} M (at the air—sea interface, on the outer surfaces of exuding phytoplankton cells and in dense blooms) then interactions between trace metals and LMW ligands are likely to be significant.

4.2. Macromolecular ligands

An overwhelming majority of investigations have implicated intermediate and high molecular weight (HMW; $5 \times 10^2 - 10^6$ amu) organic compounds, of uncertain structure, in organo-metallic interactions (Gjessing, 1965, 1970, 1971; Christman and Ghassemi, 1966; Barsdate and Matson, 1967; Matson et al., 1969; Rashid and King, 1969; Barsdate, 1970; Siegel, 1971; Chau, 1973; Andren and Harris, 1975; Ramamoorthy and Kushner, 1975a, b; Allen, 1976; Beneš et al., 1976; Sholkovitz, 1976; Smith, 1976; Giesy and Briese, 1977; Lock et al., 1977; Means et al., 1977; Reuter and Perdue, 1977; Wilson and Kinney, 1977; Burton, 1978; Gillespie and Vaccaro, 1978; Sugai and Healy, 1978). However, the molecular weight values reported by the above authors are only approximate since they have invariably been obtained using the techniques of gel chromatography, ultra-filtration or ultra-centrifugation, all of which separate polydisperse mixtures according to molecular *size* and not molecular weight. Furthermore, Gjessing (1970) and Wershaw and Pickney (1971) have demonstrated that the size of solvated macromolecular ligands such as humic acids is affected by pH, ionic strength and electrostatic effects (see Section 5).

The chemical composition and structure of the HMW ligands are inextricably linked with the ~90% of the DOM which, despite determined efforts, has remained uncharacterised. Although many of the structures put forward as a representation of the complex HMW organic matter in natural waters (see Fig. 1) should also be capable of binding metal ions, there is a pressing need for a better structural characterisation of these HMW ligands.

A variety of HMW compounds are also excreted by marine plants (Hellebust, 1965; Hoyt, 1970; Khailov and Finenko, 1970; Aaronson, 1971;

---

[1] amu = atomic mass units.

Kroes, 1972), some of which appear to be involved in the binding of metal ions in the detoxification of heavy metals (Steemann- Nielsen and Wium-Anderson, 1970; Barber, 1973) and the selective uptake of essential trace metals (Johnston, 1964; Murphy et al., 1976). The active or post-mortal release of free metallo-enzymes (Reichardt et al., 1967), cofactors (Carlucci, 1970) and nucleic acids (e.g. DNA, RNA; Pillai and Ganguly, 1970) does not represent a significant component in the organic speciation of trace metals in bulk sea water.

5. ELECTROCHEMICAL AND SURFACE CHEMICAL PROPERTIES

The few and scattered reports on the electrochemical properties of organic matter in natural waters suggest that the DOM appears to possess a net negative charge (Packham, 1964; Neihof and Loeb, 1974) as well as uncharged hydrophobic moieties (Schnitzer and Khan, 1972; Boehm and Quinn, 1973). The negative charges, which arise from the dissociation of acidic functional groups located on the macromolecular components of DOM (see Fig. 1) are stabilised partly by the formation of an electric double layer at the solution interface (Parks, 1975) and partly by the delocalisation of electrons within the conjugated structures which have been postulated for complex organic matter. In addition, the quinone and free radical sites present in humic compounds extracted from marine sediments (Rashid, 1972; Riffaldi and Schnitzer, 1972) will also contribute to the mobility of electrons within these metal-binding molecules. Visser (1964) and Szilágyi (1972, 1973) have estimated the standard oxidation—reduction potential of humic substances to be in the range 0.5—0.7 V. Humic compounds have been implicated in the reduction and kinetic stabilisation of a variety of metal ions including Hg^{2+} (Alberts et al., 1974; Strohal and Huljev, 1970; Lindberg and Harris, 1974), Fe^{3+} (Gjessing, 1964; Theis and Singer, 1973; Langford and Khan, 1975; Szilágyi, 1973; Rashid, 1972), Co^{3+} (MacCarthy and O'Cinneide, 1974; Hanck and Dillard, 1977), MoO_4^- (Szilágyi, 1967) and VO_2^{2+} (Szalay and Szilágyi, 1967). Further, humic acid can act as an extracellular substrate for the transport of electrons in biochemical redox reactions (Schindler et al., 1976). Johnston (1964) has argued that the biological availability of ferric ions to phytoplankton is, in part, regulated by the redox poise and kinetic lability of the Fe^{3+}-chelates.

Although Neihof and Loeb (1974), using microelectrophoretic measurements of DOM-coated particles in sea water, have demonstrated the anionic character of marine DOM, Marchand (1974) and Pillai et al. (1971) have observed that interactions with certain metals will result in the formation of neutral or cationic species, which can be separated by ion exchange chromatography. Mantoura (1976), Sugimura et al. (1978a, b) and Montgomery and Santiago (1978) have made use of the hydrophobic or amphipathic properties of these complexes by extracting them on the hydrophobic

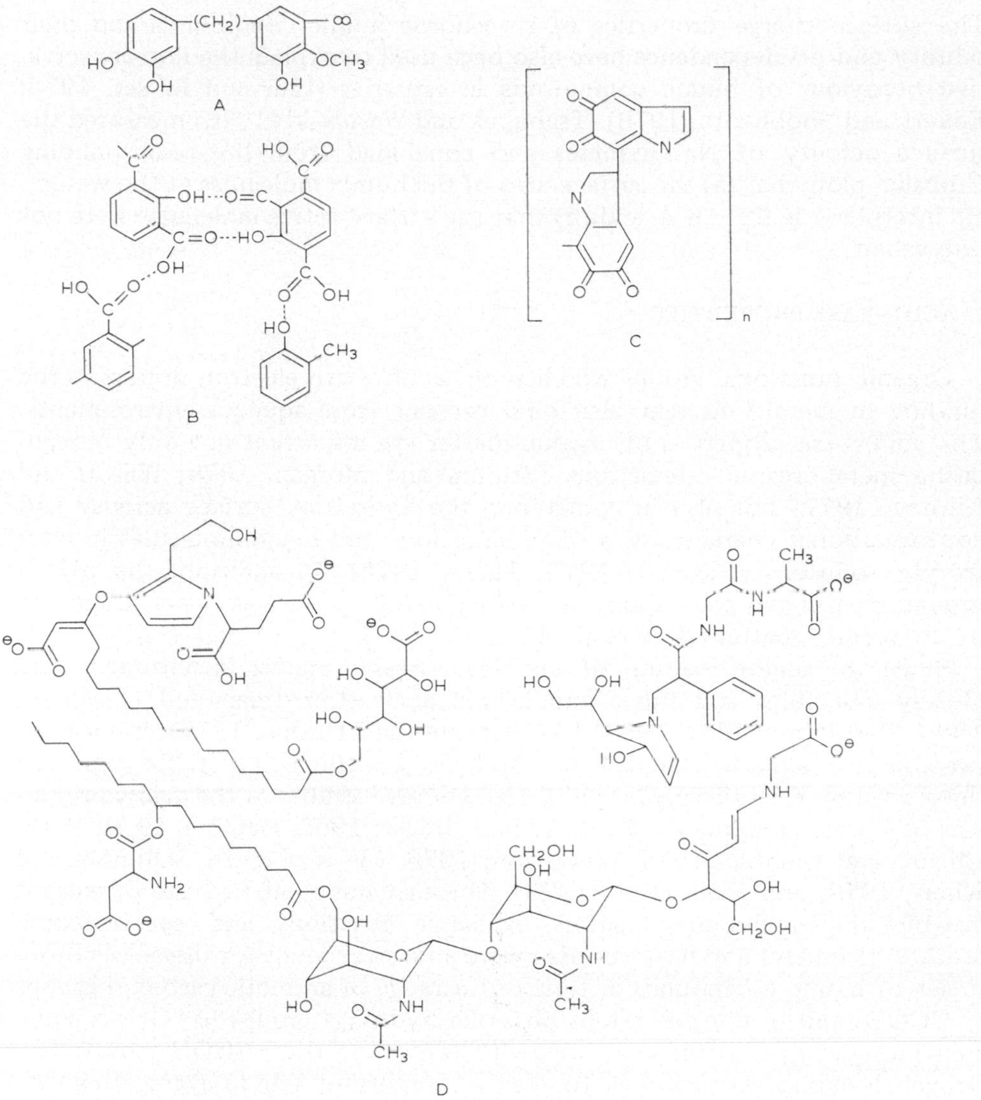

Fig. 1. Hypothetical structures of humic compounds from (A) lake water (Christman, 1970), (B) soil fulvic acids (Schnitzer and Khan, 1972), (C) melanin residue of marine humus (Duursma, 1965) and (D) sea-water humic substances (Gagosian and Stuermer, 1977).

Amberlite XAD-2 resins in order to determine "organically bound" iron, copper, zinc. Trace concentrations of surface active organic compounds dissolved in sea water have been determined by cathodic stripping voltammetry (Batley and Florence, 1976a) and through the suppression of the polarographic maxima for the reduction of oxygen (Nürnberg and Valenta, 1975).

The surface charge properties of river-borne humic compounds and their salinity and pH dependence have also been used to explain the non-conservative behaviour of humic compounds in estuaries (Hair and Basset, 1973; Eckert and Sholkovitz, 1976). Tschapek and Wasowski (1976) measured the surface activity of Na^+ humates and concluded from the corresponding Gibbsian plots that (a) the surface area of the humic molecules at the water air interphase is 62—66 Å and (b) that the surface active molecules were not polyvalent.

6. ACID—BASE PROPERTIES

Organic functional groups which serve as effective electron donors in the binding of metal ions may also bind protons from aqueous environments. The acid—base properties of organic matter are important not only in regulating metal-organic interactions (Stumm and Morgan, 1970; Wilson and Kinney, 1977) but also in controlling the hydration, surface activity and conformational characteristics of organic ions and macromolecules in electrolyte solutions (Eagland, 1975; Parks, 1975). In addition, the pH of organic-rich inland waters may also be regulated by the acid- base properties of the organic matter (Beck et al., 1974).

Since the major fraction of dissolved organic matter in natural waters closely resembles soil fulvic and humic acids (Christman and Ghassemi, 1966; Rashid and King, 1969, 1971; Reuter and Perdue, 1977), it has been possible for aquatic chemists to capitalise on the considerable expertise developed by the soil scientists in their extensive studies of the acidic properties of humic compounds (Pommer and Breger, 1960; Gamble, 1970, 1972; Cheam and Gamble, 1974; Stevenson, 1976; see reviews by Schnitzer and Khan, 1972, and Flaig et al., 1975). Through the combined use of elegant site-blocking techniques, cation exchange reactions and spectroscopic studies, Schnitzer and his associates were able to account for the acidic properties of humic compounds in terms of a range of aromatic carboxyl groups (—COOH) and to a lesser extent phenolic hydroxyl groups (ϕ—OH) [1], some of which occupied *ortho* positions with respect to the —COOH groups. Hydrogen bonding was assumed to play an important role in aggregating oxidised lignin residues of soil humics into polyelectrolyte structures. As far as the acidic properties are concerned, the other structural models which have been proposed for humic and fulvic acids [e.g. polyphenolic (Christman, 1970), melanin (Duursma, 1965) and heteropolycondensate (Degens, 1970) models; see Fig. 1] are functionally analogous since they all invoke a peripheral distribution of acidic groups on the humic macromolecule. The approximately three-fold variation in total acidity (TA) which is encountered in soil

---

[1] ϕ = phenyl ring

fulvic and humic acid (see Table III) is a reflection of the molecular polydispersity of these acidic polymers. Huizenga (1977) estimated the TA of marine DOM, isolated from sea water by adsorption on charcoal columns (yield 30—70%; Kerr and Quinn, 1975) to be 11.4 ± 2.5 mM/g C. The lower value of 4.23 mM/g C reported by Stuermer and Harvey (1974) is probably less representative since it corresponds to ~5% of the DOM as isolated on XAD-2 columns. River water humics are generally enriched in total acidity (TA ~ 25 mM/g C, see Table III) relative to their less soluble, more polymerised humic progenitors found in the soils of the catchment (TA = 8—21 mM/g C). In contrast to terrestrial humic compounds, marine humic compounds appear to contain only minor proportions of phenolic hydroxyl groups — which is in agreement with the non-aromatic (lignin-free) structure of sea-water humic substances as proposed by Gagosian and Stuermer (1977).

Potentiometric titration curves of humic compounds against strong base are usually broad and ill-defined thus reflecting the diversity in acidic groups. The apparent acidity of these groups can be expressed as:

$$pK_a = -\log K_a$$

where K_a is the acid dissociation constant, and is related to their chemical nature, position and electrostatic interactions. Pommer and Breger (1960) were able to resolve a single titrimetric equivalence point centred at pH 6.8 by employing Gran functions (Gran, 1952).

The potentiometric titration of simple acids characterised by an acid dissociation constant K_a is described by the Henderson—Hasselbalch equation (Stumm and Morgan, 1970):

$$pK_a = pH - \log \frac{\alpha}{1-\alpha} \qquad (1)$$

where $pK_a = -\log K_a$ and α is the degree of dissociation ($0 < \alpha \leqslant 1.0$). However, because of the polyprotic nature and diversity of the acid groups on the humic macromolecule, Pommer and Breger (1960) and Gamble (1970) used the generalised Katchalsky—Spitnik form of eq. 1:

$$pK_a = pH - m \log \frac{\alpha}{1-\alpha} \qquad (2)$$

where m (termed the titration exponent, $m > 1$) is introduced to correct for changes in pK_a resulting from the accumulation of negative charge on the humic polyanion during the course of the titration. Pommer and Breger (1960) employed Gran functions to resolve the equivalence point and estimated values of pK_a = 6.8—7.0 and m ~ 4.8. Using a more sophisticated polyelectrolyte model for soil fulvic acid which took into account second-order effects due to (1) chemical differences within the carboxyl and phenolic hydroxyl groupings, (2) site—site interactions, and (3) buffering and dissociation of titration medium, Gamble (1972) was able to resolve at least two equivalence points centred at pK_a = 6.54 and 7.85 (0.1 M KCl) and pK_a = 4.99 and 7.63 (0.01 M KCl) and tentatively assigned them to aromatic –COOH and ϕ–OH groups. Using a modification of Bjerrum's approach, Stevenson (1976) showed that the pK_a of –COOH groups was profoundly affected by changes in ionic strength and composition of the titrating medium, while the titration constant m remained essentially constant.

Huizenga (1977) investigated the acid–base properties of twelve marine DOM samples and one combined river water sample and used the Katchalsky–Spitnik model (eq. 2) to numerically estimate pK_a, m and total number of Type 1 sites (–COOH groups), in the range pH 3–8. Although the overall

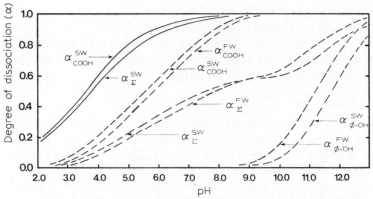

Fig. 2. The acid dissociation characteristics of dissolved organic matter, expressed in terms of the degree of dissociation (α) corresponding to the carboxyl groups (α_{COOH}), phenolic hydroxyl groups ($\alpha_{\phi-OH}$) and their normalized sum (α_Σ) where:

$$\alpha_\Sigma = \alpha_{COOH}\left(\frac{\beta}{1+\beta}\right) + \alpha_{\phi-OH}\left(\frac{1}{1+\beta}\right)$$

where β is the ratio of the total number of carboxyl groups to the number of phenolic hydroxyl groups per molecule. Calculations for fresh water (α^{FW}) and sea water (α^{SW}) organic matter are based on Huizenga's (1977) data for Sargasso Sea organics (———) using the equation:

$$pH = pK_a^{COOH} + m \log \frac{\alpha_{COOH}}{1 - \alpha_{COOH}}$$

as well as the data of Wilson and Kinney (1977) (· · · · · ·) for lake-water and sea-water (Gulf of Alaska) samples, using the equations:

$$pH = pK_i^{COOH} + \log \frac{\alpha_{COOH}}{1 - \alpha_{COOH}} + 3.30\,\alpha_{COOH}$$

$$pH = pK_i^{\phi-OH} + \log \frac{\alpha_{\phi-OH}}{1 - \alpha_{\phi-OH}} + 1.27\,(\alpha_{\phi-OH} + \beta)$$

The α coefficients 3.30 and 1.27 were determined graphically from Wilson and Kinney's (1977) data. β, ω and n values [see Wilson and Kinney (1977) for explanation of symbols] for the Gulf of Alaska sea-water sample were not provided, and these were assumed to be equal to the lake-water data provided.

TABLE III

Summary of the acid—base properties of marine, fresh-water and soil humic compounds

Sample location	Distribution of acidic functional groups (mM/g C)[1]			pKa[2]		m[3]	Method	Reference
	TA	—COOH	φ—OH	Type 1	Type 2			
Saragasso Sea	11.52	10.32	0.8	3.57	6.5	2.36	PT[4] of charcoal extracts	Huizenga (1977)
Equatorial Atlantic	9.77	9.77	nd[5]	3.73	nd	1.82		Huizenga (1977)
Pacific upwelling, Peru	9.44	8.94	0.5	3.67	6.2	1.98		
Gulf of Alaska	—[6]	—	—	3.9	9.3	—	PT of dialysis concentrates	Wilson and Kinney (1977)
Saragasso Sea	4.23	—	0.92	—	—	—	PT of XAD-2 extracts	Stuermer and Harvey (1974)
Marine sedimentary HA[7,8]	5.01	4.09	nd	—	—	—	Ba—Ca back titration	Rashid and King (1971)
Massachussetts rivers	10.43	10.43	nd	3.67	nd	1.98	PT of charcoal extracts	Huizenga (1977)
Georgia Rivers, U.S.A.[8]	25.57	18.35	7.22	4.23	8.71	—	Ba—Ca back titration	Beck et al. (1974)
Lake water	—	5.70	—	—	—	—	PT of dialysis concentrates	Wilson and Kinney (1977)
Soil FA[7]	21.63	15.15	6.48	—	—	—	Ba—Ca back titration	Schnitzer and Khan (1972)
Soil HA	13.86	6.49	7.37	—	—	—	PT of soil extracts	Gamble (1970)
Soil FA	15.44	6.30	7.14	—	—	—		
Soil HA	8.14	5.91	2.22	4.82	—	1.86	PT of chemical derivatisation	Stevenson (1976)

[1] TA = total acidity; —COOH = carboxyl groups; φ—OH = phenolic hydroxyl groups.

[2] pKa = log acid dissocation constant (Ka). Type 1 and 2 constants usually refer to the carboxyl and phenolic hydroxyl groups, but refer to authors for exact details.

[3] m is the empirical titration exponent in the modified Henderson—Hasselbach eq. 2 in Section 6.

[4] Potentiometric titration.

[5] Not detected.

[6] Not determined.

[7] Humic acid, fulvic acid.

[8] Data shown is the mean of values reported in this review.

variation was small (pK$_a$ = 3.5 ± 0.2; m = 2.1 ± 0.3; n_{COOH} = 11.4 ± 2.5 mM/ mg C), the pK$_a$ and m values in all the sub-fractions of each sample were found to be negatively correlated. This was ascribed to chemical differences in the organic matter between fractions within a sample and between fractions of different samples. In addition, the presence of highly acidic sites (low pK$_a$) was correlated with elevated number of Type 1 (—COOH) sites. In only two sub-fractions was there any evidence of a second site (Type 2) observed at pH 6.5. As a typical example, the acid—base proporties of DOM from a Saragasso Sea sample (see Table III for relevant parameters) are plotted in Fig. 2. This shows that at pH 8.0, more than 97% of the —COOH sites are dissociated, and are potentially available for interaction with metal ions. Wilson and Kinney (1977) applied Tanford's (1967) biopolymeric dissociation model in a recent investigation of lake water and marine humic compounds. The model assumes that variations in pK$_a$ for a particular acid group type can be directly related to the electrostatic interactions between protons and the macromolecule as a whole. Two intrinsic constants pK$_i^{COOH}$ and pK$_i^{\phi-OH}$ (which correspond to the acid dissociation constants for —COOH and ϕ—OH groups when there is no charge on the organic matter) were derived from the equations:

$$pK_i^{COOH} = pH - \log \frac{\alpha_{COOH}}{1 - \alpha_{COOH}} + \omega_{COOH} n_{COOH} \alpha_{COOH}$$

$$pK_i^{\phi-OH} = pH - \log \frac{\alpha_{\phi-OH}}{1 - \alpha_{\phi-OH}} + \omega_{\phi-OH} n_{\phi-OH} (\alpha_{\phi-OH} + \beta)$$

where ω is the electrostatic interaction factor, n is the number of dissociating sites per molecule, and β is the ratio n_{COOH} to $n_{\phi-OH}$. The constants obtained in this manner are shown in Table III, from which the degree of dissociation of like groups as a function of pH have been calculated (See Fig. 2). These data indicate that whereas greater than 90% of the COOH sites are dissociated at pH 8 of sea water, virtually all of the phenolic hydroxyl sites remain protonated. Both Huizenga (1977) and Wilson and Kinney (1977) show that only minor proportions (8—38%) of the acidic sites on DOM appear to be available for metal ion binding, indicating that the majority of the acidic sites are not in close proximity and therefore do not possess the polydentate characteristics necessary for chelation of metal ions. The acid—base properties of other functional groups in DOM [e.g. R—NH$_2$, heterocyclic N, R—COOH, —C(OH)—C(OH)], which may also be found in DOM are not known, but may be important in enhancing metal chelation.

7. INTERACTIONS WITH VARIOUS METALS

7.1. Alkali and alkali earth metals

Specific interactions between the alkali (Na$^+$, K$^+$, Cs$^+$) and alkali earth (Mg^{2+}, Ca^{2+}, Sr^{2+}, Ba^{2+}) metal ions and organic ligands in natural waters, have, in general, escaped the attention of aquatic chemists (see reviews by Brewer, 1975 and Wilson, 1975). This is because such interactions are weak, electrovalent and reversible in nature, and because the speciation of the abundant (0.01—0.47 M) major cations (Na^+, K^+, Mg^{2+}, Ca^{2+}) in sea water is unlikely to be affected by specific interactions with the far less abundant concentrations ($<10^{-5}$ M) of organic ligands. On the other hand, the complexing ability of organic ligands for trace metals (transition and Groups IB and IIB) is regulated by the competing effects of the weakly interacting, but very abundant, major cations (see also Section 8). Gamble (1973) using ion specific electrodes, has demonstrated that, in the absence of other metals, soil fulvic acid binds both Na^+ and K^+. In a detailed electrochemical investigation of the chelation equilibria of nitrolotriacetic acid (NTA) and Cd^{2+} in a sea-water medium, Raspor et al. (1977) found it necessary to include specific interaction terms for Na^+—NTA in order to account for the observed speciation of Cd^{2+}. Koshy et al. (1969) could not observe any binding of $^{134}Cs^+$ tracer by humic acids extracted from marine sediments, and attributed this to the presence of high concentrations of competing ions in sea water. Apart from the somewhat dated but unusual observation by Bannister and Hey (1936) of crystalline Ca-citrate and oxalate in the surface sediments of the Weddel Sea, and the occurrence of Ca-Mg salts of long-chain fatty acids in Black Sea mud (Shabarova, 1954), there do not appear to have been any *in situ* measurements of Ca^{2+} and Mg^{2+} binding by natural organics. Giesy and Briese (1977) using ultra-filtration techniques on swampy river waters from the southeast United States, concluded that Mg^{2+} and Ca^{2+} exist either as free cations or bound to small molecular weight (<500) organic compounds. Landymore and Antia (1978) have presented evidence that Ca^{2+} and Mg^{2+} chelation catalyse the photodecomposition of dihydroxylated pteridines in sea water. Experimental investigations of organic chelation of ^{60}Co and ^{65}Zn by leucine, quinoline-2-carboxylic acid and EDTA (Duursma, 1970) and of Cd^{2+} by NTA (Raspor et al., 1977) were found to agree with theoretical predictions based on stability constants provided chelation of Ca^{2+} and Mg^{2+} were also taken into account. Conditional stability constants for Mg^{2+} and Ca^{2+} complexes with soil fulvic acid (Schnitzer and Hansen, 1970) and with humic compounds extracted from a variety of natural waters (Mantoura et al., 1978) are generally low (log K $<$ 3.5) and the complexes are kinetically labile (Mantoura, 1976). The affinity between alkali earth cations and nucleotides (Bowen, 1966) makes it likely that the trace concentration of *dissolved* ATP in the deep sea is chelated with Mg^{2+} ions. The possibility of organic binding of Sr^{2+} in sea water (Angino et al., 1966) has been recently discounted as an analytical artifact (Carpenter, 1972). Using ^{133}Ba radiotracer and ion-exchange chromatography, Desai et al. (1969) have concluded that solubilisation interactions between Ba^{2+} ions and DOM containing 0.3 mg l^{-1} carbohydrates and an unspecified amount of various amino acids, are the cause for barium super-saturation in sea water. Although Mg^{2+}

containing chlorophyll *a* is thought to rapidly dechelate when the pigment is released upon the death of photosynthetic cells (Baker and Smith, 1977; Pocklington, 1977) many metal-containing enzymes do not dechelate since they retain their activity in the extracellular environment of sea water (e.g. Co^{2+}-vitamin B_{12}, Carlucci (1970); phosphatases, Reichardt et al. (1967)).

Finally, it should also be pointed out that the alkali and alkali earth metal ions also participate in important non-specific interactions. These include: (1) acting as counter ions in the electrical double layer stabilisation of macromolecular and colloidal organo-metallic complexes (Shapiro, 1964; Parks, 1975; Eckert and Sholkovitz, 1976) and (2) in regulating the electrolyte properties and ionic strength (I) of natural waters (Whitfield, 1975) which, in turn, will affect the activity coefficients and conditional stability constants (K_c) of organo-metallic complexes. For example, log K_c values for Cu^{2+}-, Pb^{2+}- and Cd^{2+}-humate complexes were found to decrease linearly with $\sqrt{I}$ (Stevenson, 1976) or simply with I (Schnitzer and Khan, 1972).

7.2. Vanadium, chromium and manganese

The association of vanadium and chromium with sedimentary organic matter has been used to explain the enrichment of these elements in shales (Baker and Smith, 1977). It is not certain whether the selective uptake of vanadium by ascidians and tunicates involves organic intermediates. The anionic nature of the principal soluble species of V and Cr in sea water ($H_2VO_4^-$; HVO_4^{2-}; $Cr(OH)_3^0$; CrO_4^{2-}; Brewer, 1975) does not favour interactions with organic ligands, which are negatively charged (see Section 5) at the pH of sea water. In contrast, Sugimura et al. (1978b) have presented data showing that more than 80% of V dissolved in sea water is recoverable on the hydrophobic macroreticular resin, Amberlite XAD-2, and have assigned this fraction as "organically bound vanadium".

The chemistry of manganese in natural waters is complicated (Stumm and Morgan, 1970; Brewer, 1975) and in general is dominated by redox processes. Although organo-manganese complexes in sea water had been postulated (Rona et al., 1962; Slowey and Hood, 1971) no such interactions have been detected in either sea water (Stumm and Morgan, 1970; Marchand, 1974) or organically rich sedimentary pore waters (Elderfield and Hepworth, 1975; Nissenbaum and Swaine, 1976; Krom and Sholkovitz, 1978). The conditional stability constants of Mn^{2+} complexes with a variety of soil and natural water organics are only marginally higher than the Mg^{2+} and Ca^{2+} complexes (log K = 2.1—4.4; Table IV) and considerably lower than other transition metals. The explanation lies in the electronic characteristics of Mn^{2+} bonding. Gamble et al. (1977) used electron spin resonance spectrometry to show that Mn^{2+} is bound by weak electrostatic and hydrogen bonding as $Mn(H_2O)_6^{2+}$ to donor groups on fulvic acid. The splitting of the $3d^5$ high spin electron configuration of Mn^{2+} by interacting ligands does *not* result in any

TABLE IV

Conditional stability constants for organo-metallic complexes derived from different environments

tal	Organic ligand	Log K_c and experimental conditions [2]	Reference
2+	soil FA [1]	1.9 (3.0, 0.1, 1.0); 2.15 (5.0, 0.1, 1.0)	Schnitzer and Hansen (1970)
	lake water HC	3.67 (8.0, 0.02, —)	} Mantoura et al. (1978)
	sea water	3.73 (8.0, 0.02, —)	
2+	soil FA	2.65 (3.0, 0.1, 0.8); 3.35 (5.0, 0.1, 0.9)	Schnitzer and Hansen (1970)
	lake water HC	3.83 (8.0, 0.02, —)	} Mantoura et al. (1978)
	sea water HC	3.86 (8.0, 0.02, —)	
2+	soil FA	2.3 (3.0, 0.1, 1.0); 3.7 (5.0, 0.1, 1.0)	Schnitzer and Hansen (1970)
		3.3 (6.35, 0.01, —) [3]	Gamble et al. (1977)
	lake water HC	2.5 (7.5, 0.1, —) [4]	Wilson and Kinney (1977)
		4.6 (8.0, 0.02, —)	} Mantoura et al. (1978)
	sea water HC	4.5 (8.0, 0.02, —)	
3+	soil FA	5.6 (6.0, 1.0, —)	Malcolm et al. (1970)
		5.1 (3.5, 0.1, —); 5.8 (5.0, 0.1, —)	Schnitzer (1969)
2+	soil FA	7.6 (1.7, 0.0, —); 6.1 (1.7, 0.1, —)	Schnitzer and Hansen (1970)
2+	soil FA	6.6 (6.0, 1.0, —)	Malcolm et al. (1970)
		2.9 (3.0, 0.1, 1.0); 4.2 (5.0, 0.1, 1.0)	Schnitzer and Hansen (1970)
	lake water HC	4.8 (8.0, 0.02, —)	} Mantoura et al. (1978)
		4.8 (8.0, 0.02, —)	
2+	soil FA	3.2 (3.0, 0.1, 1.0); 4.2 (5.0, 0.1, 1.0)	Schnitzer and Hansen (1970)
	lake water HC	5.2 (8.0, 0.02, —)	} Mantoura et al. (1978)
	sea water HC	5.5 (8.0, 0.02, —)	
2+	soil FA	3.3 (3.0, 0.1, 1.0); 4.0 (5.0, 0.1, 1.0)	Schnitzer and Hansen (1970)
		5.8 (3.5, 0.1, —); 8.7 (5.0, 0.1, —)	Schnitzer (1969)
		3.2 (3.0, —, —); 4.4 (5.0, —, —)	Cheam and Gamble (1974)
	soil HA	16.8 (6.8, —, 2.98); 10.2 (6.8, —, 1.14) [5]	Ernst et al. (1975)
		6.2 (6.8, —, —); 5.08 (6.8, —, —) [5]	Guy and Chakrabarti (1976)
	peat HA	8.9 (4.9, 0.0, —) [6]	Stevenson (1976)
	fresh-water HC	5.0 (6.0, 0.1, 1.0); 4.8 (6.0, 0.1, 1.0)	Buffle et al. (1977)
	Smith lake HC	3.68 (3.0, 0.1, —) [4]	Wilson and Kinney (1977)
	Ottawa River HC	3.70 (7.3, —, —,)	Ramamoorthy and Kushner (1975a,b)
	lake HC	9.23 (8.0, 0.02, —)	} Mantoura et al. (1978)
	sea water HC	9.30 (8.0, 0.02, —)	
2+	soil FA	2.3 (3.0, 0.1, 1.0); 3.7 (5.0, 0.1, 1.0)	Schnitzer and Hansen (1970)
	soil HA	5.0 (6.8, 0.01, 4.0)	Guy and Chakrabarti (1976)
	lake HA	5.2 (8.0, 0.02, —)	} Mantoura et al. (1978)
	sea water HC	5.3 (8.0, 0.02, —)	
2+	soil FA	3.0 (5.0, 0.1, —); 3.6 (6.0, 0.1, —)	Cheam and Gamble (1974)
		6.9 (4.9, 0.0, —) [6]	Stevenson (1976)
		5.0 (6.8, 0.01, 3.0)	Guy and Chakrabarti (1976)
	Ottawa River HC	3.7 (7.3, —, —)	Ramamoorthy and Kushner (1975a,b)
	lake HC	4.6 (8.0, 0.02, —)	} Mantoura et al. (1978)
	sea water HC	4.7 (8.0, 0.02, —)	

TABLE IV (continued)

Metal	Organic ligand	Log K_c and experimental conditions [2]	Reference
Hg^{2+}	soil FA	5.0 (3.0, 0.1, −); 5.0 (4.0, 0.1, −)	Cheam and Gamble (1974
	Ottawa River HC	6.1 (7.3, −, −)	Ramamoorthy and Kushne
	lake HC	19.7 (8.0, 0.02, −)	(1975a,
	sea water HC	18.1 (8.0, 0.02, −)	} Mantoura et al. (1978)
Pb^{2+}	soil FA	2.6 (3.0, 0.1, 1.0); 4.1 (5.0, 0.1, 1.0)	Schnitzer and Hansen (197
	soil HA	8.7 (4.9, 0.0, −) [6]	Stevenson (1976)
		7.8 (6.8, −, 1.1) [5]; 14.8 (6.8, −, 2.5) [5]	Ernst et al. (1975)
		6.5 (6.8, 0.01, 2) [5]; 5.3 (6.8, 0.01, 6) [5]	Guy and Chakrabarti (197
	fresh water HC	6.3 (6.7, 0.1, 1.0)	Buffle et al. (1977)
	Ottawa River HC	4.0 (7.3, −, −)	Ramamoorthy and Kushne
		24 (-, −, −)	Barsdate and (1975a,
			Matson (1967)

[1] The following abbreviations are used: FA = fulvic acid; HA = humic acid; HC = humic compounds, i.e. undifferentiated FA and HA.

[2] Sequence of stability data is presented as: logarithm of conditional stability constant, log K_c (pH, ionic strength, metal ligand binding stoichiometry). Where several values of log K_c are reported, only the mean value is included.

[3] Log K_c value derived from mean value of $\overline{K}_c$ (differential equilibrium function) and $[H^+] = 10^{-6.35}$ where log $K_c = \overline{K}_c [H^+]$. See Gamble et al. (1977).

[4] Log K_c value refers to the *intrinsic* stability constant. See Wilson and Kinney (1977) for details.

[5] The two values of log K_c refer to two types of binding between the metal and organic ligand.

[6] Log K_c as a function of ionic strength has been derived.

crystal field stabilisation energy (Mackay and Mackay, 1969; Stumm and Morgan, 1970). Furthermore, competition by the abundant Mg^{2+} and Ca^{2+} ions in sea water is likely to displace any organically bound Mn^{2+}, which in turn will either form inorganic Mn(II) species (e.g. $MnCO_3$, $Mn(OH)_2$, etc.) or more likely will oxidise to kinetically inert and insoluble Mn(IV) species (Stumm and Brauner, 1975).

7.3. Iron

The association of iron with organic matter in natural waters has significant geochemical and biological consequences, and as a result has been widely investigated. The reviews by Kester et al. (1975a, b) and Millero (1975) on the physical chemistry, redox reactions and inorganic complexes of Fe in marine systems are useful background to this section. The ideas of organic speciation and colloidal forms of Fe in sea water were introduced as far back as 1928 (Harvey, 1928; Gran, 1931, 1933; Cooper, 1935, 1937,

1948a) in order to explain the elevated concentration of "dissolved" Fe (10^{-7} M) in relation to the expected solubility of Fe^{3+} ion ($\sim 10^{-19}$ M) in sea water. Since then, an impressive number of investigations have demonstrated the existence of organo-Fe interactions not only in sea water (Koshy et al., 1969; Marchand, 1974; Sugimura et al., 1978a) but also in marine sediments (Rashid, 1971, 1972; Rashid and Leonard, 1973; Nissenbaum and Swaine, 1976; Picard and Felbeck, 1976), interstitial waters (Nissenbaum et al., 1971; Elderfield and Hepworth, 1975; Krom and Sholkovitz, 1978), estuarine waters (Sholkovitz, 1976; Boyle et al., 1977; Sugimura et al., 1978a, b) and fresh waters (Gjessing, 1964; Shapiro, 1964; Beck et al., 1974; Perdue et al., 1976; Giesy and Briese, 1977; Means et al., 1977). However, there is a wide range of mechanisms which could account for those interactions and these would include complexation, adsorption and inclusion reactions.

Complexation reactions between Fe^{2+} and Fe^{3+} ions and humic ligands have been extensively studied by Schnitzer and his associates (Schnitzer and Khan, 1972). Electron spin resonance and Mössbauer spectroscopy were combined with chemical treatments of soil humic compounds to obtain information on oxidation states and site symmetries of Fe^{3+} binding (Senesi et al., 1977). Two (or possibly three) different Fe^{3+}-binding sites were found and these displayed different stabilities and coordination for Fe^{3+}. The conditional stability constants of soluble complexes formed between Fe^{2+} and Fe^{3+} ions and fulvic acid (see Table IV) were determined using ion exchange and continuous variation techniques (Schnitzer and Hansen, 1970). It was necessary to determine these values in acidic media in order to avoid the formation of polynuclear complexes of Fe^{3+}. Since the ionisation of metal-binding functional groups on fulvic and humic acids is likely to be partially suppressed at these acidic conditions (see Section 6) the effective Fe^{3+} stability constants at the pH of natural waters are likely to be considerably higher (log $K_{Fe^{3+}} \sim 12$–14) than those shown in Table IV. In addition Fe^{2+}-humic complexes do not follow the Irving-Williams stability series (Schnitzer and Khan, 1972) but exhibit higher stability constants than predicted from the behaviour of other transition metals. In studies of the inorganic speciation of Fe, Kester et al. (1975a) concluded that the log K for Fe^{3+}—organic complexes must be at least 10^{18} times greater than the stability of the corresponding Mg^{2+} complexes, in order that Fe^{3+}—organic complexes be significant in sea water.

Perdue et al. (1976) observed an excellent linear correlation between the combined concentrations of Fe and Al and DOM in the Satilla River system (southeast U.S.A.). Such a correlation is likely to result from the formation of dissolved organic complexes rather than the adsorption of DOM onto colloidal iron hydroxide particles. Furthermore, using vapour pressure osmometry, Perdue et al. (1976) determined the number-average molecular weight of the DOM to be 1296, thus enabling them to derive an (Fe + Al): DOM stoichiometric binding ratio of 0.8–1.1, in agreement with the values for soil fulvic acid (Schnitzer and Khan, 1972). In contrast, Shapiro (1964), Boyle et al. (1977) and Burton (1977) have proposed a colloidal peptization mechanism involving hydrated ferric oxide particles coated with negatively changed humic molecules in order to explain the coagulative and macromolecular properties of "organic iron" in fresh water and estuaries. Although organic substances can aid markedly in the formation of stable colloidal dispersions of $Fe(OH)_3$ or $Fe-OOH$ (Stumm and Morgan, 1970), Murray and Gill (1978) found no evidence to associate either the precipitation of river-borne iron in Puget Sound or the oxidation kinetics of Fe^{2+} in the underlying pore water with organic interactions. The ultraviolet irradiation of sea water, which has been successfully used for the determination of "organic copper" (Foster and Morris, 1971) gave no consistent pattern for iron in the Menai Straits (Morris, 1974). This probably arises because the scavenging action of colloidal particles removes any photolytically released Fe^{3+} ions.

Fresh-water DOM which has been size fractionated using either ultrafiltration (Allen, 1976; Giesy and Briese, 1977) or gel chromatography (Ghassemi and Christman, 1968; Means et al., 1977) appears to possess different Fe-complexing properties. Theis and Singer (1973), using kinetic and spectrophotometric techniques, have shown that humic compounds can stabilise ferrous ion through the formation of kinetically inert Fe^{2+} organic complexes. Under certain conditions, the individual rate constants for reactions in a multicomponent system consisting of Fe^{2+}, Fe^{3+}, Fe-bound by fulvic acid, and colloidal hydrated iron oxide may be resolved using stop-flow kinetic analysis (Langford et al., 1977). The tendency of Fe to aggregate with organic detritus in sea and lake waters has also been examined by Akiyama (1971) using histological and stereological methods. Sugimura et al. (1978a) were able to recover 80—90% of the dissolved iron from sea water on a hydrophobic resin (Amberlite XAD-2), and have described this fraction as "dissolved organic iron". However, in similar experiments conducted on lake water, Mantoura (1976) has shown that colloidal Fe particles may be trapped within the macroreticular structure of this resin, and has cautioned against its use for differentiating organic from inorganic forms of metals in natural waters.

The stimulatory effects of synthetic chelators or humic acids combined with Fe on the growth of phytoplankton in organic-free sea water or synthetic media (Johnston, 1964; Prakash and Rashid, 1968; Barber and Ryther, 1969; Davies, 1970) have led to the hypothesis that sea water contains organic chelators of high specificity for Fe and that these are functionally analogous to EDTA. However, the manner in which these Fe-chelates act is uncertain (Giesy, 1976). The addition of biosynthesised Fe complexes such as ferrisaccharides, oxyhaemoglobin (Gran, 1931) and deferriferrioxamine B (Barber et al., 1971) does not appear to stimulate the growth of phytoplankton. Jackson and Morgan (1978) developed numerical models to examine three mechanisms claimed to enhance the supply of iron via chela-

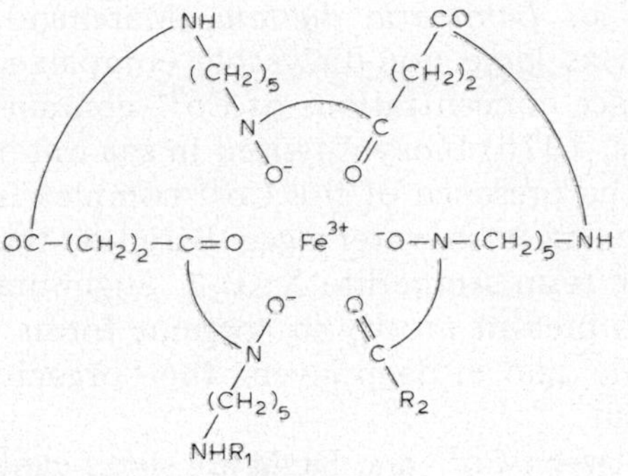

Ferrioxamine B ($R_1 = H$, $R_2 = CH_3$)
Ferrioxamine D_1 ($R_1 = CH_2CO$, $R_2 = CH_3$)
Ferrioxamine G ($R_1 = H$, $R_2 = CH_2CH_2COOH$)

Fig. 3. Structures of some trihydroxamate siderochromes which are thought to play a primary role in the extra-cellular transport of iron to biological cells. (Prelog, 1964; Neilands, 1973).

tion: these included transport through the cell membrane, ligand exchange at the cell surface, and increased supply of iron to the cell surface by dissociation of a chelate. None of these mechanisms could account for the observed effects of chelator variations on growth rate. However, in a series of bioassay experiments employing ^{55}Fe radio tracers, Murphy et al. (1976) have shown that iron deprivation of the blue-green alga *Anabaena flos-aqua* in Lake Ontario leads to the production of Fe-specific hydroxamate chelators having a molecular weight of less than 1000. This complex is a member of a recently recognised group of biosynthesised Fe-chelating ligands called siderochromes or exochelins which act as carrier molecules for the transportation of Fe across membranes (see Fig. 3) (Neilands, 1973; Macham and Ratledge, 1975). The stability constants of some Fe^{3+} trihydroxamate siderochromes may be as high as 10^{32} (Neilands, 1973). The eventual molecular characterisation of these chelates would represent the first organic ligand in natural waters to be identified.

7.4. Cobalt and nickel

The formation of "reversible" organic complexes of cobalt in sea water is unlikely because of the unfavourable mass action effects of: (a) the low concentration (8×10^{-10} M) of dissolved Co in the ocean (Brewer, 1975) and (b) the moderate stability constants of Co^{2+} chelates relative to Mg^{2+} and Ca^{2+} analogues (see Table IV; Sillén and Martell, 1971; Mantoura et al., 1978). Radiochemical experiments employing γ-emitting ^{60}Co tracer show that Co^{2+} in sea water is not chelated by additions of either 1 μg l^{-1} leucine (Duursma,

1970) or the decomposition products of *Laminaria digitata* (Marchand, 1974), and that very little was present as large non-dialysable complexes (Barsdate, 1970). Although the ultra-trace concentrations of Co^{2+} containing vitamin B_{12} (0.5—2.0 ng l^{-1}; Williams, 1975) biosynthesised in sea water represent less than 0.2% of the cobalt, the presence of this Co^{2+} complex is essential for the growth of many marine and fresh-water algae. Based on the extractability of Co on the hydrophobic resin Amberlite XAD-2, Sugimura et al. (1978b) have concluded that Co is present mostly in "organic forms" in the surface waters of the North Pacific, and in deep layers, the "organic forms" decrease to 30—40% of total cobalt.

There is evidence that interactions between Co^{2+} and DOM are significant in humic-rich environments such as pore waters (Elderfield and Hepworth, 1975; Nissenbaum and Swaine, 1976) and fresh waters (Barsdate and Matson, 1967; Means et al., 1977). Mantoura (1976) used a multi-metal, multi-ligand equilibrium speciation model and calculated that <10% of Co^{2+} is bound to lake-water fulvic acid. The kinetically non-labile nature of Co^{3+} complexes with humic compounds (MacCarthy and O'Cinneide, 1974) has been used as a basis for a polarographic determination of metal complexation capacity of natural waters (Hanck and Dillard, 1977).

Similar considerations apply to nickel, which, although it is more abundant (3×10^{-8} M; Brewer, 1975) in sea water, tends to form organic complexes that are only marginally more stable than those of cobalt (see Table IV; Mantoura et al., 1978). Morris (1974) determined the seasonal variation in the concentrations of ionic and organically associated nickel in the Menai Straits, using an ultraviolet photolytic procedure. The organic nickel fraction constituted 0—30% (0—0.45 μg Ni l^{-1}) of the total Ni, with the higher concentrations occurring throughout spring and summer blooms of *Phaeocystis*. Using different techniques, similar results were obtained by Sugimura et al. (1978b) for the western North Pacific Ocean. Like vanadium, nickel is preferentially incorporated into organic matter in sediments via the central cavity of petroporphyrins during the sedimentary diagenesis of chlorophyll *a*.

7.5 Copper

Organo-copper interactions have been the most intensely examined aspect of organo-metallic chemistry in natural waters. This is mainly because copper offers several advantages over other metal systems. Copper(II) is the highest member of the Iriving—Williams series (Stumm and Morgan, 1970), forming coordination complexes of greater stability than other divalent transition metals. Its aqueous chemistry is not complicated by heterogeneous side reactions such as hydrolysis and precipitation. The concentration and chemical forms may be directly determined by modulated voltammetric techniques and it is an essential trace metal for both plants and animals as well as being a toxic pollutant.

TABLE V

The distribution of organically associated copper in natural waters and the operational schemes employed in their determination

Location and number of samples	Concentrations (μg Cu l^{-1}) [1]			Operational and analytical scheme [2]	Reference
	organic Cu	total Cu	% organic Cu		
Florida Current (91)	<2 [3]	0–24 (8)	>80 [3]	CO,S	Alexander and Corcoran (1967)
Gulf of Mexico (5)	0.10–0.45 (0.20)	0.3–4.6 (1.5)	8–56 (23)	SE,NAA	Slowey et al. (1967)
East Pacific (20)	0.00–0.45 (0.19)	0.44–4.70 (1.31)	0–28 (16)	UV,S	Williams (1969)
Arctic Red Tide (5)	0.0–18.9 (7.8)	1.3–20 (9.5)	0–95 (49)	CO,ASV	Holm-Hansen et al. (1970)
Coastal Cape Cod (2)	1.6–2.8 (2.2)	2.3–4.4 (3.4)	64–70 (67)	UV,ASV	Fitzgerald (1971) cited by Siegel (1971)
Menai Straits, Wales (50)	0.3–0.8 (0.5)	2.4–3.6 (3.0)	9–30 (16)	UV,S	Foster and Morris (1971)
Menai Straits, Wales (63)	0.1–1.2 (0.3)	0.6–3.7 (1.7)	10–61 (18)	UV,XRF	Morris (1974)
East Atlantic (21)	<0.02 [3]	0.09–0.23 (0.19)	≤10 [3]	UV,SE,AAS	Moore and Burton (1976)
Sea water ? (1)	0.17 [4]	0.46	37	UV,ASV	Batley and Florence (1976b)
West Pacific (11)	0.8–0.9 [3]	1.1–1.2 [3]	83–90 [3]	XAD,AAS	Sugimura et al. (1978a)
Port Hacking Estuary, N.S.W. (8)	0.01–0.67 (0.18) [3]	0.17–1.60 (0.55)	5–54 (28)	UV,ASV	Batley and Gardner (1978)
Yarra Estuary (6)	0–22 (7)	2.6–29 (12)	0–74 (13)	UV,ASV	Blutstein and Smith (1978)
Puerto Rican Estuary (4)	16.7–48.1 (30.2)	25–69 (47)	64–70 (67)	XAD,AAS	Montgomery and Santiago (1978)
Ontario Lakes (16)	4–43 (19)	4–46 (21)	55–100 (95)	CO,ASV	Chau et al. (1974)
Inland waters (3)	0.1–0.2 (0.15)	0.2–0.4 (0.3)	32–63 (50)	UV,ASV	Florence (1977)
Filtered sewage (6)	3.4–7.5 (5.6)	17–30 (23)	56–89 (74)	UV,ASV	Gardiner and Stiff (1975)

[1] The ranges in the concentrations and percentage reported, and their respective means.

[2] The following abbreviations are used: UV = ultraviolet oxidation; CO = chemical oxidation; SE = solvent extraction; XAD = adsorption on Amberlite XAD-2 resin; AAS = atomic absorption spectrophotometry; S = visible spectrophotometry; XRF = X-ray fluorescence spectroscopy; NAA = neutron activation analysis.

[3] Results of individual samples unavailable.

[4] Organically associated copper assumed to equal to fractions ML2 + ML4; see Batley and Florence (1976b) for details.

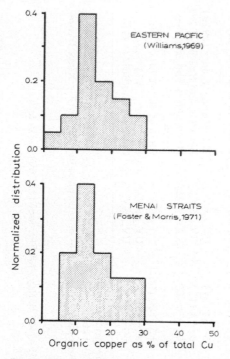

Fig. 4. Normalized distribution of organically associated copper expressed as a percentage of the total copper in the eastern Pacific Ocean (Williams, 1969) and the Menai Straits (Foster and Morris, 1971).

The distribution of organically associated copper (Cu_{org}) in marine and fresh waters and the operational schemes employed in their determination are summarised in Table V. In general, despite the various methods employed, a consistent average of ca. 20% of the total copper in sea water appears to be in the organic form (see Fig. 4), with progressively higher values in estuarine and inshore environments. The considerably higher averages of greater than 80% reported by Alexander and Corcoran (1967) and Sugimura et al. (1978b) may be linked partly with procedural blanks and partly with the unproven assumption that XAD-2 resins selectively extract the organic forms of copper from sea water (Mantoura, 1976; Montgomery and Santiago, 1978; Sugimura et al., 1978b). There is some indication that high values of Cu_{org} in surface sea water may be associated with concomitant blooms of phytoplankton (Foster and Morris, 1971; Morris, 1974) and protociliates (Holm-Hansen et al., 1970). However, the fact that no correlation could be found between Cu_{org} and DOC (Williams, 1969; Foster and Morris, 1971) may indicate (a) that there is some specificity in the organic binding of copper and (b) that these ligands constitute a small but variable proportion of the DOC. It is not known whether these chlorophyll *a* associated variations in Cu_{org} are due to stress-induced exudation of detoxifying ligands (Stolzberg and Rosin, 1977; Barber, 1973) or represent a post-mortal release of endogenous organo-copper complexes. The ability of added copper to form hydrophobic mycellar complexes with dissolved long-chain fatty acids (Tréguer et al., 1972) lends support to Slowey et al.'s (1967) hypothesis that chloroform extractable copper fractions correspond to associations with phospholipid, aminolipid, and porphyrin components of the DOM.

The biological significance of the various forms of copper is enigmatic. On the one hand, traces of copper are required as fundamental constituents of vital metallo-enzymes and proteins in which copper(II) is firmly bonded via the prosthetic groups such as the porphyrin (e.g., ascorbic oxidase, molluscan haemocyanin) or flavin residues (e.g. nitrite reductase; Bowen, 1966). Copper concentration factors of up to 3×10^4 relative to sea water have been reported for phytoplankton (Brewer, 1975). On the other hand, at the concentrations of copper normally encountered in coastal sea water (2×10^{-8} M), the ionic form is toxic to many species of phytoplankton (Steemann Nielsen and Wium-Anderson, 1970; Saifullah, 1978) and bacteria (Gillespie and Vaccaro, 1978). Cupric ion activities of 10^{-9} M or less are deleterious to the dinoflaggellate *Gonyaulax tamerensis* (Anderson and Morel, 1978) the estuarine diatom *Thalassiosira pseudonanna* and the green alga *Nannochloris atomus* (Sunda and Guilland, 1976). It is the cupric ion activity that correlates most strongly with copper toxicity. Thus the presence of even traces of copper chelating ligands in sea water would be significant in the ecology of natural systems (Barber, 1973; Johnston, 1964; Jackson and Morgan, 1978). There is no doubt that the mechanisms of membrane transport of essential copper and detoxification of excess copper (Foster, 1977; Jackson and Morgan, 1978) must be intimately related not only to the quantity and forms of copper in solution, but also in their thermodynamic and kinetic properties.

The conditional stability constants (log K_c) obtained for copper with humic compounds extracted from soils and natural waters are invariably greater than those for other transition metals (see Table IV). This is expected from the enhanced levels of crystal field stabilisation energy which result from the splitting of the $3d^9$ electronic orbitals on Cu^{2+} by an octahedral field (Mackay and Mackay, 1969). The divergence in the values of log K_c shown in Table IV, may, in part, have arisen from intrinsic variations in the copper-binding properties of the various humic samples. However, these deviations may also be explained in terms of the different experimental conditions employed (pH, ionic strength, temperature, for example) and the assumptions made in the calculations. For example, an increase in the pH will enhance the availability of dissociated binding sites (see Section 6) which are then free to participate in further complexation of copper and thus result in higher values of log K_c. Thermodynamic calculations of copper speciation which include humic complexes indicate that copper humate species may be a major form of copper in fresh and estuarine waters, and to a lesser extent in sea water (Mantoura et al., 1978; see also Section 8). Scatchard plots of copper complexation data show that the binding of copper by fulvic acid proceeds via two types of sites possessing significantly different intrinsic stability constants (Mantoura and Riley, 1975; Guy and Chakrabarti, 1976). These may correspond to the salicylate and phthalate type bidentate linkages postulated to be the metal-binding groups on soil humic compounds (see Figs. 1 and 5; Schnitzer and Khan, 1972b; Gamble and Schnitzer, 1973). Ernst et al. (1975) using a combination of differential pulse anodic stripping voltammetry (DPASV) and differential pulse polarography (DPP) were able to overcome most of the problems associated with humic ligands' non-ideal electrochemical behaviour (Brezonik et al., 1976; O'Shea and Mancy, 1976) and thus estimate the stability constants of the two copper-binding sites (see Table IV). DPASV was also used by Chau et al. (1974) to show, by analogy to the electrochemical behaviour of copper in the presence of various well-characterised ligands, that lake waters contained complexes of copper possessing log K_c values exceeding 10. Duinker and Kramer (1977) also using DPASV, concluded that the partially non-labile properties of ionic copper spiked into samples of North Sea water was due to the formation of organic complexes. Cupric ion potentiometry has been used to derive further values of log K_c (Cheam and Gamble, 1974), to assess the effects of natural organics on the non-Nernstian nature of cupric activity in sea water (Williams and Baldwin, 1976) and more recently, to even derive the molecular weights and the copper-binding stoichiometry of riverine and lacustrine organic matter (Buffle et al., 1977).

Organic complexation of copper is particularly enhanced in the organic-

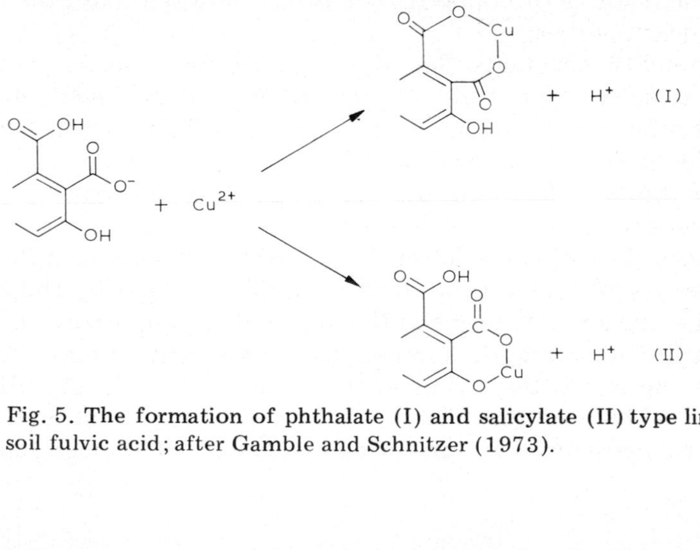

Fig. 5. The formation of phthalate (I) and salicylate (II) type linkages between Cu^{2+} and soil fulvic acid; after Gamble and Schnitzer (1973).

rich environments of sedimentary pore waters (Elderfield and Hepworth, 1975; Nissenbaum and Swaine, 1976; Sugai and Healy, 1978) and at the air— sea interface (Liss, 1975). The role of humic acids of marine origin in complexing copper has been examined by Rashid (1971) and Rashid and Leonard (1973).

7.6. Zinc and cadmium

The chemistry of zinc and cadmium in natural waters appears to be dominated by inorganic reactions. The principal species of zinc and cadmium in sea water are Zn^{2+}, $ZnCl^+$, $Zn(OH)_2^0$ (Zirino and Yamamoto, 1972) and $CdCl_2^0$ (Stumm and Brauner, 1975) although the predominance of other species of zinc have also been proposed (Bernhard et al., 1975). Early experimental studies showing the presence of non-dialysable, oxidisable and extractable forms of zinc in sea water (Rona et al., 1962; Slowey and Hood, 1971) have recently been superceded by more meaningful and quantitative techniques employing polarography (Barić and Branića, 1967; Maljković and Branića, 1971; Piro et al., 1973; Bernhard et al., 1975; Florence, 1977; Raspor et al., 1977) and radiochemistry (Duursma, 1970; Piro et al., 1973; Marchand, 1974). These studies confirm that in the absence of synthetic organic ligands these metals form only inorganic species which are kinetically inert to exchange reactions with radio tracers (Bernhard et al., 1975). However, Morris (1974) has observed that the destruction of organic matter by ultraviolet irradiation results in ca. 10% increase in zinc. Kester (1975) has hypothesised that stable metallo-enzymes of Zn such as carbonic anhydrase and carboxypeptidase may account for the "non-equilibrium" characteristics of zinc in marine waters.

Cadmium has been used in several experimental studies of metal chelation in sea waters constituted with synthetic ligands such as EDTA (Maljković and Branića, 1971) and NTA (Raspor et al., 1977). The conditional stability constants of zinc and cadmium complexes with soil humics and natural water organics are not high (see Table IV). Nevertheless, such interactions have been invoked to explain the enrichment of zinc in marine and estuarine sedimentary pore waters (Elderfield and Hepworth, 1975; Nissenbaum and Swaine, 1976).

7.7. Mercury and lead

Significant proportions of mercury dissolved in natural waters may be associated with organic matter (Fitzgerald and Lyons, 1973; Andren and Harris, 1975; Fitzgerald, 1975; Ramamoorthy and Kushner, 1975a, b). Although the neurotoxic alkyl mercury compounds [e.g. $HgCH_3^+$, $Hg(CH_3)_2$] are known to be present in marine organisms and in many inshore anoxic sediments (Jernelöv, 1974), they are unstable in oxygenated conditions and have not been detected in sea water (Andren and Harris, 1975). Jewett et al. (1975) have found that in the presence of acetate ions, Hg^{2+} was photolysed by sunlight or mild ultraviolet radiation to form transient methyl mercury compounds. Ultraviolet irradiation has been used to demonstrate the presence of other forms of organo-mercury associations amounting to ~50% of the total mercury in coastal waters of northeastern United States (Fitzgerald and Lyons, 1973; Fitzgerald, 1975). Andren and Harris (1975) size-fractionated a range of coastal and estuarine water samples and found mercury to be strongly associated with low molecular weight (<500) organic matter possessing similar properties to soil humic compounds. Strohal and Huljev (1970) also showed that no other cation could displace mercury once it was bound to humic acid isolated from the Adriatic Sea. The irreversible nature of mercury bound to humic compounds may be due to the presence of sulphur residues (Schnitzer and Khan, 1972) or to charge transfer reactions which, in the limiting case, will lead to the reduction of Hg^{2+} to volatile elemental mercury (see also Section 5; Alberts et al., 1974). Conditional stability constants of Hg^{2+} with humic compounds from natural waters are high (see Table IV) and have been incorporated into a multi-ligand multi-metal equilibrium model for speciation of mercury in natural waters (Mantoura et al., 1978). Other investigations of interactions between mercury and humic compounds include ion-selective potentiometry of mercuric activity in river water (Ramamoorthy and Kushner, 1975a, b) and synthetic media (Cheam and Gamble, 1974), gel filtration of humic bound ^{203}Hg tracer (Mantoura, 1976) and the effects of humic acids on the electrodeposition of mercury on silver gauze (Millward and Burton, 1975).

Lead is a non-conservative trace ($<10^{-10}$ M) constituent of ocean waters being particularly susceptible to adsorption and natural coprecipitation reactions (Brewer, 1975). In inshore and coastal waters it may also be associated with high molecular weight (Barsdate and Matson, 1967; Buffle et al., 1976; Sugai and Healy, 1978) and colloidal organic matter (Florence, 1977; Giesy and Briese, 1977; Batley and Gardner, 1978; Blutstein and Smith, 1978). Studies on specific interactions of lead with organics have almost entirely concentrated on humic compounds using mainly voltammetric (Barsdate and Matson, 1967; Ernst et al., 1975; Buffle et al., 1976) and potentiometric techniques (Stevenson, 1976; Ramamoorthy and Kushner, 1975a). Although these studies have provided information on stability constants (see Table IV), dissociation rates of Pb—humate complexes and their electro- and surface-active properties, there is as yet no evidence for the existence of soluble organic complexes of lead in sea water. Turner and Whitfield (1979) have recently reviewed the chemical speciation of lead in sea water and considered the effect of added complexing agents (e.g. EDTA, NTA, citric acid, humic acid) on the thermodynamic equilibria and electrochemical availability of lead. They have recommended more rigorous criteria for the detection of non-labile species based on electrokinetic approach using rotating-disc electrodes.

8. MODELS

A comprehensive treatment of organo-metallic interactions in natural waters cannot be obtained by simply examining one metal or one ligand in isolation, but must take into account the competitive effects of other inter-acting components. Thus, an organic ligand will interact with *all* cations and similarly each cation will interact with all other organic and inorganic ligands. This ensemble of interdependent interactions may be represented by a set of non-linear thermodynamic equations, the solution of which results in the equilibrium speciation for all the components in the system. The ensuing calculations are laborious and complex and are best processed by numerical iterative procedures. Dyrrsen et al. (1968) and Morel and Morgan (1972) have developed general purpose programs which may be specially

TABLE VI

A compilation of organic ligands that have been incorporated into equilibrium models of trace-metal speciation

Ligand [1]	Aquatic systems and references [2]		
	fresh water	sea water	culture media
Acetic acid		9, 11	
Glutamic acid		11	
Glycine	8	4, 9, 11	
Cysteine	8		
Leucine		3,	
Histidine		17	
Phthalic acid		11	
Tartaric acid		11	
Citric acid	4, 7, 8, 12	4, 11, 16	
Salycilic acid		4	
QCA		3	
NOC		4	
EDTA	4, 13	1, 3, 6, 13, 16, 17	2, 17, 18
NTA	7, 8	4, 14, 16	17
TRIS			18, 19
Humic acids	5, 10, 15, 20	15, 16	

[1] QCA = quinoline-2-carboxylic acid; NOC = nocardamine; EDTA = ethylenediamine tetra acetic acid; NTA = nitrilotriacetic acid; TRIS = tris(hydroxymethylamine) methane.
[2] References for Table VI: 1 = Spencer (1958); 2 = Johnston (1964); 3 = Duursma (1970); 4 = Stumm and Morgan (1970); 5 = Stiff (1971); 6 = Maljković and Branića (1971); 7 = Lehrman and Childs (1973); 8 = Morel et al. (1973); 9 = Dyrrsen and Wedborg (1974); 10 = Gardiner (1974); 11 = Stumm and Brauner (1975); 12 = Elder (1975); 13 = Gardiner (1976); 14 = Raspor et al. (1977); 15 = Mantoura et al. (1978); 16 = Whitfield and Turner (1980); 17 = Jackson and Morgan (1978); 18 = Anderson and Morel (1978); 19 = Sunda and Guillard (1976); 20 = Wilson and Kinney (1977).

adapted to model the speciation of trace metals in the presence of different organic ligands (Stumm and Morgan, 1970; Morel et al., 1973; Stumm and Brauner, 1975; Mantoura et al., 1978). The versatility of these programs allows the inclusion of gaseous and solid dissolution reactions and the formation of mixed ligand and polynuclear complexes. However, in the often-encountered limiting case when the concentrations of all competing ligands exceed the metal ion of interest (or *vice versa*), it is possible to construct much simpler models using Ringbom's theory of side reaction coefficients (Elder, 1975; Kester et al., 1975b, Gardiner, 1976; Whitfield and Turner, 1980).

The range of organic ligands which have been examined using equilibrium models are listed in Table VI. Although the results of individual ligands cannot be examined in the space allowed, there are some important principles which emerge from these studies. The extent of trace-metal complexation with organic ligands in multicomponent systems is a function of (1) the *availability* of deprotonated metal-binding sites on organic ligands as controlled by pH, and (2) the binding *specificity* of these sites for trace metals relative to the displacing action of the major cations and the competing effects of inorganic ligands. The addition or removal of *any* component in

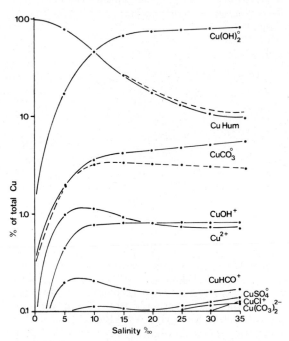

Fig. 6. The equilibrium speciation of copper(II) as a function of salinity along a model estuary. For other details, see Mantoura et al. (1978).

the equilibrium system will have a *synergic* effect on the chemical form and distribution of all other interacting components in the system. For example, the addition of Fe^{3+} or Cu^{2+} ions to sea-water medium in productivity studies (Johnston, 1964; Anderson and Morel, 1978), will lead to a redistribution of *all* trace metals and hence any observed change in productivity is not necessarily caused by a change in the availability of Fe^{3+} or toxicity of Cu^{2+} to the cell (Stumm and Brauner, 1975; Jackson and Morgan, 1978).

A quantitative impression of multiple interactive systems can be gained from Figs. 6 and 7 which depict the effect of variation in salinity on the speciation of copper in the presence of humic compounds (Mantoura et al., 1978). It is clear from Fig. 5 that the decomplexation mechanism of humic bound copper originating from acidic river systems is associated with the gradual increase in the concentration of hydroxy and carbonate ligands cor-

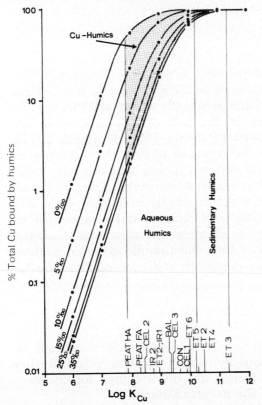

Fig. 7. The extent of copper humic interactions, as a function of the magnitude of the stability constant, at various salinities. The annotation on the horizontal axis portrays the range of stability constants found for various natural samples [see Mantoura et al. (1978) for further details].

responding to the increasing pH of more saline waters, rather than displacement effects of the surge in calcium and magnesium concentration at low salinities (as is the case for Zn^{2+}, Mn^{2+}, Co^{2+}, Ni^{2+}). The way in which the nature of humic compounds affects the degree of copper complexation at different salinities and in differing environments is illustrated in Fig. 7.

The recent extension of these thermodynamic models to include the kinetics and mechanisms of organo-metallic interactions has made it possible (1) to quantify the electrochemical availability of these metal complexes to voltammetric systems (Whitfield and Turner, 1980); (2) to examine diffusion and dissociation models for the transport of chelated iron to biological cells (Jackson and Morgan, 1978); and (3) to estimate the significance of adsorptive and convective removal processes on the equilibrium speciation of metals in natural waters (Lehrman and Childs, 1973). Thus both equilibrium and dynamic models have become an indispensable tool in the identification of the important chemical forms and critical reaction pathways of interactive elements in aquatic environments.

9. CONCLUSION

Despite the great number of independent experimental studies which show, directly or indirectly, the existence of organo-metallic interactions in natural waters, it has not been possible to actually isolate and identify the resultant complexes, mainly because of the lack of information on the molecular nature of ~90% of the DOM (Williams, 1975; Ehrhardt, 1977). However, the use of this as a basis for argueing *against* the existence of organo-metallic interactions in natural waters (Stumm and Morgan, 1970; Pocklington, 1977), in spite of all the evidence assembled here, is unwarranted. The increasing multidisciplinary nature of organo-metallic chemistry makes it likely that significant conceptual and experimental advances will be attained in the near future. The physico-chemical and structural characterisation of organic ligands derived either from the ubiquitous humic pools or from specific biological systems is an important and exciting direction for research in this field.

10. ACKNOWLEDGEMENTS

I wish to thank Drs. A.W. Morris, D. Turner, A.R.D. Stebbing and M. Whitfield for much helpful advice and discussion; my wife, Olivia, for her patience and encouragement; Mrs. V. Martin and Mrs. J. Milton for deciphering my manuscript without the aid of the Rosetta Stone.

11. REFERENCES

Aaronson, S., 1971. The synthesis of extracellular macromolecules and membranes by a population of the phytoflagellate *Ochramonas danica*. Limnol. Oceanogr., 16: 1—9.

Akiyama, T., 1971. Microscopic determination of iron-organic aggregates in sea and lake waters. Geochem. J. Jap., 27: 207—212.

Alberts, J.J., Schindler, J.E., Miller, R.W. and Nutter, D.E. Jr., 1974. Elemental mercury evolution mediated by humic acid. Science, 184: 895—896.

Alexander, J.E. and Corcoran, E.F., 1967. Distribution of copper in tropical sea water. Limnol. Oceanogr., 12: 236—242.

Allen, H.L., 1976. Dissolved organic matter in lake water: characteristics of molecular weight size fractions and ecological implications. Oikos, 27: 64—70.

Anderson, D.M. and Morel, F.M.M., 1978. Copper sensitivity of *Gonyaulax tamarensis*. Limnol. Oceanogr., 23: 283—295.

Andren, E.J. and Harris, R.C., 1975. Observation on the association between mercury and organic matter dissolved in natural waters. Geochim. Cosmochim. Acta, 39: 1253—1257.

Angino, E.E., Billings, G.K. and Anderson, N.R., 1966. Observed variations in the strontium concentration in sea water. Chem. Geol., 1: 145—153.

Bada, J.L. and Lee, C., 1977. Decomposition and alteration of organic compounds dissolved in sea water. Mar. Chem., 5: 523—534.

Baker, E.W. and Smith, G.D., 1977. Fossil porphyrins and chlorins in deep ocean sediments: In: T.F. Yen (Editor), Chemistry of Marine Sediments. Ann Arbor Press, Ann Arbor. Mich., pp. 73—99.

Bannister, F.A. and Hey, M.H., 1936. Report on some crystalline components of the Weddell Sea deposits. Discovery Rep., 13: 60—69.

Barber, R.T., 1973. Organic ligands and phytoplankton growth in nutrient rich sea water. In P.C. Singer (Editor), Metals and Metal Organic Interactions in Natural Waters. Ann Arbor Press, Mich., pp. 321—338.

Barber, R.T. and Ryther, J.H., 1969. Organic chelators: factors affecting primary production in the Cromwell current upwelling. J. Exp. Mar. Biol. Ecol., 3: 191—199.

Barber, R.T., Dugdale, R.C., MacIsaac, J.J. and Smith, R.L., 1971. Variations in phytoplankton growth associated with the source and conditioning of upwelling water. Inv. Pesq., 35: 171—193.

Barić, A. and Branića, M., 1967. Polarography of sea water: I, ionic state of cadmium and zinc in sea water. J. Polarogr. Soc., 13: 4—8.

Barsdate, R.J., 1970. Transition metal binding by large molecules in high latitude waters. In: D.W. Hood (Editor), Symposium on Organic Matter in Natural Waters. Inst. Mar. Sci., Univ. Alaska, Occ. Publ., 1: 485—491.

Barsdate, R.J. and Matson, W.R., 1967. Trace metals in arctic and sub-arctic lakes with reference to the organic complexes of metals. In: B. Aberg and F.P. Hungate (Editors), Radioecological Concentration Processes. Pergamon Press, London, pp. 711—719.

Batley, G.E. and Florence, T.M., 1976a. Determination of the chemical forms of dissolved cadmium, lead and copper, in sea water. Mar. Chem., 4: 347—363.

Batley, G.E. and Florence, T.M., 1976b. A novel scheme for the classification of heavy metal species in natural waters. Anal. Lett., 9: 373—388.

Batley, G.E. and Gardner, D., 1978. A study of copper, lead and cadmium speciation in some estuarine and coastal marine waters. Estuarine Coastal Mar. Sci., 7: 59—70.

Beck, K.C., Reuter, J.H. and Perdue, E.M., 1974. Organic and inorganic geochemistry of some coastal plain rivers of the south eastern United States. Geochim. Cosmochim. Acta, 38: 341—364.

Beneš, P., Gjessing, E.T. and Steinnes, E., 1976. Interaction between humus and trace elements in fresh waters. Water Res., 10: 711—716.

Bernhard, M., Goldberg, E.D. and Piro, A., 1975. Zinc in sea water. In: E.D. Goldberg (Editor), The Nature of Sea Water. Dahlem Konferenzen, Berlin, pp. 43—68.

Blutstein, H. and Smith, J.D., 1978. Distribution of species of Cu, Pb, Zn and Cd, in a water profile of the Yarra River Estuary. Water Res., 12: 119—125.

214

Boehm, P.D. and Quinn, J.G., 1973. Solubilization of hydrocarbons by the dissolved organic matter in sea water. Geochim. Cosmochim. Acta, 37: 2459—2478.

Bohling, H., 1972. Dissolved amino acids in surface water of the North Sea near Helgoland, concentration changes during summer 1970. Mar. Biol., 16: 281—290.

Bowen, H.J.M., 1966. Trace elements in biochemistry. Academic Press, London, 241 pp.

Boyle, E.A., Edmond, J.M. and Sholkovitz, E.R., 1977. The mechanism of iron removal in estuaries. Geochim. Cosmochim. Acta, 41: 1313—1324.

Brewer, P.G., 1975. Minor elements in sea water. In: J.P. Riley and G. Skirrow (Editors), Chemical Oceanography, I. Academic Press, London, 2nd ed., pp. 415—496.

Brezonik, P.L., Brauner, P.A. and Stumm, W., 1976. Trace metal analysis by anodic stripping voltammetry: effect of sorption by natural and model organic compounds. Water Res., 10: 605—612.

Buffle, J., Greter, F.L., Nembrini, G., Paul, J. and Haerdi, W., 1976. Capabilities of voltammetric techniques for water quality control problems. Z. Anal. Chem., 282: 339—350.

Buffle, J., Greter, F.L. and Haerdi, W., 1977. Measurement of complexation properties of humic and fulvic acids in natural waters with lead and copper ion-selective electrodes. Anal. Chem., 49: 216—222.

Burton, J.D., 1978. Behaviour of some trace chemical constituents in estuarine waters. Pure Appl. Chem., 50: 385—393.

Cahn, R.S., 1974. An Introduction to Chemical Nomenclature. Butterworths, London, 4th ed.

Carlucci, A.F., 1970. Vitamin B_{12}, thiamine and biotin. Bull. Scripps Inst. Oceanogr., 17: 23—32.

Carpenter, J.H., 1972. Problems in applications of analytical chemistry to oceanography. In: W.W. Meinke and J.K. Taylor (Editors), Analytical Chemistry: Key to Progress on National Problems. Natl. Curr. Stand. (U.S.) Spec. Publ., 351: 393—419 (Washington, D.C.).

Chau, Y.K., 1973. Complexing capacity of natural water: its significance and measurement. J. Chrom. Sci., 11: 579.

Chau, Y.K. and Lum-Shue-Chan, K., 1974. Determination of labile and strongly-bound metals in lake waters. Water Res., 8: 383—388.

Chau, Y.K. and Wong, P.T.S., 1976. Complexation of metals in natural waters. In: R.W. Andrew, P.V. Hodson and D.E. Konasewich (Editors), Toxicity of Metal Forms in Natural Water. Proceedings of a Workshop held in Duluth, Minn., October 1975, pp. 187—197.

Chau, Y.K., Gächter, R. and Lum-Shue-Chan, K., 1974. Determination of the apparent complexing capacity of lake waters. J. Fish. Res. Board Can., 31: 1515—1519.

Chave, K.E., 1970. Carbonate—organic interactions in sea water. In: D.W. Hood (Editor), Symp. Organic Matter in Natural Waters, 1968. Inst. Mar. Sci., Univ. Alaska, Occ. Publ., No. 1: 373—397.

Cheam, V., 1973. Chelation study of copper II: fulvic acid system. Can. J. Soil. Sci., 53: 377—382.

Cheam, V. and Gamble, D.S., 1974. Metal fulvic acid chelation equilibrium in aqueous $NaNO_3$ solution. Can. J. Soil. Sci., 54: 413—417.

Christman, R.F., 1970. Chemical structures of colour producing organic substances in water. In: D.W. Hood (Editor), Symp. on Organic Matter in Natural Waters, 1968. Inst. Mar. Sci., Univ. Alaska, Occ. Publ., No. 1: 181—198.

Christman, R.F. and Ghassemi, M., 1966. Chemical nature of organic colour in water. J. Am. Water Works Assoc., 58: 723—741.

Clem, R.G. and Hodgson, A.T., 1978. Ozone oxidation of organic sequestering agents in water prior to the determination of trace metals by anodic stripping voltammetry. Anal. Chem., 50: 102—110.

Cooper, L.H.N., 1935. Iron in the sea and in marine plankton. Proc. R. Soc. Lond., B 118: 419—438.

Cooper, L.H.N., 1937. Some conditions governing the solubility of iron. Proc. R. Soc. Lond., B 124: 299—307.

Cooper, L.H.N., 1948a. Some chemical considerations on the distribution of iron in the sea. J. Mar. Biol. Assoc. U.K., 27: 314—321.

Cooper, L.H.N., 1948b. The distribution of iron in the waters of the western English Channel. J. Mar. Biol. Assoc. U.K., 27: 279—313.

Davey, E.W., Morgan, M.J. and Erickson, S.J., 1973. A biological measurement of the copper complexation capacity of seawater. Limnol. Oceanogr., 18: 993—997.

Davies, A.G., 1970. Iron, chelation and the growth of marine phytoplankton: I Growth kinetics and chlorophyll production in cultures of the euryhaline flagellate *Dunaliella tertiolecta* under iron limiting conditions. J. Mar. Biol. Assoc. U.K., 55: 65—86.

Davison, W., 1978. Defining the electroanalytically measured species in a natural water sample. J. Electroanal. Chem., 87: 395—410.

Dawson, R. and Duursma, E.K., 1974. Distribution of radioisotopes between phytoplankton, sediment and sea water in a dialysis compartment system. Neth. J. Sea Res., 8: 339—353.

Degens, E.T., 1970. Molecular nature of nitrogenous compounds in sea water and recent marine sediments. In: D.W. Hood (Editor), Organic Matter in Natural Waters. Inst. Mar. Sci., Univ. Alaska, Occ. Publ., No. 1: 77—106.

Desai, M.V.M., Koshy, E. and Ganguly, A.K., 1969. Solubility of barium in sea water in the presence of dissolved organic matter. Curr. Sci., 38: 107—108.

Duinker, J.C. and Kramer, C.J.M., 1977. An experimental study on the speciation of dissolved zinc, cadmium, lead and copper in River Rhine and North Sea water by differential pulsed anodic stripping voltammetry. Mar. Chem., 5: 207—228.

Duursma, E.K., 1961. Dissolved organic carbon, nitrogen and phosphorous in the sea. Neth. J. Sea Res., 1: 1—147.

Duursma, E.K., 1965. The dissolved organic constituents of sea water. In: J.P. Riley and G. Skirrow (Editors), Chemical Oceanography, 1. Academic Press, 1st ed., London, pp. 433—475.

Duursma, E.K., 1970. Organic chelation of ^{60}Co and ^{65}Zn by leucine in relation to sorption by sediments. In: D.W. Hood (Editor), Symposium on Organic Matter in Natural Waters. Inst. Mar. Sci., Univ. Alaska, Occ. Publ., 1: 387—397.

Dyrrsen, D. and Wedborg, M., 1974. Equilibrium calculations of the speciation of elements in sea water. In: E.D. Goldberg (Editor), The Sea, 5. Wiley—Interscience, New York, N.Y., pp. 181—195.

Dyrrsen, D., Jagner, D. and Wanglin, F., 1968. Computer Calculation of Ionic Equilibria and Titration Procedures. Almquist and Wiksell, Stockholm, 250 pp.

Eagland, D., 1975. Nucleic acids, peptides and proteins. In: F. Franks (Editor), Water: A Comprehensive Treatise, 4. Plenum Press, London, pp. 305—516.

Eckert, J.M. and Sholkovitz, E.R., 1976. The flocculation of iron, aluminum and humates from river water by electrolytes. Geochim. Cosmochim. Acta, 40: 847—848.

Ehrhardt, M., 1977. Organic substances in sea water. Mar. Chem., 5: 307—316.

Eichhorn, G.L., 1975. Organic ligands in natural systems. In: E.D. Goldberg (Editor), The Nature of Seawater. Dahlem Konferenzen, Berlin, pp. 245—262.

Elder, J.F., 1975. Complexation side reactions involving trace metals in natural water systems. Limnol. Oceanogr., 20: 96—102.

Elderfield, H. and Hepworth, A., 1975. Diagenesis, metals and pollution in estuaries. Mar. Poll. Bull., 6: 85—87.

Ernst, R., Allen, H.E. and Mancy, K.H., 1975. Characterisation of trace metal species and measurement of trace metal stability constants by electrochemical techniques. Water Res., 9: 969—979.

216

Fitzgerald, W.F., 1975. Mercury analysis in seawater using cold-trap preconcentration and gas phase detection. Adv. Chem. Ser., 147: 99—109.

Fitzgerald, W.F. and Lyons, W.B., 1973. Organic mercury compounds in coastal waters, Nature, 242: 452—453.

Flaig, W., Beutelspacher, H. and Rietz, E., 1975. Chemical composition and physical properties of humic substances. In: J.E. Gieseking (Editor), Soil Components, I. Springer, Berlin, pp. 1—211.

Florence, T.M., 1977. Trace metal species in fresh waters. Water Res., 11: 681—687.

Florence, T.M. and Batley, G.E., 1977. Determination of copper in seawater by anodic stripping voltammetry. J. Electroanal. Chem., 75: 791—798.

Foster, P.L., 1977. Copper exclusion as a mechanism of heavy metal tolerance in a green alga. Nature, 269: 322—323.

Foster, P. and Morris, A.W., 1971. The seasonal variations of dissolved ionic and organically associated copper in the Menai Straits. Deep-Sea Res., 18: 231—236.

Gächter, R. and Lum-Shue-Chan, K., 1973. Complexing capacity of the nutrient medium and its relation to the inhibition of algal photosynthesis by copper. Schw. Z. Hydrol., 35: 252—261.

Gagosian, R.B. and Stuermer, D.H., 1977. The cycling of biogenic compounds and their diagenetically transformed products in seawater. Mar. Chem., 5: 605—632.

Gamble, D.S., 1970. Titration curves of fulvic acid: the analytical chemistry of a weak acid polyelectrolyte. Can. J. Chem., 48: 2662—2669.

Gamble, D.S., 1972. Potentiometric titration of fulvic acid: equivalence point calculations and acidic functional groups. Can. J. Chem., 50: 2680—2690.

Gamble, D.S., 1973. Na^+ and K^+ binding by fulvic acid. Can. J. Chem., 1: 3217—3222.

Gamble, D.S. and Schnitzer, M., 1973. The chemistry of fulvic acid and its reactions with metal ions. In: P.C. Singer (Editor), Trace Metals and Metal Organic Interactions in Natural Waters. Ann Arbor Science, Mich., pp. 265—302.

Gamble, D.S., Schnitzer, M. and Skinner, D.S., 1977. Mn(II)—fulvic acid complexing equilibrium measurements by electron spin resonance spectrometry. Can. J. Soil Sci., 7: 47—53.

Gardiner, J., 1974. The chemistry of cadmium in natural waters: a study of cadmium complex formation using the cadmium specific ion electrode. Water Res., 8: 23—30.

Gardiner, J., 1976. Complexation of trace metals by ethylenediaminetetra acetic acid (EDTA) in natural waters. Water. Res., 10: 507—514.

Gardiner, J. and Stiff, M.J., 1975. The determination of lead, copper and zinc in ground water, estuarine water and sewage effluent by anodic stripping voltammetry. Water Res. 9: 517—523.

Ghassemi, M. and Christman, R.F., 1968. Properties of the yellow organic acids of natural waters. Limnol. Oceanogr., 13: 583—597.

Giesy, J.P. Jr., 1976. Stimulation of growth in Scenedesmus obliquus (chlorophyceae) by humic acids under iron limited conditions. J. Phycol., 13: 172—179.

Giesy, J.P. Jr. and Briese, L.A., 1977. Metals associated with organic carbon extracted from Okefenokee swamp water. Chem. Geol., 20: 109—120.

Gillespie, P.A. and Vaccaro, R.F., 1978. A bacterial bioassay for measuring the copper chelation capacity of sea water. Limnol. Oceanogr., 23: 543—548.

Gjessing, E.T., 1964. Ferrous iron in water. Limnol. Oceanogr., 9: 272—274.

Gjessing, E.T., 1965. Use of sephadex gel for estimation of molecular weight of humic substances in natural water. Nature, 208: 1091—1092.

Gjessing, E.T., 1970. Ultrafiltration of aquatic humus. Environ. Sci. Tech., 4: 437—438.

Gjessing, E.T., 1971. Effect of pH on the filtration of aquatic humus using gels and membranes. Schw. Z. Hydrol., 33: 592—600.

Gjessing, E.T., 1973. Gel and ultra-membrane filtration of aquatic humus: a comparison of two methods. Schw. Z. Hydrol. 35: 286—294.

Gran, G., 1952. Determination of the equivalence point in potentiometric titrations. Analyst, 77: 661—671.

Gran, H.H., 1931. On the condition for the production of plankton in the sea. Rapp. P.-V. Reun. Cons. Int. Explor. Mer, 75: 37.

Gran, H.H., 1933. Studies on the biology and chemistry of the Gulf of Maine. Biol. Bull., 64: 159—182.

Guy, R.D. and Chakrabarti, C.L., 1976. Studies of metal organic interactions in model systems pertaining to natural waters. Can. J. Chem., 54: 2600—2611.

Haberman, J.P., 1971. Polarographic determination of traces of nitrilotriacetic acid in water samples. Anal. Chem., 43: 63—67.

Hair, M.E. and Bassett, C.R., 1973. Dissolved and particulate humic acids in an east coast estuary. Estuarine Coastal Mar. Sci. 1: 107—111.

Hanck, K.W. and Dillard, J.W., 1977. Determination of the complexing capacity of natural water by Co(III) complexation. Anal. Chem., 49: 404—409.

Harvey, H.W., 1928. Biological Chemistry and Physics of Sea Water. Cambridge University Press, Cambridge.

Hedges, J.I., 1977. Association of organic molecules with clay minerals in aqueous phases. Geochim. Cosmochim. Acta, 41: 1119—1124.

Hellebust, J.A., 1965. Excretion of some organic compounds by marine phytoplankton. Limnol. Oceanogr., 10: 192—206.

Hobbie, J.E., Crawford, C.C. and Webb, K.L., 1968. Amino acid flux in an estuary. Science, 159: 1463—1464.

Holm-Hansen, O., Strickland, J.D.H. and Williams, P.M., 1966. A detailed analysis of biologically important substances in a profile off Southern California. Limnol. Oceanogr., 11: 548—561.

Holm-Hansen, O., Taylor, F.J.R. and Barsdate, R.J., 1970. A ciliate red tide at Barrow, Alaska. Mar. Biol., I: 37—46.

Hoyt, J.W., 1970. High molecular weight algal substances in the sea. Mar. Biol., 7: 93—99.

Huizenga, D.L., 1977. Protonation Characteristics of Dissolved Organic Matter in Sea Water. Thesis, University of Rhode Island, Providence, R.I., 86 pp.

IUPAC, 1973. Nomenclature in Organic Chemistry, section D. Oxford University Press, Oxford, 149 pp.

Jackson, G.A. and Morgan, J.J., 1978. Trace metal—chelator interactions and phytoplankton growth in sea water media and theoretical analysis and comparison with reported observations. Limnol. Oceanogr., 23: 268—282.

Jernelöv, A., 1974. Heavy metals, metalloids and synthetic organics. In: E.D. Goldberg (Editor), The Sea, 5. Marine Chemistry. Wiley—Interscience, New York, N.Y., pp. 799—815.

Jewett, K.L., Brinckman, F.E. and Bellama, J.M., 1975. Chemical factors influencing metal alkylation in water. In: T.M. Church (Editor), Marine Chemistry in the Coastal Environment. Am. Chem. Soc. Symp., Ser. 18: 304—318.

Johnston, R., 1963. Sea water, the natural medium of phytoplankton. I. General features. J. Mar. Biol. Assoc. U.K., 43: 427—456.

Johnston, R., 1964. Sea water, the natural medium of phytoplankton. II. Trace metals and chelation, and general discussion. J. Mar. Biol. Assoc. U.K., 44: 87—109.

Kerr, R.A. and Quinn, J.G., 1975. Chemical studies on the dissolved organic matter in sea water. Isolation and fractionation. Deep-Sea Res., 22: 107—116.

Kester, D.R., 1975. Chemical speciation in sea water. Group Report. In: E.D. Goldberg (Editor), The Nature of Seawater. Dahlem Konferenzen, Berlin. pp. 17—41.

Kester, D.R., O'Conner, T.P. and Byrne, R.H. Jr., 1975a. Solution chemistry, solubility and adsorption equilibria of iron, cobalt and copper in marine systems. Thalassia Yugosl., 11: 121—134.

218

Kester, D.R., Byrne, R.H. Jr. and Liang Yu-Jean, 1975b. Redox reactions and solution complexes of iron in marine systems. In: T.M. Church (Editor), Marine Chemistry in Coastal Environment. Am. Chem. Soc. Symp., Ser. 18: 56—79.

Khailov, K.M. and Finenko, Z.Z., 1970. Organic macromolecular compounds dissolved in sea water and their inclusion in the food chain. In: J.H. Steele (Editor), Marine Food Chains. Oliver and Boyd, Edinburgh, pp. 6—18.

Koshy, E., Desai, M.V.M. and Ganguly, A.K., 1969. Studies on organo-metallic interactions in the marine environment: I Interaction of some metallic ions with dissolved organic substances in sea water. Curr. Sci., 38: 555—558.

Kroes, H.W., 1972. Growth interactions between *Chlamydomonas globosa* and *Chlorococcum*: the role of extracellular products. Limnol. Oceanogr., 12: 423—432.

Krom, M.D. and Sholkovitz, E.R., 1978. On the association of iron and manganese with organic matter in anoxic marine pore waters. Geochim. Cosmochim. Acta, 42: 607—611.

Kunkel, R. and Manahan, S.E., 1973. Atomic absorption analysis of strong heavy metal chelating agents in water and waste water. Anal. Chem., 45: 1465—1468.

Landymore, A.F. and Antia, N.J., 1978. White light promoted degradation of leucopterin and related pteridites dissolved in sea water, with evidence for the involvement of complexation from major divalent cations in sea water. Mar. Chem., 6: 309—325.

Langford, C.H. and Khan, T.R., 1975. Kinetics and equilibrium of binding of Fe^{3+} by a fulvic acid: a study of stepped flow method. Can. J. Chem., 53: 2979—2984.

Langford, C.H., Kay, R., Quance, G.W. and Khan, T.K., 1977. Kinetic analysis applied to iron in a natural water model containing ions, organic complexes, colloids and particles. Anal. Lett., 10: 1249—1260.

Lehrman, A. and Childs, C.W., 1973. Metal-organic complexes in natural waters: control of distribution by thermodynamic, kinetic and physical factors. In: P.C. Singer (Editor), Trace Metals and Metal Organic Interactions. Ann Arbor, Mich., pp. 201—236.

Lindberg, S.E. and Harris, R.C., 1974. Mercury-organic matter association in estuarine sediments and interstitial water. Environ. Sci. Tech., 8: 459—462.

Liss, P.S., 1975. Chemistry of the sea surface microlayer. In: J.P. Riley and G. Skirrow (Editors), Chemical Oceanography, II. Academic Press, 2nd ed., London, pp. 193—243.

Lock, M.A., Wallis, P.M. and Hynes, H.B.N., 1977. Colloidal organic carbon in running waters. Oikos, 29: 1—4.

MacCarthy, P. and O'Cinneide, S., 1974. Fulvic acid II: Interaction with metal ions. J. Soil Sci., 25: 429—437.

Macham, L.P. and Ratledge, C., 1975. A new group of water soluble iron binding compounds from mycobacteria: the exochelins. J. Gen. Microbiol., 89: 379—382.

Mackay, K.M. and Mackay, R.A., 1969. Introduction to Modern Inorganic Chemistry. Intertext, London, 258 pp.

Malcolm, R.L., Jenne, E.A. and McKinley, P.W., 1970. Conditional stability constants of a North Carolina soil fulvic acid with Co^{2+} and Fe^{3+}. In: D.W. Hood (Editor), Symp. Organic Matter in Natural Waters. Inst. Mar. Sci., Univ. Alaska, Occ. Publ., No. 1: 479—483.

Maljković, D. and Branića, M., 1971. Polarography of sea water. II. Complex formation of cadmium with EDTA. Limnol. Oceanogr., 16: 779—785.

Mantoura, R.F.C., 1976. Humic Compounds in Natural Waters and Their Complexation with Metals. Thesis, University of Liverpool, Liverpool, 249 pp.

Mantoura, R.F.C. and Riley, J.P., 1975. The use of gel filtration in the study of metal binding by humic acids and related compounds. Anal. Chim. Acta, 78: 193—200.

Mantoura, R.F.C., Dickson, A. and Riley, J.P., 1978. The complexation of metals with humic materials in natural waters. Estuarine Coastal Mar. Sci., 6: 387—408.

Marchand, M., 1974. Considerations sur les formes physico-chimiques du cobalt, manganese, zinc, chrome et fer dans une eau de mer enrichie ou non de matière organique. J. Cons., Cons. Int. Explor. Mer., 32: 130—142.

Martell, A.E., 1971. Principles of complex formation. In: S.J. Faust and J.V. Hunter (Editors), Organic Compounds in Aquatic Environments. Dekker, New York, N.Y., pp. 239—263.

Martin, D.F. and Martin, B.B., 1973. Implications of metal organic compounds in red tide outbreaks. In: P.C. Singer (Editor), Trace Metals and Metal Organic Interactions in Natural Water. Ann. Arbor, Mich., pp. 339—362.

Matson, W.R., Allen, H.E. and Reksham, P., 1969. Trace—Metal Organic Complexes in the Great Lakes. Presented before the Division of Water, Air and Waste Chemistry, A.C.S., Minnesota, 1969.

Means, J.J., Crerar, D.A. and Amster, T.L., 1977. Application of gel filtration chromatography to evaluation of organo-metallic interactions in natural waters. Limnol. Oceanogr., 22: 957—965.

Millero, F.J., 1975. The state of metal ions in sea water. Thalassia Jugosl., 11: 53—84.

Millward, G.E. and Burton, J.D., 1975. Association of mercuric ions and humic acid in sodium chloride solution. Mar. Sci. Comm., 1: 15—26.

Montgomery, J.R. and Santiago, R.J., 1978. Zinc and copper in "particulate" forms and "soluble" complexes with inorganic or organic ligands in Guanajibo River and coastal zone, Puerto Rico. Estuarine Coastal Mar. Sci., 6: 111—116.

Moore, R.M. and Burton, J.D., 1976. Concentration of dissolved copper in the eastern Atlantic Ocean 23°N to 47°N. Nature, 264: 241—243.

Morel, F. and Morgan, J.J., 1972. A numerical method for computing equilibria in aqueous chemical systems. Environ. Sci. Tech., 6: 58—67.

Morel, F., McDuff, R.E., Morgan, J.J. and Keck, W.M., 1973. Interactions and chemostasis in aquatic systems: role of pH, pE, solubility and complexation. In: P.C. Singer (Editor), Trace Metals and Metal Organic Interactions. Ann Arbor, Mich., pp. 157—200.

Morris, A.W., 1974. Seasonal variation of dissolved metals in inshore waters of the Menai Straits. Mar. Poll. Bull., 5: 54—59.

Murphy, T.P., Lean, D.R.S. and Nalewajko, C., 1976. Blue green algae: their excretion of iron selective chelators enables them to dominate other algae. Science, 192: 900—902.

Murray, J.W. and Gill, G., 1978. The geochemistry of iron in Puget Sound. Geochim. Cosmochim. Acta, 42: 9—19.

Neihof, R. and Loeb, G., 1974. Dissolved organic matter in sea water and the electric charge of immersed surfaces. J. Mar. Res., 32: 5—12.

Neilands, J.B., 1973. Microbial iron transport compounds. In: G.L. Eichhorn (Editor), Inorganic Biochemistry, 1. Elsevier, Amsterdam, pp. 167—202.

Nissenbaum, A. and Swaine, D.J., 1976. Organic matter — metal interactions in recent sediments: the role of humic substances. Geochim. Cosmochim. Acta, 40: 809—816.

Nissenbaum, A., Baedecker, M.J. and Kaplan, I.R., 1971. Studies on dissolved organic matter from interstitial water of a reducing marine fjord. In: H.R. von Gaertner and H. Wehner (Editors), Advances in Organic Geochemistry. Pergamon, London, pp. 427—440.

Nürnberg, H.W. and Valenta, P., 1975. Polarography and voltammetry in marine chemistry. In: E.D. Goldberg (Editor), The Nature of Seawater. Dahlem Konferenzen, Berlin, pp. 87—136.

O'Shea, T.A. and Mancy, K.H., 1976. Characterisation of trace metal organic interactions by anodic stripping voltammetry. Anal. Chem., 48: 1603—1607.

Packham, R.F., 1964. Studies of organic colour in natural water. Proc. Soc. Water Treat. Exam., 13: 316—334.

Parks, G.A., 1975. Adsorption in the marine environment. In: J.P. Riley and G. Skirrow (Editors), Chemical Oceanography, I. Academic Press, 2nd ed., London, pp. 241—308.

Perdue, E.M., Beck, K.C. and Reuter, J.H., 1976. Organic complexes of iron and aluminum in natural waters. Nature, 260: 418—420.

Picard, G.L. and Felbeck, G.T. Jr., 1976. The complexation of iron by marine humic Acid. Geochim. Cosmochim. Acta, 40: 1347—1350.

Pillai, T.N.V. and Ganguly, A.K., 1970. Nucleic acids in dissolved constituents of sea water. Curr. Sci., 30: 501—504.

Pillai, T.N.V., Desai, M.V.M., Matthew, E., Ganapathy, S. and Ganguly, A.K., 1971. Organic materials in the marine environment and the associated metallic elements. Curr. Sci., 40: 75—81.

Piro, A., Bernhard, M., Branica, M. and Verzi, M., 1973. Incomplete exchange reactions between radioactive ionic zinc and stable natural zinc in sea water. In: Radioactive Contamination of the Marine Environment. I.A.E.C., Vienna, pp. 29—45.

Pocklington, R., 1977. Chemical processes and interactions involving marine organic matter. Mar. Chem., 5: 479—496.

Pommer, A.M. and Breger, I.A., 1960. Potentiometric titration and equivalent weight of humic acid. Geochim. Cosmochim. Acta, 20: 30—44.

Prakash, A. and Rashid, M.A., 1968. Influence of humic substances on the growth of marine phytoplankton: dinoflagellates (Gonyaulax). Limnol. Oceanogr., 13: 598—606.

Prelog, V., 1964. Iron containing compounds in micro-organisms. In: E. Gross (Editor), Iron Metabolism. Springer-Verlag, Berlin, pp. 73—83.

Provasoli, L., McLaughlin, J.J.A. and Droop, M.R., 1957. The development of artificial media for marine algae. Arch. Mikrobiol., 25: 392—428.

Ramamoorthy, S. and Kushner, D.J., 1975a. Heavy metal binding sites in river water. Nature, 256: 399—401.

Ramamoorthy, S. and Kushner, D.J., 1975b. Heavy metal binding components of river water. J. Fish. Res. Board Can., 32: 1755—1766.

Rashid, M.A., 1971. The role of humic acids of marine origin and their different molecular weight fractions in complexing di- and trivalent metals. Soil Sci., 111: 298—306.

Rashid, M.A., 1972. Quinone content of humic compounds isolated from marine environments. Soil Sci., 113: 181.

Rashid, M.A., 1974. Absorption of metals on sedimentary and peat humic acids. Chem. Geol., 13: 115—123.

Rashid, M.A. and King, L.H., 1969. Molecular weight distribution measurements on humic and fulvic acids fractions from marine clay on the Scotian Shelf. Geochim. Cosmochim. Acta, 33: 147—151.

Rashid, M.A. and King, L.H., 1971. Chemical characteristics of fractionated humic acids associated with marine sediments. Chem. Geol., 7: 37—43.

Rashid, M.A. and Leonard, J.D., 1973. Modifications in solubility and precipitation behaviour of various metals as a result of their interactions with sedimentary humic acid. Chem. Geol., 11: 89—97.

Raspor, B., Valenta, P., Nürnberg, H.W. and Branica, M., 1977. The chelation of cadmium with NTA in sea water as a model for the typical behaviour of trace metal chelates in natural waters. Sci. Total Environ., 9: 87—109.

Reichardt, W., Overbeck, J. and Steubing, L., 1967. Free dissolved enzymes in lake water. Nature, 216: 1345—1347.

Reuter, J.H. and Perdue, E.M., 1977. Importance of heavy metal—organic matter interaction in natural water. Geochim. Cosmochim. Acta, 41: 326—334.

Riffaldi, R. and Schnitzer, M., 1972. Electron spin resonance spectrometry of humic substances. Soil Sci. Soc. Am. Proc., 36: 301—305.

Riley, J.P., 1975. Analytical chemistry of sea water. In: J.P. Riley and G. Skirrow (Editors), Chemical Oceanography, 3. Academic Press, London, 2nd ed., pp. 193—477.

Riley, J.P. and Chester, R., 1971. Introduction to Marine Chemistry. Academic Press, London, 465 pp.

Rona, E., Hood, D.W., Muce, L. and Buglia, B., 1962. Activation analysis of manganese and zinc in sea water. Limnol. Oceanogr., 7: 201—206.

Saifullah, S.M., 1978. Inhibitory effects of copper on marine dinoflaggelates. Mar. Biol., 44: 299—308.

Schindler, J.E., Williams, D.J. and Zimmerman, A.P., 1976. Investigations of extracellular electron transport by humic acids. In: J.E. Nriagu (Editor), Environmental Biogeochemistry, 1. Ann Arbor Science, Mich., pp. 109—115.

Schnitzer, M., 1969. Reactions between fulvic acid, a soil humic compound and inorganic soil constituents. Soil Sci. Soc. Am. Proc., 33: 75—81.

Schnitzer, M. and Hansen, E.H., 1970. Organo-metallic interactions in soils: 8. An evaluation of methods for the determination of stability constants of metal—fulvic acid complexes. Soil Sci., 109: 333—340.

Schnitzer, M. and Khan, S.U., 1972. Humic Substances in the Environment. Dekker, New York, N.Y., 327 pp.

Senesi, N., Griffith, S.M., Schnitzer, M. and Townsend, M.G., 1977. Binding of Fe^{3+} by humic materials. Geochim. Cosmochim. Acta, 41: 969—976.

Shabarova, N.T., 1954. Usp. Sov. Biol. 37, 203. Quoted by Saxby 1969 In: Rev. Pure Appl. Chem., 19: 131—150.

Shapiro, J., 1964. Effect of yellow organic acids on iron and other metals in water. J. Am. Water Works. Assoc., 56: 1062—1082.

Sharp, J.H., 1973. Size classes of organic carbon in sea water. Limnol. Oceanogr., 18: 441—447.

Sholkovitz, E.R., 1976. Flocculation of dissolved organic and inorganic matter during the mixing of river water and sea water. Geochim. Cosmochim. Acta, 40: 831—845.

Siegel, A., 1971. Metal organic interactions in the marine environment. In: S.D. Faust and J.V. Hunter (Editors), Organic Compounds in Aquatic Environments. Dekker, New York, N.Y., pp. 265—289.

Siegel, A. and Degens, E.T., 1966. Concentration of dissolved amino acids from saline waters by ligand exchange chromatography. Science, 151: 1098—1101.

Sillén, L.G. and Martell, A.E., 1971. Stability constants of metal—ion complexes. Supplement No. I. The Chemical Society, London, Spec. Publ. No. 25, 865 pp.

Singer, P.C., 1973. Trace Metals and Metal Organic Interactions in Natural Waters. Ann Arbor, Mich., 380 pp.

Slowey, J.F. and Hood, D.W., 1971. Copper, manganese and zinc concentrations in Gulf of Mexico water. Geochim. Cosmochim. Acta, 35: 121—138.

Slowey, J.F., Jeffrey, L.M. and Hood, D.W., 1967. Evidence for organic complexed copper in sea water. Nature, 214: 377—378.

Smith, R.G. Jr., 1976. Evaluation of combined applications of ultrafiltration and complexation capacity techniques to natural waters. Anal. Chem., 48: 74—76.

Spencer, C.P., 1958. The chemistry of ethylenediamine tetra acetic acid in sea water. J. Mar. Biol. Assoc. U.K., 37: 127—144.

Steemann Nielsen, E. and Wium-Anderson, S., 1970. Copper ions as poison in the sea and in fresh water. Mar. Biol., 6: 93—97.

Stevenson, F.J., 1976. Stability constants of Cu^{2+}, Pb^{2+} and Cd^{2+} complexes with humic acids. Soil Sci. Soc. Am. J., 40: 665—672.

Stiff, M.J., 1971. The chemical states of copper in polluted fresh waters and a scheme of analysis to differentiate them. Water Res., 5: 585—599.

Stolzberg, R.J. and Rosin, D., 1977. Chromatographic measurement of submicromolar strong complexing capacity in phytoplankton media. Anal. Chem., 49: 226—230.

Strohal, P. and Huljev, D., 1970. Investigation of mercury pollutant interaction with humic acid by means of radio tracers. In: Nuclear Techniques in Environmental Pollution. I.A.E.A., Vienna, pp. 439—446.

Stuermer, D.H. and Harvey, G.R., 1974. Humic substances from sea water. Nature, 250: 480—481.

Stuermer, D.H. and Harvey, G.R., 1977. The isolation of humic substances and alcohol soluble organic matter from sea waters. Deep-Sea Res., 24: 303—309.

Stumm, W. and Brauner, P.A., 1975. Chemical Speciation. In: J.P. Riley and G. Skirrow (Editors), Chemical Oceanography, 1. Academic Press, 2nd ed., London, pp. 173—239.

Stumm, W. and Morgan, J.J., 1970. Aquatic Chemistry: An introduction Emphasizing Chemical Equilibria in Natural Waters. Wiley—Interscience, New York, N.Y., 583 pp.

Sunda, W. and Guillard, R.R.L., 1976. The relationship between cupric ion activity and the toxicity of copper phytoplankton. J. Mar. Res., 34: 511—529.

Sugai, S.F. and Healy, M.L., 1978. Voltammetric studies of the organic association of copper and lead in two Canadian inlets. Mar. Chem., 6: 291—308.

Sugimura, Y., Suzuki, Y. and Miyake, Y., 1978a. The dissolved organic iron in sea water. Deep-Sea Res., 25: 309—314.

Sugimura, Y., Suzuki, Y. and Miyake, Y., 1978b. On the chemical forms of minor metallic elements in the ocean. J. Oceanogr. Soc. Jap., 34: 93—96.

Szalay, A. and Szilágyi, M., 1967. The association of vanadium with humic acids. Geochim. Cosmochim. Acta, 31: 1—6.

Szilágyi, M., 1967. Sorption of molybdenum by humus preparation. Geochem. Int., 4: 1165—1167.

Szilágyi, M., 1972. The geochemical role of standard potential in reactions between humic substances and metals. Geochem. Int., 9: 402—406.

Szilágyi, M., 1973. The redox properties and the determination of the normal potential of the peat—water system. Soil Sci., 115: 434—437.

Tanford, C., 1967. Physical Chemistry of Macromolecules. Wiley, New York, N.Y., 710 pp.

Theis, T.L. and Singer, P.C., 1973. The stabilisation of ferrous iron by organic compounds in natural water. In: P.C. Singer (Editor), Trace Metals and Metal Organic Interactions in Natural Waters. Ann Arbor, Mich., pp. 303—320.

Tréguer, P., Le Corre, P. and Courtot, P., 1972. A method for determination of the total dissolved free fatty acid content of sea water. J. Mar. Biol. Assoc. U.K., 52: 1045—1055.

Tschapek, M. and Wasowski, C., 1976. The surface activity of humic acid. Geochim. Cosmochim. Acta, 40: 1343—1345.

Turner, D.R. and Whitfield, M., 1979. The electrodeposition of trace metal ions from multi-ligand systems. II. Calculations of the electrochemical availability of lead at trace levels in sea water. J. Electroanal. Chem., 103: 61—79.

Visser, S.A., 1964. Oxidation reduction potential and capillary activities of humic acids. Nature, 204: 581.

Wershaw, R.L. and Pickney, D.J., 1971. Association and dissociation of humic acid fractions as a function of pH. Geol. Surv. Res. Pap., 750-D: 216—218.

Whitfield, M., 1975. Sea water as an electrolyte solution. In: J.P. Riley and G. Skirrow (Editors), Chemical Oceanography, 1. Academic Press, 2nd ed., London, pp. 44—171.

Whitfield, P.H. and Lewis, R.G., 1976. Control of the biological availability of trace metals to a calanoid copepod in a coastal fjord. Estuarine Coastal Mar. Sci., 4: 255—266.

Whitfield, M. and Turner, D.R., 1980. Theoretical studies of the chemical speciation of lead in sea water. In: M. Branica and Z. Konrad (Editors), Lead Occurrence, Fate and Pollution in the Marine Environments: International Experts Discussion, Rovinj, Yugoslavia, 1977. Pergamon, London, 364 pp. (in press).

Williams, P.J. Le B., 1975. Biological and chemical aspects of dissolved organic material in sea water. In: J.P. Riley and G. Skirrow (Editors), Chemical Oceanography, 2. Academic Press, London, 2nd ed., pp. 301—363.

Williams, P.M., 1961. Organic acids in Pacific Ocean waters. Nature, 189: 219—220.

Williams, P.M., 1969. The association of copper with dissolved organic matter in sea water. Limnol. Oceanogr., 14: 156—158.

Williams, P.M. and Baldwin, R.J., 1976. Cupric ion activity in coastal sea water. Mar. Sci. Comm., 2: 161—181.

Wilson, D.E. and Kinney, P., 1977. Effects of polymeric charge variations on the proton metal ion equilibria of humic materials. Limnol. Oceanogr., 22: 281—289.

Wilson, T.R.S., 1975. Salinity and the major elements of sea water. In: J.P. Riley and G. Skirrow (Editors), Chemical Oceanography, 1. Academic Press, 2nd ed., London, pp. 365—413.

Zirino, A. and Yamamoto, S., 1972. A pH-dependent model for the chemical speciation of copper, zinc, cadmium and lead in sea water. Limnol. Oceanogr., 17: 661—671.

Zsolnay, A., 1977. Inventory of non-volatile fatty acids and hydrocarbons in the oceans. Mar. Chem., 5: 465—475.

Chapter 8

CHEMICAL TELEMEDIATORS IN THE MARINE ENVIRONMENT

M.J. GAUTHIER and M. AUBERT

1. INTRODUCTION

Each individual marine organism, considered from an autoecological standpoint, has its own characteristics and requirements which control its behaviour within the species and in the environment. It must adjust to the physico-chemical conditions of the medium in which it lives and to the presence of other organisms in order to make up a balanced functional whole, the ecosystem.

Marine ecosystems, like terrestrial biological communities, are mainly considered to depend, as regards their structure and functioning, upon the trophic and sexual relationships between individuals, either "horizontally" between individuals at the same level, or "vertically", down through the various levels of the food chains.

Such a concept would indicate that the marine environment owes its long-term stability to a dynamic equilibrium within a highly complex biological network. The functioning of biosystems is largely governed by their structure, i.e. the qualitative and quantitative distribution of individuals within biocoenoses as well as by the relationships of organisms with each other and with the environment. Therefore, it is essential to discover how these innumerable connections work in order to analyse the biological equilibrium of the marine environment.

A more recent approach to the understanding of ecosystems is the study of interactions between members of the biotic communities by way of chemical substances which intervene in the homeostasis of the systems without necessarily having an energetic alimentary value. These substances inform the organisms of certain characteristics of the ambient medium, which may or may not be favourable to their growth and/or reproduction.

This concept of chemical ecology was initially developed from works carried out in the terrestrial field, which lent themselves more easily to experimentation (Alexander, 1971; Pasteels, 1972). In the marine sphere, the hypothesis of an intra- and interspecific control by "external metabolites" was first postulated forty years ago by Lucas (1938, 1947, 1955, 1961), and then adopted more recently by Nigrelli (1958a, b), Fontaine (1970) and Todd et al. (1972). The theory was generalised by Aubert (1971) following a number of works carried out at the C.E.R.B.O.M., essentially in the field of microbiology (Aubert et al., 1967, 1970b, 1972a; Aubert and Pesando, 1971, 1974).

The various types of chemical mediators have been classified on the basis of very different criteria: ecological function of the message, its meaning for the emitting or receptor species, mode of emission, and effect on individuals of the same or different species, etc.

Florkin (1965) has defined ecomones as "the molecular factors, specific or non-specific, exercising an action on the constitution and the persistence of a biotical community". This definition corresponds to the "ectohormones" of Bethe (1932) and the "telergones" of Kirschenblatt (1962). Depending on whether or not the receiving or emitting individuals belong to the same species, ecomones have been described as *pheromones* (Karlson and Butenandt, 1959; Karlson and Luscher, 1959) or *allomones* (Brown et al., 1970). Both types of mediators have been accorded various names: hemiohormones (Bethe, 1932), sociohormones (Pickens, 1932) and homotelergones (Kirschenblatt, 1962) for the former, and alloiohormones (Bethe, 1932), heterotelergones (Kirschenblatt, 1962), coactones (Florkin, 1965) and allelochemics (Whittaker, 1970) for the latter. Brown et al. (1970) have distinguished a sub-group among the allomones, namely kairomones (or blaptones, according to Bethe, 1932), which evoke a positive reaction from the receiving species but an indifferent or negative reaction from the emitting species.

Florkin (1965) has proposed the term *exoactones* for those mediators emitted in normal circumstances and *endoactones* where emission only occurs as a result of "trauma".

This semantic profusion gives some idea of the variety and complexity of the phenomena under study. In 1971, Aubert simultaneously widened and simplified the concept by naming "telemediators", these being the chemical substances produced by marine animals or plants and liberated into the medium, affecting the behaviour or biological functions of the same species or of other species.

In the marine environment, chemical communications are naturally facilitated by the vectorial function of the water, which ensures widespread distribution whatever the site or type of mediator. They play a significant role in widely different fields, both at the intraspecific (sexual behaviour, trailing, recognition, migration, alarm) and interspecific (nutrition, predation, defence, pseudo-social relationships, commensalism) levels.

An analogy may be perceived between this concept of function regulation and the functioning of higher organisms in which internal equilibrium is controlled by hormonal regulators. As will be seen on closer investigation, the analogy should be restricted to telemediators with a very low threshold of activity, which cannot be used as metabolites.

Studies have now reached a stage where it seems appropriate to proceed with an accurate analysis of the data gathered *in situ* and *in vitro* in order to establish the status of our knowledge in this field, and to note the unknowns and gaps that merit further study.

2. THE BASIC PRINCIPLES OF COMMUNICATION VIA CHEMICAL MEDIATORS

The biological interactions mentioned above rest on one fundamental biological phenomenon: chemoreception. This non-visual mode of perception appears to be particularly well developed in marine organisms, indeed it compensates for the lack of visual facility commonly encountered in the turbid or dark marine environment. The importance of chemoreception is further demonstrated by the often substantial increase in activity of chemoreceptor organs in marine organisms.

Remote sensing between two organisms through chemical mediation involves a sequence of actions and reactions in which each partner plays an active role (Fig. 1). The organism producing the telemediator must synthesise it, either spontaneously or in an induced way, then release it into the medium during its active growth (or following cell lysis). Next, the mediator is conveyed to the remote receiving organism, which it should reach without

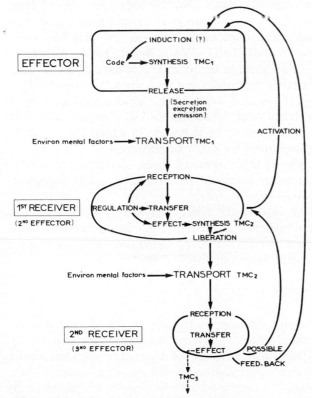

Fig. 1. General scheme of biological interactions between individuals within biocenoses by the way of chemical telemediators (*TMC*).

being degraded or chemically modified (or possibly after such modification). Its concentration must remain equal to, or higher than, its threshold of activity. The receptor detects it, with or without absorption, and reacts by modifying its own metabolism or behaviour. In extreme cases, this may even cause the death of the receptor. In more complex situations, a second mediator will be synthesised and released into the medium.

In a general sense, communication will be "positive" if the mediator enhances growth or leads to favourable modification of the behaviour of the receiving organism. Such a supply may lead to commensalism, proto-cooperation and eventually symbiosis. Interaction is said to be "negative" when the mediator is toxic in itself (antibiotic, antiseptic or toxic) or brings about a lethal metabolic or behavioural modification in the receiver.

Theoretically, this approach is useful in the study of marine biological interactions and presents a number of explanations. It involves a knowledge of complex microecological phenomena, some aspects of which are now understood. A pressing problem is that of chemotaxis (chemoreception in marine organisms). Telemediator-activity threshold and turnover in the natural environment also requires investigation. Additionally one should take into account the physical context in which the phenomena take place. For instance, it is maintained that most bacteria in the sea are adsorbed onto solid or particulate substrates (Zobell, 1937, 1943; Wood, 1963; Aubert et al., 1975b). In this case, the relationships occur in the sphere of a few fractions of a millimetre. Most benthic organisms are thus sharply dependent on the substrate: this emphasises the significance of the biochemical processes which cause micro-organisms to colonise substrates and so compete with one another. A further point of interest would be the ways in which they modify the substrate prior to their replacement by other organisms.

In this field, the literature shows that, in many cases, mediation processes are involved in the maintenance of intra- or inter-specific relationships in microbial, algal or animal populations.

Although comparatively few cases have been analysed, the evidence too often being obtained *de vitro* rather than *de situ*, it can be demonstrated that such relationships do exist between marine organisms and are indeed quite likely to occur in the natural environment.

3. CHEMICAL MEDIATION BETWEEN BACTERIA AND ALGAE

The influence of free chemical mediators in the life cycles of marine organisms, and the inter-specific competition amongst micro-organisms has only recently been studied. Paradoxically, because of their size and metabolic activity, it is the lower forms (bacteria, yeasts, unicellular algae) which liberate into the medium the greatest quantity of simple organic metabolites. It is likely therefore that these micro-organisms represent a higher echelon of telemediation phenomena in the marine environment.

3.1. Interactions between bacteria

The inter-dependency of bacterial species has been described for a number of cases. One example is the close dependency of sulphate-reducing bacteria on the presence of an associated heterotrophic microflora which supply the bacteria with additional growth factors and favour reduction of the medium (Cahet, 1965, 1966; LeGall and Postgate, 1974). Moreover, the production of hydrogen sulphide binds them to sulpho-oxidisers within the ecosystem (called "sulphuretum" by Baas-Becking, 1925).

However, the best examples of a direct relationship between species through chemical substances are to be found in the field of microbial antagonisms. Many studies have described antibiotic and antiseptic substances synthesised by marine bacteria (Gauthier, 1969b), and also their production of exo- or endocellular bacteriolytic enzymes (Mitchell and Nevo, 1964), which are active toward both bacteria and fungi (Mitchell and Wirsen, 1968). Therefore, an aggregate of lytic, antibiotic and toxic actions should be considered in order to explain the majority of microbial antagonisms, observed *in vitro* or assumed to occur in the natural environment. The assumption that these actions are the same *in situ* is borne out by parallel observations carried out in the laboratory and at sea. For example, Sieburth (1967, 1968) was able to observe, between pseudomonads and arthrobacters living in Narragansett Bay, a natural antagonism which he verified in experiments using cultures of the two bacterial groups (Fig. 2). It has also been demonstrated that the polyanionic antibiotics produced by several *Alteromonas* (Gauthier et al., 1975b), which modify bacterial respiration (Gauthier, 1976), do exist *in situ*, since similar substances were isolated out of the polyanionic carbohydrates recovered from immersed artificial substrates. In this particular case, the overall pattern of the sequence of mediation reactions taking place between the organism, producing the antibiotic, and the receiver (marine or terrestrial bacterium) is generally accepted.

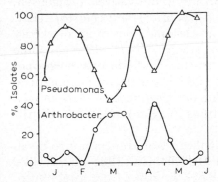

Fig. 2. Apparent inverse relationship between pseudomonads and arthrobacters in Narragansett Bay (Rhode Island), during January to June 1964 (from Sieburth, 1967).

It is likely that these antagonisms contribute to the sea's ability to purify itself of terrestrial micro-organisms, which are less competitive than the indigenous organisms (Gauthier, 1973). This implies that bacteria possess chemoreceptive and chemotactic properties (Adler, 1966, 1969; Fogel et al., 1971). These two ethological characteristics result from a series of secondary chemical mediations in the receiving organism. In bacteria, for instance, they involve the detection of the attractant or repellant by cell-wall receptors, a transmission of the stimulus into the receiving cell via molecules such as methionine (as S-adenosyl-methionine), and subsequently a chemotactic response which brings about an inversion of the flagellum rotation — right-handed for repulsion, left-handed for attraction (Larsen et al., 1974).

As demonstrated by Chet et al. (1971), this phenomenon almost certainly plays a part in causing some predatory bacteria (*Pseudomonas*) to be attracted by exudates from their prey (fungi, algae). Thus, it probably accounts for bacterial biodegradation in the sea, being a means of supporting predation, or more generally feeding, by keeping the predator close to its food source. Predation is also dependent on the predator's ability to produce enzymes that can degrade the prey (Mitchell, 1971).

3.2. Interactions between bacteria and algae

Many studies of microbial ecology have described the enhanced growth of bacterial populations due to organic substances produced by planktonic algae. The growth elements may be nutrients, vitamins or phytohormones (Bentley, 1958, 1960; Augier, 1972). They are usually released into the medium after lysis of the algal cells during the senescence of the culture or natural population. They may also be antiseptics or antibiotics: 20—25% of the Mediterranean diatoms produce such inhibitors (Aubert and Gauthier, 1967). Their chemical nature is variable: fatty acids (Pesando, 1972; Gauthier et al., 1978b), nucleosides (Aubert et al., 1970a), complex lipoproteins (Ulitzur and Shilo, 1970), chlorophyll derivatives (Jørgensen, 1962), acrylic acid (Sieburth, 1959, 1961a, b), peptides and *gelbstoff* (Berland et al., 1972c). Usually, these inhibiting substances act more markedly upon gram-positive aerobic bacteria, such as staphylococci, although some are also active toward gram-negative organisms and various anaerobes (Aubert and Gauthier, 1967; Aubert et al., 1968; Aubert and Gambarotta, 1972). A few are extremely active, such as the one produced by *Asterionella notata*, with minimum inhibitory concentrations lying around 0.005 μg ml^{-1} (Gauthier, 1969a).

The release of such substances by phytoplankton would seem not to be an experimental artifact; according to the data of Hellebust (1965), Nalewajko (1966), Samuel et al. (1971) and Thomas (1971), 1—20% of the total photo-assimilated carbon is released into the medium by natural phytoplankton populations. Although the importance of algal excretion can probably be reduced in healthy cells (Duursma, 1963; Sharp, 1977), these exudates can be highly significant, both as regards the nutrition of bacteria in the sea and the control of their development, depending on whether the mediator enhances or inhibits growth. In the former case, Bell and Mitchell (1972) suggest the formation of a "phycosphere" in the immediate vicinity of each planktonic alga, made up of bacteria which are attracted and fixed by chemotaxis close to the cell supplying them with growth elements (which are the actual attractants). These authors showed that the chemotactic response is dependent of the age of the algae and becomes significant in senescent cells only. They have neglected, however, the possibility of a counter-activity, whereby the algae produce repellants due to their own anti-biotic activity which is maximal during the growth phase. A similar phenom-enon was documented by Sieburth (1968) in order to explain the phyto-plankton repulsion of bacterial attachment by the excretion of acidic com-ponents produced by diatoms (Fig. 3).

This inhibiting activity on the part of the algae and the role it plays in the natural environment were established by the observations of Sieburth and Pratt (1962) on the inhibition of coliforms and *Vibrio* in Narragansett Bay, following the blooms of the diatom *Skeletonema costatum*. Similarly, studies carried out by Aubert and co-workers (1967—1975) showed that bio-occretions of some diatoms (*Asterionella japonica, Asterionella notata,*

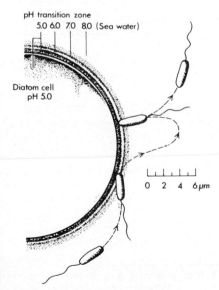

Fig. 3. Hypothetical mechanism of phytoplankton repulsion of bacterial attachment by excretion of acidic cell constituents forming an external pH transition zone into the alka-line seawater (from Sieburth, 1968).

Chaetoceros lauderi, Skeletonema costatum, Nitzschia spp.) contribute to the biological purification of the marine environment and are responsible for the destruction of terrestrial bacteria in the open sea.

More recently, Rieper (1976) showed that the growth of a fraction of the bacterial population in the Schlei Fjord (Baltic Sea) is enhanced by actively growing cells of *Chlorella* spp., whereas other bacteria are inhibited by the same algae at the termination of their spring bloom (Fig. 4). In the late summer a symbiotic or mutualistic relationship exists between other bacteria and the alga *Microcystis aeruginosa*.

Moreover, a significant number of seaweed species produce different antibiotic substances; fatty acids, acrylates, polyphenols, terpenoids, chlorophyllides, sulphonated or brominated compounds, peptides (Roos, 1957; Lewin, 1962; Sieburth, 1964, 1968; Hornsey and Hide, 1974; Khaleafa et al., 1975). Most of these antibiotics were found experimentally in algal extracts, and are also associated with gram-positive and gram-negative bacteria. Evidence for their activity in the natural environment has also been established by a few *in situ* observations. Thus, according to Sieburth (1968), "the phaeophyte tannins are sometimes excreted in amounts sufficient to discolour inshore waters. Such concentrations have inhibitory activity for bacteria, algae and animal forms". Algal tannins are responsible for the absence of fouling on *Sargassum* fronds (Sieburth and Conover, 1965).

In the case of mediators produced by bacteria, action is twofold. The relationship between the bacterium and the alga may be positive. For instance, many lower algal forms exhibit enhanced growth in the presence of vitamins or growth factors produced by micro-organisms (Ericson and Lewis, 1953; Burkholder, 1959; Gutvieb et al., 1973). Nevertheless, some

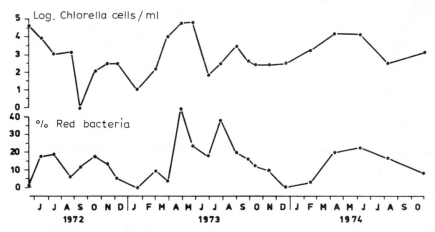

Fig. 4. Comparison of the percentage of red pigmented bacteria with the concentration of *Chlorella* sp. in the Schlei Fjord (western Baltic Sea) (from Rieper, 1976).

studies suggest that bacteria may be both the greatest producers and the greatest consumers of vitamins, in which case they would be competing with algae for their own secretions (Provasoli, 1963). In a similar way, certain epiphytic microfloras may exert a morphogenetic influence on seaweeds (Provasoli and Pintner, 1964).

The relationship may be negative if the bacterial mediator is toxic. It has been demonstrated that bacteria can produce substances with a high anti-algal activity (Berland et al., 1972a, b, 1973). It is not likely, however, that this phenomenon would actually take place in the natural environment, because of the low concentrations of inhibitors around algal cells. Berland and Maestrini (1969) also showed that the diatom *Nitzschia ascicularis*, inhibited by antibiotics from different *Vibrio* and *Pseudomonas*, in turn activates the synthesis of those inhibitors through a process similar to induction. It is one of the few cases of this type of direct feedback to be observed.

In the case of seaweeds, it should be noted that epiphytic bacteria may act as a filter, modifying both the exogenous substances absorbed by the algal thallus, and the secondary metabolites excreted by it. According to Sieburth (1968), "in this way, the inhibitory influence of seaweed population on a different, susceptible species, may be reduced".

3.3. Inter-algal relationships

"Limiting factors" in the regulation of phytoplanktonic populations, such as nutrients, light, temperature and salinity, have been traditionally evoked to explain species succession in the natural environment. Their combined influence probably accounts for major regulation in the open sea, provided that "normal" conditions are not disturbed to any great extent. Algal secondary metabolites, however, undoubtedly influence succession of phyto-plankters, although "it seems likely that, unless a metabolite is produced in prodigious amounts, its influence on succession is readily tempered and applied through interaction with growth factors" (Smayda, 1963).

In other cases metabolites from algae are involved in interactions between algal species or populations both *in situ* and *in vitro*. For phytoplankton, the best-known phenomena in this field concern the antagonism between diatoms and dinoflagellates, resulting in alternate growth of populations. Through a series of experiments carried out *in vitro*, Pincemin (1971) showed that the growth of the diatom *Asterionella japonica* was inhibited by the presence of dinoflagellates such as *Coolia monotis* or *Peridinium trochoideum*. Similar occurrences were reported for other algae: Pratt (1966) observed an antagonism between the diatom *Skeletonema costatum* and the xanthophyte *Olisthodiscus luteus*, both in the natural environment and in mixed cultures of these species. The dominance of *Skeletonema* was explained by its shorter generation time, whereas that of *Olisthodiscus* was

234

due to the production of an extra-cellular toxic metabolite (yellow phenolic material).

In a different connection, Aubert et al. (1967) have observed that, at some periods of the year and especially during spring, diatoms cease to synthesise antibacterial substances, a phenomenon which corresponds with the peak density of dinoflagellate populations. *In vitro* it has been shown that the proximity of *Prorocentrum micans* inhibits production of one of the antibacterial substances in *Asterionella japonica* (Aubert et al., 1970b). The substance released by the dinoflagellate which accounted for that effect was found to be proteinaceous with a very low threshold of activity: 10^{-9} M (Aubert and Pesando, 1971; Aubert et al., 1972a).

The multi-component biological system is more complex. *Prorocentrum micans* produces two antagonistic mediators (Fig. 5): one enhances the synthesis of carotenoid pigments in *A. japonica* and *Chaetoceros lauderi*, thus protecting their antibiotic fatty acids from *in vivo* photo-activation, and the second acts more probably on the bacterial cell, increasing the lethal effect of these antibiotics (Gauthier et al., 1978a).

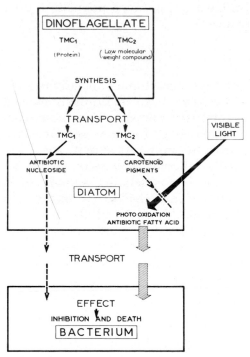

Fig. 5. Interactions between three marine microorganisms: a dinoflagellate (*Prorocentrum micans*), a diatom (*Asterionella japonica*) and a bacterium (marine or terrestrial), involving four chemical telemediators (*TMC*): *TMC1* and *TMC2* are secondary mediators, the antibiotics of *A. japonica* being primary mediators.

On closer examination of the interactions between terrestrial bacteria, diatoms and dinoflagellates, it was discovered that their populations fluctuated. Bacteria which proliferate in waters rich in organic matter will release growth mediators within one day, for instance, vitamin B_{12}, which enhances diatom growth. In their turn, the diatoms release antibiotic substances, such as fatty acids and polysaccharides, which inhibit terrestrial bacteria. The process would result in an invasion of the environment by diatoms if the latter were not inhibited by dinoflagellates over a cycle of some thirty days. Vitamin B_{12} helps the dinoflagellates grow and give off mediators which impede diatom proliferation. It should be noted that relationships between two phytoplanktonic groups are not always negative: Pincemin (1971) described an increase in the growth rate of *Peridinium trochoideum* in the presence of exudates from the diatom *Asterionella japonica*.

This difference in effect of such negative and positive mediations can partly explain the sequence of algal populations in distrophic bodies of water on the Mediterranean coast (Gauthier et al., 1975a). A similar phenomenon was described by Keating (1977, 1978) for eutrophic lakes. In the case of seaweeds, most relationships between higher algae and their algal epiphytes are based upon the production of secondary metabolites released by the thallus. According to Sieburth (1968), two factors may influence the association: excretion of essential metabolites, leading to the selection of obligate algal parasites, and repulsive substances, which prevent attachment of potential parasites.

4. CHEMICAL INTERACTIONS BETWEEN BACTERIA OR ALGAE AND MARINE ANIMALS

4.1. Relationships between bacteria and marine animals

A few observations establish the influence of bacterial components on animals, mainly protozoans. The ingestion of bacteria by protozoans may be controlled by secretions or constituent compounds of their preys. Unicellular organisms may even be able to choose the bacteria they feed on, avoiding such species as *Chromobacterium* or *Serratia*, which contain toxic or repulsive products (see Paoletti, 1964). Cyclic AMP (cyclic adenosine monophosphate), a well-known intra-cellular mediator, may play a role outside the cell, monitoring certain chemical communication systems. Chassy et al. (1969) observed that the slime mould *Dictyostelum discoideum* was attracted, in oligotrophic conditions, by C—AMP released by the bacteria on which it feeds.

On the other hand, several observations have dealt with the antibacterial activity of marine animals: alcyonarians (Burkholder and Burkholder, 1958; Ciereszko, 1962), molluscs (Prescott et al., 1962; Li et al., 1965) and sponges (Nigrelli et al., 1959). Their antibiotics are not selective and inhibit

236

both gram-positive and gram-negative bacterial strains. In addition to their activity on the microbial flora of host animals, they are probably involved in the self-purification of these organisms in polluted areas.

4.2. Relationships between algae and animals

Although a number of studies have been devoted to the purely trophic relationship between phytoplankton and zooplankton, there are few data to be found concerning the behaviour of, and competition between, these two groups through free chemical substances.

Some examples, however, suggest possible antagonisms, indicating an ecologically significant control. Fontaine (1970) reported an antagonism between certain zooplanktonic populations which excludes permanent coexistence of the two populations. The author assumed that phytoplankton release certain substances and this release, in connection with photosynthesis, would tend to drive zooplankton outside the phytoplankton mass. Conditions of alternating sunlight and darkness produce a rhythmical defence mechanism in the phytoplanktonic biomass and hence nutritional control of the zooplankton. This is merely an assumption, since the mediators released by planktonic algae have yet to be isolated or analysed. In spite of the fact

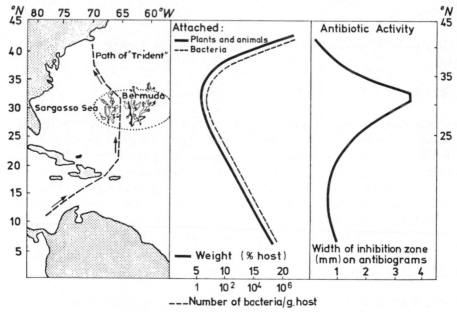

Fig. 6. Antibiotic activity of *Sargassum* weeds (*S. natans*, *S. fluitans*) on their incrusting plants, animals or epiphytic bacteria; inhibition was maximal near the Sargasso Sea gyre, the area of optimal growth (from Sieburth, 1962).

that only biological evidence exists, lacking biochemical confirmation, this phenomenon is a good example of animal exclusion, according to the theory developed by Hardy (1935).

Despite the relative lack of experiments carried out in this field, a few convincing observations suggest that a similar control takes place between phytoplankton and a number of higher organisms, e.g. molluscs. Lucas (1961) reported that oyster larvae feed on a few algae only, the growth of which depends on the presence of nutrients and toxic substances produced by other algae (Davis and Guillard, 1958; Hutner et al., 1958; McLaughlin, 1958). Similarly, the pumping rate of sea water by oysters is directly connected to the rate of some carbohydrates released by phytoplankton (Collier et al., 1953). Furthermore, Allison and Cole (1935) observed a relationship between the feeding activity of barnacles and the local phytoplankton density.

Secondary metabolites from higher algae may also influence marine animals, mainly planktonic larvae which are highly sensitive to toxic agents. A natural antagonism was observed by Conover and Sieburth (1964) between *Sargassum* seaweeds and their animal epiphytes (Fig. 6), due to release of tannins from the algae (Sieburth and Conover, 1965). For tide pools in Narragansett Bay, these authors (1966) also described a periodical decay of *Balanus balanoides* nauplii in waters containing tannins produced by the brown alga *Rafsia verrucosa*.

5. CHEMICAL COMMUNICATION BETWEEN HIGHER ORGANISMS

The improvement in experimental facilities, often related to a direct economic interest, has led to an increase in studies on the interactions between marine animals involving chemical substances. A number of taxonomic groups have been studied as a result of this interest. It has become classical to consider separately the *intra-specific* interactions or regulations, which strongly influence the sexual biology and the social evolution of numerous species, and the *inter-specific* or trans-specific interactions within the dynamic equilibrium of biocoenoses. Although arbitrary on many points, this clear differentiation has the merit of facilitating analysis of data in the literature.

5.1. Intra-specific interactions

The mediators involved in this type of interaction are the pheromones, which act in various ways: they may (1) cause a momentery behavioural modification, immediate but potentially reversible (releaser pheromone); (2) induce a series of modifications at the nervous and endocrinous levels following prolonged stimulation (primer pheromones); or (3) lead to a phenotypic modification of the receiving individual when they act at certain critical periods (imprinting pheromones).

The most important ecological role is probably played by the sexual mediators, which control the sequence of sexual processes for a number of marine animals. Other biological activities, such as alarm, trailing and individual or social recognition, also depend on this kind of mediator.

5.1.1. Intra-specific sexual mediators

Certain sexual pheromones can stimulate emission of these same pheromones by other mature individuals of the same species: this phenomenon has been described for the starfish *Acanthaster plancii* (Beach et al., 1975) and the annelid *Platynereis dumerilii* (Boilly-Marer, 1974). The presence of eggs or sperm also stimulates, in certain ascidians, the secretion of a gonadotropin by the neural gland (Mackie and Grant, 1974). The attraction of male gametes by female gametes, once they are released into the water, also depends on chemical mediating substances. This fact has been described for numerous species, belonging to various taxa: hydroids and medusa (Miller, 1972), and sea-squirts (Miller, 1975).

A similar phenomenon occurs at the level of adult individuals, whose meeting and copulation in reproductive periods are ensured by several pheromones. The majority of studies concern the crustaceans, where sexual attraction depends on ecdysones, also responsible for moulting, and for most of these animals, copulation is indeed associated with the moulting period (Patel and Crisp, 1961). A few species have been the subjects of detailed investigations: namely *Homarus americanus* (Atema and Engstrom, 1971; Atema and Gagosian, 1973), *Pachygrapsus crassipes*, *Cancer antennarius*, *Cancer anthonyi* (Kittredge et al., 1971). In all cases, chemical attraction induces sexual behaviour but suppresses aggression and the instinct to feed in both partners (McLeese, 1973). Consequently crustecdysones function at the same time as endohormones (moult) and telemediators (coupling). A similar chemical sexual attraction has been described. for the planktonic copepods *Eurythemora affinis*, *E. herdmani* and *Pseudodiaptomus coronatus* (Katona, 1973), *Calanus pacificus* and *Pseudocalanus* sp. (Griffiths and Frost, 1976). The phenomenon is not, however, restricted to crustaceans. The same has been described for the mollusc *Littorina littorea* (Dinter, 1974) and different fishes (Heiligenberg, 1976), although probably influenced by different pheromones.

Sex determination can also be regulated by free chemical mediators. This is particularly the case for the worm *Bonellia fuliginosa*, whose larvae produce only male parasites if they develop on an adult female. Sex is determined by the production of bonelline, a pheromone manufactured by this female (Mackie and Grant, 1974).

For fishes, certain mediating substances released into the medium by female individuals are operative during the pre-mating period, either in causing the meeting of conspecific individuals at spawning-time, as with *Salmo gairdneri* (Newcombe and Hartman, 1973), or in promoting the nuptial display of *Bathygobius soporator* (Tavolga, 1956).

5.1.2. Intra-specific non-sexual mediators

Some individual or social behavioural patterns other than sexual phenomena, in marine animals belonging to the same species, are regulated by chemical signals, e.g. alarm, social recognition, tracking and migration.

5.1.2.1. Phenomena of alarm.

The pheromones involved here are usually emitted by attacked or injured individuals, and represent a system of defence within the species against predators (Atema and Stenzler, 1977). With the cyprinoid *Phoxinus phoxinus*, the "fright" substance, identified as isoxanthopterin (Pfeiffer and Lemke, 1973), is actively secreted by the skin of alarmed individuals and induces flight in other individuals of the same species. It is interesting to note that this pterin is also produced by copepods and ascidians (Momzikoff, 1973).

Many examples of mediating activity of this type have been described for fishes, both *in situ* and in the laboratory (Von Frisch, 1938, 1941; Pfeiffer, 1963). The alarm mediator is emitted passively by injured individuals. The phenomenon has been found in the most diverse groups of invertebrates, classical examples being the molluscs. Thus the mud snail *Nassarius obsoletus* can emit into the water a substance with a molecular weight higher than 100,000, which is thermostable and present in its blood and tissues, and which causes a strong alarm reaction in conspecifics (Atema and Burd, 1975; Atema and Stenzler, 1977) (Fig. 7). The signal is endowed with a relative specificity: *N. obsoletus*, *N. vibex* and *N. trittatus* are highly alarmed by the crushing of one of their own species, but they react less strongly to that of the two other species. The crushing of foreign species (*Littorina, Melampus*) has no alarming effect, with the exception of *Urosalpinx cinerea* which must therefore contain a non-specific mediator (Stenzler and Atema, 1977).

Equally interesting is that with *Nassarius* species, the alarm effect is maintained *in situ* over many hours by adsorption of the mediator onto the sediments (Atema and Stenzler, 1977). An analogous behaviour pattern, expressed by rapid burying in the sediment, has been described for the mollusc *Aplysia brasiliana* (Aspey and Blankenship, 1976). This effect is also displayed by sessile organisms, such as the sea anemone *Anthopleura elegantissima*, which following an attack, emits a conspecific alarm pheromone, anthopleurin (Howe and Sheikh, 1975; Howe, 1976a). In this case the action of the mediator can be blocked by another biological substance, L-prolin (Howe, 1976b).

5.1.2.2. Phenomena of recognition.

It has been known for some time that pheromones are responsible for the schooling of fishes (McFarland and Moss, 1967). With the yellow bullhead, *Italurus natalis*, recognition takes place by way of a mediator present in the mucus of the skin (Todd et al., 1967). In the same way, *Hemichromis bimaculatus* recognises its offspring due to a pheromone liberated by them (Kühme, 1963). These phenomena protect the

240

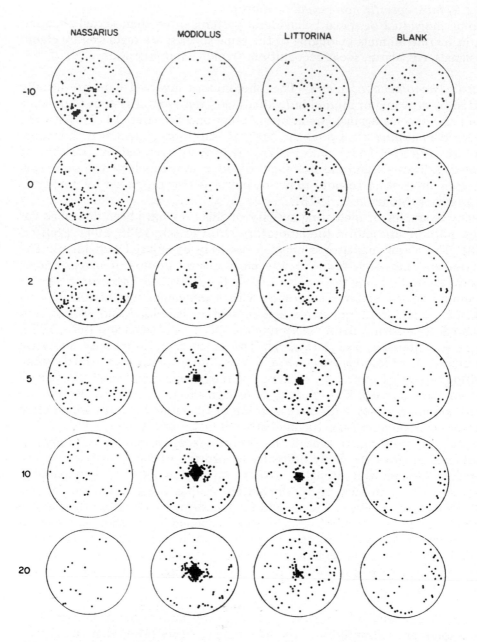

Fig. 7. Responses of the mud snail *Nassarius obsoletus* to four stimuli: crushed *Nassarius obsoletus*, *Modiolus demissus*, *Littorina littorea*, and blank (no stimulus). Dots indicate individual snails. Each circle (radius 35 cm) shows the superimposed sum of four replicate trials. The snails were counted at −10, 0, 2, 5, 10 and 20 min. Stimulus introduction was at time 0 min, in the center of the circle (from Atema and Burd, 1975).

perenniality of the species, and thus the maintenance of a pseudo-social structure promoting defence and nutrition.

5.1.2.3. Phenomena of tracking and migration. With benthic motile invertebrates, the spacial positioning of individuals in the environment in relation to shelter and food sources, and their possible grouping, may be realised by means of trailing substances only perceptible to individuals of the same species. Such mediators are deposited on the substratum mixed with superficial mucus. This is especially the case for the gastropods *Bursatella leachii pleii* (Lowe and Turner, 1976) and *Siphonaria alternata* (Cook and Cook, 1975). Pratt (1976) showed also that sated oyster drills (*Urosalpinx cinerea*) are able to signal their hunting success to conspecifics by way of a chemical signal.

The migrations of fishes probably have an analogous explanation, the waters being marked on a much larger scale. In the case of *Salmo alpinus*, Doving et al. (1974) have described pheromones working as homing devices, actively secreted by the skin of the fishes. According to the work of Hara et al. (1965), the electro-encephalographic activity of the olfactory bulbs of salmon during the spawning period is greatly increased in water originating from the river of birth, average for different waters crossed by the animal during its catadromous migration, and nil for the waters situated out of the zone of migration. A comparable phenomenon has been described for *Onchorynchus kisutch* (Cooper et al., 1976). It may be deduced, therefore, that these animals retain the memory of a sequence of olfactory stimuli at very low concentrations during their catadromous migration. The existence of molecular signals in the migration zones, which permanently trace the water and thus mark out specific migratory channels, may also be postulated. Such a concept of migration implies the emission of a specific pheromone for each population of salmon, which indicates the diversity and specificity of these chemical signals.

The origins of these substances remain controversial. According to Nordeng (1971), they are produced by youngsters during their descent to the sea. According to Atema et al. (1973), the alewife *Alosa pseudoharengus* is attracted to its river of reproduction by certain organic acids or bases of low molecular weight (less than 1000) which may derive from bacterial decomposition of organic matter found only in that river.

5.2. Inter-specific interactions

The inter-specific mediators, or allomones, may work in one of two directions: some are attractant, or allelostimulant, whilst others are repellent, or allelopathic. This distinction is not always clear, since the effects of such chemical mediation varies according to the receiving species within natural populations. It may, however, be retained for clarity of account.

5.2.1. Attractant allomones

These play a fundamental role in the nutritional activity of organisms by creating an attractant specific bond between prey and predator. They are also implicated in certain commensal or symbiotic relationships and appear to be necessary for the continuance of the life cycle of several sessile invertebrates during their larval fixation.

5.2.1.1. Nutritional phenomena. A great many examples may be found in the literature of the last twenty years to show the extent of the phenomenon of chemodetection of prey by predator animals. In motile animals the allomones guide the active search for food, and induce filter feeding in sessile organisms. The mediators generally form part of the blood or tissue of the prey, but may also involve compounds emitted by the skin or excretory organs. In most of the cases studied, they are small molecules, essentially amines (primary, secondary or tertiary amines, amino acids, etc.). The phenomenon is well-known following the works of Kleerekoper and Morgensen (1963) on the attraction of lampreys by leucine methyl ester, released by fishes of the family Salmonidae, inhabiting American lakes. In the same way, a combination of amino acids (TAU-ASP-THR-SER-GLU-GLY-ALA), present in the flesh of *Tapes japonica*, induces the search for food in the eel (Hashimoto et al., 1968). Further, several fishes are attracted by arcamine (hypotauryl-2 glycine) or by strombine (methyliminodiacetic acid), produced respectively by the molluscs *Arca zebra* and *Strombus gigas* (Sangster et al., 1975). Certain sharks are attracted by substances present in animal tissues (Kleerekoper et al., 1975). Another example is particularly interesting: the yellow fin tuna, *Thunnu: albacores*, can, when visual detection of prey is impeded, locate its prey by means of certain amino constituents representing in unknown ways the "chemical search image" of the prey. These amines may be detected to concentrations of 10^{-11} M (Atema et al., 1977).

With marine invertebrates, examples of chemodetection in the trophic field are numerous. One of the first studies concerns the crab *Limulus*; its chemoreceptor organs react strongly to aqueous extracts of bivalves (Barber, 1956). With coelenterates, reduced glutathione (GLU-CYS-GLY) frequently induces a feeding response which has been found for *Hydra littoralis* (Loomis, 1955) and the Portuguese man-of-war *Physalia physalis* (Lenhoff and Schneidermann, 1959). Some coelenterates respond to other amines: *Cyphastrea ocellina* to 1-proline (Mariscal, 1971), *Montastrea cavernosa* to glutamic acid (Lehman and Porter, 1973), *Palythoa psammophilia* to proline-glutathione synergy (Lenhoff and Lindstedt, 1974). The attractant role of biological amines may be found in various taxa: coelenterates (Lenhoff and Lindstedt, 1974; Walder, 1973), echinoids (Valentincic, 1973; Reimer and Reimer, 1975) and crustaceans (Case et al., 1960). Conversely, for certain species, the trophic attractant mediator appears to be macromolecular. This is the case for *Asterias forbesi* (Heeb, 1973) and *Nassarius obsoletus*

(Carr et al., 1974), both of which are attracted to their prey by proteins. In the case of *N. obsoletus*, chemodetection of prey has been observed *in vitro* and in the natural environment (Atema and Burd, 1975) (Fig. 6).

Specificity of chemoreception in predators generally is very pronounced. With the actinia *Boloceroides* sp., valine is a powerful activator of the feeding response, whereas leucine has no action in this sense and isoleucine inhibits this behaviour (Lindstedt, 1971). A further example of stereospecificity in detection is furnished by the crustacean *Homarus gammarus*, which reacts to a complex mixture of laevo-rotatory amino acids, but shows no response to the same elements in their dextro-rotatory form (Mackie, 1973).

It should be emphasised that the specificity of attraction of the predator to its prey is not fixed: to a certain extent, it would appear that some species can be conditioned to respond to organisms other than their habitual prey, when the latter disappear. This is the case for the tuna *Thunnus albacores* (Atema et al., 1977), and the starfish *Asterias rubens* (Castilla, 1972).

5.2.1.2. Commensalism. The root causes of commensalism or symbiosis in organisms, in fresh water or in the sea, have long been investigated simply from the viewpoint that partners may derive food or protection from the association. In fact, it would seem that there exists between them a "molecular" bond established through recognition mediators, perhaps produced by the host and perceived by the commensal or parasite. Thus the polychete annelid *Arctonoe fragilis* recognises, in experiments, water which has passed over the body of its host *Evasteria troschelli* (Davenport, 1950; Hickok and Davenport, 1957). In the same way, *Arctonoe pulchra*, a symbiont of limpets, star fishes and holothuria, recognises its hosts by an allomone (Dimock and Davenport, 1971).

5.2.1.3. Settlement. The fixation of the larvae of benthic animals does not take place randomly, but is sometimes guided by the presence of immerged substrata of organic substances deposited by adults previously fixed there. Thus, in the case of *Balanus balanoides*, the cyprid larvae attach themselves to surfaces where earlier barnacles have deposited a complex mixture of mucopolysaccharides and proteins associated to nucleic acids (Crisp, 1974). This "mediator" is not specific: it is produced by several crustaceans and has therefore received the name arthropodin. A similar phenomenon was described by Nott (1973) for the annelid *Spirorbis spirorbis*.

5.2.2. Repulsive or allelopathic allomones

The term embraces many aggression, defence or attack substances, released into the medium by various species, and of varied chemical composition: poisons, toxins, venoms, antibiotics, antigerminatives, antimitotics, etc. They would appear to play a basic regulatory role in certain ecosystems, e.g. coral reefs (Burkholder, 1973).

Typical examples are furnished by the benthic invertebrates, which provide particularly suitable material for experimentation. A number of flight or burying reactions initiated by the detection of chemical signals emitted by predators have been described. Thus, saponins and steroidal glucosides produced by the starfish *Marthasterias glacialis* cause a violent flight reaction in molluscs (Mackie and Grant, 1974). In the same way, choline esters produced by the hypobranchial glands of predatory gastropods repel other molluscs: urocyanilcholine (*Tritonalis erinacea, Urosalpinx cinerea, Thais lapillus*) (Laurenson, 1970), senecioylcholine (*Thais floridana*) and acrylyl-choline (*Buccinum undatum*) (Whittaker, 1960). The same is described for *Asterias rubens*, where flight is caused by a complex mixture of organic compounds emitted by the stellerid *Solaster papposus*. The mollusc *Mercenaria mercenaria* rapidly buries itself in the sediment in the presence of *Asterias forbesi* (Doering, 1976).

On the other hand, the production of a defence mediator in an animal can be provoked by the approach of, or contact with, a predator. This is true of the gastropod opisthobranch *Berthellina citrina*, which releases acidic defensive secretions when touched by some sea anemones, fishes or crustaceans (Marbach and Tsurnamal, 1973). For *Octopus vulgaris*, defence is ensured by simultaneous flight and ejection of ink, a clouding mixture which contains orthoquinones (8-hydroxy 4-quinolone for *Octopus dofleini*; Siuda, 1974). This product temporarily inhibits the sense of smell of the moray eel (McGinitie and McGinitie, 1968). Spinochromes, derived from juglone and naphtazarine, may play a similar role with echinoderms and crustaceans, as with the crab *Pachygrapsus crassipes* which is repulsed by these compounds (Kittredge et al., 1974).

In addition to the nutritional aspect, the role of a few allomones can also be related to behavioural inhibition. It is known, for example, that the anadromous migration of salmon can be inhibited by the presence of serine from the human skin (Idler et al., 1956) or by certain substances from hatcheries (Sutterlin and Gray, 1973). The feeding behaviour of some echinoderms (*Ophiaderma brevicaudum, O. apressum, O. cinereum* and *O. rubicundum*), induced by niacine and glycine from prey, can also be inhibited by ornithine and hydroxyproline, possibly released by other organisms (Reimer and Reimer, 1975).

6. DISCUSSION —PURPOSE OF TELEMEDIATORS IN THE NATURAL ENVIRONMENT

No matter which biological phenomena are considered, the examples show that there are several stages in telemediation mechanisms, corresponding to the increasing complexity of biological systems:

(a) A primary mechanism, where the mediator is synthesised by an organism or species, and simply controls or modifies the metabolism or behaviour of another organism.

(b) A secondary mechanism, in which the mediator is synthesised by an organism to modify the behaviour of another organism. The latter, in turn, releases a primary mediator which controls the functions of a third organism. This longer sequence would involve a greater number of organisms, from several species.

If the mechanism becomes cyclic, feedback may occur, which may be either positive or negative ecologically, depending on whether it speeds or slows a biological process. It is positive where one species grazes on another and the latter activates the synthesis of the inhibiting compound in the former; some bacteria produce antibiotics in the presence of the alga *Nitzschia ascicularis*. The self-regulation system may be more complex: one species may destroy another which was inhibiting a third one, which may then grow again and may even enhance the first growth. Negative feedback occurs where one species helps another to grow by its secretions and the latter then inhibits the former. For instance, the vitamins (especially B_{12}) released by bacteria in sewage enhance the development of some bacterial and phytoplanktonic species: they are in turn destroyed by the antibiotic secretions of those species. A further example of mediation of this kind is the inhibition of the feeding response in *Ophioderma* (Reimer and Reimer, 1975).

It has been observed that telemediators, intra- or inter-specific, can act on receptor species whether or not they are directly utilised in the metabolism of the species. In micro-organisms, molecules are often integrated as metabolites (growth-promoting or growth-limiting substances), whereas with higher animals or plants, most act only as a signal and at much weaker concentrations. However, the distinction is factitious: the site of molecular action of mediators in metazoans, although macroscopically localised at the level of a sensory receptive organ, is in the last analysis intracellular, and at this level it is possible that mediators will be found to have a similar action, whatever the organism or function concerned. It is equally possible that a relay may exist between the outer telemediator and its intracellular site of action, using endomediators such as cyclic AMP. There are no data available to support an argument on this point.

Mediating molecules appear to be highly specific to receiving organisms or biological functions, and are sometimes extraordinarily discriminative. Reactions produced in receiving species are, on the other hand, variable: immediate and limited to the synthesis of a particular specific compound (antibiotic produced by a diatom, for instance), or extending through chain-reaction until, exceeding cellular limits, they affect several organs, or even the receiving organism as a whole, leading to a global modification of its behaviour.

A major problem posed by telemediation processes in marine environments is that of the concentrations in which these molecules act on biological systems. This notion is directly tied to the turnover of these substances, i.e. their biodegradation following discharge into the medium. A mediating substance in its active state will be particularly highly concentrated if it is produced in great quantity, and if its rates of degradation and diffusion in the aqueous phase are slow.

The acuity of chemoreceptive organs tends to compensate for the weakness of emission of the substances. On the other hand, degradation may play a fundamental ecological role. Chemically, repulsive orthoquinones in the ink of the octopus lose their potency in sea water by progressive polymerisation. Biologically, numerous telemediators are excellent growth factors for micro-organisms. Degradation or its converse, stability, are necessary properties in mediation: an instantaneous alarm reaction supposes a fleeting mediation and therefore a rapid degradation of the mediator, but where the constancy of migration is fundamental to the perenniality of certain species, the compounds must be more stable or permanently produced. As to diffusion, this is obviously weaker in the absence of current, and when the molecular weight is high the effect of dilution can be retarded by adsorption onto substrates, which is the case for alarm or tracking pheromones in benthic invertebrates. Micro-organisms are also worthy of attention in this connection.

The problems of concentrations of mediators give rise to arguments which seek to minimise the significance of telemediation phenomena in microbial populations; except in specific cases, where organisms are present in very high numbers (dystrophic waters, red tides), the activity of messenger molecules alone might be supposed. The practical problems involved in estimating the actual concentrations of these substances around receiving cells are great, and would require measuring amounts of substances released, rates of diffusion (different according to chemical nature) and, most important of all, chronological stability. Since competition mainly occurs on the surface of submerged substrates, it would be erroneous to reason quantitatively on such active substances as if the medium were homogeneous. Actually, the active substance concentration is probably much higher in the micro-environments, or microniches, where most phenomena are likely to occur, and here further qualitative and quantitative observations are needed.

7. CONCLUSION — CYBERNETIC CONCEPT OF THE MARINE BIOLOGICAL EQUILIBRIUM

The concept of a marine universe where equilibrium is controlled by interaction between organisms suggests an analogy with cybernetic structures (Margalef, 1963, 1967; Aubert, 1971). Each marine organism may be considered as a separate functional unit, a module characterised by particular properties which govern its actions. Units are linked, and are thus involved in the functioning of complex systems in which they intervene locally, channelling the flow of energy and matter to a fixed extent and in a given direction. The analogy is still more pronounced when feedbacks are considered, which control various sectors of the system. Margalef (1967) showed that species and populations control one another in order to maintain a highly diversified network, corresponding to the most mature systems, parallel to the most complex logical circuits. In either case, the stability and plasticity of the populations are greater when the elements and their interconnections are numerous. In addition, as with cybernetic systems, ecological phenomena as a whole are closely dependent on time, since control preferentially occurs during the slow phases of the process. In this connection, the speed of metabolism and growth of biological elements play a key role in the adaptation time of a population: bacteria can "respond" in a few hours, whilst latency is higher in algae (a few days), and is *a fortiori* in animals (several months).

In addition, the fragility of telemediation mechanisms should be considered. In a series of studies of biological systems controlled by chemical mediators, Aubert et al. (1972b, 1975a) were able to demonstrate that various chemical pollutants bring about changes in the structure of the biological components of sea water, resulting in either the modification or the destruction of the mediating metabolites or "signals".

According to Atema and Stein (1974), small quantities of crude oil (oil : sea water ratio 1 : 10^5) constitute a noxious smell for *Homarus americanus*, depressing chemical excitability and appetite and increasing the time taken to find food. The hydrosoluble fraction of this petroleum had no effect, contrary to the observations of Takahashi and Kittredge (1973), who found that the feeding and sexual behaviour of the crab *Pachygrapsus crassipes* is strongly inhibited by this fraction. Kerosene and its branched chain-cyclic fraction stimulate, however, the search for food in the lobster, whilst the mud snail *Nassarius obsoletus* is inhibited in its search for prey by a dilute sea-water extract of kerosene (1 ppb, mainly benzene and naphthalenes). The attraction of the flatworm *Bdelloura candida* to its host *Limulus polyphemus* is inhibited by $HgCl_2$, $FeCl_2$ and by some detergents (1—100 ppm) (Atema et al., 1973) (Table I).

These phenomena could effect a pronounced ecological drift, varying according to the activity of the pollutants. From a cybernetic point of view, it may be imagined that these exogenous attacks may compel some portions of the ecosystem to change course as their functions are altered. The result is extensive imbalance which brings about a reduction in the diversity and density of the biological network through the disappearance of a number of interactions, a phenomenon usually observed in dystrophic polluted areas. Beyond a certain threshold, the destruction of a maximum number of interactions results in complete disorganisation of the system. Thus, apart from the purely fundamental aspect of research in this field, it is desirable that correlative studies be carried out on the action of chemical pollutants which may alter the marine equilibrium.

The difficulty of study in this field, which accounts for the scarcity of *in situ* observations, mainly results from the extraordinary complexity of eco-

TABLE I

Effects of different petroleum fractions on marine animal behaviour (from Atema, 1976)

Petroleum fraction	Conc.	Exposure time	Effects of behaviour	Type effect [2]
Nassarius obsoletus				
No. 2 fuel oil	50 ppm	1 h	acutely toxic: interferes w. posture, locomotion	1
	10 ppm	1 h	increased alarm response: faster burial	3
	1 ppm	1 h	increased feeding attraction; slightly increased alarm response	2 + 3
Solubles of No. 2	50 ppb?	1 h	increased alarm response	3
	10 ppb?	1 h	increased alarm response	3
	1 ppb?	1 h	slightly increased alarm response	3
	0.1 ppb?	1 h	no effect	4
	0.01 ppb?	1 h	no effect	4
Solubles of kerosene (largely polar aromatics)	1. 4 ppb	15 min	blocks or reduces feeding attraction and upstream movement	3
Homarus americanus				
Kerosene	10 ppb range	min—h	attraction, feeding; depressed activity, stop feeding for days	1 + 3 + 2
br-c of kerosene [1]			attraction, feeding, aggression, alarm, depressed activity	2 (+ 3 + 1)
pol-ar of kerosene [1]			attraction, repulsion, increased activity	3 + 2
str-al of kerosene [1]			no effect; depressed activity, stop feeding for days after contact	4 (+ 1 + 2)
La Rosa crude	10 ppm	1—5 days	delayed feeding; change in chemosensory movements	3
Solubles of La Rosa crude	10 ppb	1—5 days	no effect	4

[1] br-c: branched cyclic fraction; pol-ar: polar aromatic fraction; str-al: straight-chain-aliphatic fraction.
[2] Effects: 1 = acutely toxic; 2 = attractant, feeding; 3 = repellent, noxious; 4 = no effect.

systems in the open sea, where the diversity of biological networks is extremely high. It is likely that a deeper knowledge of the biochemical inter-actions between individuals of marine communities will be acquired more easily by investigating eutrophicated coastal ecosystems which have three research advantages, these being fewer species, hence fewer interactions, a greater cell concentration, and ease of access.

In any case, it may be assumed that many organic substances dissolved in sea water (which can be isolated for analysis) have some more or less specific biological purpose. They deserve systematic research with a view to discovering their possible ecological significance. In fact, whatever the degree of complexity of the studied ecosystems, research to discover new examples of telemediation would appear to be the most promising path towards comprehension of the dynamic equilibrium in which marine populations have long maintained themselves.

8. REFERENCES

Adler, J., 1966. Chemotaxis in bacteria. Science, 153: 708—716.

Adler, J., 1969. Chemoreceptors in bacteria. Science, 166: 1588—1597.

Alexander, M., 1964. Biochemical ecology of soil microorganisms. Annu. Rev. Microbiol., 18: 217—252.

Alexander, M., 1971. Biochemical Ecology of Microorganisms. In: Microbial Ecology. Wiley—Interscience, New York, N.Y., pp. 361—392.

Allison, J.B. and Cole, W.H., 1935. Behaviour of the barnacle *Balanus balanoides* as correlated with the planktonic content of the seawater. Mt. Desert Inst. Biol. Lab. Bull., pp. 24—25.

Aspey, W.P. and Blankenship, J.E., 1976. *Aplysia* behavioural biology. II. Induced burrowing in swimming *Aplysia brasiliana* by burrowed conspecifics. Behav. Biol., 17: 301—312.

Atema, J., 1976. Sublethal effects of petroleum fractions on the behavior of the lobster, *Homarus americanus*, and the mud snail, *Nassarius obsoletus*. In: M. Wiley (Editor), Estuarine Processes, 1. Uses, Stresses and Adaptation to the Estuary. Academic Press, London, pp. 302—312.

Atema, J. and Burd, G.D., 1975. A field study of chemotactic responses of the marine mud snail, *Nassarius obsoletus*. J. Chem. Ecol., 1: 243—251.

Atema, J. and Engstrom, D.G., 1971. Sex pheromone in the lobster *Homarus americanus*. Nature, 232: 261—263.

Atema, J. and Gagosian, R.B., 1973. Behavioral responses of male lobsters to ecdysones. Mar. Behav. Physiol., 2: 15—20.

Atema, J. and Stein, L.S., 1974. Effects of crude oil on the feeding behaviour of the lobster, *Homarus americanus*. Environ. Pollut., 6: 77—86.

Atema, J. and Stenzler, D., 1977. Alarm substance of the marine mud snail *Nassarius obsoletus*: biological characterization and possible evolution. J. Chem. Ecol., 3: 173—187.

Atema, J., Boylan, D.B., Jacobson, S. and Todd, J., 1973. The importance of chemical signals in stimulating behaviour of marine organisms: effects of altered environmental chemistry on animal communication. In: G. Glass (Editor), Bioassay Techniques in Environmental Chemistry. Ann. Arbor Science, Mich., pp. 177—197.

Atema, J., Holland, K. and Ikehara, W., 1977. Chemical Search Image: Olfactory Responses of Yellowfin Tuna (*Thunnus albacores*) to Prey Odors. Unpublished Report.

Aubert, J. and Gambarotta, J.P., 1972. Etude de l'action antibactérienne d'espèces phytoplanctoniques marines vis-à-vis de germes anaérobies. Rev. Int. Océanogr. Méd., 25: 39—48.

Aubert, J., Pesando, D. and Thouvenot, H., 1968. Action antibiotique d'extraits planctoniques vis-à-vis de germes anaérobies. Rev. Int. Océanogr. Méd., 10: 259—266.

Aubert, J., Belaich, J.P., Fernex, F., Pouthier, J. and Pesando, D., 1975b. Behaviour of bacteria discharged with particulates in the sea. In: E.A. Pearson and E. de Fraja Frangipane (Editors), Marine Pollution and Marine Wastes Disposal. Pergamon Press, London, pp. 111—124.

Aubert, M., 1971. Télémédiateurs chimiques et équilibre biologique océanique. 1re Partie: Théorie générale. Rev. Int. Océanogr. Méd., 21: 5—16.

Aubert, M. and Gauthier, M.J., 1967. Origine et nature des substances antibiotiques présentes dans le milieu marin. 8e Partie: Etude systématique de l'action antibactérienne d'espèces phytoplanctoniques vis-à-vis de germes telluriques aérobies. Rev. Int. Océanogr. Med., 5: 63—67.

Aubert, M. and Pesando, D., 1971. Télémédiateurs chimiques et équilibre biologique océanique. 2e Partie: Nature chimique de l'inhibiteur de synthèse de l'antibiotique produit par une Diatomée. Rev. Int. Océanogr. Méd., 21: 17—26.

Aubert, M. and Pesando, D., 1974. Médiateurs biochimiques et équilibre biologique de la mer. Rev. Int. Océanogr. Méd., 35—36: 195—212.

Aubert, M., Aubert, J., Gauthier, M.J. and Pesando, D., 1967. Etude de phénomènes antibiotiques liés à une efflorescence de Péridiniens. Rev. Int. Océanogr. Méd., 6—7: 43—52.

Aubert, M., Pesando, D. and Gauthier, M.J., 1970a. Phénomènes d'antibiose d'origine phytoplanctonique en milieu marin. Substances antibactériennes produites par une Diatomée, Asterionella japonica (Cleve). Rev. Int. Océanogr. Méd., 28—29: 69—76.

Aubert, M., Pesando, D. and Pincemin, J.M., 1970b. Médiateurs chimiques et relations interespèces. Mise en évidence d'un inhibiteur de synthèse métabolique d'une Diatomée produit par un Péridinien (Etude "in vitro"). Rev. Int. Océanogr. Méd., 17: 5—21.

Aubert, M., Pesando, D. and Pincemin, J.M., 1972a. Télémédiateurs chimiques et équilibre biologique océanique. 4e Partie: Seuil d'activité de l'inhibiteur de la synthèse d'antibiotique produit par une Diatomée. Rev. Int. Océanogr. Méd., 25: 17—22.

Aubert, M., Gauthier, M.J., Donnier, B., Pesando, D., Pincemin, J.M. and Barelli, M., 1972b. Effets des pollutions chimiques vis-à-vis de télémédiateurs intervenant dans l'écologie microbiologique et planctonique en milieu marin. Rev. Int. Océanogr. Méd., 28: 129—166.

Aubert, M., Gauthier, M.J. and Pesando, D., 1975a. Effets des pollutions chimiques vis-à-vis de télémédiateurs chimiques intervenant dans l'écologie microbiologique et planctonique en milieu marin. 2e Partie. Rev. Int. Océanogr. Méd., 37—38: 69—88.

Augier, H., 1972. Contribution à l'étude biochimique et physiologique des substances de croissance chez les Algues. Thesis, University of Marseille, Marseille, 323 pp.

Baas-Becking, L.G.M., 1925. Studies on the sulfur bacteria. Ann. Bot., 39: 613.

Barber, S.B., 1956. Chemoreception and proprioception in Limulus. J. Exp. Zool., 131: 51—73.

Beach, D.H., Hanscomb, N.J. and Ormond, R.F.G., 1975. Spawning pheromone in crown-thorns starfish. Nature, 254: 135—136.

Bell, W. and Mitchell, R., 1972. Chemotactic and growth responses of marine bacteria to algal extracellular products. Biol. Bull. Mar. Biol. Lab., 143: 265—277.

Bentley, J.A., 1958. Role of plant hormones in algal metabolism and ecology. Nature, 181: 1499—1502.

Bentley, J.A., 1960. Plant hormones in marine phytoplankton, zooplankton and seawater. J. Mar. Biol. Assoc. U.K., 39: 433—447.

Berland, B.R. and Maestrini, S.Y., 1969. Study of bacteria associated with marine algae in culture. II. Action of antibiotic substances. Mar. Biol., 3: 334—335.

Berland, B.R., Bonin, D.J. and Maestrini, S.Y., 1972a. Are some bacteria toxic for marine algae? Mar. Biol., 12: 189—193.

Berland, B.R., Bonin, D.J. and Maestrini, S.Y., 1972b. Etude des relations algues—bactéries du milieu marin. Possibilité d'inhibition des algues par les bactéries. Tethys, 4: 339—348.

Berland, B.R., Bonin, D.J., Cornu, A.L., Maestrini, S.Y. and Marino, J.P., 1972c. The antibacterial substances of the marine alga: *Stichochrysis immobilis* (Chrysophyta). J. Phycol., 8: 383—392.

Berland, B.R., Bonin, D.J. and Maestrini, S.Y., 1973. Study of bacteria inhibiting marine algae: a method of screening which use gliding algae. Mar. Biol. Mar. Oceanogr., 3: 1—10.

Bethe, A., 1932. Vernachlässigte Hormone. Naturwissenschaften, 20: 177—192.

Boilly-Marer, Y., 1974. Etude expérimentale du comportement nuptial de *Platynereis dumerilii* (Annelida: Polychaeta): chemoréception, emission des produits génitaux. Mar. Biol., 24: 167—179.

Brown, W.L., Eisner, T. and Whittaker, R.H., 1970. Transpecific chemical messengers and ecologic aspects. Adv. Chemoreception, 1: 35—49.

Burkholder, P.R., 1959. Vitamin-producing bacteria in the sea. In: M. Sears (Editor), Int. Oceanogr. Congress. Amer. Assoc. Adv. Sci., Washington, D.C., pp. 912—913. (preprints).

Burkholder, P.R., 1973. The ecology of marine antibiotics and coral reefs. In: O.A. Jordan and R. Endean (Editors), Biology and Geology of the Coral Reefs. Vol. 2, Biology: 1, Academic Press, New York, N.Y., pp. 117—182.

Burkholder, P.R. and Burkholder, L.M., 1958. Antimicrobial activity of horny corals. Science, 127: 1174—1175.

Cahet, G., 1965. Contribution à l'étude des eaux et des sédiments de l'Etang de Bages—Sigean, Aude. III. Réduction des composés soufrés. Vie Milieu, 16: 917—981.

Cahet, G., 1966. Substrats energétiques naturels des bactéries sulfato-réductrices. C.R. Acad. Sci. Paris, 263: 691—692.

Carr, W.E.S., Hall, E.R. and Gurin, S., 1974. Chemoreception and the role of proteins: a comparative study. Comp. Biochem. Physiol., 47: 559—566.

Case, J., Gwilliam, G.F. and Hanson, F., 1960. Dactyl chemoreceptors of brachyurans. Biol. Bull., 119: 308.

Castilla, J.C., 1972. Responses of *Asterias rubens* to bivalve prey in a Y-maze. Mar. Biol., 12: 222—228.

Chassy, B.M., Love, L.L. and Krichevsky, M.I., 1969. The acrasin activity of $3'-5'$ cyclic nucleotides. Proc. Acad. Sci. U.S.A., 64: 296—303.

Chet, I., Fogel, S. and Mitchell, R., 1971. Chemical detection of microbial prey by bacterial predators. J. Bacteriol., 106: 863—867.

Ciereszko, L.S., 1962. Chemistry of Coelenterates. III. Occurrence of antimicrobial terpenoid compounds in the zooxanthellae of alcyonarians. Trans. N.Y. Acad. Sci., Ser. II, 24: 129—183.

Collier, A., Ray, S.M., Magnitski, A.W. and Bell, J.O., 1953. Effects of dissolved organic substances on oysters. Fish. Bull. U.S., 54: 167—185.

Conover, J.T. and Sieburth, J.McN., 1964. Effects of *Sargassum* distribution on its epibiota and antibacterial activity. Bot. Mar., 6: 147—157.

Conover, J.T. and Sieburth, J.McN., 1966. Effects of tannins excreted from Phaeophyta on planktonic animal survival in tidepools. Int. Seaweed Symp., 5: 99—100.

Cook, S.B. and Cook, C.B., 1975. Directionality in the trail-following response in the pulmonate limpet *Siphonoria alternata*. Mar. Behav. Physiol., 3: 147—155.

Cooper, J.C., Scholz, A.T., Horrall, R.M., Hasler, A.D. and Madison, D.M., 1976. Experimental confirmation of the olfactory hypothesis with homing, artificially imprinted coho salmon (*Oncorhynchus kisutch*). J. Fish. Res. Board Can., 33: 703—710.

Crisp, D.J., 1974. Factors influencing the settlement of marine invertebrates larvae. In:

P.J. Grant and A.M. Mackie (Editors), Chemoreception in Marine Organisms, 1974. Academic Press, London, pp. 177—266.

Davenport, D., 1950. Studies in the physiology of commensalism. I. The polynoid genus *Arctonoë*. Biol. Bull., 98: 81—93.

Davis, H.C. and Guillard, R.R., 1958. Relative value of ten genera of microorganisms as foods for oysters and clam larvae. Fish. Bull. U.S., 58: 293—304.

Dimock, R.V. and Davenport, D., 1971. Behavioral specificity and the induction of host recognition in a symbiotic polychaete. Biol. Bull., 141: 472—484.

Dinter, I., 1974. Pheromonal behaviour in the marine snail *Littorina littorea* Linnaeus. Veliger, 17: 37—39.

Doering, P.H., 1976. A burrowing response of *Mercenaria mercenaria* (Linnaeus, 1758) elicited by *Asterias forbesi* (Desor, 1848). Veliger, 19: 167—175.

Doving, K.B., Nordeng, H. and Oakley, B., 1974. Single unit discrimination of fish odours released by char (*Salmo alpinus* L.) populations. Comp. Biochem. Physiol., 47: 1051—1063.

Duursma, E.K., 1963. The production of dissolved organic matter in the sea, as related to the primary gross production of organic matter. Neth. J. Sea Res., 2: 85—694.

Ericson, L.E. and Lewis, L., 1953. On the occurrence of vitamin B 12 factors in marine algae. Ark. Kemi, 6: 247—442.

Florkin, M., 1965. Approches moléculaires de l'intégration écologique. Problèmes de terminologie. Bull. Acad. R. Belg., 51: 239—248.

Fogel, S., Chet, I. and Mitchell, R., 1971. Chemotactic responses in marine bacteria. Bacteriol. Proc. G., 31.

Fontaine, M., 1970. Introduction à l'Océanologie. Conférence Congrès de l'A.F.A.S., Paris.

Gauthier, M.J., 1969a. Activité antibactérienne d'une Diatomée marine: *Asterionella notata* (Grun.). Rev. Int. Océanogr. Méd., 15—16: 103—171.

Gauthier, M.J., 1969b. Substances antibactériennes produites par les bactéries marines. 1re Partie: Etude systématique de l'activité antagoniste de souches bactériennes marines vis-à-vis de germes telluriques aérobies. Rev. Int. Océanogr. Méd., 15—16: 41—60.

Gauthier, M.J., 1973. Antagonismes microbiens en milieu marin. Influence sur la microécologie et l'auto-épuration du domaine benthique. In: S. Genovese (Editor), Atti del 5e Colloquio Internazionale di Oceanographia Medica, pp. 623—633.

Gauthier, M.J., 1976. Modification of bacterial respiration by a polyanionic antibiotic produced by a marine *Alteromonas*. Antimicrob. Agents Chemother., 9: 361—366.

Gauthier, M.J., Breitmayer, J.Ph. and Aubert, M., 1975a. Etude des facteurs responsables de dérives écologiques en milieu méditerranéen côtier. Proc. 10e Europ. Mar. Biol. Symp., Ostende, 2: 271—283.

Gauthier, M.J., Shewan, J.M., Gibson, D. and Lee, J.V., 1975b. Taxonomic position and seasonal variations in marine neritic environment of some gram-negative bacteria. J. Gen. Microbiol., 87: 211—218.

Gauthier, M.J., Bernard, P. and Aubert, M., 1978a. Modification de la fonction antibiotique de deux Diatomées marines, *Astrionella japonica* Cleve et *Chaetoceros lauderi* Ralfs par un dinoflagellé, *Prorocentrum micans* (Ehrenberg). J. Exp. Mar. Biol. Ecol., 33: 37—50.

Gauthier, M.J., Bernard, P. and Aubert, M., 1978b. Production d'un antibiotique lipidique photosensible par la diatomée marine *Chaetoceros lauderi* Ralfs. Ann. Microbiol., 129B: 63—70.

Griffiths, A.M. and Frost, B.W., 1976. Chemical communication in the marine planktonic copepods *Calanus pacificus* and *Pseudocalanus* sp. Crustaceana, 30: 1—8.

Gutvieb, L.G., Benzhitski, A.G. and Lebedeva, M.N., 1973. Synthesis of biologically active substances of vitamin B 12 group in the bacterioneuston of the Tropical Atlantic. In: S. Genovese (Editor), Atti del 5e Colloquio Internazionale di Oceanographia Medica, pp. 161—175.

Hara, T.J., Ueda, K. and Gorbman, A., 1965. Electroencephalographic studies of homing salmon. Science, 149: 884—885.

Hardy, A.C., 1935. Observations on the uneven distribution of oceanic plankton. Discovery Rep., 11: 511—538.

Hashimoto, Y., Konosu, S., Fusetani, N. and Nose, T., 1968. Attractants for eels in the extracts of shortnecked clam. I. Survey of constituents eliciting feeding behaviour by the omission test. Bull. Jap. Soc. Sci. Fish, 34: 78—83.

Heeb, M.A., 1973. Large molecules and chemical control of feeding behaviour in the starfish Asterias forbesi. Helgol. Wiss. Meeresunter., 24: 425—435.

Heiligenberg, W., 1976. Chemical stimuli and reproduction in fish. Experientia, 32: 1091.

Hellebust, J.A., 1965. Excretion of some organic compounds by marine phytoplankton. Limnol. Oceanogr., 10: 192—206.

Hickok, J.F. and Davenport, D., 1957. Further studies in the behaviour of commensal polychaetes. Biol. Bull., 113: 397—406.

Hornsey, I.S. and Hide, D., 1974. The production of antimicrobial compounds by British marine algae. I. Antibiotic producing marine algae. Br. Phycol., 9: 353—351.

Howe, N.R., 1976a. Behaviour of sea anemones evoked by the alarm pheromone anthopleurine. J. Comp. Physiol., 107: 67—76.

Howe, N.R., 1976b. Proline inhibition of a sea anemone alarm pheromone response. J. Exper. Biol., 65: 147—156.

Howe, N.R. and Sheikh, Y.M., 1975. Anthopleurine: a sea anemone alarm pheromone. Science, 189: 386—388.

Hutner, S.H., Cury, A. and Baker, H., 1958. Microbiological assays. Anal. Chem., 30: 849—867.

Idler, D.R., Fagerlund, U.H.M. and Mayoh, H., 1956. Olfactory perception in migrating salmon. I. L-Serine, a salmon repellent in mammalian skin. J. Gen. Physiol., 39: 889—892.

Jørgensen, E.G., 1962. Antibiotic substances from cells and culture solutions of unicellular algae with special reference to some chlorophyll derivates. Physiol. Plant., 15: 530—545.

Karlson, P. and Butenandt, A., 1959. Pheromones (Ectohormones) in insects. Ann. Rev. Entomol., 4: 39—43.

Karlson, P. and Luscher, M., 1959. Pheromones: a new term for a class of biologically active substances. Nature, 183: 55—56.

Katona, S.K., 1973. Evidence for sex pheromones in plankton copepods. Limnol. Oceanogr., 18: 574—583.

Keating, K.I., 1977. Allelopathic influence on blue-green bloom sequence in a eutrophic lake. Science, 196: 885—887.

Keating, K.I., 1978. Blue-green algal inhibition of diatom growth: transition from mesotrophic to eutrophic community structure. Science, 199: 971—973.

Khaleafa, A.F., Kharboushman, M.A.M., Metwalli, A., Moshen, A.F. and Serwi, A., 1975. Antibiotic action from extracts of some seaweeds. Bot. Mar., 18: 163—165.

Kirschenblatt, J., 1962. Terminology of some biologically active substances and validity of the term "pheromone". Nature, 195: 916—917.

Kittredge, J.S., Terry, M. and Takahashi, F.T., 1971. Sex pheromone activity of the molting hormone, crustecdysone, on male crabs. Fish Bull., 69: 337—343.

Kittredge, J.S., Takahashi, F.T., Lindsey, J. and Lasker, R., 1974. Chemical signals in the sea: marine allelochemics and evolution. Fish Bull., 72: 1—11.

Kleerekoper, H. and Morgensen, J., 1963. Role of olfaction in the orientation of Petromyzon marinus. I. Response to a single amine in prey's body odor. Physiol. Zool., 36: 347—360.

254

Kleerekoper, H., Gruber, D. and Mathis, J., 1975. Accuracy of localization of a chemical stimulus in flowing and stagnant water by the nurse shark, *Ginglymostoma cirratum*. J. Comp. Physiol., 98: 257—275.

Kühme, W., 1963. Chemisch ausgelöste Brutpflege und Schwarmverhalten bei *Hemichromis bimaculatus* (Pisces). Z. Tierpsychol., 20: 688—704.

Larsen, S.H., Reader, R.W., Kort, E.N., Tso, W.W. and Adler, J., 1974. Change in direction of flagellar rotation is the basis of chemotactic response in *Escherichia coli*. Nature, 249: 74—77.

Laurenson, D.F., 1970. Behavioral and Physiological Studies of the Escape Response Elicited in Members of the Trochidae by Thaid and Starfish Predators. Thesis, University of Auckland, Auckland, 128 pp.

Legall, J. and Postgate, J.R., 1974. The physiology of sulfate-reducing bacteria. In: A.H. Rose and T.W. Tempest (Editors), Advances in Microbial Physiology, 10. Academic Press, London, pp. 81—113.

Lehman, J.T. and Porter, J.W., 1973. Chemical activation of feeding in the Carribean reef-building coral *Montastrea cavernosa*. Biol. Bull., 145: 140—149.

Lenhoff, H.M. and Lindstedt, K.J., 1974. Chemoreception in aquatic invertebrates with special emphasis on the feeding behaviour of Coelenterates. In: P.T. Grant and A.M. Mackie (Editors), Chemoreception in Marine Organisms. Academic Press, London, pp. 143—176.

Lenhoff, H.M. and Schneidermann, H.A., 1959. The chemical control of feeding in the Portuguese man-of-war, *Physalia physalis* L. and its bearing on the evolution of the Cnidaria. Biol. Bull., 116: 452—460.

Lewin, R.A., 1962. Physiology and Biochemistry of Algae. Academic Press, New York, N.Y., 929 pp.

Li, C.P., Prescott, B., Eddy, N., Caldes, G., Green, W.R., Martino, E.C. and Young, A.M., 1965. Antimicrobial activity of paolins from clams. Ann. N.Y. Acad. Sci., 130: 374—382.

Lindstedt, K.J., 1971. Valine activation of feeding in the sea anemone *Boloceroïdes*. In: H.W. Lenhoff, L. Muscatine and L.V. Davis (Editors), Experimental Coelenterate Biology. University of Hawaii Press, Honolulu, Hawaii, pp. 92—99.

Loomis, W.F., 1955. Glutathione control of the specific feeding reactions of hydra. Ann. N.Y. Acad. Sci., 62: 209—228.

Lowe, E.F. and Turner, E.L., 1976. Aggregation and trailfollowing in juvenile *Bursatella leachii pleii*. Veliger, 19: 153—155.

Lucas, C.E., 1938. Some aspects of integration in plankton communities. J. Cons., Cons. Perm. Explor. Mer., 13: 309—322.

Lucas, C.E., 1947. The ecological effects of external metabolites. Biol. Rev., 22: 270—295.

Lucas, C.E., 1955. External metabolites in the sea. Papers in marine biology and oceanography. Deep-Sea Res., 3 (Supp.): 139—148.

Lucas, C.E., 1961. On the significance of external metabolites in ecology. Symp. Soc. Exp. Biol., 15: 190—206.

McFarland, W.N. and Moss, S.A., 1967. Internal behaviour in fish schools. Science, 156: 260—262.

McGinitie, G.E. and McGinitie, N., 1968. Natural History of Marine Animals. McGraw-Hill, New York, N.Y., 2nd ed., 523 pp.

Mackie, A.M., 1973. The chemical basis of food detection in the lobster *Homarus gammarus*. Mar. Biol., 21: 103—108.

Mackie, A.M. and Grant, P.T., 1974. Interspecific and intraspecific chemoreception by marine invertebrates. In: P.T. Grant and A.M. Mackie (Editors), Chemoreception in Marine Organisms. Academic Press, London, pp. 105—142.

McLaughlin, J.J.A., 1958. Euryhaline chrysomonads: Nutrition and toxigenesis in *Prymnesium parvum* (with notes on *Isochrysis galbana* and *Monochrysis lutheri*). J. Protozool., 5: 75—81.

McLeese, D.W., 1973. Chemical communication among lobsters (*Homarus americanus*). J. Fish. Res. Board Can., 30: 775—778.

Marbach, A. and Tsurnamal, M., 1973. On the biology of *Berthellina citrina* (Gasteropoda—Opisthobranchia) and its defensive acid secretion. Mar. Biol., 21: 331—339.

Margalef, R., 1963. Ecologie marine, nouvelles vues sur de vieux problèmes. Ann. Biol., Sér. 4, 2: 3—16.

Margalef, R., 1967. El ecosistema. In: Ecologia marina. Fundacion La Salte de Ciencias Naturales, Caracas, pp. 377—453.

Mariscal, R.N., 1971. The chemical control of the feeding behaviour in some Hawaiian corals. In: H.W. Lenhoff, L. Muscatine and L.V. Davis (Editors), Experimental Coelenterate Biology, University of Hawaii Press, Honolulu, Hawaii, pp. 100—118.

Miller, R.L., 1972. Gel filtration of the sperm attractants of some marine Hydrozoa. J. Exp. Zool., 182: 281—298.

Miller, R.L., 1975. Chemotaxis of spermatozoa of *Ciona intestinalis*. Nature, 254: 244—245.

Mitchell, R., 1971. Role of predators in the reversal of imbalances in microbial ecosystems. Nature, 230: 257—258.

Mitchell, R. and Nevo, Z., 1964. Decomposition of structural polysaccharides of bacteria by marine microorganisms. Nature, 205: 1007—1008.

Mitchell, R. and Wirsen, Z., 1968. Lysis of non-marine fungi by marine microorganisms. J. Gen. Microbiol., 52: 335—351.

Momzikoff, A., 1973. Identification de la riboflavine et de l'isoxanthoptérine parmi les substances fluorescentes d'*Ascidiella aspersa* (O.F. Muller 1776). Arch. Zool. Exp. Gen., 114: 603—609.

Nalewajko, C., 1966. Photosynthesis and excretion in various planktonic algae. Limnol. Oceanogr., 11: 1—10.

Newcombe, C. and Hartman, G., 1973. Some chemical signals in the spawning behaviour of rainbow trout (*Salmo gairdneri*). J. Fish Res. Board. Can., 30: 995—997.

Nigrelli, R.F., 1958a. Dutchman's baccy juice or growth-promoting and growth-inhibiting substances of marine origin. Trans. N.Y. Acad. Sci., Ser. II, 20: 248—262.

Nigrelli, R.F., 1958b. Biochemistry and pharmacology of compounds derived from marine organisms. Ann. N.Y. Acad. Sci., 90: 1—335.

Nigrelli, R.F., Jakowska, S. and Calventi, I., 1959. Ectyonin, an antimicrobial agent from the sponge, *Microciona prolifera* (Verill.). Zoologica, 44: 173—176.

Nordeng, H., 1971. Is the local orientation of anadromous fishes determined by pheromones? Nature, 233: 411—413.

Nott, J.A., 1973. Settlement of the larvae of *Spirorbis spirorbis* L. J. Mar. Biol. Assoc. U.K., 53: 437—453.

Paoletti, A., 1964. Micro-organisms pathogènes dans le milieu marin. In: Proc. Symp. Pol. Mar. Microorg. Prod. Petr., C.I.E.S.M.M., pp. 133—184.

Pasteels, J.M., 1972. Ecomones: messages chimiques des écosystèmes. Ann. Soc. R. Zool. Belg., 103: 103—117.

Patel, B. and Crisp, D.J., 1961. Relation between the breeding and moulting cycles in Cirripedes. Crustaceana, 2: 89—107.

Pesando, D., 1972. Etude chimique et structurale d'une substance lipidique antibiotique produite par une Diatomée marine, *Asterionella japonica*. Rev. Int. Océanogr. Méd., 25: 49—70.

Pfeiffer, W., 1963. Alarm substances. Experientia, 19: 113—123.

Pfeiffer, W. and Lemke, J., 1973. Untersuchungen zur Isolierung und Identifizierung des

256

Schreckstoffes aus der Haut der Elritze, *Phoxinus phoxinus* (L.). J. Comp. Physiol., 82: 407—410.

Pickens, A.L., 1932. Observations on the genus *Reticulitermes* Holmgren. Pan-Pac. Entomol., 8: 178—202.

Pincemin, J.M., 1971. Télémédiateurs chimiques et équilibre biologique océanique. 3e Partie: Etude "in vitro" de relations entre populations phytoplanctoniques. Rev. Int. Océanogr. Méd., 22—23: 165—196.

Pratt, D.M., 1966. Competition between *Skeletonema costatum* and *Olisthodiscus luteus* in Narragansett Bay and in culture. Limnol. Oceanogr., 11: 447—455.

Pratt, D.M., 1976. Intraspecific signalling of hunting success or failure in *Urosalpinx cinerea* Say. J. Exp. Mar. Biol. Ecol., 21: 7—9.

Prescott, C.P., Li, B., Jahnes, W.G. and Martino, E.C., 1962. Antimicrobial agents from molluscs. Trans. N.Y. Acad. Sci., Ser. II, 24: 504—509.

Provasoli, L., 1963. Organic regulation in phytoplankton fertility. In: M.N. Hill (Editor), The Sea, 2. Wiley—Interscience, New York, N.Y., 165—219.

Provasoli, L. and Pintner, I.J., 1964. Symbiotic relationships between microorganisms and seaweeds. Am. J. Bot., 51: 681.

Rieper, M., 1976. Investigations on the relationships between algal blooms and bacterial populations in the Schlei Fjord (Western Baltic Sea). Helgol. Wiss. Meeresunters., 28: 1—18.

Reimer, R.D. and Reimer, A.A., 1975. Chemical control of feeding in four species of tropical ophiuroïds of the genus *Ophioderma*. Comp. Biochem. Physiol., 51: 915—927.

Roos, H., 1957. Untersuchungen über das Vorkommen antimikrobieller Substanzen in Meeresalgen. Kiel Meeres., 13: 41—58.

Samuel, S., Shaw, N.M. and Fogg, G.F., 1971. Liberation of extracellular products of photosynthesis by tropical phytoplankton. J. Mar. Biol. Assoc. U.K., 51: 793—798.

Sangster, A.W., Thomas, S.E. and Tingling, N.L., 1975. Fish attractants from marine invertebrates: arcamine from *Arca zebra* and strombine from *Strombus gigas*. Tetrahedron, 31: 1135—1137.

Saunders, G.W., 1957. Interrelations of dissolved organic matter and phytoplankton. Bot. Rev., 23: 389—410.

Sharp, J.H., 1977. Excretion of organic matter by marine phytoplankton: Do healthy cells do it? Limnol. Oceanogr., 22: 381—399.

Sieburth, J.McN., 1959. Antibacterial activity of antarctic marine phytoplankton. Limnol. Océanogr., 4: 419—424.

Sieburth, J.McN., 1961a. Acrylic acid, an antibiotic principle in *Phaeocystis* blooms in antarctic waters. Science, 132: 676—677.

Sieburth, J.McN., 1961b. Antibiotic properties of acid, a factor in the gastrointestinal antibiosis of polar marine animals. J. Bacteriol., 82: 72—79.

Sieburth, J.McN., 1962. Biochemical Warfare among Microbes of the Sea. Honors Lecture. University of Rhode Island, Providence, R.I., 13 pp.

Sieburth, J.McN., 1964. Antibacterial substances produced by marine algae. Dev. Ind. Microbiol., 5: 124—134.

Sieburth, J.McN., 1966. The influence of algal antibiosis on the ecology of marine microorganisms. In: M.R. Droop and E.J.F. Wood (Editors), Advances in Microbiology of the Sea. Academic Press, New York, N.Y., pp. 63—94.

Sieburth, J.McN., 1967. Inhibition and agglutination of arthrobacters by pseudomonads. J. Bacteriol., 93: 1911—1916.

Sieburth, J.McN., 1968. Observations in planktonic bacteria in Narragansett Bay, Rhode Island: a résumé. Bull. Misaki Mar. Biol. Inst., Kyoto Univ., 12: 49—64.

Sieburth, J.McN. and Conover, J.T., 1965. *Sargassum* tannin, an antibiotic which retards fouling. Nature, 208: 52—53.

Sieburth, J.McN. and Pratt, D.M., 1962. Anticoliform activity of sea water associated with the termination of *Skeletonema costatum* blooms. Trans. N.Y. Acad. Sci., Ser. II, 24: 498—501.

Siuda, J.F., 1974. Chemical defence mechanisms of marine organisms. Identification of 8-hydroxy-4-quinolone from the ink of the giant octopus, *Octopus dofleini* Martini. Lloydia, 37: 501—503.

Smayda, T.J., 1963. Succession of phytoplankton, and the ocean as an holocoenotic environment. In: C.H. Oppenheimer (Editor), Symp. Mar. Microbiol. Thomas, Springfield, Ill., pp. 260—274.

Stenzler, D. and Atema, J., 1977. Alarm response of the marine and mud snail, *Nassarius obsoletus*: specificity and behavioral priority. J. Chem. Ecol., 3: 159—171.

Sutterlin, A.M. and Gray, R., 1973. Chemical basis for homing of Atlantic salmon (*Salmo salar*) to a hatchery. J. Fish. Res. Board Can., 30: 985—989.

Takahashi, F.I. and Kittredge, J.S., 1973. Sublethal effects of the water soluble components of oil: chemical communication in the marine environment. In: D.G. Ahearn and S.P. Meyers (Editors), The Microbial Degradation of Oil Pollutants. Louisiana State University, New Orleans, La., pp. 105—125.

Tavolga, W.N., 1956. Visual, chemical and sound stimuli as cues in the sex discriminatory behaviour of the gobiid fish *Bathygobius soporator*. Zoologica, 41: 49—64.

Thomas, J.P., 1971. Release of dissolved organic matter from natural populations of marine phytoplankton. Mar. Biol., 11: 311—323.

Todd, J.H., Atema, J. and Bardach, J.E., 1967. Chemical communication in social behaviour of a fish, the yellow bullhead (*Ictalurus natalis*). Science, 158: 672—673.

Todd, J.H., Atema, J. and Boylan, D.B., 1972. Chemical communication in the sea. Mar. Tech. Soc. J., 6: 54—56.

Ulitzur, S. and Shilo, M., 1970. Procedure for purification and separation of *Prymnesium parvum* toxins. Biochem. Biophys. Acta, 201: 350—263.

Valentincic, T., 1973. Food finding and stimuli to feeding in the sea star *Marthasterias glacialis*. In: Seventh European Symposium on Marine Biology, Texel, 11—16 September, 1972. Neth. J. Sea. Res., 7: 191—199.

Von Frisch, K., 1938. Zur Psychologie des Fisch-Schwarmes. Naturwissenschaften, 26: 601—606.

Von Frisch, K., 1941. Ueber einen Schreckstoff der Fischhaut und seine biologische Bedeutung. Z. Vergl. Physiol., 29: 46—145.

Walder, G.L., 1973. The feeding response of *Pachycerianthus fimbriatus* (Ceriantharia). Comp. Biochem. Physiol., 44: 1085—1092.

Whittaker, R.H., 1970. The biochemical ecology of higher plants. In: E. Sondheimer and J.B. Simeone (Editors), Chemical Ecology. Academic Press, New York, N.Y., pp. 43—78.

Whittaker, V.P., 1960. Pharmacologically active choline esters in marine gastropods. In: Biochemistry and Pharmacology of Compounds Derived from Marine Organisms. Ann. N.Y. Acad. Sci., 90: 695—705.

Wood, E.F.J., 1963. Heterotrophic microorganisms in the oceans. Oceanogr. Mar. Biol., Annu. Rev., 1: 197—222.

Zobell, C.E., 1937. The influence of solid surfaces on the physiological activities of bacteria in seawater. J. Bacteriol., 33: 86.

Zobell, C.E., 1943. The effect of solid surfaces upon bacterial activity. J. Bacteriol., 46: 39—56.

Chapter 9

ORGANIC SEA SURFACE FILMS

K.A. HUNTER and P.S. LISS

1. INTRODUCTION

No physical chemist would be surprised to learn that films of natural organic matter can sometimes form spontaneously on the surface of the sea. A present-day synopsis of the organic compounds already identified in seawater (Table I) shows that well-known surface active species such as planktonic lipids and proteins, for example, exist in measurable amounts in seawater, especially in regions of high primary production ("organic-rich" waters). It can also be seen from the Table that some 90% of the dissolved organic matter (DOM) in seawater remains to be properly characterised. Some or all of this uncharacterised DOM represents the so-called seawater yellow substance, *Gelbstoffe*, marine humus or "gunk" which is presently the subject of much attention. Within this uncharacterised fraction of the DOM there appears to be plenty of surface-active material, such as the humic and fulvic acids which can be isolated from seawater by adsorption onto a macro-reticular hydrophobic resin (e.g., Amberlite XAD-2, Mantoura and Riley, 1975), which phenomenon is itself an expression of activity for a hydrophobic, organic-polymer surface. These acids can represent up to 20% of the DOM (Stuermer and Harvey, 1977). From the point of view of molec-

TABLE I

A synopsis of presently known average organic composition of seawater, after Williams (1975)

Component	Concentration as mg C m^{-3} (ppb)
Vitamins	0.0065
Total fatty acids	5
Urea	5
Total free sugars	10
Total carbohydrates	200
Total free amino acids	10
Total combined amino acids	50
Dissolved organic carbon	500—2000
Dissolved organic nitrogen	75—230 (as mg N m^{-3})

ular size, dialysis and ultrafiltration studies have shown that a sizeable portion, at least 10%, of the DOM is macromolecular (Sharp, 1973; Ogura, 1974; Wheeler, 1976), an almost certain indicator of surface-active properties.

Much of our present understanding of this complex, uncharacterised DOM comes from analogies made with their freshwater counterparts such as soil derived humic, fulvic and hymatomelanic acids. Soil humic acids, for example, are well known to be surface active (Tschapek and Wasowski, 1976). There is currently uncertainty as to whether terrestrial humics can enter the marine environment directly via river inflow or whether they are largely precipitated in the estuarine zone. Both Williams (1975) and Head (1976), who have reviewed this topic, favour estuarine removal although two recent studies, in which the behaviour of freshwater DOM on mixing with seawater has been examined, indicate that less than 20% of the inflowing organics are precipitated during estuarine mixing (Sholkovitz, 1976; Moore et al., 1979). As far as particular details of molecular structure and functional groups are concerned the analogy between freshwater humus and marine humus is probably very limited. Structural studies of marine humus suggest a considerably more aliphatic nature than is found for the terrestrial counterparts, arising perhaps from extensive incorporation of planktonic lipids during molecular formation (Stuermer and Payne, 1976; Stuermer and Harvey, 1978). Furthermore, at least some of the marine humus is thought to derive from condensation and degradation reactions taking place *in situ* in the marine environment between proteinaceous, carbohydrate material and phenolic precursors (Sieburth and Jensen, 1968) in much the same way as lignin, cellulose and higher plant proteins enter into the formation of soil humic substances (Flaig et al., 1975). The macromolecular nature (condensation/polymerisation reactions) and the presence together of hydrophobic (lipid/protein precursors) and hydrophilic (polar functional groups) molecular segments in such organic "brews" will lead almost inevitably to surface active properties in at least part of the uncharacterised DOM. Indeed, it will be seen that organic material of this type probably represents the most important group of film-forming substances in the sea under normal conditions. Experimental support for *in situ* formation of such material comes from a recent study by Hedges (1978) in which it was found that amino acids and sugars will participate in condensation reactions to form melanoidins (polymers closely resembling natural humic substances) which exhibit considerable surface activity for clay mineral particles.

Jarvis (1967) placed samples of fresh seawater in a film balance tray, swept the surface free of existing films and, using film pressure and surface potential measurements, followed the build-up (i.e. adsorption) at the air/water interface of organic films from the bulk seawater. He found that seawater normally contains natural surface active organic compounds which will adsorb at the air/water interface, and that bubbling the seawater as opposed to simple stirring was particularly effective in bringing about the transfer to the interface. Jarvis also concluded that most of the seawater samples contained insufficient film-forming substances to allow the formation of closely packed surface films.

More information about the ubiquitous presence in seawater of natural organic surfactants has been obtained by the use of seawater/solid interfaces than from studies of the interface between seawater and air. For practical reasons, it is usually very much simpler to study the adsorption of organic matter from solution onto solid surfaces, where a variety of powerful techniques such as electrocapillarity, electrical double-layer capacitance measurements, electrophoresis and ellipsometry can be used to study the progress of adsorption and the nature of the adsorbed layer. Neihof and Loeb (1972, 1974) and Loeb and Neihof (1975, 1977) have demonstrated by electrophoresis and ellipsometry that a wide variety of solid surfaces become covered by a strongly adsorbed film of polymeric acids upon exposure to seawater. Hunter (1977) found the same type of effect and has shown by electrophoretic studies at different pH and metal ion concentrations that phenolic and carboxylic groups are probably responsible. This adsorbed organic material seems likely to represent an important part of the natural surfactants in seawater and, as such, will adsorb at the air/sea interface as well.

2. PHYSICAL PROPERTIES OF SEA SURFACE FILMS

A very useful indicator of the existence of a coherent organic film is the presence of slicks on the water surface. These are visible because with a coherent surface film, capillary waves (wavelength <1.7 cm) are rapidly damped out leading to a change in the light-reflecting properties of the water surface. Under these circumstances, the slick can be seen as a floating patch with a silvery sheen, particularly if it is observed very near to the water level, and can be easily distinguished from the ruffled surface appearance of those parts of the water not covered by slicks. Blanchard (1963) has remarked that slick-covered areas of the water surface become strikingly apparent during a shower of rain. The concentric, expanding capillary ripples caused by raindrops striking the water surface are rapidly damped out by the organic film within the slick-covered regions, giving rise to a considerably smoother surface appearance than is seen in the slick-free regions. One of the authors has noted a similar phenomenon in a sheltered, coastal area of the North Sea.

The earliest reported observations of natural films on the sea made use of this simple indicator, and much of this work is described in the papers, and references cited therein, of Ewing (1950), Stommel (1951) and LaFond (1959). When surface slicks are seen, they are usually aligned in long, parallel streaks that may or may not be oriented with respect to the wind (Blanchard,

1963). Some of these slicks mark regions of surface convergence driven by internal waves and thus may have no particular orientation with respect to the wind (Ewing, 1950; Dietz and LaFond, 1950). Other appear to mark convergence zones that are largely wind-driven, such as in the helical vortex water circulation pairs envisaged by Langmuir (1938). In this case there is an obvious alignment with the wind direction, but regardless of their orientation, it is clear that sea slicks are almost always banded in nature, pointing to the presence of alternate regions of upwelling and convergence with the greatest amount of film material to be found in the convergent or downwelling areas. Furthermore, it is possible that the maintenance, if not the genesis, of Langmuir cells is in part due to differential wind stress on the surface corresponding to the alternate regions of high and low concentration of film-forming material, an idea originally suggested by Welander (1963).

Garrett and Bultman (1963) and Garrett (1967a) have shown by laboratory studies that insoluble monolayer films will damp out capillary waves on a water surface with an exponential decay of wave amplitude with distance from the wave source given by:

$$a = a_0 \exp(-kx)$$

where a is amplitude, k is the **capillary wave damping coefficient** and x is distance. The damping coefficient k **goes through a maximum** value at a film pressure of about 1×10^{-3} N m^{-1}, being much smaller at lower film pressures. It appears from these studies that a film pressure of at least 1×10^{-3} N m^{-1} is needed for effective damping of capillary waves (N = Newton).

2.1. Film pressure measurements

The surface tension of clean seawater, properly called the air/seawater interfacial tension, at a salinity of 35‰ and a temperature of $10°C$ is approximately 75×10^{-3} N m^{-1}. The adsorption of surface active species at the air/seawater interface will lower this interfacial tension, the magnitude of the decrease being equal to the positive film pressure exerted by the adsorbed film. Measurements of film pressures made at sea have generally employed the spreading drop technique devised by Adam (1937). In this method drops of oil of known spreading pressure are added to the water surface. The pressure of the sea surface film is taken to be equal to that of the calibrated oil for which the applied drop neither spreads nor contracts. Table II is a compilation of such measurements made *in situ* taken from Hunter and Liss (1977). Significant film pressures were observed in all of the studies, but it must be pointed out that almost all of these measurements have been made in coastal waters or areas of similarly high biological production. Even under these conditions, where the supply of surfactants exuded, for example, by planktonic blooms or beds of sessile algae is expected to be elevated, film pressures in excess of a few milliNewtons per metre are relatively uncom-

TABLE II

In situ measurements of film pressures at the sea surface, from Hunter and Liss (1977)

Investigators	Area	Condition of water surface	Number of results	Film pressure $(10^{-3} \text{ N m}^{-1})$
Lumby and Folkard (1956)	Monaco Bay	ruffled smooth	13 16	2—2.5 2—22.5
Sieburth and Conover (1965)	Sargasso Sea	rippled slicked	9 9	1—2 1—4.5
Garrett (1965)	Chesapeake Bay	slicked	—	3—12
Sturdy and Fischer (1966)	Kelp Beds, S. California	—	—	2—24
Barger et al. (1974)	Mission Beach, S. California (1 km offshore)	rippled slicked	150 18	1 or less 1—23

mon. In the comprehensive study of Barger et al. (1974) made at Mission Beach, San Diego, Calif., 150 out of 168 measurements were below the value of $1 \times 10^{-3} \text{ N m}^{-1}$, which is the usual limit of detection assumed for the spreading drop method. No visible slicks were apparent when the great majority of these 150 measurements were taken, an observation which is not surprising since a film pressure of $1 \times 10^{-3} \text{ N m}^{-1}$ represents the practical limit for the onset of capillary wave damping, at least as determined by studies in the laboratory using synthetic surfactants. Most reports of film pressure measurements at sea comment on the correspondence between high film pressures ($>1 \times 10^{-3} \text{ N m}^{-1}$) and the presence of visible slicks.

There are no measurements available for typical ocean situations, but it seems unlikely that slicks with film pressures significantly greater than $1 \times 10^{-3} \text{ N m}^{-1}$ will be at all common on the surface of the open ocean. Firstly, the availability of organic material in these areas of relatively low primary production should be much reduced relative to coastal waters and, secondly, wind and wave conditions in the open sea are almost certainly too severe for the formation of visible slicks except under very calm and therefore unusual conditions. Goldacre (1949) made film pressure measurements on the surface of lake waters which are of special interest here as some of the findings, though obtained for inland waters, can be extrapolated to the oceanic situation. He found organic films (mainly protein and lipo-protein) on all of more than fifty natural bodies of water examined, with a few film pressure values in excess of $30 \times 10^{-3} \text{ N m}^{-1}$. Furthermore, at wind speeds as low as 6 knots ($\simeq$3 m s^{-1}), the gradient of film pressure in the direction of the wind was of the order of 3×10^{-3} N m^{-2}. Thus, the pressure established across any extensive slick, e.g., 20 m or more in length, will amount to 60×10^{-3} N m^{-1} or more, quite sufficient to collapse the film into a foam patch or surface scum. Blanchard (1963) comments on the difficulties he experienced when trying to cover a small harbour with surfactant when moderate winds were blowing.

A wind speed of about 6 knots (2—3 on the Beaufort Scale) is not at all high for the open ocean, where velocities of 70 knots or more have been recorded (Blanchard, 1963; Kanwisher, 1963). Although this in itself suggests that visible slicks should be a relatively rare phenomenon on the open sea, it should not be taken to mean that under these conditions the ocean surface is clean, i.e., completely free of adsorbed organic surfactants. It will be seen in the next section that the film pressure versus area behaviour of most organic surfactants, including sea surface film material, is such that a very small change in surface concentration (i.e., amount of film organics per unit area) leads to a rapid change of film pressure in the region of film pressure values which are accessible by the spreading drop method (1×10^{-3} N m^{-1} and upwards). Thus, film pressures which are lower than this, such as are found in the majority of cases where visible slicks are absent, do not necessarily indicate that the coverage of the water surface by film material is negligible.

2.2. Force/area behaviour of surface film organics

Another physical property of organic films which can be measured, though not easily at sea, is their force/area behaviour. The film pressure versus molecular area behaviour of a wide variety of known organic surfactants has been studied over the last fifty years or more. This work needs no detailed treatment here, but several general points about the physical behaviour of such films will be made as an aid to understanding the results available for natural marine surfactants. More detailed information is to be found in any good textbook of surface chemistry such as those of Rideal (1930) and Adamson (1976).

There are several ways of classifying the force/area curves of different organic surfactants. A convenient one for the present purposes is shown in Fig. 1, where four types of idealised monolayer behaviour are depicted.

The general appearance of a *solid-type* monolayer is that of a rigid, dense skin on the water surface. Their pressure/area curves are characterised by a broad region above a certain limiting area per molecule where the film pressure is very low, probably less than 0.1×10^{-3} N m^{-1}, and a region below this limiting area where the film pressure increases abruptly in a more or less linear way with very little change in the area per molecule. Most fatty acids and alcohols exhibit this type of film behaviour at low enough temperatures

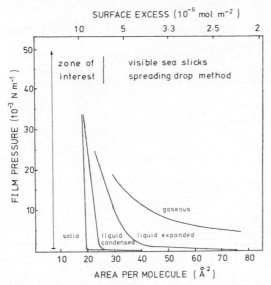

Fig. 1. Idealised pressure area diagrams for four types of organic monolayer.

or when the alkyl chain is sufficiently long, e.g. >14—16 carbon atoms. In the presence of metal ions, fatty acids readily form soap films which for divalent ions such as Ca^{2+}, Mg^{2+} and the heavy metal ions are almost always solid condensed films. Some surfactants form liquid-like monolayers which can be of the *liquid-condensed* type, where the pressure/area behaviour resembles that of a solid film but much more compressible, or of the *liquid-expanded* type for which film compressibility is even higher. The distinction between these two types can be made largely on the basis of a first-order transition from the liquid-expanded phase to the *gaseous* type of film. Gaseous monolayers frequently obey an equation of state of the form:

$$P\sigma = kT$$

where P is film pressure, σ is the area/molecule, k is Boltzmann's constant and T is the thermodynamic temperature, which is identical in form to the equation of state for an ideal gas. Like the analogous three-dimensional gas, this type of monolayer is observed at sufficiently low film pressures with most surfactant films, but it is only with relatively water-soluble surface active compounds that such a film persists up to film pressures greater than about 5×10^{-3} N m^{-1}.

The film pressure at which a gaseous film changes, or condenses, into a liquid-expanded type of film is the equivalent of a saturated vapour pressure for the surface phases. Table III gives such saturated vapour pressures at 15°C for a number of simple surface active compounds. It can be seen that

TABLE III

Saturated vapour pressures (SVP/10^{-3} N m^{-1}) of some organic surfactants at 15°C, from
Adamson (1976)

Substance	SVP/10^{-3} N m^{-1}
Tridecylic acid	0.31
Myristic acid	0.20
Pentadecylic acid	0.11
Palmitic acid	0.039
Margaric nitrile	0.11
Tetradecyl alcohol	0.11
Ethyl margarate	0.10
Ethyl stearate	0.033
Lauric acid	above critical temp.
Dodecyl alcohol	above critical temp.
Ethyl palmitate	above critical temp.

the transition between the liquid and gaseous states takes place at a film
pressure which is much smaller than the limit of detection of the spreading
drop method used for the measurement of film pressures on the sea, at
least for simple lipids where the carbon chain length is more than twelve
atoms long. Since in visible sea surface slicks the film pressure is usually
greater than 1×10^{-3} N m^{-1}, surface slicks will be of the condensed type
(liquid or solid) unless they contain an appreciable fraction of more water-
soluble surfactants which more readily form films of the expanded, gaseous
variety. At the other extreme, the usual surface state of the sea, even in sheltered coastal areas, is a marked absence of visible slicks and film pressures
probably much less than 1×10^{-3} N m^{-1}. Thus the ambient surface films on
the sea may be largely of the gaseous, expanded type. Fig. 1 shows that at
low film pressures, the more expanded the type of monolayer the greater
the reduction in surface tension at a given coverage of the surface.

A second aspect of the film properties of typical known surfactants,
which is of relevance to the *in situ* film pressure measurements, concerns the
usefulness of the film pressure as a measure of the amount of organic material present in a surface film. For solid and liquid-condensed films, the film
pressure is almost independent of area/molecule in the region of interest, i.e.,
those pressures above 1×10^{-3} N m^{-1}. For a liquid-expanded film, Fig. 1 suggests that the area/molecule might change by a factor of 2 or 3 if the film
pressure was increased 30-fold from this lower limit to 30×10^{-3} N m^{-1}.
Finally for gaseous films, which may not in fact exist on the sea in this film
pressure range, the film pressure should be approximately linear with the
amount of organic material present per unit of surface. Only in this extreme
case, therefore, is the film pressure data a reliable indicator of the amount
of natural surfactant molecules at the water surface. For all of the other

monolayer types, the film pressure measurements *in the range* 1×10^{-3} N m^{-1} and above, the region of visible sea slicks, considerably overestimate the importance of coherent films and are blind to what might be a uniform and ambient coverage of the sea surface at lower pressures. It is therefore very important to develop a satisfacotry method for the measurement *in situ* of very low film pressures, down to 0.01×10^{-3} N m^{-1} or less.

The pressure/area behaviour of sea surface film material can be measured by spreading some of it on a clean seawater surface contained in a tray. The tray is equipped with a movable barrier, often automated, and an instrument for measuring the film pressure. Barger et al. (1974) have performed experiments of this kind using surface film samples collected with the Garrett (1965) surface microlayer sampling device (Section 3.2) at the same time as their *in situ* film pressure measurements summarised in Table II were made. The results (Fig. 2) show that considerable compression of the film must take place before there is any significant increase in film pressure. It is really necessary to know the area per molecule or the effective molecular weight in order to compare these curves with those of known monolayers, but the region of increasing film pressure appears to correspond quite closely to the same regions for liquid-condensed and liquid-expanded films in Fig. 1; i.e., condensed but relatively fluid surface films.

The low slope of the force area curves at the higher film areas probably

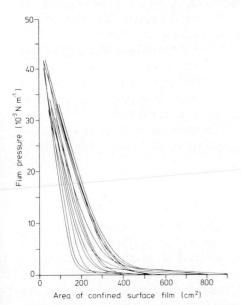

Fig. 2. Force area curves for untreated microlayer samples collected 1 km offshore from Mission Beach, San Diego (Barger et al., 1974). Reproduced with permission of the authors and publisher.

represents a squeezing out of the film of more soluble, polar species as it is compressed into a compact layer. Once this point is reached, at a film pressure of about 1×10^{-3} N m^{-1}, further compression yields a more compact film for which the pressure changes abruptly with decrease in film area. It should be emphasised that it is very difficult to say anything more specific about such surface films since their chemical composition is so heterogeneous.

The fact that the natural film materials had to be considerably compressed in order to form a coherent film implies that in the samples examined by Barger et al. (1974) there was insufficient material available to form such a film. Nevertheless, it is apparent from the approximate area changes in Fig. 2 that the concentration of strongly surface active species that remain in the film at high pressure is only a factor of 4 to 8 lower than is expected in visible sea slicks. If the soluble species are added to these, the concentration of organic material at the surface in these samples may not be a great deal lower than in the slicks.

This gradual squeezing out of more soluble, polar species as the film is compressed in the laboratory may have its analogy in the natural system. It has been already noted that slick formation occurs at film pressures of 1×10^{-3} N m^{-1} and upwards, frequently in convergent zones where the film is being compressed by movement of the surface water into the convergent zone. Thus a great many polar surface active species may be squeezed out of the surface film as slick formation progresses, giving the possibility of changes in film composition with film pressure and, in particular, a profound difference in the composition of visible slicks and ambient films of very low pressure. Coherent monolayer formation will be a function not only of the supply to the surface of surface active material, but also of the homogeneity of the film-forming substances; highly mixed films containing molecules of widely different chemical type not readily forming compact surface films (La Mer, 1962). Thus, taking what is probably a rather simple-minded view, it may be reasoned that changes of film composition with increasing film pressure may be in the direction of an increasing simplification (decrease in heterogeneity) of this composition.

2.3. Bubbles and foams

Bubbles are produced in the ocean by the action of wind and waves. Whitecaps, which cover on average 3% of the ocean surface, consist of a dense raft of bubbles with a size spectrum centred on about 200 μm diameter ranging from perhaps 50 μm up to 1 mm or two (Blanchard, 1963). The surface of a bubble is also a miniature air/sea interface analogous to the sea surface proper. Not only will the chemistry of bubbles be similar, therefore, to that of surface films, but it seems very plausible that surfactants picked up by the surfaces of rising bubbles will be transported to the surface and be a major source of surface film material (Liss, 1975).

Blanchard and Syzdek (1974) have shown that the ejection height of the top jet drop from bubbles bursting in a sample of seawater from Long Island Sound decreases rapidly with the age of the bubble, i.e., the time for which the bubble is exposed during its rise to the surface, until an age of about 10 sec (Fig. 3). Since the kinetic energy of the jet drop comes from the surface energy of the original bubble (Blanchard, 1963), the ejection height of this drop increases with the bubble surface energy. Fig. 3 shows that organic surfactants present in the seawater sample rapidly adsorb on the surface of the bubble and decrease its surface energy, reaching an apparently complete coverage after 10 sec.

Garrett (1967b) has studied the effect of organic films on the surface of water upon the stability of air bubbles which rise up and reach the film-covered surface, i.e. the time for which they persist before bursting. The lifetime of air bubbles at the surface of organic-free water is essentially zero. The bubble ruptures by thinning of the bubble cap as liquid drains out of the film formed between the air inside the bubble cap and the atmosphere. At the top of the cap, this water drainage creates a local deficiency of surface tension at both air/water interfaces. If the bubble surface or the water surface was covered by a fluid organic monolayer, this can flow into the surface tension "hole" and, by stabilising a film of water beneath, oppose the thinning of the bubble cap and prolong its lifetime dramatically. At high film pressure, however, most insoluble surfactants such as fatty acids and alcohols condense into solid-type films which are not very fluid and cannot rapidly repair sudden gaps in the film surface tension. For surfactants of this type, Garrett (1967b) found that at moderate film pressures where the films on the water surface were fluid, bubble lifetimes were increased from zero to

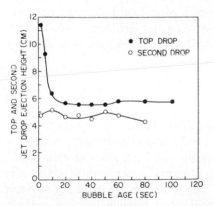

Fig. 3. Ejection height of top and second jet drop as a function of age of the air bubble rising through Long Island Sound seawater (Blanchard and Syzdek, 1974). Reproduced with permission of the authors and publisher.

3--20 sec, depending on the surfactant used. At higher film pressures, bubble lifetimes returned to zero as the films became more condensed and solid-like. It was found that the lifetimes of bubbles bursting at the surface of natural seawater samples were quite a bit higher than was found with fluid synthetic monolayers, e.g., 60—120 sec for one sample. Evidently natural seawater surfactants are very effective in stabilising thin water films such as bubble film caps. MacIntyre (1974) has noted the unusual stability of oceanic bubble films, particularly the observation that they rarely thin sufficiently to give interference colours (<1 μm thick).

Similar differences in the surfactant nature of soluble and insoluble species can be found in studies of foam stability. Broadly speaking, insoluble or solid-type surfactants do not efficiently stabilise foams (Garrett, 1967b). On the other hand, water-soluble surfactants such as the alkyl sulphonates, phosphates and tri-alkyl ammonium salts, all common detergents, form prolific foams. Wilson (1959) noted the accumulation of albuminoid-N compounds in coastal sea foams, while Southward (1953), who found a variety of planktonic and benthic organisms in such foams, concluded that proteins were probably responsible for the foam stability. Wilson and Collier (1972) have observed the production of such foam-stabilising agents in appreciable quantities by various marine organisms such as diatoms.

3. THE ORGANIC CHEMICAL COMPOSITION OF SURFACE FILMS

There has been considerable controversy over what types of organic compounds compose sea surface films (Blanchard, 1974). Part of this controversy has arisen from interpretation of the methods of chemical analysis used, and the relevant aspects of this will be discussed in the Sections 3.3. to 3.6. which are concerned with different types of organic matter. A second part of the controversy comes from the methods of sampling the surface films. Since the sampling technique is, in fact, a practical definition of what constitutes a sea surface film, it is more central to the problem. This important aspect of the chemical composition studies will therefore be discussed first of all in the next two sections.

3.1. Sampling methods

Blodgett (1934, 1935) and Langmuir and Blodgett (1937) developed many years ago a simple method for the removal of organic monolayers from a water surface. If a small plate of any hydrophilic material such as clean glass is raised up through a monolayer of barium stearate spread on a water surface, then, as illustrated in Fig. 4, the film clings to the hydrophilic surface with its hydrophobic hydrocarbon chains oriented outwards. The surface of such a film covered plate is now hydrophobic, and provided the plate was removed sufficiently slowly so that the film emerged completely dry, it can

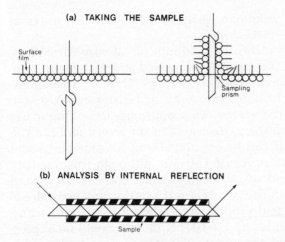

(a) TAKING THE SAMPLE

Surface film

Sampling prism

(b) ANALYSIS BY INTERNAL REFLECTION

Sample

Fig. 4. Removal of an organic fatty acid monolayer on a water surface by a rising hydrophilic plate after the method of Blodgett (1934, 1935) and its analysis by multiple attenuated internal reflection (Baier, 1970). Reproduced with permission of the author.

be replaced in water without the film being removed. On a second passage into the film-covered water surface, the plate will pick up a second monolayer of barium stearate, this time with the hydrocarbon chains back-to-back. The process can be continued for some time so that up to 100 layers may be deposited on the plate with careful work (Blodgett, 1935). Instead of removing the plate very slowly, the film can be consolidated on the plate surface by careful drying before the next dip. Otherwise, it will be removed to the water surface upon subsequent dipping.

Baier (1970, 1972, 1976) and Baier et al. (1974) have used this sampling method, mostly with only a single deposited film, for organic films on natural bodies of water. A germanium prism $5 \times 2 \times 0.1$ cm maintained at a high state of surface cleanliness is used, machined in such a way that infrared spectroscopy of the adsorbed film can be carried out by the method of multiple attenuated internal reflections (Fig. 4). In addition, measurements of the film thickness and refractive index can be made by ellipsometry, and the surface potential using a vibrating reed electrometer. The contact angle of a range of standard organic liquids is also measured, allowing calculation of the critical surface tension intercept. According to the work of Zisman (1964) this parameter can be highly characteristic of the type of surface present.

Unfortunately, the Baier germanium prism does not collect a sufficient quantity of organic material to allow the use of chemical methods of analysis other than IR spectroscopy. There are other methods which do sample a sufficiently large area of surface, but these are for the most part devices which collect a thin layer of seawater adjacent to the interface. They are therefore thought to collect an unknown quantity of subsurface water as well.

Garrett (1965) has described a screen of 0.14-mm metal wire for collecting a surface layer of this kind. The screen is immersed and withdrawn horizontally through the water surface, whereupon small films of a thickness approximately equal to the wire diameter adhere by surface tension forces to the gaps or cells in the mesh of the screen. This water can be subsequently drained into a sample bottle. From the surface area of the screen and the volume of water delivered per dip, it can be calculated that the apparent water layer thickness obtained is of the order of 150 μm, although this quantity depends quite a lot on the method of use, the mesh size and the material of which the screen is made. The efficiency of Garrett's device, made of Monel metal, for the sampling of oleic acid monolayers (a reasonably fluid surfactant) was 75% down to film pressures of 1×10^{-3} N m^{-1}. It would be expected on the basis of this result that the screen will effectively sample visible sea slicks. The less than 100% efficiency arises because on the first dip of the screen, oleic acid adsorbs on the mesh wires and is not removed by draining. This deactivates the screen material toward further such adsorption, however, so that at all other times it appears to sample only with the void area of the screen. We have no idea of the efficiency for surface material at low film pressures, but Hunter (1977) has found that the efficiency of sampling very dilute radioactively labelled palmitic acid films is of the order of 10%.

By using the screen a number of times, several litres of surface layer sample can be collected in 0.5–1 h. At sea, the device is probably the most practical of all the samplers of this general type and is the most frequently used. There must be some doubt, however, as to whether it collects selectively a sufficiently thin and unmixed water layer which includes a representative part of the interface, especially under rough conditions. This is, to a greater or lesser extent, a shortcoming of most of the sampling methods.

Harvey and Burzell (1972) have described what must be one of the simplest methods for collecting surface films. A clean glass plate (20 × 20 cm) is immersed vertically through the sea surface and withdrawn at a rate of $\simeq$20 cm s^{-1}. A water film of thickness 60–100 μm adheres to the plate and can be scraped off both sides with a neoprene wiper blade. Superficially, this device resembles the Blodgett method already described, but the rate of removal of the plate and its size are much greater. Nonetheless, compressed surface monolayers can usually be removed in this way, emerging very wet and easily scraped off. At low film pressures, however, the glass plate used in the manner described by Harvey and Burzell (1972) was found to have an efficiency of not much more than 10% (Hunter, 1977). There is no doubt that it does work very well at sea, e.g., for the harvesting of surface-dwelling organisms (neuston) as found by Harvey and Burzell (1972) or total organic matter (Barker and Zeitlin, 1972; Section 3.3). A clean glass plate used in the manner described above will remove a water film in the absence of organic films simply by the wetting of the glass and viscous retention of a water film (it can be shown by simple calculation that such a film should be about 100 μm thick). Because of this successful operation of the device probably depends both on the surface effect of organic film transfer described earlier, and on the ability of an organic film to stabilise a thin water layer, perhaps a few μm thick, around itself as the larger water film is lifted by the plate.

Harvey (1966) has designed another surface film sampler which depends in a similar way on the trapping of thin water films on a hydrophilic surface by viscous forces. This device consists of a rotating drum with a hydrophilic ceramic coating which is attached to the front of a small boat or raft that moves slowly through the water. A water film of thickness 60—100 μm adheres to the drum and is scraped off into a sample bottle using a fixed wiper blade. Under average conditions, the sampling rate is about one-third of a litre per minute, so that large sample volumes can be collected equivalent to many m^2 of water surface.

Larsson et al. (1974) describe the use of a teflon plate (highly hydrophobic) for the adsorption of hydrophobic film materials such as lipids. The plate is perforated with a large number of conical holes that reduce the water/air contact area when the plate is touched to the water surface. The presence of these holes means that the device entrains physically a good deal of seawater as well, which from the surface area of the device is equivalent to a water film of 50—100 μm. Owing to the nature of this device's construction, however, it is doubtful that this water comes from a genuinely unmixed layer adjacent to the interface.

Garrett and Barger (1974) have described the use of a similar adsorption sampler made of thin teflon sheet that will collect surface lipids with very little entrained seawater. The sheet of teflon is held by clips in a circular holder and is touched on to the water surface. After sample collection, the teflon sheet can be cut up with scissors and placed in a Soxhlet apparatus for extraction.

Morris (1974) has described a further, very ingenious, method for collecting surface films. A large polythene funnel ($\simeq$20 cm diameter) is immersed in seawater beneath the surface and withdrawn vertically with its stem sealed, thus isolating a section of the surface $\simeq$300 cm^2 in area. The seawater is then allowed to drain away slowly so that the organic film remains attached to the walls of the funnel which are then rinsed with a small volume of chloroform/methanol mixture.

Finally mention should be made of a very different kind of surface layer sampler which has not yet been used for the study of organic film constituents. This is the bubble microtome as described by MacIntyre (1968). The jet drop released by a bubble bursting at a film-covered surface consists of material originally within a distance of 0.05% of the bubble diameter from the air/water interface. For 1-mm bubbles, for example, this represents a surface "cut" of about 0.5 μm, considerably thinner than many of the devices already mentioned. A seagoing realisation of this principle, the Bubble Interfacial Microlayer Sampler (BIMS), has been described by Fasching et al. (1974) and used by Piotrowicz (1977) for a study of trace metals in coastal waters. A considerable amount of organic chemical analysis of samples obtained with this device is to take place in the near future within the Searex (Sea—Air Exchange Programme) project.

3.2. The concept of the sea surface microlayer

It is evident that many of the sampling devices described in the previous section in fact collect thin layers of water adjacent to and including the air/sea interface itself. These type of surface film samples have become known as surface microlayers (Liss, 1975). Although such layers are in fact operationally defined by the devices used for their collection, there is the understanding that something happens to the properties of ordinary, bulk seawater at and near the air/sea interface which is contained in such microlayer samples. In this way, the microlayer becomes a real phenomenon in much the same way that the particulate state, understandable in a common-sense sort of way, is also usually defined by operational methods such as membrane filtration.

Thus, the presence of a discontinuity in properties near or at the air/sea interface can be inferred from a difference in properties between the bulk seawater and the microlayer sample which contains the interfacial region. For an organic surfactant adsorbed at or near the sea surface, for example, this will manifest itself by a higher concentration or *enrichment* in the microlayer sample. There is, nevertheless, the vexed question of how much of a screen microlayer sample (of typical thickness $\simeq 200$ μm) consists of a real concentration anomaly near the interface and how much consists of seawater of bulk composition. Since there have been a number of differing and intractable opinions voiced in this regard, it is timely to examine closely what can be said, in terms of known surface chemistry, about the structure of the microlayer.

To some authors, it seems to be apparent that the microlayer consists of a surface monolayer of adsorbed organic matter of thickness $\simeq 20$ Å diluted by a vast excess, $10^4 - 10^5$ times as much, subsurface seawater. It should be noted that a monolayer thickness of this magnitude applies mostly to monolayers of simple surfactants such as fatty lipids. Water-soluble surfactants of the "wet" variety (MacIntyre, 1974) can form monolayer films of much greater thickness, with hydrophobic parts of the molecule attached to the interface and hydrophilic parts extending by as much as $\simeq 1$ μm into the aqueous phase. The results of Baier et al. (1974), to be discussed shortly, show that films in a dry state on germanium prisms used for the Blodgett (1934, 1935) type of sampling method have thicknesses determined by ellipsometry of between 100 and 300 Å. These films could therefore be considerably expanded in solution, reaching dimensions of up to 1 μm. Despite the fact that surface films could be considerably thicker than 20 Å, it is clear that if the microlayer consists only of these films even 1 μm thick, there is still considerable dilution in a screen microlayer sample of thickness 200 times as great.

The observation of an enrichment in concentration of a given species in a microlayer sample compared to the subsurface water taken at least a few centimetres below the surface in no way proves, however, that the excess quantity of this species came from a surface monolayer, be it 20 Å or 1 μm. How might such an enrichment arise in other ways? A thin, diffusion boundary layer [for example, use of a one-film model in studies of gas transfer across the air/sea interface yields diffusion layer thicknesses averaging about 50 μm (Peng et al., 1979)] is an example of a totally different phenomenon which has all the external characteristics of a film enriched with the diffusing species. Such a layer may very well arise at the sea surface in the presence of a monolayer which stabilises the water film adjacent to the interface. Thus water-soluble surfactants may be present in such a stabilised boundary layer adjacent to the surface film to an extent which, under agitated conditions, could represent a major part of the excess material present. It will be seen in the ensuing sections that the major part of the surfactants present in the microlayer are of the polymeric, largely water-soluble or water-dispersed type which can lead to the properties conjectured here. In this regard, it is of interest to note that monolayer chemistry is almost always studied by spreading films of very insoluble surfactants onto the surface of water, when techniques such as the use of very short-range isotopes can be used to show that the organic material remains within a very short distance of the macroscopic water surface. The same methods do not work easily for more soluble species, which are, nonetheless, believed to adsorb at the surface, since they also occur in the bulk solution. In fact, the studies of the adsorption of such species from solution have been made using surface samplers equivalent in principle to those used to sample the microlayer (McBain and Humphreys, 1932). If the ideas outlined here prove to be correct then they imply that the slice of the surface taken by a Garrett-type sampler (≃200 μm) may not dilute, by admixture with subsurface water, the near-surface excess concentration as substantially as is sometimes assumed.

Since it is tempting to think of a still water surface as a mathematical plane, it is instructive at this point to examine what the surface of water looks like on a molecular level, i.e., as seen by surface film material. Within an area of 20 Å^2, the cross-section of a water molecule, there are 3×10^5 bombardments per second by other water molecules. Thus, the surface is in a state of immense turbulence on the molecular level. Adamson (1976) notes that the disturbance of the water structure as a result of this turbulence may be felt at distances as much as 100 Å from the interface. Under a suitable magnification, the interface itself would appear as a fuzzy blur, with the properties of water, air and adsorbed surface active species smeared out over a considerable distance. It is apparent that as the physical dimensions of the layer we wish to discuss become similar to those of this molecular smear, macroscopic parameters such as film thickness and concentration cease to have any definable meaning.

Surface chemists, who are used to these sorts of problems, have defined a quantity called the surface excess, a measure of surface concentration per unit area which can be related to macroscopic, measurable thermodynamic variables such as the change in interfacial tension. The surface excess, denoted Γ, of a soluble surfactant is defined as the excess amount per unit area present in a finite section through the surface (i.e., including some of each phase) compared to the amount that would be present in an identical section of the aqueous bulk phase containing the same number of moles of water as the surface section. It can be shown that such a definition implies the existence of a plane such that the excess of water present in the "fuzzy" air phase above is balanced by the depleted amount of water in the "fuzzy" water phase below. The surface excess of the water is thus taken as zero. If this plane is taken as the zero of a depth scale into the bulk solution and $c(x)$ is the profile of concentration of a surface-adsorbed species, it can be shown that:

$$\Gamma = \int_0^\infty [c(x) - c_b] \, dx$$

where $c_b = c(x = \infty)$, i.e., in the bulk solution. It is important to note that because this is a definite integral, information about the integrand, i.e., the concentration profile or its thickness, cannot be obtained from measurements of the surface excess. It is fortunate, however, that Γ does not depend on the value of this thickness if the difference between $c(x)$ and c_b vanishes within a distance much less than the surface film cut off or sampled in order to measure Γ. Thus for microlayer samples of thickness $\simeq 200 \ \mu$m, for which this is most likely true, the surface excess can be easily shown to be given by:

$$\Gamma = (c_\mu - c_b) \, d$$

where d is the microlayer thickness and c_μ is the concentration in the microlayer.

The microlayer surface excess, defined by the simple formula given above, is perhaps the best variable for describing what is present in excess in a microlayer sample compared with the same amount of subsurface water as a result of surface processes. It is, in particular, almost certainly independent of the thickness of the sample obtained with one or other sampling device used. Thus different samplers should give the same results in terms of surface excess when the concentration in the microlayers taken from the same piece of seawater surface will differ, ostensibly as a result of differing dilutions of the hypothetical very thin surface film with subsurface waters.

Alternatively, since the Garrett (1965) screen type of sampler is used in the majority of microlayer studies, the simple excess concentration $c_\mu - c_b$ can be used to gain an idea of the quantity of material present as a result of surface effects. In addition, the ratio of microlayer and subsurface concentrations, c_μ/c_b, variously referred to as the fractionation ratio or the enrichment factor is often used to gain an idea of the magnitude of the enrichment in the microlayer compared to the bulk concentration. Both of these quantities will obviously depend on the type of sampler used, i.e. the thickness of the microlayer taken. An advantage in calculating enrichment factors is that, since the microlayer concentration is normalised to the subsurface concentration, values for substances of widely differing bulk concentration should be directly comparable. In contrast, surface excesses calculated for various chemicals cannot be so simply compared since Γ is a function of the integrated *difference* between surface and bulk concentrations.

Daumas et al. (1976) have compared the operation of the rotating drum and screen types of samplers in the field. A detailed examination of their results shows clearly that the surface excesses of different substances analysed in microlayers collected with each device on the same water surface are not equal but vary considerably. There are even cases where some species are enriched in one type of microlayer and depleted in another. This is not so difficult to understand; the very wide diversity in surface concentrations in the same patch of water at the same time is well-known (e.g., Parker and Zeitlin, 1972). This type of surface heterogeneity should be borne in mind when the results of different studies are compared. It is possible, nevertheless, that some of this variation arises from a disturbance of the original surface film by one or both of the sampling devices, i.e., they lose some of the film or its components in an unknown and irregular way.

3.3. Total organic matter in the surface microlayer

Analytical determinations of specific organic compounds and classes of compounds should be referred to the total concentrations of organic material in the samples. These can take the form of organic carbon analysis (DOC and POC), or, for nitrogenous and phosphatic organic compounds, organic nitrogen (DON and PON) and organic phosphorus (DOP and POP). Numerous workers have measured the concentrations of these total organic matter indicators in surface microlayer samples obtained mostly with the screen type of sampling device, comparing the results with concentrations in subsurface waters. Tables IV and V give the mean results for such determinations for dissolved and particulate species respectively.

It can be seen in Table IV that DOC, DON and DOP are enriched in the microlayer by a factor of 1.5—3 on average, depending on the sampling site

TABLE IV

Concentrations of dissolved organic material expressed as dissolved organic carbon (DOC), dissolved organic nitrogen (DON) and dissolved organic phosphorus (DOP) in mg m^{-3} (ppb) for surface microlayer (M) and subsurface water (Sub) samples

Investigators	Sampler used	DOC		DON		DOP		Sampling area
		M	Sub	M	Sub	M	Sub	
Williams (1967)	screen	3150	1050	440	140	23	12	Peru
Nishizawa (1971)	screen	1420	920					0°, 155°W
Sieburth et al. (1976)	screen	1730	1100					North Atlantic
Barker and Zeitlin (1972)	glass plate	18 400 [1]	1800 [1]					Hawaii, deep water

[1] Total organic carbon (TOC) = DOC + POC.

and investigator concerned. Not all of the authors carefully report the state of the sea surface at the time of sampling, but in the study of Williams (1967), five samples were taken between 100 and 500 km from the coast of Peru when there were no visible slicks present. A single sample, included in the averages of Table IV, was obtained within a heavy slick associated with

TABLE V

Concentrations of particulate organic material, expressed as particulate organic carbon (POC) and particulate organic nitrogen (PON), in mg m^{-3} (ppb), for surface microlayer (M) and subsurface (Sub) samples

Investigators	Sampler used	POC		PON		Sampling area
		M	Sub	M	Sub	
Williams (1967)	screen	980	140	310	30	coastal Peru
Nishizawa (1971)	screen	420	50	390	70	Equator, 155°W
Daumas et al. (1976)	screen rotating drum	560 940	350 350	50 90	40 40	Brusc, north Mediterranean
Daumas et al. (1976)	screen rotating drum	14 000 >29 000	2600 2600	1200 6000	400 400	Étang de Berre Marseille (highly polluted)

a diatom bloom 26 km from the coast. Although this sample had the highest DOC, DON and DOP levels in the microlayer, these concentrations were not much higher than the averages of Table IV. It appears, therefore, that enrichment of these species in the microlayer is a reasonably constant and widespread phenomenon, occurring even when visible slicks are absent and the film pressure is probably less than 1×10^{-3} N m^{-1}. This is in agreement with the ideas developed in Section 2.2 where it was suggested that wide changes in film pressure above this limit and the presence of visible slicks are not necessarily associated with a very much greater surface concentration of natural surfactants.

For DOC, it can be seen that the results of Williams (1967), for example, show an extra 2.1 g m^{-3} DOC in the microlayer. If the thickness of the water film obtained with the screen device is taken to be $\simeq 200$ μm, the surface excess of DOC can be calculated as $2.1 \times 200 \times 10^{-6} = 4.2 \times 10^{-4}$ g m^{-2}. A reasonable lower limit to take for the molecular weight of this extra organic material in the surface film is that of a relatively short-chain acid or alcohol with $\simeq 14$ carbon atoms, equivalent to $\simeq 170$ g mole^{-1} carbon. Using this minimum value, the area per molecule in the ambient type of films sampled by Williams (1967) can be calculated as $> \simeq 170/4.2 \times 10^{-4} \times 6.02 \times 10^{23} = > 70$ Å^2. It can be seen from Fig. 1 that for all surface film types except gaseous films, such an area per molecule has no effect on the surface tension of seawater, as measured by the spreading drop method, or on the damping of capillary waves. Moreover, only relatively water-soluble surfactants remain in the gaseous state at film pressures of $\simeq 10^{-3}$ N m^{-1}.

Table V indicates that particulate organic species are considerably more enriched in the surface microlayer than the corresponding dissolved species. The ratio of microlayer and subsurface concentrations for POC and PON falls in the range 6–10 for the oceanic samples, but is a little lower at Brusc in the north Mediterranean. The results for Brusc and also for the heavily polluted site Étang de Berre near Marseille, where POC and PON concentrations in the microlayer are extremely high (in the presence of a heavy, continuous slick), show clearly that microlayer concentrations for the rotating drum type of sampler are higher than those for the Garrett screen, suggesting that the drum effectively collects a thinner water film. This result is also borne out by other chemical analyses reported by Daumas et al. (1976) for the same samples. Phosphorus was also measured in the particulate phase by Williams (1967) but the results are not given in the Table. For all of the oceanic samples, microlayer POP was no more than a few mg m^{-3}, with undetectably small quantities in the subsurface waters. In the plankton bloom already mentioned subsurface POP was still undetectable, but over 80 mg m^{-3} was found in the microlayer, indicating an unusually high enrichment.

Taking typically 50–70% carbon in organic matter, the results in Table IV show that any specific class of organic compounds amenable to chemical analysis will have to have concentrations of the order of several g m^{-3} in the microlayer on top of what is normally present in the seawater in order to account for the DOC enrichments found at the surface. This is a figure typical of unslicked oceanic and shallow waters at all times, but higher values are expected for productive waters in the presence of heavy slicks. Table I showed that fatty acids, classically known as natural surfactants, have concentrations in oceanic water of the order of a few mg m^{-3}. Thus, if fatty acids and their lipid derivatives account for all of the excess organic material in the microlayer, then they should be enriched there by factors of the order of 1000 over the bulk water. It will be seen in the next section that the observed enrichments of fatty acids in the microlayer are in fact never more than a few percent of this value.

It is more difficult to think of particulate organic material even as a mixture of simple organic compounds. Such simple compounds as are present are most likely adsorbed or absorbed, and will therefore be amenable to solvent extraction (e.g., lipids). The rest undoubtedly consists of polymeric material such as cell wall detritus and polymerised simpler structures, which sort of material is more suited to chemical class analysis (polysaccharide, protein, etc.) usually after hydrolytic or enzymatic degradation into simpler forms.

3.4. Lipids in the surface microlayer

An obvious group of organic compounds to investigate in the microlayer is the lipids, since this group contains most of the well-known natural surfactants of simple molecular structure such as the long-chain fatty acids, alcohols and their esters. A number of workers have made analysis for these particular lipids in microlayer and subsurface seawater samples by gas chromatography (Garrett, 1967c; Duce et al., 1972; Quinn and Wade, 1972; Larsson et al., 1974; Marty and Saliot, 1974; Daumas et al., 1976; Marty et al., 1979). Since the significance of these results in terms of the total amount of organic matter in surface films depends to some extent on the methods of analysis used, it would be useful at this point to present a few essential aspects of lipid chemistry.

Lipids are defined as organic substances soluble in solvents regarded as fat-like, i.e., chloroform, carbon tetrachloride, benzene, diethyl ether, petroleum ether, etc., and which are related actually or potentially to esters of the long-chain fatty acids. A simple classification scheme which illustrates the role of these fatty acids in lipid chemistry is given in Table VI. It can be seen that most combined lipids consist of esters with long-chain alcohols, sterols, vitamins or, most commonly, glycerol and its substituted derivatives, viz., RCOOM where R = alkyl chain and M is an alcoholic moiety.

In animal and vegetable storage tissue, for example, triglycerides (a combination of one glycerol and three fatty acids) comprise the major part of the lipid material and are themselves made up of about 90% by weight of

TABLE VI

A simple classification of lipid compounds, after Deuil (1951)

Name	Description	Hydrolysis products	
		lipid	water-soluble
Simple combined lipids			
1. Neutral fats	glycerol esters (glycerides) of the fatty acids	3RCOOH	glycerol
2. Waxes	a. esters of straight-chain higher alcohols, e.g., C_{14}, C_{16} and C_{18}	RCOOH alcohol	
	b. cholesterol esters	RCOOH cholesterol	
	c. Vitamin A, D esters	RCOOH vitamins	
Complex combined lipids			
1. Phospholipids	a. lecithins	2RCOOH	glycerol, H_3PO_4, choline
	b. phosphatidyl lipids, e.g., of ethanol-amine, serine, inositol, etc.	2RCOOH	glycerol, H_3PO_4, ethanol-amine, etc.
	c. phosphatidic acids	2RCOOH	glycerol, H_3PO_4
2. Glycolipids	galactolipids	RCOOH	galactose
	glucolipids	RCOOH	glucose
Derived lipids			
Hydrocarbons, free fatty acids, alcohols, sterols		not changed by hydrolysis remain in lipid fraction	

the fatty acids in their esterified form. Thus, the fatty acids are initially at least the most important component of lipid mixtures. In a medium such as seawater, hydrolysis of the lipid esters such as triglycerides is expected to be important (Larsson et al., 1974), so that the components of combined lipids should also be present in the lipid fraction, i.e., free fatty acids, long-chain alcohols and sterols.

All lipid analyses are preceded by isolation of the total lipid material by solvent extraction with a solvent such as $CHCl_3$. As far as seawater and the microlayer are concerned it cannot be stressed too strongly that this method of isolation specifically excludes water-soluble organic compounds from whatever form of organic analysis that follows. The lipid extract is normally then hydrolysed with aqueous acid or base which breaks the ester linkages, releasing the constituents of the combined lipids. Water-soluble components such as glycerol and the phosphoric acid of phospholipids are removed automatically into the aqueous phase at this point. Thus, the original lipid molecules lose part of their weight which, as mentioned above, is not very large for triglycerides but becomes significant for esters of larger alcohols. As a variant on the sample work-up to this point, it is also possible to separate the major lipid classes before hydrolysis by a method such as thin layer chromatography (TLC) thus obtaining the fatty acid contribution to each separate class.

Before analysis by gas liquid chromatography (GLC), the fatty acids are usually converted to their more volatile methyl esters by methanolysis, e.g., MeOH with a catalyst such as BF_3. At this point, many workers make a TLC separation of the methylated fatty acids from other lipid components, particularly the hydrocarbons. The GLC analysis of fatty acid methyl esters is a straightforward enough procedure to warrant no special comment here. The most commonly used method of calibration for the analysis is the addition of one of the odd carbon-number fatty acids at the first stage of sample work up. Thus this type of analytical scheme yields eventually a concentration in the original sample of each identified fatty acid which serves as a general measure of all lipids containing that particular acid, including the free acid itself.

Garrett (1967c) was the first to use a scheme such as that outlined above for the analysis of fatty acids in screen microlayer samples and subsurface seawater, although he used no internal standards and did not, therefore, report concentrations of the fatty acids in each type of sample, although some indications of the fatty acid levels compared with the total amount of lipid material found were given. In retrospect, this proved to be a most unfortunate omission, since it has led a great many workers to assume that simple lipids such as the fatty acids and alcohols found by Garrett (1967c) were the major film components.

Lipid material was extracted from the seawater and screen samples by co-precipitation with ferric hydroxide. Back in the laboratory, the precipitate was acidified and extracted with chloroform. After hydrolysis and methylation of the extract with BF_3/MeOH, the relative proportions of fatty acids and alcohols in each of the samples was determined by GLC. A distribution of fatty acids centred on $C_{16:0}$ [i.e., $CH_3(CH_2)_{14}COOH$, palmitic acid] typical of planktonic lipids was found as illustrated in Fig. 5a for a typical surface sample. Garrett (1967c) noted that for some samples, the lipid extract was dominated by a high molecular weight material, apparently hydrocarbon, which was not surface active.

Other workers, such as Quinn and Wade (1972), extracted the aqueous samples directly with chloroform and also separated hydrocarbons and fatty acid esters after methanolysis by TLC. The French group of workers (Marty and Saliot, 1974; Daumas et al., 1976; Marty et al., 1979) on the other hand, filtered the aqueous samples before extraction of the filtrate with chloroform to obtain the dissolved lipids. The filter and particulate material

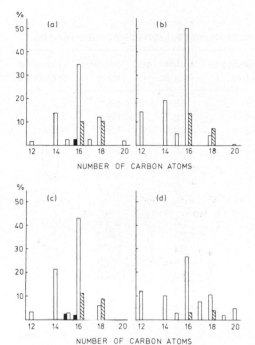

Fig. 5. Percent by weight distributions of total fatty acids as a function of the number of carbon atoms in the fatty acid chain. (a) Garrett (1967c), sample G-10, gulf of lower California. (b) Marty et al. (1979), East Tropical Atlantic. (c) Quinn and Wade (1972), North Atlantic. (d) Marty and Saliot (1974), Mediterranean Sea. □ = n-carboxylic acids; ■ = branched chain aliphatic compounds; ▨ = unsaturated fatty acids.

were extracted in a Soxhlet apparatus with a benzene/methanol mixture and analysed in a similar way for the concentrations of particulate fatty acids. These authors have also separated hydrocarbons by TLC after methanolysis.

Fig. 5b, c and d show percentage by weight distributions for typical microlayer samples obtained by three of the groups of workers mentioned in the previous paragraph. It appears that these distributions, together with the typical result found earlier by Garrett (1967c) (Fig. 5a), show properties expected of planktonic lipids; a high even-to-odd carbon number predominance, a small amount of branched structures and a carbon number range from C_{12} to C_{20} centred around C_{16} (Morris, 1974; Stuermer and Harvey, 1978).

Four groups of workers have independently reported total fatty acid measurements in microlayer and subsurface seawater samples. Their results are collected in Table VII. In surface slicks of Narragansett Bay, adjudged by visual observation to be heavy, continuous but moderate and light respectively, Quinn and Wade (1972) found that microlayer fatty acid concentra-

284

TABLE VII

Total fatty acid concentrations in surface microlayer and subsurface seawater samples

Investigator	Area	Microlayer $(mg\ m^{-3})$	Subsurface $(mg\ m^{-3})$	Surface state [1]	Quantity measured
Quinn and Wade (1972)	Narragansett Bay	128	36	HS	total
		94	62	CS	fatty
		64	20	LS	acids
Quinn and Wade (1972)	New York Bight	53	18	NVS	total
		173	200	NVS	fatty
		46	26	NVS	acids
		25	9	NVS	
		11	8	NVS	
Quinn and Wade (1972)	North Atlantic	223	25	NVS	total
		91	61	NVS	fatty
		81	24	NVS	acids
		79	53	NVS	
		51	13	NVS	
		90	44	NVS	
		91	11	NVS	
		78	32	NVS	
Marty and Saliot (1974)	Roscoff,	11	1.7		dissolved
	Mediprod 15	95	6.1		total fatty
	Mediprod 17	57	5.0		acids
	Mediprod 28	172	11		
Marty and Saliot (1974)	Roscoff	12	2.6		
	Mediprod 15	25	4.4		particulate
	Mediprod 17	103	22		total fatty
	Mediprod 28	40	10.3		acids
Daumas et al. (1976)	Brusc, north Mediterranean	5.3	2.2	NVS	dissolved
		9.9	2.8	NVS	particulate
		15.2	5.0	NVS	total fatty acid
Daumas et al. (1976)	Étang de Berre, Marseille	23.9	10.9	HS	dissolved
		1463	123	HS	particulate
		1487	134	HS	total
Marty et al. (1979) [1] (mean of 3 samples)	eastern tropical Atlantic	8.9	3.8	NVS	dissolved
		11.1	1.9		particulate
		20.0	5.7		total

[1] HS = heavy slick; CS = continuous, moderate slick; LS = light, broken slick; NVS = no visible slicks.

tions increased with the intensity of the slick. Similarly, in a polluted and heavily slicked area near Marseille, the Étang de Berre, Daumas et al. (1976) found extremely high fatty acid concentrations in the microlayer, most of these appearing in the particulate phase. In a more remote and unpolluted site at Brusc in the north Mediterranean, the same workers found much lower fatty acid concentrations both in the microlayer and in subsurface waters, values which seem to be typical also of some remote oceanic areas such as the eastern Tropical Atlantic (Marty et al., 1979) and for Roscoff in coastal northwestern France near the English Channel (Marty and Saliot, 1974). While some of the fatty acid concentrations in New York Bight and the North Atlantic (Quinn and Wade, 1972) are also at this generally low level, the majority are somewhat higher, both in the microlayer and in subsurface waters. The fatty acid concentrations in both phases are also higher for the three Mediterranean (Mediprod) samples obtained on a sector between Nice and Calvi in Corsica (Marty and Saliot, 1974).

The three sets of results given by the French groups of workers show that there can be a considerable variability in the relative proportions of dissolved and particulate fatty acids. The data of Marty and Saliot (1974) suggest that the dissolved fatty lipids are considerably more enriched in the microlayer (average factor $\simeq 12.3$) compared with the particulate fatty lipids ($\simeq 4.7$), but in the Étang de Berre and the eastern Tropical Atlantic, the opposite is observed. For total fatty acids, the enrichment factor in the microlayer in non-slicked areas is normally between 1.5 and 8, averaging around 4.3. It can be readily appreciated that this sort of enrichment is very much less than what is required to account for the enrichment of total organic matter already discussed.

Owing to the wide variability of fatty acid concentrations in Table VII, it is difficult to choose typical values for comparison with the DOC and POC results. For simplicity, the geometric mean concentration of all the results obtained in non-slicked areas will be used. The values calculated are 63 mg m^{-3} for the microlayer and 19 mg m^{-3} for subsurface water, giving an average enrichment of 3.3 times in the surface layer. It will be further assumed that half of the lipid fatty acids are present each as dissolved and particulate species. Thus, the extra quantity of dissolved fatty acids in the microlayer is, on average, 22 mg m^{-3}. If these fatty acids are present mostly as free fatty acids and triglycerides as expected by Larsson et al. (1974), then the fatty acid molecules themselves will make up most of the lipid weight. Taking the centre of the fatty acid distributions in Fig. 6, i.e., C_{16} or palmitic acid, as the average molecular size of the lipids, it can be readily calculated that the lipid carbon content is of the order of 190 g mole^{-1} or approximately 80% of the molecular weight. Thus an excess of 22 mg m^{-3} of fatty lipids contributes about 17 mg m^{-3} to the total concentration of DOC, i.e. only 0.8% of the DOC excess normally found in the microlayer (Table IV, results of Williams, 1967). Similarly, fatty acid esters with alcohols much larger than glycerol that are expected and found in seawater, e.g., the fatty alcohols and the sterols, will only increase the lipid contribution as indicated by the fatty acid concentrations to a few percent of the DOC at most. Indeed, it is apparent that the molecular weight of the alcoholic part of the lipid combination must be in the vicinity of $100/0.8) \times 240$ (240 is the MW of the C_{16} acid) in order to account for the enrichment of DOC by the presence solely of fatty acid esters. This is a MW of the order of 30,000, that of a genuine macromolecule. Such a macrospecies will have at least a thousand carbon atoms, as well as about fifty nitrogens and one phosphorus per molecule in order to account for the DON and DOP enrichments at the same time. The important points to be made concerning this calculation are:

(1) Such a molecule is by no means a simple lipid, nor need it be lipid in order to be surface active.

(2) If it is surface active, as it must be to explain the DOC results, it is unlikely to have this property conferred on it by the relatively small fatty acid part of the molecule. Indeed, the presence of the fatty acid in conjunction with such a large, hypothetical species is probably only accidental.

A second calculation of interest can be made using the average values for the fatty acid concentrations. The excess of 22 mg m^{-3} fatty acid in a microlayer of thickness $\simeq 200$ μm is equivalent to a surface excess of 4.4 μg m^{-2}. Taking again the average fatty acid chain length as 16 carbon atoms (MW $\simeq$ 240), the area per lipid molecule can be calculated as $\simeq 9000$ Å^2, much too great to have any appreciable effect on the seawater surface tension by itself, even as a film in the gaseous form. The gaseous equation of state indicates that for a surface excess of 4.4 μg m^{-2}, the film pressure contribution should be only 0.04×10^{-3} N m^{-1}.

After the lipid group of simple surfactants, the only species in organic chemistry well known for their surface active properties are macromolecules or polymers. If, therefore, as the preceding calculations and experimental work to be discussed in the next two sections suggest, the enrichment of DOC, DON, etc., in the microlayer comes essentially from the enrichment of macromolecular species, it would seem unlikely that the dilute concentrations of fatty lipids discussed above would remain free in a molecular sense, but would rather become adsorbed/absorbed by the overwhelmingly greater quantity of macrospecies. Wallace (pers. comm., 1976) has noted from foam flotation studies that the fatty acids in bulk seawater are not so readily stripped by foaming as fatty acids added to the seawater, suggesting that much of the fatty material in bulk seawater is not free but adsorbed by other, less readily stripped organic material. It is possible, therefore, that fatty lipids observed in the microlayer, at least in ambient films (non-slicked), arrive there attached to other species, i.e., independently of their own surface activity. It should be noted in this connection that the term *free fatty acid* used frequently in lipid chemistry and therefore in the various publications concerned with microlayer lipids, means fatty acids that are not esterified, and says nothing at all about the dissolved and/or adsorbed state of the molecules in seawater. Nor is such a distinction easily possible when solvent extraction is used.

In slicked areas such as Narragansett Bay, or where microlayer concentrations of fatty acids are very high (first result for New York Bight), the excess of dissolved lipids seems to be 5—10 times higher than the averages used in the above discussion. Thus under these conditions, the lipids may constitute 5—10% of the enriched DOM. This may be true for the Mediprod samples, for example, although DOC measurements for microlayer samples from the Mediterranean are not at present available. In contrast, in the most obviously slicked area of all the studies presented in Table VII, at the Étang de Berre, the total fatty acid microlayer concentrations and the microlayer excesses are at least an order of magnitude higher than in most of the other studies, but over 98% of this material is particulate and does not contribute directly to the film properties. The dissolved fatty acids on the other hand, are present at levels not significantly different from the apparently remote or unproductive sites. Even the enormous fatty acid concentrations in the particulate phase are only a small fraction of the POC and thus total organic matter present in excess in the microlayer.

The results for POC enrichment in the microlayer given in Table V show that for non-slicked, open ocean sites between 300 and 800 mg m^{-3} of extra POC exists in screen microlayer samples. Taking 500 mg m^{-3} as a reasonable average, the same kind of numerical comparison as was made for the dissolved species shows that particulate fatty lipids of the simple type contribute only a few percent of the excess POC found in the microlayer. This is not very surprising in one sense, since simple planktonic lipids should not be present as pure lipid solids. Although most of the fatty acids in the microlayer of the Étang de Berre are particulate, this is in conjunction with a much greater quantity of other particulate organic matter.

The only other components of the lipid class to receive attention in microlayer studies are the hydrocarbons. Quinn and Wade (1972) reported hydrocarbon analyses by GLC for Narragansett Bay and the North Atlantic for the same samples as were used in their fatty acid studies. In Narragansett Bay, 5.9 mg m^{-3} of total hydrocarbons were found in the subsurface water, with 40% more in the microlayer, not as great an enrichment as was found for the fatty lipids. In the North Atlantic, between 9 and 50 mg m^{-3} of subsurface hydrocarbons were found, again with a relatively low enrichment of between 0.6 and 2.4 times subsurface levels in the microlayer. Daumas et al. (1976) found that the microlayer enrichment of dissolved total hydrocarbons at Brusc, north Mediterranean was equally small, but reached 6—11 at times in the particulate phase. In the heavy slick at the Étang de Berre, dissolved hydrocarbons were enriched by six times and particulate species by 150—600 times, an unusually high enrichment. Marty et al. (1978) have studied the microlayer and subsurface concentrations of hydrocarbons in the eastern

Tropical Atlantic. They found that a little less than 80% of the hydrocarbons were aliphatic, being enriched by a factor of $\simeq 1.8$ in the microlayer. For n-alkanes, which were 9% of the total alkanes, the enrichment was also small at $\simeq 2.7$, while for aromatic hydrocarbons (13% of the total) enrichment was much more important, averaging 16.5. Studies made of the fatty acids at the same time showed that the hydrocarbons were never the major part of the lipid material analysed. In contrast, Morris (1974) found that surface films in the Mediterranean consisted of over 85% hydrocarbon, the rest of the lipid material being fatty acid lipids. However, since the amounts of lipid film material recovered by Morris (1974) were in the range of 40--230 mg m^{-2}, i.e., 100 to 1000 times higher than the total organic matter excess that occurs in the microlayer under non-slicked conditions, it is clear that multiple-layer oil slicks were all that was sampled in these studies. In the heavily slicked Étang de Berre study, for example, the surface excess of total organic matter is probably less than 10 mg m^{-2}. This emphasises the danger in thinking that grossly obvious slicks are all that is important in the surface chemistry of the sea, since this attitude leads to a predisposition for a surface state that is not at all typical of the ambient sea surface. This aspect will be made even clearer in the discussion of Section 3.6, where surface films from all types of water bodies ranging from the open ocean to areas covered by obvious oil slicks, studied using the monolayer sampling method of Baier (1970), will be considered.

3.5. Carbohydrates and proteins in the surface microlayer

Table I showed that carbohydrates are an important part of the dissolved organic matter in subsurface seawater. Sieburth et al. (1976) have measured the concentrations of mono- and polysaccharides in screen microlayer and subsurface seawater samples from Block Island Sound and the North Atlantic. They find on average that 21% of the microlayer DOC and 16% of the subsurface DOC is accounted for by total carbohydrates. Table VIII gives their average results for nine samples, all of which were collected on slick-

TABLE VIII

Dissolved carbohydrate carbon concentrations in microlayer and subsurface samples from non-slicked areas, after Sieburth et al. (1976)

Quantity	Microlayer (mg C m^{-3})	Subsurface (mg C m^{-3})	Excess (mg C m^{-3})	Percent of excess DOC
Monosaccharides	156	81	75	12
Polysaccharides	199	99	100	16
Total carbohydrates	355	180	175	28
DOC	1730	1100	630	100

free water surfaces. It is clear that monosaccharides, polysaccharides and total DOC are enriched in the microlayer to similar extents, by a factor of 2, 2.3 and 1.6, respectively. In these samples, an average of 28% of the DOC present in excess in the microlayer can be accounted for by carbohydrate, 12% monosaccharides and 16% polysaccharides. Since monosaccharides are not themselves surface active to any important extent, it seems that in the microlayer at least, the excess monosaccharides must be bound to other molecules.

Daumas et al. (1976), within their diverse set of analyses, made measurements of proteins and total carbohydrate in the particulate phase for microlayer and subsurface samples from the Étang de Berre (the polluted site) and Brusc in the north Mediterranean. Their results are given in Table IX together with some quantities calculated with a view to assessing the importance of proteins and total carbohydrates in the surface enrichment of POC and PON. For the Brusc region, where there were no visible slicks, it can be seen that there is an enrichment of proteins and carbohydrates by factors of about 1.3 for the screen and 3.3 for the drum device, i.e., approximately comparable with the results for POC and PON given in Table V, and for fatty acids (Table VII, screen). The POC content of the proteins and carbohydrates combined (P + CHO) has been calculated by assuming 60% carbon by

TABLE IX

Particulate carbohydrates and protein in screen and rotating drum microlayer samples and subsurface seawater, after Daumas et al. (1976)

Sample taken	Carbohydrate (CHO)	Protein (P)	P + CHO as POC	Percent of POC	Protein N as PON	Percent of PON
Étang de Berre						
Microlayer D [1]	17 000	10 000	16 200 [2]	<56	2000 [3]	34
Microlayer S	4000	1430	3260	23	290	24
Subsurface	1290	880	1300	50	180	45
Excess D	15 710	9120	14 900	<50	1820	33
Excess S	2710	550	1960	17	110	14
Brusc, north Mediterranean						
Microlayer D	410	238	389	42	48	55
Microlayer S	221	100	193	35	20	38
Subsurface	175	72	148	43	14	41
Excess D	235	166	241	41	33	65
Excess S	46	28	44	21	5.6	31

All concentrations are given in mg m^{-3} (ppb).
[1] D = rotating drum sampler; S = screen sampler.
[2] Assuming 60% carbon in both proteins and carbohydrates.
[3] Assuming a C/N ratio of 3.

weight in each of these materials. The results of this calculation suggest that these species account for an important part of the excess POC in the micro-layer, between 21 and 41%. By further assuming a C/N weight ratio of 3 for the proteinaceous material, the amount of the PON accounted for by the measured proteins can be estimated. It can be seen that for the drum micro-layer sample, 65% of the observed excess of PON can be accounted for by proteins in this way. For the screen sample, the PON fraction is correspond-ingly 31%, still a significant fraction. The same calculation made for the pol-luted site (Étang de Berre) gives similar results; between 17 and 50% of the POC and 14—33% of the PON accounted for by proteins and carbohydrates.

There are, unfortunately, no studies to date of the dissolved protein con-tent of microlayer samples. With the recent development of many sensitive techniques for the analysis of amino-acid mixtures in seawater using liquid chromatography and fluorescence detectors (e.g., Dawson and Pritchard, 1978), it should be relatively simple to analyse for combined amino acids after hydrolysis of the microlayer samples. Analyses of free amino acids in the microlayer seem not to have been performed to date either, but since considerable degradation of surface-adsorbed proteins may take place as a result of UV irradiation, this may be a fruitful area for future research.

3.6. Studies by Baier et al. (1974) using the germanium prism method

As described in Section 3.1, Baier et al. (1974) have used a machined germanium prism to sample surface organic monolayers in the manner devel-oped earlier by Blodgett (1934, 1935) for fatty acid films. The basis of their analytical method, infrared spectroscopy by the technique of internal reflec-tions inside the machined prisms, is only qualitative but serves very well to examine the chemical nature of the surface organics. It is well known that when an internal reflection prism made of a material with a sufficiently high index of refraction such as germanium is used, the internal reflection IR spectrum obtained suffers no band distortion or band shift when compared to conventional transmission spectra of the same substance (Barr and Flournoy, 1969).

Samples are obtained by dipping the prism in and out of the water surface attached to a fishing line, when a surface film of organic matter deposits on the prism surface as in Fig. 4. IR spectra are then obtained using conven-tional internal reflection methods normally before and after leaching of water-soluble film components with distilled water. As a result of examining a formidable number of surface film samples throughout the world, Baier et al. (1974) find it possible to separate the IR spectra obtained and the chemi-cal structures deduced therefrom into three groups:

(1) *Ambient air/sea interfacial films*, present under all sea conditions when visible slicks are absent and the film pressure is substantially less than 3×10^{-3} N m^{-1}, were characterised by IR bands at 3400, 1650 and 1100 cm^{-1}

which correlate with the presence of extensively hydroxylated components such as the remnants of carbohydrate material. Bands at 3300 cm^{-1} and 1530 cm^{-1} are also routinely observed and indicate the presence of protein, as well as bands at 1440 cm^{-1} indicating carboxyl groups. The latter are especially prominent in the ambient films on freshwater bodies. A typical spectrum of an ambient sea surface film is given in Fig. 6a. Arrows indicate absorption bands mentioned in the text. It can be seen from this spectrum that the characteristic C—H stretching vibration bands at 2950 and 2850 cm^{-1}, found in all fatty planktonic lipids, are completely absent from this spectrum. There seems to be general agreement on the assignment of absorption bands to functional groups given by Baier et al. (1974), although their statement that the carbohydrate and proteinaceous character of the ambient films suggests that the organic material falls into the categories which biochemists call glycoproteins and proteoglycans, is more speculative. However, it is hard to escape from their conclusion that the dissolved organic materials discussed under the familiar headings of *Gelbstoffe*, yellow matter, humic substances, biochemical phenolics, algal exudates or gunk have essentially the same type of properties, i.e., the property of film formation lies with some part of the uncharacterised fraction of seawater DOM.

(2) *Obvious sea slicks*, where the film pressure exceeds $\simeq 3 \times 10^{-3}$ N m^{-1} and capillary waves are damped, as well as surface scums and foam patches, all exhibit IR spectra with the same essential features as ambient films but often considerably enhanced. In addition, minor C—H bands at 2950, 2850 cm^{-1} are often found (Fig. 6b). Thus the "glycoprotein/proteoglycan" material appears to dominate even in visible slicks, while the differences between slicks and ambient films are largely quantitative and not qualitative. It can be noted that such behaviour was predicted on the basis of film pressure/area behaviour in Section 2.2. Simple planktonic lipids do not appear to be an important constituent of natural sea slicks on the basis of the IR evidence, which agrees with the comparisons of microlayer analyses for these compounds with DOC and POC made in Section 3.4.

Neither ambient films nor natural slicks appear to have an appreciable thickness. The organic film thickness for both types, determined in the dry state by ellipsometry, was found to be in the range 100—300 Å. Surface potential measurements, again performed in the dry state, always gave moderately negative results (—100 to – 200 mV), suggesting the presence of acidic functional groups in the film organic material, e.g., phenolic-OH and carboxyl. In addition, the critical surface tensions of the dried organic films as determined by the method of Zisman (1964) were always found to be in the range 30 to 40 $\times$ 10^{-3} N m^{-1} consistent with a high degree of surface oxygenation and a moderately high surface energy for an organic surface. Simple lipid surfaces dominated by hydrocarbon structures normally have critical surface tensions in the range 22 to 28 $\times$ 10^{-3} N m^{-1} (Rosoff, 1969). Thus the critical surface tension measurements are not consistent with an important lipid content in the surface films.

(3) *Polluted sea slicks* were found to be characterised by variable and often very appreciable lipid content as evidenced by C—H absorption bands at 2950 and 2850 cm^{-1} (Fig. 6c). Hydrocarbons and highly esterified oils dominated the surface films of water associated with shipping, this type of activity frequently giving rise to very thick (>100 nm) surface films. The surface films studied by Morris (1974) were also very thick, between 5 and 30 μm, accounting easily for his observation that surface films consist mostly of hydrocarbons and fatty lipids — such films are not at all typical of the sea surface, especially in the open ocean. Baier et al. (1974) find that in contrast to the ambient film and natural slick material, the critical surface tension of polluted surface films is about 26×10^{-3} N m^{-1}, typical of hydrocarbon surfaces.

Garrett (pers. comm.) has criticised the IR technique of Baier, suggesting that it is insensitive to the presence of fatty lipids and that perhaps it collects only subsurface organic matter since the fatty compounds are not seen in the ambient films and are apparently minor components of natural slicks. Possible evidence against the second objection comes from the experiments of Loeb and Neihof (1977) where the uptake of seawater organic material by a polished platinum plate indicates that adsorption, although rapid initially, continues for up to 20 h. This could be taken to mean that in the few seconds the germanium prism, used by Baier, remains underwater it will adsorb little or no organic matter from the subsurface water. The first question is really a matter of the relative sensitivity of the internal reflection method for lipids containing hydrocarbon chains in the presence of the "glycoprotein/ proteoglycan" material found in the surface films. The C—H bands at 2950, 2850 cm^{-1}, although sharp and relatively intense, can be easily masked by the broad and intense —OH absorption band centred on 3400 cm^{-1}. A calibration of the method could be simply performed by depositing successive monolayers of simple fatty lipids, perhaps radioactively labelled, on top of sea surface film samples and comparing the quantity of fatty lipid added with the amounts needed to show visible absorption bands. As a simple experiment of this kind, Garrett has shown that if stearic acid monolayers are sampled by a germanium prism already covered with a natural sea surface film, the C—H bands of the stearic acid only begin to become visible at quite

---

Fig. 6. Internal reflection IR spectra of three surface film types. (a) IR spectrum typifying the surface films which occupy air/sea interfaces at all times, obtained in Pacific Ocean, 5 km seaward of Balboa, Panama at slack high water (Fig. 1e, Baier et al., 1974, p. 576). (b) IR spectrum typical of natural sea slicks, obtained within a slick of many square miles coverage off Central California Coast in Pacific Ocean (Fig. 2b, Baier et al., 1974, p. 579). (c) IR spectrum typifying surface films of polluted seawater harbours New York City Harbour (Fig. 8c, Baier et al., 1974, p. 592). Arrows indicate absorption bands mentioned in the text. Reproduced with permission of the authors and publisher.

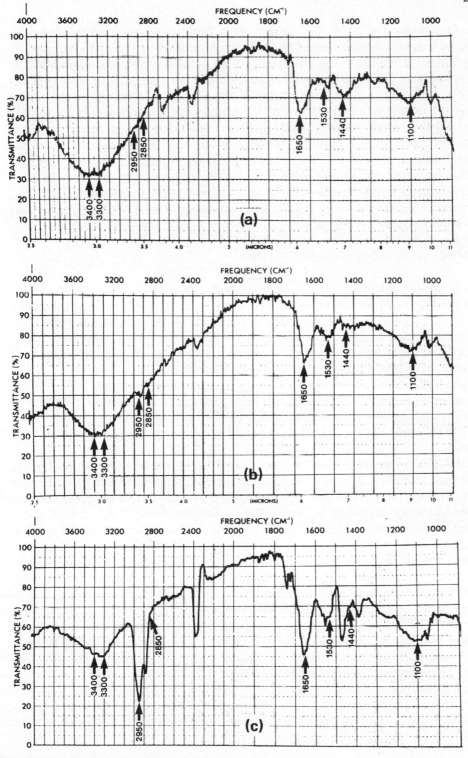

high film pressures, 20×10^{-3} N m^{-1}. This finding can be used to obtain an order of magnitude estimate of the detection limit of the Baier technique for fatty lipids. If we assume a thickness of 20 Å for the stearic acid and 200 Å for the natural film this suggests that the lipid must constitute very approximately 10% of the film (making no allowance for differences in molecular weight or degree of compression of the films) before it is detected by the Ge prism/IR technique. In Section 3.4, it was concluded that in compressed sea slicks the fatty lipids might represent about this percentage of the total organic matter present in excess in microlayer samples. Thus, it appears that on the basis of this simple criterion, the Baier prism dip/IR method does detect fatty lipids in surface films when these begin to be significant, i.e., it appears to give a reliable indication of the importance of these simple and well-known organic surfactants.

4. PROPERTIES OF ORGANIC SEA FILMS

Contrary to intuitive expectation, organic films on the surface of the sea in the absence of petroleum pollution do not consist of the classically known simple surfactants such as the fatty acids and their esters of planktonic origin. Instead, such films consist for the most part of complex polymeric material with a high degree of hydroxylation, carboxylation and proteinaceous content, with the simple lipids accounting for at most a few percents of ambient sea films and perhaps a little more in compressed natural slicks. There are no qualitative differences in this composition in and out of such slicks, as evidenced by the IR spectral results as well as DOC, DON measurements, etc. Fatty lipids do appear, however, to become more important in natural slicks.

Surface films polluted by petroleum hydrocarbons are dominated by hydrocarbons and highly esterified oils and can be many hundreds of molecular layers thick. This arises because hydrocarbons are not themselves surface active but only insoluble in water. Thus hydrocarbon films do not easily spread because intermolecular forces between the molecular layers are too large.

Thus the surface chemistry of the ocean consists essentially of the chemistry of that part of the uncharacterised and complex part of the DOM in seawater which is surface active. Apart from other effects, this can lead to the entrainment of trace elements in the surface layer by complex formation with the surface active polymers and their enrichment in the microlayer (Barker and Zeitlin, 1972; Duce et al., 1972; Piotrowicz et al., 1972, Hunter, 1977) and possible enrichment of the atmospheric aerosol (Duce et al., 1972, 1976), at least near the ocean surface (Chesselet et al., 1976).

The hydrophobic humic substances which can be extracted from bulk seawater using the hydrophobic resin Amberlite XAD-2 (Mantoura and Riley, 1975) will be part of the complex, surface active material in the DOM that should be important in the formation of organic sea films (Hunter and Liss, 1977). The lower molecular weight fraction of this material, at least, has recently been subjected to a considerable effort at structural identification (Stuermer and Payne, 1976; Gagosian and Stuermer, 1977; Stuermer and Harvey, 1978).

5. ACKNOWLEDGEMENTS

We thank Patrick Buat-Ménard for his comments on an early version of the manuscript and Norman Sheppard and John Hedges for discussions on the interpretation of IR spectra. During the period of this work the first author was supported by awards from the Royal Society, initially a studentship from the Rutherford Memorial Committee and subsequently a postdoctoral fellowship under the European Exchange Scheme.

6. REFERENCES

Adam, N.K., 1937. A rapid method for determining the lowering of surface tension of exposed water surfaces, with some observations on the surface tension of the sea and some inland waters. Proc. R. Soc. Lond., B122: 134—139.

Adamson, A.W., 1976. Physical Chemistry of Surfaces. Wiley, New York, N.Y., 698 pp.

Baier, R.E., 1970. Surface quality assessment of natural bodies of water. Proc. 13th Conf. Great Lakes Res., pp. 114—127.

Baier, R.E., 1972. Organic films on natural bodies of water: their retrieval, identification and modes of elimination. J. Geophys. Res., 77: 5062—5075.

Baier, R.E., 1976. Infra-red spectroscopic analysis of sea fog water residues, ambient atmospheric aerosols and related samples collected during the USNS "Hayes" cruise off the coast of Nova Scotia, Canada 29 July—12 August, 1975. Calspan Corp. Internal Rep., VA-5788-M-2, 45 pp.

Baier, R.E., Goupil, D.W., Perlmutter, S. and King, R., 1974. Dominant chemical composition of sea surface films, natural slicks and foams. J. Rech. Atmos., 8: 571—600.

Barger, W.R., Daniel, W.H. and Garrett, W.D., 1974. Surface chemical properties of banded sea slicks. Deep-Sea Res., 21: 83—89.

Barker, D.R. and Zeitlin, H., 1972. Metal ion concentrations in the sea surface microlayer and size-separated atmospheric aerosol samples in Hawaii. J. Geophys. Res., 77: 5076—5086.

Barr, J.K. and Flournoy, P.A., 1969. Internal reflection spectroscopy. In: B. Carroll (Editor), Physical Methods in Macromolecular Chemistry, 1. Dekker, New York, N.Y., pp. 109—164.

Blanchard, D.C., 1963. The electrification of the atmosphere by particles from bubbles in the sea. Progr. Oceanogr., 1: 73 -202.

Blanchard, D.C., 1974. International symposium on the chemistry of sea/air particulate exchange processes: Summary and conclusions. J. Rech. Atmos., 8: 509—513.

Blanchard, D.C. and Syzdek, L.D., 1974. Importance of bubble scavenging in the water to air transfer of organic material and bacteria. J. Rech. Atmos., 8: 529—540.

Blodgett, K.B., 1934. Monomolecular films of fatty acids on glass. J. Am. Chem. Soc., 56: 495.

Blodgett, K.B., 1935. Films built by depositing successive monomolecular layers on a solid surface. J. Am. Chem. Soc., 57: 1007—1022.

Chesselet, R., Buat-Ménard, P., Lesty, M. and Jehanno, C., 1976. Heavy metals in oceanic microlayer-derived aerosols. Abstracts of papers, Joint Oceanographic Assembly, Edinburgh, September, 1976. FAO, Rome.

Daumas, R.A., Laborde, P.L., Marty, J.C. and Saliot, A., 1976. Influence of sampling method on the chemical composition on water surface films. Limnol. Oceanogr., 21: 319—326.

Dawson, R. and Pritchard, R.G., 1978. The determination of α-amino acids in seawater using a fluorimetric analyser. Mar. Chem., 6: 27—40.

Deuil, H., 1951. Lipids, 1: Chemistry. Interscience, New York, N.Y., pp. 2—6.

Dietz, R.S. and LaFond, E.C., 1950. Natural slicks on the ocean. J. Mar. Res., 9: 69—76.

Duce, R.A., Quinn, J.G., Olney, C.E., Piotrowicz, S.R., Ray, B.J. and Wade, T.L., 1972. Enrichment of heavy metals and organic compounds in the surface microlayer of Narragansett Bay. Science, 176: 161—163.

Duce, R.A., Hoffman, G.L., Ray, B.J., Fletcher, I.S., Wallace, G.T., Fasching, J.L., Piotrowicz, S.R., Walsh, P.R. and Hoffman, E.J., 1976. Trace metals in the marine atmosphere: sources and fluxes. In: H.L. Windom and R.A. Duce (Editors), Marine Pollutant Transfer. D.C. Heath and Co., Lexington, Mass., pp. 77—119.

Ewing, G., 1950. Slicks, surface films and internal waves. J. Mar. Res., 9: 161—187.

Fasching, J.L., Courant, R.A., Duce, R.A. and Piotrowicz, S.R., 1974. A new surface microlayer sampler utilising the bubble microtome. J. Rech. Atmos., 8: 649—652.

Flaig, W., Beutelspacher, H. and Rietz, E., 1975. Chemical composition and physical properties of humic substances. In: J.E. Gieseking (Editor), Soil Components, 1. Springer-Verlag, Berlin, pp. 1—211.

Gagosian, R.B. and Stuermer, D.H., 1977. The cycling of biogenic compounds and their diagenetically transformed products in seawater. Mar. Chem., 5: 605—623.

Garrett, W.D., 1965. Collection of slick forming materials from the surface of the sea. Limnol. Oceanogr., 10: 602—605.

Garrett, W.D., 1967a. Damping of capillary waves at the air—sea interface by oceanic surface active material. J. Mar. Res., 25: 279—291.

Garrett, W.D., 1967b. Stabilisation of air bubbles at the air—sea interface by surface active material. Deep-Sea Res., 14: 661—672.

Garrett, W.D., 1967c. The organic chemical composition of the ocean surface. Deep-Sea Res., 14: 221—227.

Garrett, W.D. and Barger, W.R., 1974. Sampling and determining the concentration of film-forming organic constituents of the air—water interface. Naval Res. Lab. Memorandum Rep., 2852: 13 pp.

Garrett, W.D. and Bultman, J.D., 1963. Capillary-wave damping by insoluble organic monolayers. J. Colloid Sci., 18: 798—801.

Goldacre, R.J., 1949. Surface films on natural bodies of water. J. Anim. Ecol., 18: 36—39.

Harvey, G.W., 1966. Microlayer collection from the sea surface: a new method and initial results. Limnol. Oceanogr., 11: 608—614.

Harvey, G.W., and Burzell, L.A., 1972. A simple microlayer method for small samples. Limnol. Oceanogr., 17: 156—157.

Head, P.C., 1976. Organic processes in estuaries. In: J.D. Burton and P.S. Liss (Editors), Estuarine Chemistry. Academic Press, London, pp. 53—91.

Hedges, J.I., 1978. The formation and clay mineral reactions of melanoidins. Geochim. Cosmochim. Acta 42: 69—76.

Hunter, K.A., 1977. The chemistry of the sea surface microlayer. Thesis, University of East Anglia, Norwich, 363 pp.

Hunter, K.A. and Liss, P.S., 1977. The input of organic material to the oceans: air—sea interactions and the organic chemical composition of the sea surface. Mar. Chem., 5: 361—379.

Jarvis, N.L., 1967. Adsorption of surface active material at the air—sea interface. Limnol. Oceanogr., 12: 213—221.

Kanwisher, J., 1963. On the exchange of gases between the atmosphere and the sea. Deep-Sea Res., 10: 195—207.

LaFond, E.C., 1959. Surface slicks off Mission Beach. J. Geophys. Res., 64: 691.

La Mer, V.K., 1962. Preface, In: V.K. La Mer (Editor), Retardation of Evaporation by Monolayers: Transport Processes. Academic Press, New York, N.Y., pp. vii—xvii.

Langmuir, I., 1938. Surface motion of water induced by wind. Science, 87: 119—123.

Langmuir, I. and Blodgett, K.B., 1937. Built-up films of barium stearate and their optical properties. Phys. Rev., 51: 317—347.

Larsson, K., Odham, G. and Sodergren, A., 1974. On lipid surface films on the sea. I. A simple method for sampling and studies of composition. Mar. Chem., 2: 49—57.

Loeb, G.I. and Neihof, R.A., 1975. Marine conditioning films. In: R.E. Baier (Editor), Applied Chemistry at Protein Interfaces. A.C.S Adv. Chem. Series, 145: 319—335.

Loeb, G.I. and Neihof, R.A., 1977. Adsorption of an organic film at the platinum—sea-water interface. J. Mar. Res., 35: 283—291.

Liss, P.S., 1975. The chemistry of the sea surface microlayer. In: J.P. Riley and G. Skirrow (Editors), Chemical Oceanography, 2. Academic Press, London, 2nd ed., pp. 193—243.

Lumby, J.R. and Folkard, A.R., 1956. Variation in the surface tension of seawater in situ. Bull. Oceanogr. Inst. Monaco, 1080: 1—19.

McBain, J.W. and Humphreys, C.W., 1932. The microtome method of the determination of the absolute amount of adsorption. J. Phys. Chem., 36: 300—311.

MacIntyre, F., 1968. Bubbles: a boundary layer microtome for micron-thick samples of a water surface. J. Phys. Chem., 72: 589—592.

MacIntyre, F., 1974. Non-lipid related possibilities for chemical fractionation in bubble film caps. J. Rech. Atmos., 8: 515—527.

Mantoura, R.F.C. and Riley, J.P., 1975. The analytical concentration of humic substances in natural waters. Anal. Chim. Acta, 76: 97—106.

Marty, J.C. and Saliot, A., 1974. Étude chimique comparée du film de surface et de l'eau de mer sous-jacente: acides gras. J. Rech. Atmos., 8: 561—570.

Marty, J.C., Saliot, A. and Tissier, M.J., 1978. Hydrocarbures aliphatiques et polyaromatiques dans l'eau, la microcouche de surface et les aerosols marins en Atlantique tropical est. C.R. Acad. Sci. Paris, Ser. D, 286: 833—836.

Marty, J.C., Saliot, A., Buat-Ménard, P., Chesselet, R. and Hunter, K.A., 1979. Relationship between the lipid compositions of marine aerosols, the sea-surface microlayer and subsurface water. J. Geophys. Res., 84: 5707—5716.

Moore, R.M., Burton, J.D., Williams, P.J. Le B. and Young, M.L., 1979. The behaviour of dissolved organic material, iron and manganese in estuarine mixing. Geochim. Cosmochim. Acta, 43: 919—926.

Morris, R.J., 1974. Lipid composition of surface films and zooplankton from the eastern Mediterranean. Mar. Pollut. Bull., 5: 105—109.

Neihof, R.A. and Loeb, G.I., 1972. Surface charge of particulate matter in seawater. Limnol. Oceanogr., 17: 7—16.

Neihof, R.A. and Loeb, G.I., 1974. Dissolved organic matter in seawater and the electric charge of immersed surfaces. J. Mar. Res., 32: 5—12.

Nishizawa, S., 1971. Concentration of organic and inorganic material in the sea surface skin at the equator 155°W. Bull. Plankton. Soc. Jap., 18: 42—44.

Ogura, N., 1974. Molecular weight fractionation of dissolved organic matter in coastal seawater by ultrafiltration. Mar. Biol., 24: 305—312.

Peng, T.H., Broecker, W.S., Mathieu, G.G., Li, Y.H. and Bainbridge, A.E., 1979. Radon evasion rates in the Atlantic and Pacific Oceans as determined during the GEOSECS program. J. Geophys. Res., 84: 2471—2486.

Piotrowicz, S.R., 1977. Studies of the Sea to Air Transport of Trace Metals in Narragansett Bay. Thesis, University of Rhode Island, Providence R.I., 170 pp.

Piotrowicz, S.R., Ray, B.J., Hoffman, G.L. and Duce, R.A., 1972. Trace metal enrichment in the sea surface microlayer. J. Geophys. Res., 77: 5243—5254.

Quinn, J.G. and Wade, T.L., 1972. Lipid measurements in the marine atmosphere and the sea surface microlayer. In: E.D. Goldberg (Editor), Baseline Studies of Pollutants in the Marine Environment. National Science Foundation, Washington, D.C., pp. 633—663.

Rideal, E.K., 1930. An Introduction to Surface Chemistry. Cambridge University Press, Cambridge.

Rosoff, M., 1969. Surface chemistry and polymers. In: B. Carroll (Editor), Physical Methods in Macromolecular Chemistry. Dekker, New York, N.Y., pp. 1—108.

Sharp, J.H., 1973. Size classes of organic carbon in seawater. Limnol. Oceanogr., 18: 441—447.

Sholkovitz, E.R., 1976. Flocculation of dissolved organic and inorganic matter during the mixing of river water and seawater. Geochim. Cosmochim. Acta, 40: 831—845.

Sieburth, J.McN. and Conover, J.T., 1965. Slicks associated with the Trichodesmium blooms in the Sargasso Sea. Nature, 205: 830—831.

Sieburth, J.McN. and Jensen, A., 1968. Studies of algal substances in the sea. I. Gelbstoffe (humic material) in terrestrial and marine waters. J. Exp. Mar. Biol. Ecol., 2: 174—189.

Sieburth, J.McN., Willis, P., Johnson, K.M., Burney, C.M., Lavoie, D.M., Hinga, K.R., Caron, D.A., French, F.W., Johnson, P.W. and Davis, P.G., 1976. Dissolved organic matter and heterotrophic microneuston in the surface microlayers of the North Atlantic. Science, 194: 1415—1418.

Southward, A.J., 1953. Sea foam. Nature, 172: 1059—1060.

Stommel, H., 1951. Streaks on natural water surfaces. Weather, 6: 72—74.

Stuermer, D.H. and Harvey, G.R., 1977. The isolation of humic substances and alcohol soluble organic matter from seawater. Deep-Sea Res., 24: 303—309.

Stuermer, D.H. and Harvey, G.R., 1978. Structural studies on marine humus: a new reduction scheme for carbon skeleton determination. Mar. Chem., 6: 55—70.

Stuermer, D.H. and Payne, J.R., 1976. Investigation of seawater and terrestrial humic substances with 13-C and 1-H nuclear magnetic resonance. Geochim. Cosmochim. Acta, 40: 1109—1114.

Sturdy, G. and Fischer, W.H., 1966. Surface tension of slick patches near kelp beds. Nature, 211: 951—952.

Tschapek, M. and Wasowski, C., 1976. The surface activity of humic acids. Geochim. Cosmochim. Acta, 40: 1343—1345.

Welander, P., 1963. On the generation of wind streaks on the sea surface by action of a surface film. Tellus, 15: 67—71.

Wheeler, J.R., 1976. Fractionation by molecular weight of organic substances in Georgia coastal water. Limnol. Oceanogr., 21: 846—852.

Williams, P.J. Le B., 1975. Biological and chemical aspects of dissolved organic material in seawater. In: J.P. Riley and G. Skirrow (Editors), Chemical Oceanography, 2. Academic Press, London, 2nd ed., pp. 301—363.

Williams, P.M., 1967. Sea surface chemistry: organic carbon, nitrogen and phosphorus in surface films and subsurface waters. Deep-Sea Res., 14: 791—800.

Wilson, A.T., 1959. Surface of the ocean as a source of air-borne nitrogenous material and other plant nutrients. Nature, 184: 99—101.

Wilson, W.B. and Collier, A., 1972. The production of surface active material by marine phytoplankton cultures. J. Mar. Res., 30: 15—21.

Zisman, W.A., 1964. Relation of the equilibrium contact angle to liquid and solid constitution. In: R.F. Gould (Editor), Contact Angle: Wettability and Adhesion. Adv. Chem. Ser., 43: 1—51.

Chapter 10

MARINE ORGANIC PHOTOCHEMISTRY

ROD G. ZIKA

1. INTRODUCTION

The enormous amount of solar energy incident on the earth has obvious consequences in the control of its environment, through the formation and maintenance, via photosynthesis, of an oxygen-containing atmosphere and through transformation of light energy into thermal energy. Less obvious, but also of major importance, are the solar-induced processes of the atmosphere, the study of which has led to the recognition of how significant an effect mankind can play in perturbing these processes, even to his own detriment. In another vein, man has realized the useful potential of sunlight as a controlled energy source to be transformed into either thermal or electrical energy or to serve as a mechanism, when coupled with chemical oxidants or photocatalysts, to destroy or transform his refuse to less innocuous by-products. In an even more remote sense we have recognized and studied the sun's role in supplying the energy for photochemical reactions during the earth's early history which may have led to the evolution of a prebiotic oxygen-containing atmosphere and to the generation of the chemical precursors of life.

Whereas the information resulting from research in these areas has mounted steadily, there is a dearth of investigations, and especially systematic investigations, on photochemical processes occurring in sunlit seawater. The impetus for such work is certainly there in view of the demonstrated importance of sunlight on other of the earth's environments and in view of the fact that the major portion of the solar energy which penetrates the atmosphere is absorbed in the oceans. This constitutes an equatorial sea surface flux of some 6×10^{21} photons cm^{-2} day^{-1} for visible and ultraviolet wavelengths (290—700 nm). Much of this energy is utilized in photosynthesis and heating of the oceans' surface. The remainder, which is absorbed, can initiate photochemical reactions which could have significant effects upon the chemistry and biology of the oceans.

It is the purpose of this chapter to define some of the possible photochemical processes which may be affecting the nature of organic materials found in seawater and to review some of those papers which provide evidence for the occurrence of these processes in the marine environment.

2. SOME BASIC PRINCIPLES OF PHOTOCHEMISTRY

Before addressing the primary topics of this chapter it is essential that the reader be versed in some of the concepts of photochemistry and of the natural properties of seawater which make investigations in the marine system differ markedly from typical classical studies in organic photochemistry. The discussion here must be brief; for more comprehensive treatments the reader is referred to some of the many excellent texts on organic photochemistry (Calvert and Pitts, 1966; Wayne, 1970; Cowan and Drisko, 1976) and inorganic photochemistry (Balzani and Carassiti, 1970; Adamson and Fleischauer, 1975).

For a photochemical reaction to occur, a fundamental prerequisite must be fulfilled that only light which is absorbed by a system can induce a chemical reaction. The mere passage of light through the system will not produce a chemical reaction even where light is scattered by particles or molecules. In the case of Rayleigh scattering, only the directional property of the light is altered, and for Raman scattering, only a small energy transfer in a vibrational mode occurs. The significance of this in the oceans is that only those photons of sunlight which are originally entrained and escape by backscattering into the atmosphere will not be absorbed by some component of seawater. The number of photons in incident sunlight backscattered both from the sea surface and from internal scattering is dependent on conditions such as the sea state, time of day, and particle load in the water column and is, therefore, very dependent on local conditions (Jerlov, 1968). Averaged over the world's oceans, this should amount to a 10—20% loss of incident solar radiation. Therefore, some 80—90% of the visible and ultraviolet radiation incident on the oceans is available to initiate chemical reactions. It does not follow, however, that all of this light will initiate photoreactions, since most of it is utilized in photophysical processes.

The Jablonski diagram (Fig. 1) is an illustration of the possible events occurring during a photophysical process (e.g., fluorescence, phosphorescence, or energy transfer). The ordinate direction in the diagram is characteristic of increasing energy; the horizontal direction has no physical significance. Electronic states are represented by the heavy horizontal lines and the light lines represent vibration energy levels. Different electronic states are designated by the symbols S_0, S_1, and S_2 for the ground state and the first and second singlet excited states (electron spins paired), respectively, and triplet states T_0 and T_1 (electron spins unpaired) are designated in a similar way. The absorption step usually results in the direct formation of a singlet excited state, but a low probability transition of $S_0 \rightarrow T$ also exists. Generally, the triplet state is reached by the isoenergetic transformation of intersystem crossing (ISC) from $S_1 \rightarrow T$. The radiative processes of fluorescence and phosphorescence shown in Fig. 1 are primary processes and describe the sequence of steps of a molecule in going from its ground state through the excited

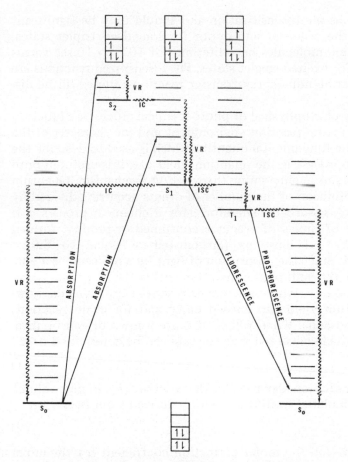

Fig. 1. Jablonski diagram showing absorption and subsequent photophysical modes of excited state decay. Abbreviations are as follows: IC = internal conversion; ISC = intersystem crossing; VR = vibrational relaxation; S = singlet; and T = triplet. The subscripts (i.e., 0, 1, and 2) indicate ground, first, and second states, respectively, and the arrows in the boxes indicate electron spins.

singlet and triplet states back to ground state. In a photochemical primary process the same course is followed, except that during the lifetime of the excited species in either a singlet or triplet state, a chemical alteration occurs resulting in the formation of a new ground state product. In certain instances the new product is reactive and proceeds either intramolecularly or intermolecularly to undergo further reactions in the system. Such reactions of products initially resulting from photo-excited molecules are referred to as secondary reactions. In certain situations the excited species does not chemically react, but instead serves as a donor and through energy transfer excites another molecule (the acceptor) which becomes the reactive species.

This process is known as photosensitization and should only be significant in dilute solutions of the acceptor and donor for long-lived triplet states, which for typical organic molecules have lifetimes of 10^{-5} to 10 sec versus only 10^{-6} to 10^{-9} sec for excited singlet states. Photosensitized reactions are also often used to describe non-energy transfer reactions; these will be discussed in Section 4.2.1.

The efficiency of any photophysical or photochemical process is a function of both the properties of the reaction environment and the character of the excited state species. The fundamental quantity which is used to describe the efficiency of any photoprocess is the quantum yield (ϕ); it is useful in both quantifying the process and in elucidating the reaction mechanism. Quantum yield has the general definition of the number of events occurring divided by the number of photons absorbed. Therefore, for a chemical process ϕ is defined as the number of moles of reactant consumed or product formed divided by the number of einsteins (an einstein is equal to 6.02×10^{23} photons) absorbed. Since the absorption of light by a molecule is a one-quantum process, then the sum of the quantum yields for all primary processes occurring must be one. Where secondary reactions are involved, however, the overall quantum yield can exceed unity and for chain reactions reach values in the thousands. When values of ϕ are known or can be measured for a specific photochemical reaction the rate can be determined from:

$$-[dA]/dt = \phi_A I_A \tag{1}$$

The amount of actinic radiation absorbed by the reactant (I_A) in unit volume and unit time can be calculated from the equation derived from Beer's law:

$$I_A = I_0 (1 - e^{-2.3\epsilon cd}) \tag{2}$$

if the incident intensity (I_0), the molar extinction coefficient (ϵ), the molar concentration (c), and the pathlength (d) are known. The intensity is usually expressed as einsteins cm^{-2} sec^{-1}.

Like thermal reactions, photochemical reactions have energy thresholds below which the reaction is not energetically feasible. The energy equivalence (E, in kcal einstein^{-1}) of the wavelength or frequency of radiation can be determined from:

$$E = (2.859 \times 10^5)\lambda^{-1} \simeq (9.537 \times 10^{-4})\nu \tag{3}$$

where λ is the wavelength of the radiation in angstroms and ν is the frequency in cycles sec^{-1}. The energies available for the wavelengths found in sea-surface sunlight are shown in Fig. 2. Although the potential effect of sunlight-induced marine photochemical processes is often acknowledged in the literature, the consideration has been confined to the immediate surface film layer. This conclusion is apparently based on the impression that it is only short-ultraviolet radiation which has sufficient energy to promote chemical reactions. It is apparent, however, that sufficient energy is available in sun-

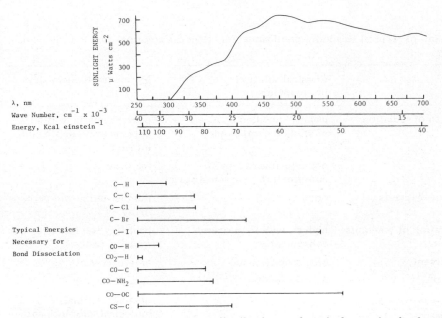

Fig. 2. The sea surface solar energy distribution and typical energies for bond dissociation.

light to promote various possible reactions at all wavelengths in the ultraviolet and visible region (Fig. 2). The entire photic zone, therefore, represents a region of potential photochemical reactivity.

3. SEAWATER AS A PHOTOCHEMICAL REACTION MEDIUM

For the most part, studies in classical solution photochemistry have been conducted under conditions where experimental variables are limited to as few as possible in order to obtain the most unambiguous information from the experiment. Operationally, the experimentalist sets the conditions which are best suited to the particular problem. This approach certainly has its place in marine photochemistry when one attempts to elucidate a particular process, but caution must be used in extrapolating the results of studies where considerable deviation from prevailing natural conditions have been employed. Extensive liberties have been taken with regard to this point, with the result that the literature now contains many references which were done under the guise of environmental photochemistry, which probably have little or no relevance with regard to the natural environment. It is probably not of value to belabor this point further; instead some aspects of seawater which make it a unique photochemical reaction medium will be considered.

A comparison of the conditions usually chosen in classical solution

304

TABLE I

Comparison of typical reaction conditions for classical solution and marine photo-chemistry

Property	Classical solution photochemistry	Marine photochemistry
Wavelength of radiation	usually specified by ϵ_{max} of studied compound	polychromatic, width dependent on depth and location
Solvent	non-aqueous except in coordination photochemistry	seawater
Number of reactants	one	number unknown, perhaps many
Concentration of reactants and products	high enough to measure conveniently	probably too low to measure easily
Phases present	one, homogeneous	heterogeneous
Oxygen	usually avoided	always present
Competing processes	avoided	possibly thermal, biological, physical, and other photochemical
Reaction rate	significant conversion, μ second to hours	environmentally significant conversion, hours to years
Variability of reaction medium (solvent and reactants)	no	unknown, probably yes

organic photochemistry versus those prevailing in the oceans' photic zone are shown in Table I. To a certain degree the variability of the reaction conditions is limited by the requirement that authenticity is best assured by reproducing the naturally imposed conditions as closely as possible. This immediately puts limits on the radiation (near-ultraviolet and visible wavelengths), the temperature (0—30°C), the solvent (seawater), and concentration of oxygen (4.5×10^{-4} to 2.0×10^{-4} M). Each of these parameters of the reaction environment is relatively easy to measure and control; the same is not true for other potentially important variables in the system.

3.1. Light attenuation in seawater

On a sunny day, a square meter of the ocean may have as much as one kilowatt of solar power impinging on it. Approximately 95% of this enters the water column; of this about one half, which consists primarily of the infrared region, is absorbed by molecules in the upper one meter, and is converted to rotational, translational and vibrational molecular motion. The remainder of the radiation is composed mostly of the visible (400—700 nm) and the near-ultraviolet (290—400 nm). The total attenuation of light in sea-water (Jerlov, 1968) is described by:

$$c = a_w + a_p + a_o + a_i + s_w + s_p \qquad (4)$$

where a_w = absorption by water; a_p = absorption by particles; a_o = absorption by dissolved organic constituents; a_i = absorption by dissolved inorganic constituents; s_w = scattering by water; s_p = scattering by particles. Light scattering has no direct photochemical consequence other than to alter the intensity, primarily as the result of backscattering out through the surface, which results in a 5—7% loss of the radiation entering the water column and varies for molecular scattering as λ^{-4}. The remainder of the light entering the water column is accounted for by absorption by water, particles, and dissolved inorganic and organic constituents.

The classical photochemistry, measuring the reactant's absorption spectrum, gives fundamental information about the characteristics of the reaction system. The absorption spectrum of seawater, however, is quite unrevealing of detail and generally appears in the region between 290 and 700 nm as a gradually decreasing continuous band with its maximum in the ultraviolet. With the exception of water, considerable variation exists for the absorbing components, and this shows up dramatically as regional differences in the attenuation of sunlight by seawater (Fig. 3). These differences result mainly from attenuation by dissolved organic materials and particles and are especially marked between regions of low and high productivity or where strong terrestrial influences exist. Contrary to the popular opinion that marine photochemical reactions initiated in the ultraviolet region are restricted to the immediate vicinity of the surface film, Fig. 3 clearly shows that a significant fraction of the incident ultraviolet penetrates deeply into the water column, especially in ocean regions.

The principal absorber in these oceanic regions may be water. For most purposes in spectroscopy and photochemistry, water is considered to be transparent to near-ultraviolet and visible radiation. In the oceans water becomes a major absorber of light energy where long light pathlengths exist. It is difficult to accurately quantify the absorbance, because of the lack of agreement in determining extinction coefficients for pure water from 300 to 500 nm (Hale and Querry, 1973). Since no primary photochemical processes are known for water above 300 nm, it probably does not represent a major contribution to organic photochemistry, except where it is involved in secondary reactions initiated by other absorbers.

The inorganic dissolved constituents apparently represent a minor absorbing component in the visible and near-ultraviolet regions (Lenoble, 1956; Armstrong and Boalch, 1961). Of particular interest as primary absorbers are nitrate, nitrite, and the coordination compounds of many of the transition metals found in seawater. In certain instances metals can

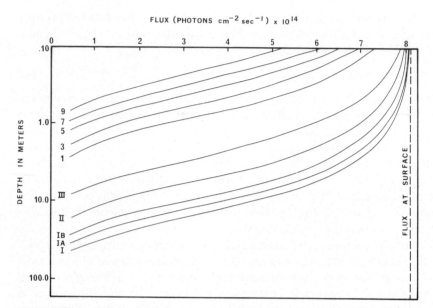

FLUX (PHOTONS cm^{-2} sec^{-1}) x 10^{14}

Fig. 3. The variation of photon flux with depth in different water types. Types I, IA, IB, II, and III represent different oceanic waters and types 1, 3, 5, 7 and 9 represent coastal waters (Jerlov, 1968). Surface light flux was calculated from data of Pettit (1932) for 10 nm spectral band centered at 350 nm. The measurement was made at latitude 32°N with the sun in the zenith.

involve organic compounds, which are otherwise transparent to the incident light, by forming organic coordination compounds with them. Some of these compounds have strong transitions well into the visible and even the near-infrared regions. Another inorganic constituent which absorbs weakly below 400 nm is hydrogen peroxide. Although limited measurements of it have been made in seawater (Van Baalen and Marler, 1966; Zika, 1978), its presence may be ubiquitous and its concentration variable and dependent to a large extent on production via photochemical and biological processes.

The strongest light-absorbing fraction in seawater is the dissolved organic material (Armstrong and Boalch, 1961). Of the known naturally occurring organic compounds, relatively few are significant absorbers above 300 nm (Table II). It is the uncharacterized fraction that is the principal absorber of light in the near-ultraviolet and the visible region out to 600 nm. Depending on whose estimates one accepts, this fraction constitutes some 65—90% of the total organic carbon present in natural seawater. The chromophoric character of this fraction may be, at least in part, the result of plankton by-products, which, through a type of Maillard reaction, have condensed to form melanoidines (Kalle, 1963). In coastal waters this fraction may contain other principal chromophoric constituents of terrigenous origin (Prakash,

TABLE II

The "dissolved" organic constituents of seawater [1]

Component	Concentration $(\mu g \, C \, l^{-1})$	Concentration [2] $(mole \, l^{-1} \times 10^{-7})$	Maximum wavelength of significant absorption [3] (nm)
Free amino acids	10	0.3	<300
Combined amino acids	50	—	<300
Free sugars	20	0.3	<300
Combined sugars	200	—	<300
Fatty acids	10	0.05	<300
Phenols	2	0.02	350
Sterols	0.2	0.0006	<300
Vitamins	0.006	0.000002	only biotin < 300
Ketones	10	0.2	350
Aldehydes	5	0.1	325
Hydrocarbons	5	0.03	aliphatic < 300 aromatic > 400
Urea	10	0.83	<300
Uronic acids	18	0.25	<300
Uncharacterized fraction	660	1.3 [4]	500—600
Total	1000	3.4	—

[1] Estimated concentration in $\mu g \, C \, l^{-1}$ (from Dawson, 1976).
[2] Based on average molecular weight for group classification.
[3] Based on typical absorption spectra for the compounds in group classification.
[4] Assumes a molecular weight of 1000 of which 50% is carbon.

1971) as well as benthic algal exudates (Khailov, 1963; Craigie and McLachlan, 1964; Sieburth and Jensen, 1969).

Heterogeneous microenvironments (e.g., particles, surface films, and micelles) constitute another absorber in natural seawater. Such microenvironments in the ocean may facilitate unique reactions, since the reactants become concentrated and light energy may be efficiently absorbed as the result of their having incorporated transition metals, portions of photosynthetic apparatus, and condensed organic chromophores. The quantification of the light energy absorbed in these microenvironments presents a formidable task, especially when they are variable in composition, in size, and relatively few in number.

This complex aggregate of light absorbers and the long pathlengths common to the marine photic zone combine to create a highly variable and difficult to simulate reaction environment with respect to the intensity and wavelength distribution of the incident radiation. Since the rate and efficiency of photochemical reactions are controlled by the intensity and wavelength, it is a common practice to impose carefully controlled limits on these parameters in classical photochemistry. The intensity, except in special instances, is held constant and the wavelength is fixed to a relatively narrow bandwidth during the course of the reaction. Limiting the wavelength of the radiation often reduces the number of reaction routes available and makes the quantum yield a constant for primary processes. In cases where secondary reactions are involved, the level of intensity directly affects the steady-state concentration of reactive transients in the solution and consequently the rates of second-order reactions. The implications of these intensity and wavelength distribution variations in the oceans are complex, hence the photochemistry observed in one region or at one depth does not necessarily extrapolate directly to another region or depth.

3.2. Composition of natural seawater

In classical organic photochemistry, reactions are usually conducted in pure organic solvents, with a single reactant present, and with the exclusion of oxygen from the system. In this way, the compositional variables which might contribute to the complexity of the system are minimized. In the marine environment it is reasonable to assume, until demonstrated otherwise, that photoreactions can involve many of the constituents present in one way or another. The "purity" of the solvent is, therefore, difficult to define since natural seawater is probably variable temporarily or spatially in all of its components. Even the major components which do not vary significantly on a proportional basis with changes in salinity, can have an appreciable effect on those photoreactions which are dependent on ionic strength or on the concentration of specific major ions. The non-conservative inorganic constituents represent an even more formidable problem, for not only do they vary in concentration, but probably also in speciation, and both of these properties are coupled to temporal and spatial features of the oceans. It will undoubtedly be suggested that the use of artificial seawater as a control blank is a way around this problem; however, an analysis for the trace-elemental composition of the reagent grade salts used in its preparation should discourage that proposed solution to the problem. When only the metals iron and manganese are considered, the concentration levels of these metals in prepared artificial seawater can be orders of magnitude higher than in natural seawater.

The organic material in seawater probably represents the major compositional problem, for it is a composite of many potential reactants (Table II) which are probably constantly undergoing changes through a combination of biological and abiotic processes. In the classical photochemical sense the situation is somewhat analogous to collecting a sample from a reaction vessel after some unknown reaction time has lapsed, during which some unknown number of reactions was active in altering an unknown number of reactants to give an unknown number of products. Marine photochemistry is further complicated by being an open system; therefore, the organic photochemical properties of any parcel of seawater in the photic zone may be ultimately controlled by the primary sources of inputs to its composition, that is fluvial, eolian, *in situ* biological production, or upwelled water. These properties might also be affected by conditioning of the constituents already there through biological, physical, and chemical processes.

3.3. Heterogeneous microenvironments

Particles, surface films, and colloids represent unique microenvironments where the composition and concentration can vary dramatically from the surrounding seawater. Since many of the substances which are considered to be part of the dissolved organic fraction of seawater are hydrophobic, they have a strong tendency to adsorb on surfaces (Zsolnay, 1977) or coalesce into particles. In the reverse sense hydrophobic materials can be solubilized by association with the dissolved organic matter in seawater (Boehm and Quinn, 1973). It has been suggested that the particulate and dissolved organic carbon are associated through a complex equilibrium (Parsons, 1975) in which the displacement is established by the concentration and nature of the organic materials involved and by processes which serve to control the forward and reverse rates (e.g., bubbles, bacteria, inorganic particles, and chemical condensation reactions). Classical organic photochemistry tends to avoid heterogeneous reactions by using pure organic solvents in which the reactants are very soluble. While this is a general goal, there has been a growing interest in the photochemistry of reactions occurring at the surface of metal compound solid phases and in micelles (Thomas, 1977). These and other perturbing polar environments, such as silica, can effect a reduction in the threshold energy and an increase in the efficiency of organic chemical reactions. If particles and other heterogeneous microenvironments are important mediators of photochemical reactions in the oceans, then they represent another variable which must be considered. For particles alone this means variations in composition and abundance for different latitudes, for coastal versus oceanic conditions, and for different oceans (Lal, 1977).

3.4. Reaction rates

To ascertain the magnitude and the mechanism of a photochemical reaction, it is necessary to determine the rate of the process (see eq. 1). In classical photochemistry, reactions which are very slow are usually considered to be unimportant and hence of little interest. In the marine environment very slow reactions can be significant if they are the only operating mechanism for a particular process or if they compete favorably with other abiotic or biotic mechanisms contributing to the same process. Remineralization of organic matter, for example, is generally attributed solely to biological routes, and for those compounds readily utilized by the biota there is likely to be an insignificant contribution from abiotic processes. Studies on the degradation of algal cultures (Otsuki and Hanya, 1972) and the natural organic fraction in seawater (Ogura, 1976) demonstrate an initial rapid microbial utilization of a significant fraction of the organic materials present, but a major dissolved refractory fraction remained. Ogura found that as much as 80—90% of the dissolved organic matter in oceanic surface water appeared in the time scale of the study to be refractory. This is supported in the ocean by a 3400 year apparent age for the dissolved organic carbon in deep water (Williams et al., 1969). Obviously, the rate constants for the processes (i.e. sedimentation and biological and abiotic degradation) acting on this fraction are small and difficult to accurately determine. If it is assumed that 5% of the net primary productivity of the oceans is the yearly contribution to the refractory dissolved organic fraction and that an average steady-state concentration of 1 mg C l^{-1} is maintained in the oceans, then the first-order rate constant for remineralization needed to maintain the steady-state concentration would be on the order of 0.002 yr^{-1}. Even if it is assumed that remineralization was solely attributed to photochemical reactions in the photic zone, the rate constant there would not be more than 0.1 yr^{-1}. This constitutes a daily average conversion for a 100-m surface layer of only 0.26 μg l^{-1} of organic carbon. Rather than suggest that this apparent biologically inert fraction is converted solely by photochemical processes, it is more realistic to conclude that a synergistic interaction involving both biological and non-biological processes is involved in which light may act primarily as the energy source for priming the refractory organic materials for further degradation.

The determination of rates for indirect photochemical pathways involving secondary chemical and biological processes may in most cases not be a tractable problem because of the many parameters involved in their determination. Even for the determination of the rates of primary reactions, which in classical photochemistry are a relatively straightforward procedure, the situation is complicated by the variability of conditions in the environment; these being primarily concerned with changes in the light field. Recently, methods have been developed which can be applied to calculating environmental direct photochemical reaction rates in the environment (Zepp and Cline, 1977; Zafiriou, 1977). Zepp and Cline have applied their method to calculation of photolysis rates for various pesticides in aquatic environments. In this method the assumption is made that ϕ is not wavelength-dependent. The rate is expressed as the first-order equation:

$$-[dA]/dt = \phi \left(\Sigma k_{a\lambda} \right) [A] \tag{5}$$

where $\Sigma k_{a\lambda}$ represents the computer-calculated sum of the rate constants for all wavelengths of sunlight that are absorbed by the reactant and $[A]$ is the concentration of reactant. Since ϕ is not always a constant at different wavelengths for some compounds, this should be determined before applying the equation. However, even if ϕ is not known, the minimum half-life ($t_{1/2}$) can be determined from:

$$t_{1/2} \leqslant 0.693/\Sigma k_{a\lambda} \qquad (6)$$

in which it is assumed that ϕ is not likely to exceed unity at the low concentrations of reactants encountered in natural waters.

4. POSSIBLE PHOTOCHEMICAL REACTIONS

The possible photochemical reactions which may be involved in the alteration of organic materials found in seawater are undoubtedly highly varied and involve both organic and inorganic constituents of seawater. Many of the potential reactions have recently been previewed (Zafiriou, 1977), and will, therefore, be mentioned only briefly here. A somewhat arbitrary division may be made into primary and secondary reactions, where primary reactions denote those involving the excited state of an organic molecule, and secondary reactions denote those involving inorganic excited species and reactive non-excited state intermediates.

4.1. Primary organic photochemical processes

Numerous possible reaction routes are known from both singlet and triplet photo-excited state manifolds of organic molecules (Fig. 4). Although the excited state of a specific compound may react by more than one chemi-

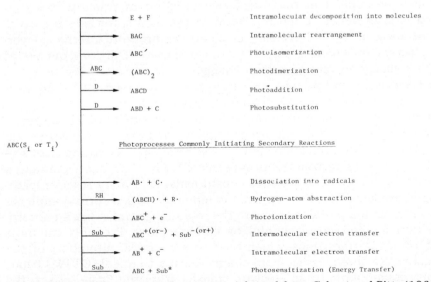

Fig. 4. Primary photochemical processes. Adapted from Calvert and Pitts (1966, p. 367).

cal process, there will usually only be one predominant reaction route observed. Theoretical prediction of the nature and efficiency of photochemical processes in organic molecules is not yet possible, but reasonable predictions based on correlations in chemical structure can often be made. However, the only sure way of determining the photochemical fate of a particular organic compound is to measure it. The number of reaction routes followed and the efficiency, especially for small molecules, can vary with changes in the energy of the absorbed radiation and with changes in the environment of the excited species. Many properties of the reaction environment can affect the nature of the reactions occurring. Second- and higher-order reactions (i.e., hydrogen-atom abstraction, photodimerization, photosubstitution, photosensitization, and intermolecular electron transfer) are primary processes which display high concentration dependence. In seawater, these processes are likely to be confined to reactions between the excited state species and major constituents, to microenvironments where the concentration is sufficiently high for them to be important, or to excited state species with sufficiently long lifetimes to allow for reasonably high collision rates of the reactants. Intramolecular reactions (i.e., dissociation into radicals, photoisomerization and intramolecular decomposition, rearrangement, and electron transfer) are far less concentration-dependent and extrapolation to environmental levels of experimental values for reaction rates determined at high concentrations can usually be made. This has been the most common recourse followed in studying environmental photochemistry of pesticides. In many instances these studies have been conducted in organic solvents, because of the limited solubility of many of these compounds in water. Caution must be applied in the direct extrapolation of primary processes determined in such media to natural water environments, for factors such as the concentration of O_2 and of heavy elements can promote internal conversion and intersystem crossing steps in excited species. The consequence of this can be a change in rate or even a change in the nature of the products observed. Likewise, some primary processes can be affected by factors such as the nature of solvent, viscosity, temperature, and pH.

In the marine environment only those organic compounds which absorb at wavelengths longer than 290 nm are candidates for primary photochemical processes. These include relatively few of the identified constituents of seawater and the uncharacterized fraction (Table II), which probably contains a host of natural and xenobiotic minor constituents. A primary reaction pathway, also, is available for those organic compounds which form complexes with certain transition metal ions where the resulting complex has an absorption transition in the near-ultraviolet or visible regions. Probably the most important of these transitions which lead to a chemical alteration of the organic ligands are those arising from charge transfer to metal (CTTM) band. The net process amounts to an electron transfer from the ligand (L) to the metal (M):

$$[M^{x+}(L_n)^{y-}]^z + h\nu \xrightarrow{\text{CTTM}} [M^{(x-1)+}(L_n)^{(y-1)-}]^z \tag{7}$$

Ensuing reactions of the oxidized ligand molecule then lead to the observed products. These processes are likely to occur for compounds containing two or more functional groups that can form relatively stable chelates with transition metals existing in their higher oxidation states, but having an accessible lower oxidation state (e.g., Fe, Cu, Cr, V, Hg, Ce, and U).

4.2. Secondary photochemical processes

While primary photochemical processes are restricted to those compounds which are excited by the direct absorption of radiation, secondary processes of one type or another can involve all organic compounds. The reactivity, the concentration, and the lifetime of secondary transient reactants will be determining factors in what types of organic compounds are susceptible to attack and what predominant modes of reactions are occurring. Because so many of the constituents of seawater may be involved, the number of possible alternative reactions occurring is potentially very complex. Therefore, only the immediate reactions of three initiating sources of secondary reactions will be discussed.

4.2.1. Organic initiators

Some of the reactions shown in Fig. 4 lead to unstable products which will undergo further secondary reactions. In some cases the products will be reactive with other organic and inorganic materials in seawater and new reactive species will be generated that can contribute to further secondary reactions. In this regard, free radicals (odd or unpaired electron species) may be one of the most important products arising from primary reactions, for once formed they continue to propagate and will only be destroyed by combination with other odd electron species. In an oxygenated solution such as seawater, a major portion of organic-free radicals generated will probably be involved in the reactions commonly associated with autoxidation:

$$\text{initiator} \rightarrow \text{X}\cdot \tag{8}$$
$$\text{X}\cdot + \text{RH} \rightarrow \text{XH} + \text{R}\cdot \tag{9}$$

$\left.\right\}$ initiation

$$\text{R}\cdot + \text{O}_2 \rightarrow \text{ROO}\cdot \tag{10}$$
$$\text{ROO}\cdot + \text{RH} \rightarrow \text{ROOH} + \text{R}\cdot$$

$\left.\right\}$ propagation

$$\text{ROO}\cdot + \text{R}\cdot \rightarrow \text{ROOR} \tag{12}$$
$$2\,\text{R}\cdot \rightarrow \text{RR} \tag{13}$$

$\left.\right\}$ termination

$$2\,\text{ROO}\cdot \rightarrow \text{ROOR} + \text{O}_2 \tag{14}$$

Since organic-free radicals ($\text{R}\cdot$ and $\text{ROO}\cdot$) should be present at low steady-state concentrations in seawater, the termination reactions should be unimportant as long as a sufficient source of abstractable hydrogen atoms are present. In seawater the organic fraction should be the only available source of hydrogen atoms; consequently, both inorganic and organic free radicals should contribute to the degradation of this fraction. One of the primary products of these reactions should be organic peroxides, which will serve as further initiators of free radical production by virtue of their reactivity with other seawater constituents (e.g., metals, halide ions, and amines) and by photolysis:

$$ROOR' + h\nu \rightarrow RO\cdot + R'O\cdot \tag{15}$$

since the peroxy group is a weak chromophore at near-ultraviolet wavelengths.

Organic compounds can generate the initiators of free radical sequences through the primary photochemical processes: homolytic dissociation into radicals, hydrogen-atom abstraction, photoionization, and electron transfer reactions. The homolytic dissociation reactions are limited to compounds containing relatively weak bonds (<98 kcal), such as sulfides, peroxides, and some halides and ethers. Representatives of all of these classes of compounds are certainly present in seawater, but the limited information on the qualitative and quantitative aspects of their occurrence does not allow for an estimate of their importance in the promotion of free radical reactions. The same is true for electron transfer reactions, which may be an important photochemical process for organic transition metal complexes.

The only attempt that has been made to estimate the magnitude of a specific organic-free radical initiation process was done for hydrated electron (e_{aq}^-) production via photoionization of aromatic compounds (Swallow, 1969). By assuming that all incident sunlight below 325 nm was absorbed by organic chromophores and that the quantum yield for photoionization was one, a maximum production rate of 3×10^{15} e_{aq}^- l^{-1} sec^{-1} was obtained at the sea surface. This probably constitutes a considerable overestimate, but even if it were accurate the hydrated electron should not be an important reactant in homogeneous solution reactions with organic compounds. The reasons for this are its high reactivity with oxygen ($k = 1.88 \times 10^{10}$ $mole^{-1}$ sec^{-1} at pH 7.0) and carbon dioxide ($k = 7.7 \times 10^9$ $mole^{-1}$ sec^{-1} at pH 7.0) in the following reactions:

$$O_2 + e_{aq}^- \rightarrow O_2^-, \text{ and} \tag{16}$$

$$CO_2 + e_{aq}^- \rightarrow CO_2^- \tag{17}$$

At pH values of 8.0 and above, the unhydrolyzed carbon dioxide level is low and the predominant scavenger should be oxygen. The resulting superoxide radical anion from this reaction is a much longer lived species than the e_{aq}^- and can react both as an oxidizing and reducing agent in reactions with various organic compounds (Bors et al., 1974). However, in the absence of sufficient quantities of reactive substrates it should rapidly disproportionate:

$$2 O_2^- + 2 H^+ \rightarrow H_2O_2 + O_2 \tag{18}$$

The measurement of significant concentrations of hydrogen peroxide in seawater (Van Baalen and Marler, 1966) and the demonstration of its photochemical production (Zika, 1978) in seawater support the occurrence of such processes. If hydrogen peroxide generation is ubiquitous over the oceans then through its spontaneous degradation pathways it represents an important secondary reaction process.

The initiation of secondary reactions by oxidation—reduction or energy transfer processes of the excited state species are often combined under the heading photosensitization. Strictly speaking, however, physical chemists usually consider photosensitization to be only an energy transfer process. The transfer of energy from an excited state species to a ground state acceptor can occur as the result of direct collision or through resonance interactions between the pair which can occur at distances of up to 100 Å. In view of the typical short lifetimes of excited state species and the high dilution of dissolved organic compounds in seawater, the photosensitization of organic molecules should be of little importance except, perhaps, in microenvironments. The efficiency of the sensitizer in transferring energy to an organic molecule is further reduced by the presence of competing inorganic energy acceptors. Oxygen is the most important of these for its concentration is relatively high and its first singlet excited state ($^1\Delta_g$) is only 22 kcal mole^{-1} above ground state energy; it could be a substantial transient product even for sensitizers with low triplet excited state energies. In fact, energetically its production is possible from sensitizers activated at wavelengths as long as 700 nm. It is, therefore, not surprising that substantial generation rates of singlet oxygen in sunlight-initiated reactions have been measured in both freshwater and coastal seawater (Zepp et al., 1977).

Reactions of singlet oxygen with many organic compounds are known (Foote, 1976). Particularly susceptible are phenols, conjugated and unconjugated unsaturated compounds, and certain compounds containing the heteroatoms nitrogen, sulfur, or phosphorus. In many of these reactions unstable products such as hydroperoxides, cyclic peroxides, and phenoxy radicals result. All of these products can provide yet another source of free radicals to initiate further secondary reactions. With the exception of certain special environments, the effectiveness of singlet oxygen in initiating these processes in seawater may not be very significant, for the competing process of quenching of $^1\Delta O_2$ in aqueous solutions is very rapid (Merkel et al., 1972).

Aside from promoting energy transfer processes, there are other possible alternative routes available to photosensitizers (Fig. 5). The excited states of many photosensitizers act as oxidizing and reducing agents as the result of the change in electron distribution brought about by promotion of an elec-

316

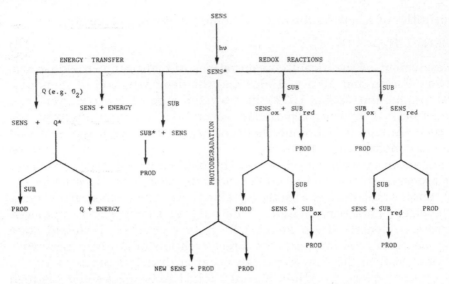

Fig. 5. Possible alternatives of an excited photosensitizer. Abbreviations are as follows: SENS = photosensitizer; PROD = product(s); SUB = substrate; Q = quencher; E = energy; and subscripts, ox = oxidized and red = reduced.

tron to a higher-energy orbital. Although the substrates reacting directly with excited sensitizer molecules can be various organic compounds, the high dilution of these in seawater probably restrict this reaction to heterogeneous environments, or to long-lived low-energy triplet excited state sensitizers. In solution the more probable reactions for photosensitizers exhibiting oxidation--reduction characteristics are the reduction of oxygen to the superoxide radical anion and the oxidation of inorganic anions. Many examples of photosensitizers which reduce oxygen exist, but the ratio of oxygen reduction to singlet oxygen formation is usually less than 1 : 100 (Kasche and Lindquist, 1965). Fewer examples of inorganic anion oxidation are known, but compounds, such as anthraquinones, have been shown to photo-oxidize chloride and carbonate ions (Kuzmin and Chibisov, 1971). Secondary reactions of these resulting reactive inorganic species, as well as reactions with the reduced or oxidized sensitizers, can because of their longer lifetimes be more important reactants with the organic fraction than is direct interaction with the triplet excited state of the sensitizer.

4.2.2. Transition metals

Besides the primary photoreactions of organic chelates, there is a variety of ways in which transition metals can initiate secondary reactions. In general, the reactions involve photoderived unstable oxidation states of the metal (e.g., Cu^{1+}, Fe^{2+}, or U^{4+}), reactive inorganic species, or reactive organic fragments, such as free radicals. The possible initial photoreactions of dissolved metal complexes which give secondary reactants can be categorized by the absorption transition into charge transfer to metal (CTTM):

$$[M^{x+}L^{y-}]^z + h\nu \rightarrow M^{(x-1)+} + L^{(y-1)-} \tag{19}$$

charge transfer to ligand (CTTL):

$$[M^{x+}L^{y-}]^z + h\nu \rightarrow M^{(x+1)+} + L^{(y+1)-} \tag{20}$$

charge transfer to solvent (CTTS):

$$M^{x+}(H_2O)_n + h\nu \rightarrow M^{x+1}(H_2O)_n + e^-_{aq} \tag{21}$$

and ion pair charge transfer (IPCT):

$$[ML]^{z+}X^- + h\nu \rightarrow [ML]^{(z-1)+} + X\cdot \tag{22}$$

Each photoreactive metal, of course, represents a separate case and the speciation of the metal ion is of primary importance in determining the nature of the reactions. These reactions, however, will generally all give free radical products and metal ion species differing by one unit charge higher or lower in oxidation state.

4.2.3. Nitrate and nitrite

Most of the inorganic anions of seawater are not subject to primary photochemical processes, because their electronic excitation energies lie well above the maximum photon energy available in sea surface sunlight. Nitrate and nitrite do, however, exhibit weak transitions with maxima at 303 nm (ϵ_{max} = 7 mole^{-1} cm^{-1}) and 355 nm (ϵ_{max} = 22.5 mole^{-1} cm^{-1}), respectively. The primary photoreactions of both of these oxyanions produce free radical products in seawater (Zafiriou and True, 1979a, b). The overall reactions are:

$$NO_3^- + H_2O + h\nu \rightarrow NO_2 + OH\cdot + OH^- \tag{23}$$

$$NO_2^- + H_2O + h\nu \rightarrow NO + OH\cdot + OH^- \tag{24}$$

and a second process for nitrate which gives nitrite and another unidentified product which may be an oxygen atom. Subsequent reactions of NO and NO_2 with organic constituents are a likely possibility for certain reactive compounds and certainly with free radicals, but the significance of these processes may be greatly reduced by competing re-oxidation and hydrolysis steps which lead to the regeneration of NO_2^- and NO_3^- (Zafiriou, 1974). The major role of NO_2^- and NO_3^- photolysis may be in the generation of the hydroxyl radical (OH$\cdot$), which is a strong oxidant. Its high reactivity with nearly all of the major inorganic anions of seawater makes its direct interaction with organic constituents small, and its major contribution should be the generation of longer lived, but less reactive free radicals like Br_2^-, CO_3^-, and HCO_3 (Zafiriou, 1974).

Recent studies demonstrate that NO_3^-, even though usually present in seawater at much higher concentrations, is probably a far less effective contributor to free radical production than NO_2^- (Zafiriou and True, 1979a, b). Based on the photochemical rate of reaction of NO_2^-, a yearly global average for OH· production of 1—10 mmole m^{-2} yr^{-1} has been estimated. For comparison to the remineralization first-order rate constant of 0.002 yr^{-1} calculated earlier, a first-order rate constant of 0.0025 to 0.025 yr^{-1} for OH· radical production is obtained when averaged over a mean ocean depth of 4000 m. Although not all of the radicals generated via NO_2^- photolysis are likely to be consumed by reactions with the refractory organic fraction, it, nevertheless, could constitute a mechanism for the initiation of alteration processes in organic constituents.

5. EXAMPLES OF MARINE PHOTOCHEMICAL PHENOMENA

Although no comprehensive systematic studies on the processes outlined above have yet been published, there are examples in the literature which have defined specific light-induced phenomena or which have explained observations by invoking undefined photochemical mechanisms. These reports are rare, however, for seldom is there any recognition of the possible involvement of photochemical reactions in investigations of the marine environment. There is little question that all of the processes which have been discussed in preceding sections are operating in sunlit ocean water. The important question is whether their effect on the environment is significant. This section of the chapter is, therefore, devoted to examining some of the ways in which the oceans could be affected by photochemical processes, by providing a cursory look at some of the scattered observations that have already been made which give credence to the significance of photochemistry in the marine environment.

5.1. Degradation and transformation of xenobiotic materials

The introduction of thousands of different xenobiotic materials into the environment in ever increasing quantities has spurred numerous investigations on their subsequent removal or alteration by natural physical, biological, chemical, and photochemical pathways. The major amount of attention to photochemical reaction pathways in natural waters has probably been given to pesticides, which as a group may constitute potentially the most deleterious class of xenobiotic materials. Many of these compounds are degraded by primary photochemical processes. Unfortunately, for reasons of limited solubility and practical aspects of product analysis, these studies have been confined primarily to organic solvents and high concentrations of reactants. Growing recognition of the possible involvement of properties common only to natural waters has stimulated a need for conducting these reactions under closely simulated environmental conditions (Rosen, 1971; Faust, 1975). Recent investigations that have been performed in natural freshwater samples reveal that the photochemical reaction rates of some pesticides are greatly increased by natural photosensitizers (Ross and Crosby, 1973, 1975; Wolfe et al., 1976; Zepp et al., 1976). The sensitizers responsible for the observed reactions have not been characterized, but the results are generally attributed to humic acids. Certainly porphyrins, carbonyl compounds, flavins, polyaromatic hydrocarbons, indoles, and numerous others, including some xenobiotics are also potential photosensitizers and are common to both fresh and marine water environments. For those sensitizers which have the combined property of behaving as surfactants, their efficiency may be greatly increased through association with insoluble pesticides. The enhanced photodecomposition of 2,3,7,8-tetrachlorodibenzo-paradioxin by the cationic surfactant 1-hexadecylpyridinium chloride is an example of this (Botre et al., 1978), and this property may be attributed to humic acids of seawater as well (Kahn and Schnitzer, 1972).

5.2. Surface microlayer reactions

In considering the various processes active in the removal of oil from the sea surface, Pilpel (1968) concluded that oxidation by micro-organisms was the most important and might proceed at rates of up to ten times faster than spontaneous chemical oxidation. In contrast to Pilpel, Baier (1972) found in experiments conducted in the field and laboratory with freshwater, that bacterial degradation was not fast enough to account for the rapid disappearance of oil films from natural waters. On the basis of results using internal-reflection infrared spectroscopy, it was concluded that the removal of oil films was facilitated by the photo-induced introduction of O_2 into the organic film (as evidenced by the appearance of ester bands in the infrared spectra) and that this process, combined with bubble breaking at the surface, provided the most effective removal mechanism.

Another mechanism for the removal of surface films was discovered by Wheeler (1972), who found that fatty acid films on seawater collapsed to form particles when exposed to near-ultraviolet radiation. The results indicated that there was an introduction of hydroperoxide groups into the parent fatty acid molecule with resultant polymerization of the products. Instead of polymerization, Timmons (1962) found that the constituents of plankton oil films were converted to smaller and more soluble fragments when exposed to artificial sunlight. Photochemically initiated solubilization appears to be a process common to some constituents of crude oil and fuel oil films as well, with low molecular weight acids, sulfoxides, and peroxides comprising some of the product soluble fraction (Burwood and Speers, 1974; Hansen, 1975; Larson et al., 1977). The rate of photo-oxidation of films of various fractions of crude oil spread on water was greatly increased by the addition of naphthalene derivatives (Klein and Pilpel, 1974; Pilpel, 1975). The naphthalene derivatives apparently acted as photosensitizers which caused an increase in solubilization of the films through a reaction mechanism involving the formation of peroxides and alcohols. A case was made for the addition of photosensitizers to oil spills at sea to accelerate their removal by sunlight photo-oxidation, but in view of the observed toxicity of photoproducts from oil layer degradation towards aquatic micro-organisms (Lacaze and Villedon, 1976; Larson et al., 1977), this approach warrants careful consideration. The use of chemical dispersants was found to magnify the toxicity by a factor of at least 10—15 (Lacaze and Villedon, 1976).

5.3. Metal—organic photo-interactions

Much speculation exists on the presence and role of chelators in natural waters, but little has been accomplished towards their characterization and virtually nothing is known about their photochemistry. There are, however, examples of the effects that metal ions can have on the photochemistry of added organic substances. A major area of interest has been in the sunlight degradation of synthetic aminopolycarboxylates in freshwater. Such investigations were prompted by concern over the increased introduction of these compounds into the environment through both commercial and domestic uses. Studies on Cu^{2+} nitrilotriacetate (NTA) (Langford et al., 1973) and Fe^{3+} NTA (Trott et al., 1972) chelates demonstrated that both of these compounds were rapidly degraded on exposure to the radiation wavelengths found in sea surface sunlight. It was proposed that the reactions proceeded by way of a CTTM transition; based on similar results for Fe^{3+} EDTA and glycine (Carey and Langford, 1973) it was concluded that this may be a general reaction:

$$RR'NCH_2 COOH + 1/2 O_2 + h\nu \rightarrow RR'NH + CO_2 + CH_2O \qquad (25)$$

for aminocarboxylate metal complexes. Further investigations on the products of Fe^{3+} +EDTA (Lockhart and Blakeley, 1975) showed that eight major products were formed, including glycine. The EDTA chelates of Na^+, Mg^{2+}, Mn^{2+}, Fe^{3+}, Co^{2+}, Cu^{2+}, Zn^{2+}, Cd^{2+}, Ni^{2+}, and Hg^{2+} were tested for photoreactivity, but only the chelates of Mn^{2+}, Fe^{3+}, and Co^{2+} were found to be photolabile. The specificity of such reactions with respect to changes in solution environment or the character of the chelator is demonstrated by the observation that in seawater only Cu^{2+} was found to promote a significant photodegradation of the aminocarboxylate glycine (Zika, 1978).

Although metals such as the alkaline earth group should not promote photo-oxidative processes via CTTM transitions, they have been implicated in the photodecomposition of pteridines in seawater (Landymore and Antia, 1978). This has been explained through a structural modification of pteridines resulting from complexation with divalent cations to give a more photolabile form of the heterocycle. The photolysis rates of all the pteridines tested were much greater than their dark reaction rates in seawater, and light-initiated degradation could be an important step in environmental recycling of this group of compounds.

The photochemical formation of covalently bonded metal—organic compounds in water has also been observed. Agaki and Takabatake (1973) found that when aqueous solutions of mercury (II) acetate were exposed to simulated sunlight, mercury alkylation occurred. In a similar investigation, methyl mercuric ion and dimethyl mercury were found to be the products resulting from the irradiation of mercury (II) acetate solutions with normal fluorescence laboratory lighting (Jewett et al., 1975) Salinity was found to be a deciding ingredient in promoting these reactions. They found similar results for thallium (I) acetate, and the acetate solutions of both metal ions gave gaseous products on extended irradiation; these were identified as ethane and CO_2.

5.4. Other observations

In a few instances, light-initiated reactions in seawater have been discovered by testing the stability of substances in light and dark controls, or from the observation that the concentration of certain constituents of seawater increase on exposure of seawater to light. The latter of these led Wilson et al. (1970) to postulate that carbon monoxide (10^{-4} ml l^{-1} day^{-1}), ethylene, and propylene (10^{-4} ml l^{-1} day^{-1}) were formed at least in part by photochemical processes in seawater. Although the production rates were small, they were significant when compared to the normal concentration of these materials in seawater. The amount produced was found to be dependent on the concentration of organic materials present. The authors do not suggest a mechanism for the formation of the observed products, but it is possible that since aldehydes have been identified as a constituent of seawater (Kamata, 1966), their direct photolysis in known reactions like:

$$CH_2O + h\nu \rightarrow H_2 + CO \tag{26}$$

$$CH_3CH_2CHO + h\nu \rightarrow CH_2 = CH_2 + CH_2O \tag{27}$$

$$CH_3CH_2CH_2CHO + h\nu \rightarrow CH_3CH = CH_2 + CH_2O \tag{28}$$

could explain the observations.

Photochemical reactions of biologically active organic compounds have also been observed. Such reactions could have a significant place in the conditioning of seawater for marine organisms. An example of this is the demonstrated photolability of the essential ingredients of phytoplankton growth: vitamin B_{12}, thiamine, and biotin (Carlucci et al., 1969). While the degradation of vitamin B_{12} and thiamine could be explained by direct photolysis the

biotin is transparent to sunlight radiation, suggesting that secondary processes must be active in this case. The same may be true for the purines, uric
acid and xanthine, which were found to photodegrade when irradiated with
light to which they should have been transparent (Antia and Landymore,
1974). On the basis of EDTA inhibition of the uric acid photolysis, trace
metals were implicated in the reaction scheme. The addition of EDTA to
natural seawater solutions of xanthine, however, resulted in an increase in
the rate of photolysis and this was advanced as evidence for the inhibitory
effect of metals on the reaction. Aside from being a good chelator, EDTA is
also a good reducing agent. It is entirely possible, therefore, that it is acting
as a free radical scavenger or a hydrogen or electron source for triplet excited
states. In any case, the influence of EDTA on the photodegradation of
xanthine and uric acid does not necessarily pin down the mechanism to one
involving trace metals.

6. CONCLUSION

The understanding of marine photochemical processes and their effects on
the organic fraction is still only rudimentary. The slow development of this
potentially important area of marine chemistry is probably the result of a
combination of factors including misconceptions about energy requirements
for photochemical reactions and the depth of penetration of actinic radiation into seawater. An examination of the light absorbers in seawater reveals
that many potential reactions are possible in the marine environment, and
initial investigations have demonstrated that significant reactions are
occurring there. The relatively few studies to date and the likelihood that
some marine photochemical processes, such as remineralization of organic
carbon, might be difficult to measure in view of their slow rates, suggest that
many more as yet unrecognized, but significant, processes are occurring. The
experimental elucidation of these processes is complicated by the simultaneous involvement of a maze of reactions and conditions which are controlling them.

7. REFERENCES

Adamson, A.W. and Fleischauer, P.D., 1975. Concepts of Inorganic Photochemistry.
    Wiley-Interscience, New York, N.Y., 439 pp.
Agaki, H. and Takabatake, E., 1973. Photochemical formation of methylmercuric com-
    pounds from mercuric acetate. Chemosphere, 3: 131—133.
Antia, N.J. and Landymore, A.F., 1974. Physiological and ecological significance of the
    chemical instability of uric acid and related purines in sea water and marine algal
    culture medium. J. Fish. Res. Board Can., 31: 1327—1335.
Armstrong, F.A.J. and Boalch, G.T., 1961. The ultraviolet absorption of seawater. J.
    Mar. Biol. Assoc. U.K., 41: 591—597.
Baier, R.E., 1972. Organic films on natural waters: Their retrieval, identification, and
    modes of elimination. J. Geophys. Res., 77: 5062—5075.

323

Balzani, V. and Carassiti, V., 1970. Photochemistry of Coordination Compounds. Academic Press, New York, N.Y., 400 pp.

Boehm, P.D. and Quinn, J.G., 1973. Solubilization of hydrocarbons by the dissolved organic matter in sea water. Geochim. Cosmochim. Acta, 37: 2457—2477.

Bors, W., Saran, M., Lengfelder, E., Spöttl, R. and Michel, C., 1974. The relevance of the superoxide anion radical in biological systems. Curr. Top. Radiat. Res. Q., 9: 247—309.

Botre, C., Memoli, A. and Alhaique, F., 1978. TCDD solubilization and photodecomposition in aqueous solutions. Environ. Sci. Technol., 12: 335—336.

Burwood, R. and Speers, G.C., 1974. Photo-oxidation as a factor in the environmental dispersal of crude oil. Estuarine Coastal Mar. Sci., 2; 117—135.

Calvert, J.G. and Pitts, J.N., 1966. Photochemistry. John Wiley, New York, N.Y., 899 pp.

Carey, J.H. and Langford, C.H., 1973. Photodecomposition of Fe (III) aminopolycarboxylates. Can. J. Chem., 51. 3665 3670.

Carlucci, A.E., Silbernagel, S.B. and McNally, P.M., 1969. Influence of temperature and solar radiation on persistence of vitamin B_{12}, thiamine, and biotin in sea water. J. Phycol., 5: 302—305.

Cowan, D.O. and Drisko, R.L., 1976. Elements of Organic Photochemistry. Plenum Press, New York, N.Y., 586 pp.

Craigie, J.S. and McLachlan, J., 1964. Excretion of colored ultraviolet-absorbing substances by marine algae. Can. J. Bot., 42: 23—33.

Dawson, R., 1976. Water soluble organic compounds in seawater. Paper presented at Symposium on "Concepts in Marine Organic Chemistry," Edinburgh, Scotland (unpublished).

Faust, S.D., 1975. Nonbiological degradation and transformations of organic pesticides in aqueous systems. In: T.M. Church (Editor), Marine Chemistry in the Coastal Environment. A.C.S. Symp. Ser. 18. Am. Chem. Soc., Washington, D.C., pp. 572—595.

Foote, C.S., 1976. Photosensitized oxidation and singlet oxygen: Consequences in biological systems. In: W.A. Pryor (Editor), Free Radicals in Biology, 2. Academic Press, New York, N.Y., pp. 85—133.

Hale, G.M. and Querry, M.R., 1973. Optical constants of water in the 200 nm to 200 μm wavelength region. Appl. Opt., 12: 555—563.

Hansen, H.P., 1975. Photochemical degradation of petroleum hydrocarbon surface films on seawater. Mar. Chem., 3: 183—195.

Jerlov, N.G., 1968. Optical Oceanography. Elsevier, Amsterdam, 194 pp.

Jewett, K.E., Brinckman, F.E. and Bellama, J.M., 1975. Chemical factors influencing metal alkylation in water. In: T.M. Church (Editor), Marine Chemistry in the Coastal Environment. A.C.S. Symp. Ser. 18. Am. Chem. Soc., Washington, D.C., pp. 304 317.

Kahn, S.U. and Schnitzer, M., 1972. The retention of hydrophobic compounds by humic acid. Geochim. Cosmochim. Acta, 35: 745—754.

Kalle, K., 1963. Über das Verhalten und die Herkunft der in den Gewässern und in der Atmosphäre vorhandenen himmelblauen Fluoreszenz. Dtsch. Hydrogr. Z., 16: 153—166.

Kamata, E., 1966. Aldehydes in lake and sea waters. Bull. Chem. Soc. Jap., 39: 1227.

Kasche, V. and Lindquist, L., 1965. Transient species in the photochemistry of eosin. Photochem. Photobiol., 4: 923—940.

Khailov, K.M., 1963. Some unknown organic substances in seawater. Dokl. Akad. Nauk S.S.S.R., 147: 1355—1357.

Klein, A.E. and Pilpel, N., 1974. The effects of artificial sunlight upon floating oils. Water Res., 8: 79—83.

Kuzmin, V.A. and Chibisov, A.K., 1971. One electron photooxidation of inorganic anions by 9,10-anthraquinone-2,6-disulforic acid in the triplet state. J. Chem. Soc. D., 23: 1559—1560.

324

Lacaze, J.C. and Villedon, O., 1976. Influence of illumination on phytotoxicity of crude oil. Mar. Pollut. Bull., 7: 73—76.

Lal, D., 1977. The oceanic microcosm of particles. Science, 198: 997—1009.

Landymore, A.F. and Antia, N.J., 1978. Whitelight promoted degradation of leucopterin and related pteridines dissolved in seawater with evidence for involvement of complexation from major divalent cations of seawater. Mar. Chem., 6: 309—325.

Langford, C.H., Wingham, M. and Sastri, V.S., 1973. Ligand photooxidation in copper (II) complexes of nitrilotriacetic acid — Implications for natural waters. Environ. Sci. Technol., 7: 820—822.

Larson, R.A. Hunt, L.L. and Blankenship, D.W., 1977. Formation of toxic products from a # 2 fuel oil by photooxidation. Environ. Sci. Tech., 11: 492—496.

Lenoble, J., 1956. Sur le rôle des principaux sels dans l'absorption ultraviolette de l'eau de mer. C.R. Acad. Sci., Paris, 242: 806—808.

Lockhart, H.B., Jr. and Blakeley, R.V., 1975. Aerobic photodegradation of Fe (III)-ethylenedinitrilotetraacetate. Implications for natural waters. Envir. Sci. Tech., 9: 1035—1038.

Merkel, P.B., Nilsson, R. and Kearns, D.R., 1972. Deuterium effects on singlet oxygen lifetimes in solutions. A new test of singlet oxygen reactions. J. Am. Chem. Soc., 94: 1030—1031.

Ogura, N., 1976. Decomposition of organic matter in seawater. Chikyu Kagaku, 10: 19—22.

Otsuki, A. and Hanya, T., 1972. Production of dissolved organic matter from dead green algal cells. I. Aerobic microbial decomposition. Limnol. Oceanogr., 17: 248—257.

Parsons, T.R., 1975. Particulate organic carbon in the sea. In: J.P. Riley and G. Skirrow (Editors), Chemical Oceanography, 2. Academic Press, New York, N.Y., pp. 301—363.

Pettit, E., 1932. Measurements of ultra-violet solar radiation. Astrophys. J., 75: 185—221.

Pilpel, N., 1968. The natural fate of oil on the sea. Endeavour, 100: 11—13.

Pilpel, N., 1975. Photo-oxidation of oil films sensitized by naphthalene derivatives. Inst. Pet., I.P. 75—007, 16 pp.

Prakash, A., 1971. Terrigenous organic matter and coastal phytoplankton fertility. In: J.D. Costlow (Editor), Fertility of the Sea, 2. Gordon and Breach, London, pp. 351—367.

Rosen, J.D., 1971. Photodecomposition of organic pesticides. In: S.D. Faust and J.V. Hunter (Editors), Organic Compounds in Aquatic Environments. Marcel Dekker, New York, N.Y., pp. 425—438.

Ross, R.D. and Crosby, D.G., 1973. Photolysis of ethylenethiourea. J. Agric. Food Chem., 21: 335—337.

Ross, R.D. and Crosby, D.G., 1975. The photooxidation of aldrin in water. Chemosphere, 4: 277—282.

Sieburth, J.M. and A. Jensen, 1969. Studies on algal substances in the sea. II. The formation of gelbstoff (humic material by exudates of phaeophyta). J. Exp. Mar. Biol. Ecol., 3: 275—289.

Swallow, A.J., 1969. Hydrated electrons in seawater. Nature, 222: 369—370.

Thomas, J.K., 1977. Effect of structure and charge on radiation-induced reaction in micellar systems. Acc. Chem. Res., 10: 133—138.

Timmons, C.O., 1962. Stability of plankton oil films to artificial sunlight. U.S. Nav. Res. Lab. Rep., No. 5774: 8 pp.

Trott, T., Henwood, R.W. and Langford, C.H., 1972. Sunlight photochemistry of ferric nitrilotriacetate complexes. Environ. Sci. Technol., 6: 367—368.

Van Baalen, C. and Marler, J.E., 1966. Occurrence of hydrogen peroxide in sea water. Nature, 211: 951.

Wayne, R.P., 1970. Photochemistry. Elsevier, New York, N.Y., 263 pp.

Wheeler, J.R., 1972. Some effects of solar levels of ultraviolet radiation on lipids in artificial sea water. J. Geophys. Res., 77: 5302—5306.

Williams, P.M., Oeschger, H. and Kinney, P., 1969. Natural radiocarbon activity of the dissolved organic carbon in the North East Pacific Ocean. Nature, 224: 256—257.

Wilson, D.F., Swinnerton, J.W. and Lamontagne, R.A., 1970. Production of carbon monoxide and gaseous hydrocarbons in seawater: relation to dissolved organic carbon. Science, 168: 1577—1579.

Wolfe, N.L., Zepp, R.G., Baughman, G.L., Fincher, R.C. and Gordon, J.A., 1976. Chemical and photochemical transformation of selected pesticides in aquatic systems. U.S. Environmental Protection Agency, Athens, G.A., rep. No. EPA-600/3-76-067.

Zafiriou, O.C., 1974. Sources and reactions of OH and daughter radicals in seawater. J. Geophys. Res., 79: 4491—4497.

Zafiriou, O.C., 1977. Marine organic photochemistry previewed. Mar. Chem., 5: 497—522.

Zafiriou, O.C. and True, M.B., 1979a. Nitrite photolysis in seawater by sunlight. Mar. Chem., 8: 9—32.

Zafiriou, O.C. and True, M.B., 1979b. Nitrate photolysis in seawater by sunlight. Mar. Chem., 8: 33—42.

Zepp, R.G. and Cline, D.M., 1977. Rates of direct photolysis in aquatic environment. Environ. Sci. Technol., 11: 359—366.

Zepp, R.G., Wolfe, N.L., Gordon, J.A. and Fincher, R.C., 1976. Light-induced transformations of methoxychlor in aquatic systems. J. Agric. Food Chem., 24: 727—733.

Zepp, R.G., Wolfe, N.L., Baughman, G.L. and Hollis, R.C., 1977. Singlet oxygen in natural waters. Nature, 267: 421—423.

Zika, R.G., 1978. An Investigation in Marine Photochemistry. Thesis, Dalhousie University, Halifax, N.S.

Zsolnay, A., 1977. Sorption of benzene on particulate material in sea water. Rapp. P.-V. Réun., Cons. Int. Explor. Mer, 171: 117—119.

Chapter 11

NATURAL HYDROCARBONS IN SEA WATER

ALAIN SALIOT

1. INTRODUCTION

Interest in the organic chemistry of the marine environment has increased considerably in the past decade, due to advances in analytical techniques such as gas chromatography/mass spectrometry enabling identification of compounds present at very low concentrations and to the development of new concepts (Ehrhardt, 1977).

The organic matter present in the sea water and the sediments shows one constant characteristic: its complexity. Two main directions in the development of marine organic geochemistry are: the compilation of an inventory of many compounds in different geographical areas, associated with the determination of physical, chemical and biological parameters, and the study of well-defined molecules which offer information as to their role, cycling and fate in the sea. Among these molecules, hydrocarbons have been extensively studied for many reasons. Biosynthesis and transformation mechanisms produce complex mixtures of compounds with great specificity depending on the organisms and the physicochemical conditions of the medium. This compound class shows remarkable stability in water and sediments, and as such, hydrocarbons or derivatives of other natural structures are suitable candidates for use as biological or geochemical markers. Research into hydrocarbons has also been intensified in order to evaluate the background levels of pollution, to estimate any increase in concentration as a result of transport and industrial activities and to predict the effects of anthropogenic hydrocarbons on physical, chemical and biological processes. Research into the role of component hydrocarbons of algae as chemical mediators is also now in progress (Jaenicke, 1977).

This chapter is concerned with natural hydrocarbons with fourteen or more carbon atoms. The distinction between volatile — low molecular weight — and non-volatile hydrocarbons ($>C_{14}$) is assumed by the differences in the analytical procedures used and the different chemical and biological roles of the two different types of compounds in the marine environment (for recent data on volatile hydrocarbons, see Swinnerton and Lamontagne, 1974; Sackett and Brooks, 1975; Scranton and Brewer, 1977).

The second section of this chapter focuses on analytical procedures commonly used with special reference to typical data obtained for hydrocarbons extracted from sea water. Section 3 attempts to summarise the occurrence of biogenic hydrocarbons in the ocean by considering the known sources: phytoplankton, benthic algae, zooplankton, bacteria and the contributions by terrestrial plants. While some recent excellent general reviews on hydrocarbons in the marine environment have appeared (Farrington and Meyers, 1975; Morris and Culkin, 1975; Andersen, 1977; Hardy et al., 1977a), this information is presented using a chemical classification of different series of hydrocarbons: aliphatic n-alkanes, regular branched isoprenoids, branched alkanes, alkenes and alicyclic cyclanes, cyclenes, and aromatics. In the last two sections, the discussion is focused on problems relating to the determination of the origin of hydrocarbons found in sea water and surface sediments.

2. EXTRACTION AND ANALYSIS OF HYDROCARBONS FROM SEA WATER AND SEDIMENTS

Techniques for the collection, storage, conservation, isolation and analysis of hydrocarbons from sea water and sediment are briefly described below and illustrated in Fig. 1. Typical data and descriptions of the techniques employed for the analysis of natural or mixed hydrocarbons from sea water or sea surface microlayers from coastal and open sea areas considered to be relatively unpolluted, are given chronologically in Table I.

2.1. Hydrocarbons from sea water

2.1.1. Sampling, storage and conservation of samples

Reviews on appropriate clean-water sampling techniques have appeared recently: Farrington (1974) and Brown et al. (1975). Great care must be taken to avoid contamination either from the sampling platform, usually a ship, or from the sampling device. A ship is associated with a floating cloud of hydrocarbons which can contaminate samples. Before arrival on station, bilge and sanitary tank pumpings should be secured. It is essential, nevertheless, to have samples available of all possible contaminants such as ship's fuel oil and paints, for later analysis to compare hydrocarbon compositions of potential contaminants with the composition of hydrocarbons isolated from the samples (Grice et al., 1972).

Collection of samples of the sea surface microlayer can be performed from a rowed dinghy, at least one mile upwind of the ship. The method usually used is the metallic steel screen technique described by Garrett (1965).

Surface samples can be taken with a stainless steel bucket from the windward side, under way at low speed, just before arrival on station. An uncontaminated sea water line, continuously flushed, is also a source for surface samples.

Profile samples are generally collected with bottles using conventional

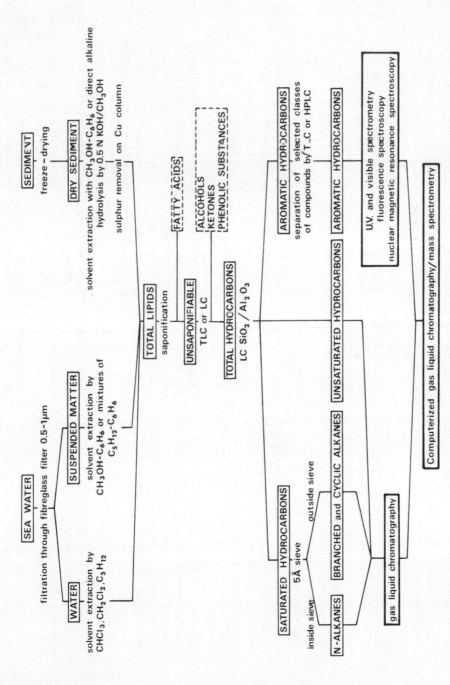

Fig. 1. Example of extraction, separation and analysis of hydrocarbons from sea water and sediments.

TABLE I

Typical data on dissolved and particulate hydrocarbons from sea water; a chronological table

Sampling, filtration, extraction and analysis techniques used	Location	Hydrocarbons studied	Range of concentrations and general comments	Reference
Collection of large water samples (~40 l); filtration on 0.45-μm Millipore; extraction at pH 3 with petroleum ether and ethyl acetate, fractionation of lipid extracts by silicic acid column chromatography (LC), thin-layer chromatography (TLC), infrared spectrophotometry (IR), gas—liquid chromatography (GLC)	Gulf of Mexico (coastal, surface waters) and Pacific, from 8 to 2780 m depth	dissolved paraffinic hydrocarbons	820—950 μg l^{-1} identification of C_{10}, C_{12}, C_{14}, C_{16} and C_{18} paraffins	Jeffrey et al. (1964)
Collection of sea surface microlayer (100—150 μm) by a metallic screen and subsurface water with Van Dorne bottles; no filtration, coprecipitation with ferric hydroxide followed by an extraction with chloroform; IR, GLC	Bay of Panama	total hydrocarbons	isolation of a hydrocarbon substance ($M = 400$), with a computed formula $C_{29}H_{52}$	Garrett (1967)
Sampling by glass or stainless steel beakers, filtration through Whatman No. 54; extraction at pH 2 with pentane, column chromatography on silica gel; GLC; mass spectrometry (MS)	Buzzards' Bay near-shore surface waters	saturated straight-chain and branched paraffins, isoprenoids and olefinic hydrocarbons	isolation of straight-chain hydrocarbons from C_{14} to C_{33} with a maximum around C_{25} to C_{28} without odd carbon predominance; characterisation of pristane, phytane, a C_{21} hexaene and squalene	Blumer (1970)
Sampling with Niskin bottles, filtration through Millipore 0.45-μm, extraction with chloroform and ethyl acetate; TLC, IR, GLC, MS	Mediterranean Sea coastal waters	dissolved hydrocarbons	mean concentration: 47 μg l^{-1} identification of C_{16}, C_{20} and C_{26} n-paraffins	Copin and Barbier (1971)
Collection of sea surface microlayer (100—150 μm) using a metallic screen and subsurface water (20 cm) using a teflon bottle; extraction after acidification with	Narragansett Bay, Rhode Island	total hydrocarbons	concentration in surface microlayer: 8.5 ± 1.7 μg l^{-1}, with an enrichment factor of 1.4 ± 0.4 with respect to the underlying	Duce et al. (1972)

Method	Location	Analyte	Results	Reference
Extraction of unfiltered samples in a continuous extractor with hexane, LC on silica gel-alumina, GLC	Gulf of Mexico	total n-paraffins	$C_{24.0}$ relative to n-alkanes on an Apiezon L column surface water concentration: $0.087-0.2$ μg l^{-1} deep water (2000 m): 0.78 μg l^{-1}	Parker et al. (1972)
Sampling with a metal bucket for coastal samples and metal bottle for deep-sea samples, filtration through Millipore 0.45 μm; extraction of large volumes (100 l), at pH 2 with chloroform; TLC, IR, GLC, MS	Caribbean Sea coastal waters: Manche Mediterranean open sea waters: equatorial Atlantic (50—4500 m)	dissolved hydrocarbons	surface water: 1.1 μg l^{-1} $46-137$ μg l^{-1} 75 μg l^{-1} $10-43$ μg l^{-1} regular distribution of hydrocarbons from C_{14} to C_{37} with a maximum around $C_{28}-C_{29}$, without predominance of odd carbon number n-paraffins	Barbier et al. (1973)
Collection of surface samples, no filtration, extraction with carbon tetrachloride; IR, MS	samples collected along two transects: U.S. Gulf coast to east coast, Caribbean to east coast	total hydrocarbons	$1-12$ μg l^{-1}; data are given concerning the composition of saturated hydrocarbon fractions in paraffins, naphthenes, aromatic hydrocarbons	Brown et al. (1973)
Collection of samples at various depths with Niskin bottles, extraction of unfiltered small volumes (1 l) with carbon tetrachloride, collection of particulate hydrocarbons using a neuston sampler; determination of hydrocarbons by fluorescence spectrophotometry.	Gulf of St. Lawrence	dissolved and particulate hydrocarbons	background level throughout the water column for dissolved hydrocarbons: $1-10$ μg l^{-1}	Levy and Walton (1973)
Sampling from 0.5 to 8 m using a glass bottle, filtration through a fine fritted Büchner (pore size ca. 5 μm); extraction with petroleum ether; GLC and gas chromatography/mass spectrometry (GC/MS)	San Francisco Bay	dissolved hydrocarbons	description of n-alkane distributions, characterisation in one sample of a sterane, lanostane $m/e = 414$	Simoneit et al. (1973b)

TABLE I (continued)

Sampling, filtration, extraction and analysis techniques used	Location	Hydrocarbons studied	Range of concentrations and general comments	Reference
Extraction of small volumes (1 l) of unfiltered sea water with hexane; purification by liquid chromatography; quantification of hydrocarbons with a micro-adsorption detector	Baltic Sea	non-aromatic hydrocarbons	data indicate that hydrocarbons originate largely *in situ* at the sediment/water interface, presumably synthesised by anaerobic bacteria	Zsolnay (1973a)
	upwelling region off West Africa	non-aromatic hydrocarbons	a significant linear correlation ($r = 0.63$) between the non-aromatic hydrocarbons and the chlorophyll *a* content in the euphotic zone, during the considered period (March, 1972) is indicated	Zsolnay (1973b)
Sampling with Niskin bottles, filtration through fibreglass filter Whatman No. 1; extraction at pH 2 of large samples (20 l) with chloroform; LC on silica gel, GLC	Gulf of Mexico (0—500 m)	dissolved non-polar hydrocarbons	traces to 75 μg l^{-1}	Iliffe and Calder (1974)
	Cariaco Trench Caribbean Sea		5 μg l^{-1} 8 μg l^{-1} the non-polar hydrocarbon fraction is characterised by relatively large amounts of *n*-alkanes between 15 and 20 carbon atoms, relatively small amounts of *n*-alkanes with more than twenty carbon atoms and an unresolved envelope extending approximately from C$_{15}$ to C$_{30}$, with the maximum between the C$_{20}$ and C$_{23}$ positions	

Method	Location	Parameter	diss. (μg l^{-1})	part. (μg l^{-1})	Remarks / Results	Reference
...layer using an aluminium-backed teflon disk; GLC, MS	Louisiana and Florida				characterisation of branched alkanes from C_{15} to C_{35} and cycloalkanes as frequently predominant components	Laseter (1974)
Collection of samples of the sea surface microlayer with a stainless steel screen and of subsurface layers with a glass flask; extraction of the unfiltered samples with chloroform or methylene chloride; TLC, GLC	Sargasso Sea	total hydrocarbons			surface microlayer: 14—559 μg l^{-1}; mean: 155 μg l^{-1} subsurface water: 13—239 μg l^{-1}; mean 73 μg l^{-1}	Wade and Quinn (1975)
Collection of sea surface microlayer samples with a metallic screen and subsurface samples with Niskin bottles or bucket; filtration through fibreglass filter Whatman GF/C; extraction at pH 2 of large samples (20 or 100 l) with chloroform; extraction of the particulate fraction with CH_3OH—C_6H_6 1:1; TLC, GLC	English Channel (coastal waters)	dissolved (diss.) and particulate (part.) n-alkanes	sea surface microlayer 17.7 subsurface water 0.1	98 0.3	n-alkanes account for 10% of the total hydrocarbons in the subsurface water and 15% in the surface microlayer; the distribution of n-alkanes is centred around C_{28}; there is an accumulation of hydrocarbons in the surface microlayer with an enrichment factor averaging 50	Marty and Saliot (1976)
	Mediterranean		sea surface microlayer 20 subsurface water 0.7—2.4	11—72 0.4		
	West African upwelling		sea surface microlayer 114 subsurface water 5.7	3.3 0.3		

TABLE I (continued)

Sampling, filtration, extraction and analysis techniques used	Location	Hydrocarbons studied	Range of concentrations and general comments	Reference
Sampling with Niskin bottles, filtration through fibreglass filter Whatman GF/C, polyethylene filter discs and Millipore 0.45 μm filters; extraction of the filters with pentane and benzene; GLC	Sargasso Sea off Bermuda (1–100 m)	particulate hydrocarbons especially tar balls	concentration of n-alkanes: 1 m: 4.38 $\mu g \, l^{-1}$; 5 m: 1.13; 25 m: 0.65; 50 m: 1.15; 100 m: 0.89	Morris et al. (1976)
Collection of surface film using a stainless steel screen and of surface water by bucket, filtration on filter paper Whatman No. 111; extraction with pentane, isolation of hydrocarbons by column chromatography on silicic acid; GLC	Celtic Sea / North Sea	dissolved n-alkanes	film: 4.3–7.7 $\mu g \, m^{-2}$; mean: 5.7 $\mu g \, m^{-2}$; film: 13–75 $\mu g \, m^{-2}$; mean: 32.8 $\mu g \, m^{-2}$; subsurface water: 0.2–0.8 $\mu g \, m^{-2}$; mean: 0.57 $\mu g \, m^{-2}$	Hardy et al. (1977b)
Sampling of surface and 50 m depth water with 4-l glass bottle; extraction of unfiltered water (12 l) with pentane; determination of aromatic hydrocarbons by fluorescence spectroscopy; LC, GLC	Bedford Basin / Scotian shelf / Gulf of St. Lawrence / Sargasso Sea	total n-alkanes	2 m: 0.065–0.590 $\mu g \, l^{-1}$; 50 m: 0.025–0.395; 2 m: 0.078–0.099; 50 m: 0.080–0.154; 2 m: 0.024–0.209; 50 m: 0.076–1.005; 2 m: 0.118; 50 m: 0.054–0.145; estimation of oil concentrations ranging from 0.2 to 9.3 $\mu g \, l^{-1}$	Keizer et al. (1977)
Collection of 20-l water samples with a tinned-steel container every month during one year, filtration on a 400-μm steel screen; extraction with chloroform; TLC, GLC	Jeddore Harbour, Nova Scotia	total n-alkanes up to C_{21}	1–10 $\mu g \, l^{-1}$ identification of pristane which represents 1/30th of the total n-alkanes and disappears during the winter	Paradis and Ackman (1977)
Sampling with a metal bottle at the	Norwegian Sea	dissolved and	total dissolved hydro-	Saliot and

through Whatman GF/C; extraction of large samples (100 l) at pH 2 with chloroform; extraction of the particles with CH_3OH-C_6H_6 1:1; TLC, GLC		hydrocarbons	identification of n-alkanes from C_{15} to C_{30} with n-C_{15} and n-C_{17} predominant	(1977)
Collection of sea surface water with glass bottles filtration on Gelman A filters; extraction of the filters with 0.1 N KOH/CH_3OH-C_6H_6 2:1; GLC	Narragansett Bay, Rhode Island	total particulate hydrocarbons	3–219 µg l^{-1} concentrations in total hydrocarbons decreasing with increasing distance from river areas	Schultz and Quinn (1977)
Sampling of surface water; extraction of small samples (1 l) with 1,2,2-trichloro-trifluoroethane; high-pressure liquid chromatography (HPLC) on silica gel; determination of hydrocarbon concentrations by measuring the heat of adsorption and desorption of eluted compounds	euphotic zone between Gulf Stream and Nova Scotia	total hydro-carbons	mean value: 4.9 ± 0.92 µg l^{-1}; the correlation coefficient (0.434) between the hydrocarbon and chlorophyll contents was significantly greater than zero at a probability level of 0.05, indicating that hydrocarbons were partly being produced by the biosphere in this region	Zsolnay (1977b)
Same as for Marty and Saliot (1976); GC/MS	eastern tropical Atlantic	polycyclic aromatic hydrocarbons	the sea surface microlayer is characterised by the predominance of phenan-threne ≥ 1.5 µg l^{-1}, the subsurface water by the predominance of phenanthrene and pyrene, 0.3 µg l^{-1} for each	Marty et al. (1978)

hydrographic means, except no lubrication of the sampling gear is permitted. Different types of bottles have been employed, but it is recommended to work with glass or metallic samplers, with good flushing characteristics. Several models have been described: 15-l rupture-disc triggered sampler made of an aluminium pressure case with a glass liner by Clark et al. (1967), 64-l and 140-l metallic devices by Bodman et al. (1961). To avoid contamination when using Niskin bottles, the rubber o-rings normally used should be replaced by viton o-rings. Before use, each Niskin must be cleaned with alternate rinses of methanol, distilled water and a more specific solvent such as carbon tetrachloride. Nevertheless, the walls of the polyvinyl chloride bottle can adsorb a considerable quantity of dissolved hydrocarbons (Gordon et al., 1974), and this constitutes a serious argument against the use of PVC bottles for analysing organic lipid material in sea water.

The water from the sampler is poured into glass bottles before extraction on board or in the laboratory. If the sample is not extracted immediately, a mineral (Hg_2Cl_2, H_2SO_4) or organic ($CHCl_3$, CCl_4) preservative may be added to prevent bacterial or algal proliferation. The bottles should be stored refrigerated in the dark.

A new promising sampling technique is described by Ehrhardt (1976) and Dawson et al. (1976), whereby the samples are extracted *in situ* without contamination or loss, using a glass column filled with a non-specific substrate lowered directly to the desired depth.

2.1.2. Filtration

Particulate hydrocarbons may be recovered from sea water by filtration with gentle vacuum or moderate over pressure to avoid rupture of cell material. Since lipid molecules, such as fatty acids have been shown to be adsorbed by membrane (0.45-μm) filters (Quinn and Meyers, 1971), it is recommended to use fibreglass filters (e.g. Whatman GF/C, 1 μm or GF/F, 0.7 μm), which can be combusted and pre-extracted with a solvent. The filtration techniques employed by various workers differ depending on the sample volumes involved (see Table I).

2.1.3. Extraction of dissolved hydrocarbons from sea water

Most of the methods used are based on the liquid—liquid extraction of lipids by a solvent; various solvents such as CH_2Cl_2, $CHCl_3$, CCl_4, nC_6H_{14}, nC_5H_{12}, petroleum ether and ethyl acetate have been tested for the recovery and employed (Table I). Recovery of a synthetic mixture of saturated hydrocarbons by extraction with petroleum ether and ethyl acetate is about 97% (Jeffrey et al., 1964). Blumer (1970) demonstrated that the extraction of dissolved lipids at pH 2 using pentane as the solvent was quantitative after four extraction steps. Parker et al. (1972), using a technique for the continuous extraction from sea water with hexane, found in control experiments that the recovery was better than 95% for n-paraffins with more than twenty carbon atoms, with serious losses of lighter paraffins. The best recovery for hydrocarbons is attributed to the extraction with chlorinated solvents. Iliffe and Calder (1974) using ^{14}C-labelled hexadecane added to 1.5 l of sea water, found that the recovery after chloroform extraction (50 ml) was 83% in the first extraction, 12% in the second, 4% in the third and 1% in the fourth. Extraction of No. 2 fuel oil, Sargasso Sea pelagic tar, or a mixture of even-carbon numbered n-alkanes (n-C_{16} to n-C_{24}) added to sea water showed that extraction with chloroform or methylene chloride gave 100% recovery of hydrocarbons compared with the internal standard (Wade and Quinn, 1975).

Another promising method involves trapping non-volatile hydrocarbons on a Bondapak C_{18} packed column as described by May et al. (1975). The recovery from sea water of internal aromatic standards is $92 \pm 2\%$ for phenanthrene, $78 \pm 17\%$ for pyrene and $58 \pm 12\%$ for benzo(a)pyrene.

The solvent containing the total lipid extract, after dessication, is evaporated under low pressure at moderate temperature ($30-40°C$) and dried under nitrogen to prevent loss of the more volatile components. The extract is stored in a refrigerator ($T < -20°C$) until analysis.

2.1.4. Extraction of particulate hydrocarbons from sea water

The filters with their retained particles can be extracted by different procedures. Morris et al. (1976, 1977) extracted the filters in a flat glass funnel with pentane; this procedure is about 90% efficient for n-alkanes. An additional extraction with benzene is also performed to dissolve any aromatic compounds. Marty and Saliot (1976), Saliot and Tissier (1977, 1978) extracted the fibreglass filters with a 1 : 1 mixture of methanol-benzene in a soxhlet apparatus for 24 h.

2.2. Hydrocarbons from sediments

2.2.1. Sampling and storage

Surface sediments are generally collected using a gravity corer, for example the short or long "Reineck" or the "Kullenberg" with a metal liner to avoid contamination of the sediment from the usual plastic liner (Debyser, 1975). The sediment is stored at very low temperature ($-20°C$) in the dark or after freeze-drying in metal or glassware, under inert atmosphere.

2.2.2. Extraction of hydrocarbons

For a review of the methods used, see Murphy (1969), Farrington and Meyers (1975), and Farrington and Tripp (1975). The methods briefly indicated here are among the most extensively applied for the extraction of lipophilic components from a sediment: (a) extraction of the sediment by placing the sample in an appropriate solvent in an ultrasonic bath; (b)

Soxhlet extraction with hexane or methanol-benzene 1 : 1 for a period of 72 h, the solvent being changed daily; (c) direct alkaline hydrolysis — the sediment is refluxed for several hours with benzene and 0.5 N methanolic KOH; (d) dissolution of rock matrix with concentrated solutions of hydrochloric and hydrofluoric acid, to release bound organic matter, followed by extraction with a solvent.

The last three methods have been examined by Farrington and Tripp (1975) for the recovery of hydrocarbons from a Narragansett Bay sediment. They conclude that there are no substantial differences in the efficiencies of the extraction methods, a finding in contrast to the extraction of fatty acids from sediments where alkaline hydrolysis yields significantly greater amounts of fatty acids (Farrington and Quinn, 1973b).

2.3. Analysis of hydrocarbons

Various analytical techniques used for hydrocarbon analyses of sea water samples, ultraviolet and fluorescence spectrophotometry (Keizer and Gordon, 1973), infrared spectrometry (Carlberg and Skarstedt, 1972), liquid column chromatography together with the determination of heat of adsorption (Zsolnay, 1977b) provide only data about a fraction of the hydrocarbons or yield a value equivalent to the total hydrocarbon concentration. Thus, we will only consider studies where hydrocarbons are precisely determined by gas—liquid chromatography and mass spectrometry.

The preliminary extraction of hydrocarbons and other compounds from sea water or sediment is followed by the isolation of hydrocarbons from the extract. At this stage a variety of analytical methods may be applied to analyse for total hydrocarbons or selected fractions.

2.3.1. Isolation of the hydrocarbons from other lipids

The total lipid extract may be subjected to removal of elemental sulphur by passage through an activated copper column (Blumer, 1957) and then to chromatographic separation on adsorbent columns or thin layer plates. For column chromatography, silicagel is used with a short alumina bed on the top of the silicagel. Both adsorbents should be partially deactivated by the addition of water (2—5%) to prevent the formation of artifacts (Blumer, 1970). Elution with a non-polar solvent such as hexane or pentane and subsequently with mixtures of non-polar and polar solvents, e.g. benzene and methanol, permits the isolation of several fractions containing saturated, unsaturated, aromatic hydrocarbons and more polar compounds (methyl esters, alcohols, acids, phenols and heterocyclic compounds). The interference from esters encountered in the isolation of aromatic hydrocarbons can be avoided prior to separation by saponification of the esters of fatty acids, which are easily removed.

Preparative chromatography on thin layers of silica in the pentane-ethyl acetate 4 : 1 system (Barbier et al., 1973) allows the isolation of the hydro-carbons which migrate with the solvent front from the remainder of the un-saponifiable substances. The hydrocarbons are extracted from the silica with ether or pentane.

Other separating techniques may be used to separate total hydrocarbons into different classes. Thus, the normal paraffins are selectively removed by 5 Å molecular sieve (Mortimer and Luke, 1967) or by urea adduction, although it is less specific than the former method. Unsaturated hydrocar-bons are separated from the saturated fraction by thin layer or column chromatography on silicic acid/AgNO$_3$.

High-pressure liquid chromatography (Hites and Biemann, 1972; Klimisch, 1973; Loheac et al., 1973; Simoneit et al., 1973b; Dong et al., 1976; Hunt et al., 1977) and gel permeation chromatography on Sephadex LH 20 (Giger and Blumer, 1974; Giger and Schaffner, 1977) are now commonly used to isolate and to analyse for aromatic hydrocarbons.

2.3.2. Gas—liquid chromatography

For a general review, in an organic geochemistry context, see for example Douglas (1969). The sensitivity of the apparatus and the resolution of the extremely complex mixtures encountered enhance the value of this tech-nique for analysis of hydrocarbons in the marine environment. The extract, liquid or solid, dissolved in a solvent of low boiling point, is injected with a leak-proof microsyringue directly through a septum into a heated vaporiza-tion block, at the top of the column. Analysis can be performed using capillary wall coated or support coated columns filled with various polar or non-polar substrates such as silicone rubbers, Apiezon L, Dexsil 300 and F.F.A.P., applying the retention index concept (Kovats, 1958).

2.3.3. Mass spectrometry and gas—liquid chromatography/computer inter-faced mass spectrometry (C/GC/MS)

The combination of gas chromatography and mass spectrometry repre-sents a powerful versatile analytical technique with the advantages of speed, sensitivity, elimination of contamination during isolation steps and adaptabil-ity to routine analysis. Such C/GC/MS have been used extensively by organic geochemists (Burlingame and Schnoes, 1969; Simoneit et al., 1973a, b; Eglinton et al., 1974; Giger et al., 1976; Giger and Schaffner, 1977; Brooks et al., 1977; among others) and permit the recall of programmes for the treatment of mass fragmentograms, and the interpretation of mass spectral records. Mass fragmentography provides rapid information about the type of compounds in a fraction and their relative distribution from the same GC/MS run, after computer reconstruction of mass chromatograms of selected key ions. Mass spectral identifications are based on recognition of fragmenta-tion patterns characteristic of particular class of compounds. Reference spec-tra of authentic standard compounds may be stored in search files; the mass spectra of unknowns may then be compared with these reference standards.

2.3.4. Interlaboratory calibration

Hydrocarbons are encountered in the marine environment as complex structures over large concentration ranges. Thus, it is necessary to compare results of analyses obtained by a number of laboratories, employing similar or different analytical procedures (Farrington et al., 1973, 1976).

3. SOURCES OF NATURAL HYDROCARBONS IN THE MARINE ENVIRONMENT; SOME PROCESSES CONTROLLING THEIR DISTRIBUTION (Fig. 2)

To understand the cycle of natural hydrocarbons in the ocean, we need to know the size and the nature of different reservoirs in the marine biota and the flux and direction of organic carbon to them. Virtually all the organic matter of internal origin in the oceans is originally derived from photosynthetic marine algae in the euphotic zone.

Many species of algae including marine and fresh-water phytoplankton have been analysed either from culture or from nature; but, they are not truly representative of the algal biomass (Whittle, 1977) and few systematic quantitative studies of hydrocarbon biosynthesis have been carried out to evaluate the lipid composition with growth cycle and season. It is difficult to estimate the marine primary hydrocarbon production with such data. Calculations by Farrington and Meyers (1975), based on the primary productivity, the average hydrocarbon concentration of phytoplankton, and assuming no recycling of hydrocarbons from primary productivity, arrive at a rate of hydrocarbon biosynthesis by marine producers of 1—10 million metric tons per year, close to the annual input rate from anthropogenic inputs (Blumer et al., 1972).

We shall examine biogenic hydrocarbons associated with marine biota, considering either major contributions or very specific compounds for different classes of aliphatic, alicyclic and aromatic hydrocarbons, derived from phytoplankton, benthic algae, macrophytes from coastal waters, zooplankton, bacteria, with reference to other important contributions such as fresh-water plankton and bacteria or higher terrestrial plants. Inputs from terrestrial origin represent only negligible quantities of organic matter. Data partially from Williams (1971), Skopintsev (1971), and Williams (1975) give annual inputs: by rain 2.2×10^{14} g carbon; by rivers 1.8×10^{14} g C; and by net primary productivity: 3.6×10^{16} g C.

These inventories and the knowledge of physical, chemical and biological processes controlling the distribution of hydrocarbons are necessary to understand the origin and the cycling of natural and anthropogenic hydrocarbons in the entire ocean; a future requirement is certainly a study of organics after their release in the water column and determination of the quantitative trophic transfer of hydrocarbons through the different reservoirs.

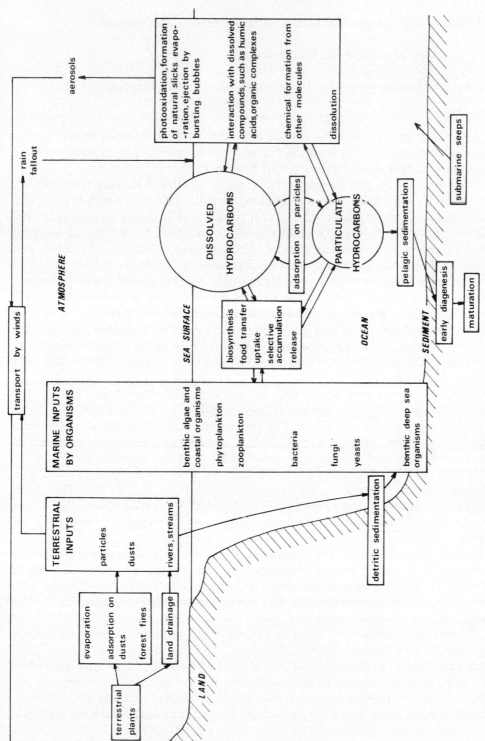

Fig. 2. Processes involved in the cycling of natural hydrocarbons in the marine environment.

3.1. Sources of natural hydrocarbons in the marine environment

3.1.1. Saturated aliphatic hydrocarbons

N-alkanes. Typical compound: *n*-heptadecane; n-C_{17}; structure:

N-alkanes are dominant constituents of natural hydrocarbons in the marine environment and their ease of analysis by gas—liquid chromatography explains that this class is relative well documented.

N-alkanes from phytoplankton: Analysis of the *n*-alkanes from phytoplankton from pure cultures or net samples from natural sea waters generally indicate the predominance of one compound (n-C_{15} or n-C_{17}) over a normal alkane distribution, usually at low level, from about C_{14} to C_{32-40} without predominance of individual alkanes. Predominant *n*-alkanes in marine and fresh-water phytoplankton species are indicated in Table II.

N-alkanes from benthic and pelagic algae: Among the saturated hydrocarbons, n-C_{15} predominates in the brown and n-C_{17} in the red algae over a lower secondary maximum in the C_{20} to C_{30} range (Table III).

N-alkanes from zooplankton: Nine monospecific zooplankton samples, including *Pseudocalanus elongatus, Evadne nordmanii, Podon leuckarti, Temora longicornis, Acartia clausii* and *Calanus finmarchicus*, show different alkane patterns which may be a reflection of differences in metabolism, in the trophic level, feeding pattern or maturity (Whittle et al., 1977). The intrinsic origin of the *n*-alkanes in zooplankton species is questionable: as reported by Murray et al. (1977) short-term incubation of a mixed zooplankton culture with ^{14}C-labelled algal diet did not show any evidence of the incorporation of ^{14}C into the *n*-alkane array of the zooplankton.

N-alkanes from bacteria: The hydrocarbons from photosynthetic and non-photosynthetic bacteria range in chain length from C_{13} to C_{31}. The major *n*-alkane is found in the range C_{17}–C_{20} (Table IV); for long-chain alkanes the amounts of odd-numbered homologues and the even-numbered ones are about equal.

The saturated hydrocarbon composition of the sulfate-reducing bacterium *Desulfovibrio desulfuricans* common in sediments, described by Davis (1968), indicates a prominent series of *n*-alkanes in the C_{25}–C_{35} range with no particular preference for odd- or even-numbered carbon chains, and in the range $<C_{25}$, a predominant C_{18}.

Terrestrial n-alkanes: N-alkanes derived from terrestrial origin are mainly associated with higher plant metabolism. The composition of higher plant cuticular waxes is characterised by a marked predominance of odd carbon number, high molecular weight *n*-alkanes from C_{23} to C_{33} (Eglinton et al., 1962; Eglinton and Hamilton, 1963; Mazliak, 1968; Caldicott and Eglinton, 1973). Plants from lower organised levels produce high amounts of lower

TABLE II

Predominant n-alkanes in marine (*) and fresh-water phytoplankton species

Species	Pre-dominant n-alkane	Total hydro-carbons (%)	References
Cyanophyceae			
Oscillatoria woronichinii *	n-C$_{15}$		Blumer et al. (1971)
Xanthophyceae			
Tribonema aequale	n-C$_{15}$		Blumer et al. (1971)
Chlorophyceae			
Dunaliella tertiolecta *	n-C$_{17}$		Blumer et al. (1971)
Derbesia tenuissima *	n-C$_{17}$		Blumer et al. (1971)
Chlorella pyrenoidosa	n-C$_{17}$	18.0	Han and Calvin (1969)
Coelastrum microsporum	n-C$_{17}$	100.0	Gelpi et al. (1970)
Rhodophyceae			
Porphyridium sp. *	n-C$_{17}$		Blumer et al. (1971)
Cyanophyceae			
Synechococcus bacillaris *	n-C$_{17}$		Blumer et al. (1971)
Phormidium luridum	n-C$_{17}$	96.0	Han and Calvin (1969)
Chlorogloca fritschii	n-C$_{17}$	87.3	Han and Calvin (1969)
Anacystis cyanea	n-C$_{17}$	87.0	Gelpi et al. (1970)
Spirulina platensis	n-C$_{17}$	70.0	Gelpi et al. (1970)
Lyngbya aestuarii	n-C$_{17}$	35.0	Gelpi et al. (1970)
Nostoc sp.	n-C$_{17}$	48.0	Gelpi et al. (1970)
Chroococcus turgidus	n-C$_{17}$	32.0	Gelpi et al. (1970)
Anacystis montana	n-C$_{17}$	11.5	Gelpi et al. (1970)
Anacystis nidulans	n-C$_{17}$	73.7	Han and Calvin (1969)
Dinophyceae			
Peridinium trochoideum *	n-C$_{17}$		Blumer et al. (1971)
Euglenophyceae			
Eutrepiella sp. *	n-C$_{17}$		Blumer et al. (1971)
Chrysophyceae			
Syracosphaera carterae *			Clark and Blumer (1967)
Bacillariophyceae			
Rhizosolenia setigera *	n-C$_{21}$		Blumer et al. (1971)
Chorophyceae			
Tetraedron sp.	n-C$_{23}$	40.0	Gelpi et al. (1970)
Bacillariophyceae			
Skeletonema costatum *	n-C$_{29}$		Clark and Blumer (1967)
Cryptophyceae			
Undetermined cryptomonad			Clark and Blumer (1967)

TABLE III

Predominant n-alkanes in marine benthic and pelagic algae.

Species	Predominant n-alkane	Total alkanes (%)	Saturated and olefinic hydrocarbons (%)	Reference
Phaeophyceae				
Ascophyllum nodosum	n-C$_{15}$	98.4		Clark and Blumer (1967)
Fucus sp.	n-C$_{15}$	94.7—96.9		Clark and Blumer (1967)
Agarum cribosum	n-C$_{15}$	28.3		Clark and Blumer (1967)
Laminaria digitata	n-C$_{15}$	77.1		Clark and Blumer (1967)
Sargassum sp. (pelagic)	n-C$_{15}$	54.9		Clark and Blumer (1967)
Sargassum sp. (pelagic)	n-C$_{15}$	56 (mean)		Burns and Teal (1973)
Chordaria flagelliformis	n-C$_{15}$		96	Youngblood et al. (1971)
Leathesia difformis	n-C$_{15}$		59	Youngblood et al. (1971)
Punctaria latifolia	n-C$_{15}$		85	Youngblood et al. (1971)
Scytosiphon lomentaria	n-C$_{15}$		38	Youngblood et al. (1971)
Chorda filum	n-C$_{15}$		38	Youngblood et al. (1971)
Laminaria agardhii	n-C$_{15}$		73	Youngblood et al. (1971)
Laminaria digitata	n-C$_{15}$		64	Youngblood et al. (1971)
Ascophyllum nodosum	n-C$_{15}$		56	Youngblood et al. (1971)
Fucus distichus	n-C$_{15}$		98	Youngblood et al. (1971)
Fucus spiralis	n-C$_{15}$		95	Youngblood et al. (1971)
Fucus vesiculosus	n-C$_{15}$		65	Youngblood et al. (1971)
Rhodophyceae				
Corallina officinalis	n-C$_{17}$	76.3		Clark and Blumer (1967)
Polysiphonia sp.	n-C$_{17}$	86.1		Clark and Blumer (1967)
Rhodymenia palmata	n-C$_{17}$	79.0		Clark and Blumer (1967)
Chondrus crispus	n-C$_{17}$	57.7—95.2		Clark and Blumer (1967)
Dumontia incrassata	n-C$_{17}$		98	Youngblood et al. (1971)
Ceramium rubrum	n-C$_{17}$		96	Youngblood et al. (1971)
Polysiphonia urceolata	n-C$_{17}$		96	Youngblood et al. (1971)
Chlorophyceae				
Chaetomorpha linum	n-C$_{17}$	35.5		Clark and Blumer (1967)
Codium fragile	n-C$_{17}$		89	Youngblood et al. (1971)

TABLE IV

Predominant n-alkanes in photosynthetic and non-photosynthetic bacteria

Bacteria	Pre-dominant n-alkane	Total hydro-carbons (%)	Reference
Photosynthetic bacteria			
Rhodospirillum rubrum	n-C_{17}	3.5	Han and Calvin (1969)
Rhodopseudomonas spheroides	n-C_{17}	42.5	Han and Calvin (1969)
Chlorobrium (sulfur bacterium)	n-C_{17}	50.0	Han and Calvin (1969)
Non-photosynthetic bacteria			
Clostridium acidiurici	n-C_{17}	50.0	Han and Calvin (1969)
Clostridium tetanomorphum	n-C_{17}	9.1	Han and Calvin (1969)
P. shermanii	n-C_{17}	13.3	Han and Calvin (1969)
Vibrio marinus	n-C_{17}	24.0	Oro et al. (1967)
Escherichia coli	n-C_{18}	27.6	Han and Calvin (1969)
Micrococcus lysodeikiticus	n-C_{19}	13.3	Han and Calvin (1969)
Desulfovibrio	n-C_{20}	16.5—34.0	Han and Calvin (1969)

homologues (n-C_{15} to n-C_{23}), with an almost equal number of odd- and even-numbered homologues. (Stransky et al., 1967).

These compounds may be introduced into the marine environment either fluvially or by aeolian transport of dusts and aerosols over continents and oceans (Simoneit, 1977a; Simoneit and Eglinton, 1977).

Regular branched alkanes: isoprenoids. Typical compound: 2,6,10,14-tetra-methylpentadecane or pristane; structure:

Isoprenoids from phytoplankton: The isoprenoid hydrocarbon pristane was found to be present in trace concentrations in different species of marine phytoplankton (Blumer et al., 1971).

Isoprenoids from benthic algae: Clark and Blumer (1967) first reported the occurrence of pristane in the Phaeophyceae *Agarum cribosum* and *Laminaria digitata* (pristane/n-C_{17} = 1 : 20 and 1 : 14, respectively). Data obtained by Youngblood et al. (1971) are consistent with this observation: pristane appears in measurable amounts only in the Phaeophyceae *Laminaria aghardii* and *Laminaria digitata*; in both the ratio of pristane to n-hepta-decane is about 1 : 20; no 2,6,10,14-tetramethylhexadecane (phytane) was detected in any sample.

Isoprenoids from zooplankton: Blumer et al. (1964) presented an excellent review of the characterisation of pristane in the marine environment.

This common isoprenoid hydrocarbon, derived probably from phytol (Avigan and Blumer, 1968), occurs in unusually high concentrations (1—3% of the body lipid) in different zooplankton, e.g. *Calanus finmarchicus, Calanus glacialis* (Blumer et al., 1963), *Calanus hyperboreus* (Conover, 1960; Blumer et al., 1963). It is present at lower concentrations in copepods, e.g. *Rhincalanus nasutus* (0.03% of total lipid), *Paraeuchaeta norvegica* (0.14%), *Pleuromamma robusta* (0.01%), *Euchirella rostrata* (0.04%), in the amphipod *Parathemisto gaudichaudii* (0.04%) and the euphausid *Nematocelis megalops* (0.09%) (Conover, 1960).

The importance of pristane was confirmed later for natural mixed plankton, (representing 21% of total paraffins) by Clark and Blumer (1967) and for zooplankton consisting of essentially *Calanus hyperboreus* by Gastaud (1977).

Isoprenoids from bacteria: Pristane may be the major hydrocarbon of some anaerobic bacteria, as found in *Pseudomonas shermanii*, representing 46.5% of the total hydrocarbons (Han and Calvin, 1969). For other photosynthetic and non-photosynthetic organisms analysed, pristane accounts for only 0.5—9.6% of the total hydrocarbons (Han and Calvin, 1969). In bacteria tested by the same authors phytane occurs at lower concentrations (0.3—2.5% of the total hydrocarbons).

Non-regular branched alkanes. Typical compound: 7-methylheptadecane; structure:

Branched alkanes from phytoplankton: The 50 : 50 mixture of 7-methylheptadecane and 8-methylheptadecane occurs uniquely in the fresh-water Cyanophyceae *Nostoc muscorum* representing 16.1% of the total hydrocarbons (Han and Calvin, 1969), *Lyngbya aestuarii* (38%) and *Chroococcus turgidus* (22%) (Gelpi et al., 1970). These methyl-branched alkanes are absent in other photosynthetic organisms, particularly bacteria. Because the methyl branches are not the *iso* and *anteiso* structures which are found in many living organisms, they are of interest in taxonomic studies.

Branched alkanes from terrestrial origin: Branched alkanes are generally minor constituents of plants, although Eglinton et al. (1962) have shown that *iso*-alkanes may constitute more than 50% of the hydrocarbon fractions. The specificity in the biosynthesis of branched paraffins in leaves is discussed by Kolattukudy (1968).

3.1.2. Unsaturated aliphatic hydrocarbons

N-alkenes. Typical compounds: 7-heptadecene; n-C_{17} : 1; structure:

Cis-3,6,9,12,15,18-heneicosahexaene (HEH); n-C_{21} : 6; structure:

Olefinic hydrocarbons may overshadow the contribution of saturated compounds of some marine organisms' hydrocarbons.

N-alkenes from phytoplankton: Unsaturated odd-carbon numbered olefins and particularly n-C_{21} : 6, derived from the corresponding docosahexaenoic acid (Lee and Loeblich, 1971), predominate largely in marine phytoplankton species (Table V). The absence of n-C_{21} : 6 in non-photosynthetic dino-flagellates (*Oxyrrhis* and *Noctiluca*) and diatoms such as *Nitzschia alba* suggests its localisation within the chloroplasts of photosynthetic organisms. The n-C_{21} : 5 and n-C_{21} : 4 homologues, tentatively identified, appear to be important hydrocarbons in dinoflagellates and euglenids (Lee and Loeblich, 1971).

Even carbon atom normal olefins are significant components of the non-photosynthetic diatom *Nitzschia alba*: the n-C_{16} : 2 represents 33.8% of total n-alkanes + n-alkenes, the n-C_{18} : 2 20.4%, the n-C_{28} : 2 9.3% (Tornabene et al., 1974). The predominance in *Nitzschia alba* of even-carbon numbered olefins, as compared with the high content of odd-carbon numbered olefinic hydrocarbons found in other algae is unique and suggests that this diatom may have an unusual biosynthetic pathway for hydrocarbons.

N-alkenes from benthic algae: n-C_{17} : 1 constitutes the bulk of n-alkanes + n-alkenes in some Chlorophyceae, such as *Enteromorpha compressa* and *Ulva lactuca* (Youngblood et al., 1971). n-C_{19} : 5 represents 62% of n-alkanes + n-alkenes in the rhodophycean alga *Porphyra leucostica* (Youngblood et al., 1971). n-C_{21} : 6 is the major hydrocarbon in the chlorophycean *Spongomorpha arcta* (60% of total n-alkanes + n-alkenes), and in the Phaeophyceae *Ectocarpus fasciculatus* (91%), *Pilayella littoralis* (98%) and *Chorda tomentosa* (47%) (Youngblood et al., 1971). n-C_{21} : 6 and n-C_{21} : 5 are also described as abundant in the phaeophycean *Fucus vesiculosus* (Halsall and Hills, 1971).

N-alkenes from zooplankton: n-C_{21} : 6 occurs in wild *Rhincalanus nasutus* and in animals raised in the laboratory on a diet containing this olefin (Blumer et al., 1970). This is probably due to a selective accumulation (not normally demonstrated by copepods) of olefinic hydrocarbons, together with triglyceride lipids. Other related species of copepods are incapable of accumulating n-C_{21} : 6, even when grown in cultures of the algae which provide *Rhincalanus nasutus* with the olefin.

Mono-olefinic hydrocarbons C_{14}, C_{19}, C_{22} and C_{30} were found to be the major components in zooplankton samples, essentially comprising of *Eucalanus* sp., *Rhincalanus nasutus*, *Nannocalanus minor*, *Euphausia americana* and *Eucalanus gibboides*, collected in the Atlantic Ocean by Gastaud (1977).

N-alkenes from bacteria: n-C_{17} ; 1 and n-C_{17} : 2 are significant components

TABLE V

Predominant n-alkenes in marine (*) and fresh-water phytoplankton species

Species	Predominant n-alkene	Total hydrocarbons (%)	Cell weight (%)	Reference
Cyanophyceae				
Synechococcus bacillaris *	n-C$_{17}$:1			Blumer et al. (1971)
Chlorophyceae				
Chlorella pyrenoidosa		80		Han and Calvin (1969); Gelpi et al. (1970)
Coccochloris elabens *	n-C$_{19}$:1	85		Winters et al. (1969)
Agmenellum quadruplicatum *		92—98		Winters et al. (1969)
Bacillariophyceae				
Chaetoceros curvisetus *	n-C$_{21}$:6	90		Lee and Loeblich (1971)
Cynlindrotheca fusiformis *	n-C$_{21}$:6	90		Lee and Loeblich (1971)
Ditylum brightwelii *	n-C$_{21}$:6	90		Blumer et al. (1970); Lee and Loeblich (1971)
Lauderia borealis *	n-C$_{21}$:6	90		Lee and Loeblich (1971)
Skeletonema costatum *	n-C$_{21}$:6	90		Blumer et al. (1970); Lee et al. (1970);

Dinophyceae				
Peridinium trochoideum *	n-C21:6	80	0.0037	Blumer et al. (1970)
Gymnodinium splendens *	n-C21:6	90		Blumer et al. (1970); Lee and Loeblich (1971)
Conyaulax polyedra *	n-C21:6	80		Lee and Loeblich (1971)
Peridinium sociale *	n-C21:6			
Cryptophyceae				
Cryptomonas sp. *	n-C21:6		0.008	Blumer et al. (1970)
Rhodomonas lens *	n-C21:6	45		Lee and Loeblich (1971)
Haptophyceae				
Isochrysis galbana *	n-C21:6		0.01	Blumer et al. (1970)
Phaeocystis poucheti *	n-C21:6		0.005	Blumer et al. (1970)
Chrysophyceae				
Cricosphaera carterae *	n-C21:6	85		Lee and Loeblich (1971)
Isocchrysis galbana *	n-C21:6	90		Lee and Loeblich (1971)
Prymnesium parvum *	n-C21:6	80		Lee and Loeblich (1971)
Botryococcus braunii	n-C29:2	50		Gelpi et al. (1970)
Chlorophyceae				
Scenedesmus quadricauda	n-C27:1	43		Gelpi et al. (1970)
Anacystis montana		35		Gelpi et al. (1970)

of the marine bacterium *Vibrio marinus* (Oro et al., 1967). n-C_{17} : 1 represents 42.5% of total hydrocarbons in *Rhodopseudomonas spheroides* and 50% in the sulphur bacterium *Chrorobium* (Han and Calvin, 1969).

N-alkenes from terrestrial sources: N-alkenes of plant waxes are generally accompanied by low amounts of a homologous series of n-alkenes ranging from 23 to 31 carbon atoms (Sorm et al., 1964; Giger and Schaffner, 1977, for reed leaves).

Branched alkenes. Typical compound: squalene, $C_{30}H_{50}$; structure:

Branched alkenes from phytoplankton: Paoletti et al. (1976) identified squalene as the major hydrocarbon (24.7%) together with the n-C_{17} in the fresh-water alga *Uronema terrestre*.

Branched alkenes from zooplankton: Phytol-derived olefinic hydrocarbons have been isolated in low concentration from marine zooplankton: (1) isomeric mono-olefins from mixed zooplankton from the Gulf of Maine by Blumer and Thomas (1965b); (2) isomeric phytadienes derived by dehydration of phytol, presumably by acid catalysis (Blumer and Thomas, 1965a); (3) C_{19} di- and tri-olefinic hydrocarbons (Blumer et al., 1969). These olefins are not present in ancient sediments and in petroleum and are therefore valuable markers in the distinction between marine hydrocarbons from organisms and from oil pollution.

Branched alkenes from bacteria: The highly unsaturated isoprenoid squalene is the major constituent of the branched and cyclic fraction in the photosynthetic bacteria *Rhodospirillum rubrum* and *Rhodomicrobium vannielii* (Han and Calvin, 1969).

3.1.3. Saturated alicyclic hydrocarbons

Typical compound: hopane; $C_{30}H_{52}$; structure:

Cyclanes from benthic algae: A minor hydrocarbon component from the Chlorophyceae *Ulva lactuca* and *Enteromorpha compressa* isolated by Youngblood et al. (1971) is an alkylated cyclopropane derivative, containing seventeen carbon atoms.

Cyclanes from bacteria: Approximately 1% of steranes and triterpanes were detected by gas chromatography/mass spectrometry in the hydrocarbon fraction of the photosynthetic bacteria *Rhodospirillum rubrum* and *Rhodomicrobium vannielii* by Han and Calvin (1969).

Derived from diploptene, diplopterol and pentacyclic polyols identified in bacteria and blue-green algae (Rohmer and Ourisson, 1976), hopanes are good indicators of microbiological activity (Rohmer, 1975).

3.1.4. Unsaturated alicyclic hydrocarbons

Cyclenes from phytoplankton: A pentacyclic triterpene, with a structure close to gammacerane, of the empirical formula $C_{30}H_{50}$, has been reported as a major constituent of some species of Cyanophyceae, including *Chroococcus turgidus* (38% of the total hydrocarbons) and *Lynghya aestuarii* (16%) (Gelpi et al., 1970).

Cyclenes from terrestrial origin: Cyclic di- and triterpenoid hydrocarbons occur in significant amounts only in higher plants and are therefore potential markers for terrigenous plant lipids. Many compounds with various structures have been isolated from terrestrial and marine sediments and described (among others: Streibl and Herout, 1969; Albrecht and Ourisson, 1971; Dastillung et al., 1977; Simoneit, 1977b, c).

3.1.5. Aromatic hydrocarbons

Typical compounds: 3,4-benzopyrene; $C_{20}H_{12}$, structure:

Retene, $C_{18}H_{18}$, structure:

Perylene, $C_{20}H_{12}$, structure:

With the exception of some minor contributions by microalgae, bacteria (reviews by Zobell, 1971; Zitko, 1975) and plants (Hancock et al., 1970), aromatic hydrocarbons are not synthesised by organisms (Grimmer and Duevel, 1970). They are generally considered as products of pyrolysis at high

352

temperatures in processes initiated by man and natural phenomena such as forest fires. *In situ* chemical aromatisation of naturally occurring compounds such as terpenoids and pigments can also generate specific aromatic hydrocarbons (Blumer, 1965).

Aromatic hydrocarbons from algae: Biosynthetic polycyclic aromatic hydrocarbon formation by the fresh-water alga *Chlorella vulgaris* was shown by Borneff et al. (1968) by demonstrating the conversion of ^{14}C tagged acetate by algal cultures. Different hydrocarbons were produced: 3,4-benzopyrene, 11,12-benzofluoranthene, 1,2,3-indenopyrene, benzo(ghi)perylene, 3,4-benzofluoranthene, 1,2-benzanthracene and fluoranthene.

Aromatic hydrocarbons from bacteria: Several species of bacteria have been shown to synthesise polycyclic aromatic hydrocarbons: the anaerobic *Clostridium putride* assimilates lipids associated with dead plankton forming benzo(a)pyrene (Lima-Zanghi, 1968); Niaussat et al. (1970) observed that *Bacillus badius* synthesises benzo(a)pyrene and perylene. The hydrocarbon 3,4-benzopyrene was isolated from phytoplankton cultures and tentatively identified by Niaussat et al. (1969); the amounts of 3,4-benzopyrene found after four months are closely associated with the bacterial activity.

More recently, the possible biosynthesis of polycyclic aromatic hydrocarbons in recent sediments was studied by Hase and Hites (1976): a mixed culture of anaerobic bacteria taken from Charles River sediment showed no evidence to suggest biosynthesis of PAH, although a bioaccumulation of PAH in the medium was established.

3.2. Processes controlling the distribution of hydrocarbons in the marine environment

The hydrocarbons released into the water and in the sediments by biosynthesis from various life forms or issued from non-marine sources are subjected to physical, chemical and biological transformations, some of these affecting the molecular structure of the more labile compounds. There have been several studies of the modification of hydrocarbons in the ocean, initiated mainly by an attempt to evaluate the persistence of crude oil in the sea. The mechanisms affecting petroleum products are closely related to those for natural hydrocarbons with one important exception: many biogenic hydrocarbons are unsaturated compounds and probably disappear rapidly in the water column. This is an important consideration in evaluating the persistent contribution of biogenic hydrocarbons to dissolved and particulate lipids of sea water and sediments.

3.2.1. Dissolution

When a hydrocarbon is released into the sea water, depending on its concentration and the physicochemical characteristics of the medium, it may be partitioned into two fractions: one soluble in the water and the second forming a particle or associated by adsorption with a mineral or an organic particle. A significant amount of organic compounds generated in surface waters rapidly leave the ocean surface via the air/sea interface either by evaporation or in association with bursting bubbles (see Hunter and Liss, chapter 9). After ejection into the atmosphere and recombination with aerosols and other particles, these compounds may be transported over considerable distances by winds.

Many of the hydrocarbons have small but measurable solubilities in water. Data obtained by Sutton and Calder (1974) indicate that normal paraffins are less soluble in sea water than in distilled water (for example 0.8 compared with 1.1 μg l^{-1} for the n-C$_{20}$). This plays a significant role in the estuarine environment. If the river water is saturated with respect to normal paraffins, salting out will occur in the fresh water/sea water mixing zone and salted-out molecules may be adsorbed onto mineral or organic particles. So, estuaries may act to limit the amount of dissolved molecules and may increase the amount of particulate organic carbon entering the ocean (Sutton and Calder, 1974).

For hydrocarbons with the same number of carbon atoms the solubility increases in the order: n-alkanes, iso-alkanes, cycloparaffins and aromatic hydrocarbons. For each homologous series of hydrocarbons the logarithm of the solubility in water is a linear function of the hydrocarbon molar volume (McAuliffe, 1966).

Several other factors may influence the solubility of hydrocarbons in sea water: water temperature, salinity and the presence of certain types of organic compounds such as fulvic or humic acids (Boehm and Quinn, 1973; Sutton and Calder, 1974, 1975; Eganhouse and Calder, 1976).

Qualitatively the effect of dissolution of a mixture of hydrocarbons is the loss of the more soluble compounds. However, the preferential solvation and the greater solubility of aromatic and heterocyclic compounds, especially those of lower molecular weight, enhances their dissipation relative to the saturates of similar molecular size. The differences in relative solubilities of the various types of hydrocarbon may decide their distribution in sediments (Farrington and Quinn, 1973a; Zafiriou, 1973).

3.2.2. Evaporation

Evaporation selectively depletes the lower-boiling components of a hydrocarbon mixture, but leads to little or no fractionation among hydrocarbons of the same volatility but belonging to different structural series. It is to be expected, that the lower-boiling aromatic hydrocarbons are removed less rapidly by evaporation at ambient temperature than the n-paraffins of the same boiling point. The higher-boiling aromatics on the other hand, should evaporate more rapidly at ambient temperature than the corresponding n-paraffins (Ehrhardt and Blumer, 1972). These phenomena have been observed in $situ$ by Blumer et al. (1973) for two light paraffinic crude oils, and by Butler (1975) for samples of pelagic tar collected near Bermuda.

354

3.2.3. Adsorption-transfer to the sediment

Studies on the association of hydrocarbons with mineral particles in saline solution (Meyers and Quinn, 1973) or with particulate material occurring naturally in sea water (Zsolnay, 1977a) have revealed that such an association with particulate matter (i.e. faecal pellets, dead and living organisms, clay minerals, silicates) is essentially responsible for removing hydrocarbons from marine waters and their incorporation into the sediments. Additionally, dissolved hydrocarbons may be adsorbed directly by the sediment; this association being controlled by the nature of the sediment and the content of indigenous matter such as humic acids (Meyers and Quinn, 1973). Thus, generally a correlation between organic content of the sediment and hydrocarbon uptake is observed.

3.2.4. Chemical processes in the water column

The reactions occurring at the sea surface and in the euphotic zone, photo-oxidation, oxidation processes and association of some hydrocarbons with organic complexes such as humic or fulvic acids (Khan and Schnitzer, 1972; Gagosian and Stuermer, 1977), tend to reduce the concentration of more labile compounds, especially the unsaturated hydrocarbons. Oxidation processes lead to the formation of alcohols, acids, alkyl and arylethers, carbonyl compounds and sulfoxides (Kawahara, 1969; Hansen, 1977).

3.2.5. Chemical processes in superficial sediments: early diagenesis

The early diagenesis of organic matter has been described in several papers (Brown et al., 1972; Farrington and Quinn, 1973b; Johnson and Calder, 1973; Eglinton et al., 1974; Cranwell, 1975; Swetland and Wehmiller, 1975; Cardoso et al., 1976; Pelet, 1977; among others). Associated with the action of meiofauna, macrofauna and micro-organisms, the formation of hydrocarbons from other related structures or the modification of pre-existent hydrocarbons by reduction, decarboxylation or aromatization is induced by adsorption phenomena and a complex series of interactions between the mineral fraction of the sediment and the organic matter. Changes in the relative distribution and carbon number dominance of the n-alkanes in the upper sediment layers are often observed which can be explained either by early diagenesis, by the bacterial activity (Johnson and Calder, 1973; Cranwell, 1976) or by a change in the lipid contribution (Cardoso et al., 1976). The characterisation of diagenetic intermediates allows some insight into the evolution pathways undergone by the organic matter; some examples illustrate this point: (a) n-alkenes found in recent sediments are probably intermediates in the transformation of alkanoic acids into n-alkanes (Cooper and Bray, 1963; Shimoyama and Johns, 1972; Debyser et al., 1977), or products occurring during the transformation of n-alcohols into n-alkanes by hydrogenation; (b) the identification of Δ^2-sterenes by Dastillung and Albrecht (1977) in various sediments suggests that these compounds are degradation intermediates of the precursor sterols, via the corresponding stanols; (c) pristane-to-phytane ratios may be indicators of the palaeoenvironmental oxicity (Brooks et al., 1969).

3.2.6. Biological transformations

The uptake, retention and release of hydrocarbons by living organisms (particularly fish and invertebrates) has received considerable attention in recent studies due to an interest in evaluating the biological effects of oil pollution (for reviews, see Stegeman and Teal, 1973; Duursma and Marchand, 1974; Neff et al., 1976; Malins, 1977; McIntyre and Whittle, 1977; and Wolfe, 1977).

Some of the more important aspects in the study of the biological cycling of hydrocarbons in the marine environment are: hydrocarbon uptake from food and water and selective accumulation, storage and release and transfer through the oceanic food web. Stegeman and Teal (1973) have suggested that the concentration and the composition of non-biogenic hydrocarbons in oysters reflect both the complete exposure history and the current exposure level, result from the interplay between uptake, release, retention, route of entry and the possible residence of hydrocarbons in multiple lipid compartments.

Discrimination between hydrocarbons occurs during uptake. Blumer et al. (1970) found that among various species of copepods, *Rhincalanus* was the only species to accumulate n-C_{21} : 6. Boutry et al. (1977) showed that the uptake of hydrocarbons by the diatom *Chaetoceros calcitrans simplex* was irregular and depended upon the age of the cultures and the composition of the medium. Bioaccumulation has been invoqued by Murray et al. (1977) to explain the enrichment of long-chain aliphatic hydrocarbons in mixed plankton.

Long residence times in organisms have a marked effect on food web transport of hydrocarbons in marine organisms. During experiments on the uptake and release of aromatic and paraffinic hydrocarbons by copepods, Corner et al. (1976) showed that for naphtalenes, the dietary route of entry was more important than direct uptake from the water. Most of the aromatic compounds were lost during depuration with half lifes of 2—3 days; however, even after a 28 days depuration period not all hydrocarbons were eliminated. Most depuration studies (field or laboratory) on bivalves, show an initial rapid discharge, with a residual small concentration of hydrocarbons remaining for much longer periods.

The role of bacteria. For a general review, see Zobell (1969, 1971), Crow et al. (1974), Yen (1975), and Bartha and Atlas (1977). Bacteria have been shown to accumulate hydrocarbons with concentration factors ranging from a few hundred to over fifty thousand for different compounds and species of microorganisms (Grimes and Morrison, 1975). They apparently store hydrocarbons in their cells in unmodified form and can produce more lipids when growing on hydrocarbons than on control media (Finnerty et al., 1973). Bacteria are capable of transforming or degrading hydrocarbons. N-alkanes are more easily degraded than the other hydrocarbon compounds; order of degradation being branched alkanes > cyclic > aromatic hydrocarbons. The rate of bacterial degradation, even for the n-alkanes, is probably much lower in the natural environment than under laboratory conditions, where aeration, nutrients, agitation and elevated temperatures are controlled.

4. HYDROCARBONS IN SEA WATER

Based on the variability of hydrocarbons among species of marine organisms, and on their chemical and biological stability, Blumer (1970) suggested that when released to the water column hydrocarbons might serve as unique markers for the various water masses of the ocean. The correlation of the hydrocarbon composition of water, particles and sediment samples with the hydrocarbon composition of the different marine biota (phytoplankton, zooplankton, benthic algae, bacteria, fungi and yeasts), however, is still far from complete. Information is lacking concerning the hydrocarbon contributions by micro-organisms, and the effects of physical, chemical and biological transformations on the organic matter in the water column and at the air/sea and sediment/water interfaces. Likewise, the transformed products of petroleum pollution have not been completely identified.

Blumer (1970) reported carbon numbers ranging from C_{14} to C_{33} with a maximum occurring between C_{25} and C_{28} and no systematic odd carbon predominance for n-alkanes in surface coastal water. He also identified diakyl and cycloalkyl branched paraffins, the isoprenoids pristane and phytane, squalene and the $C_{21} : 6$ olefin at low concentrations. These data constitute undoubtly a reference point for the inventory of hydrocarbons in sea water and the differentiation between natural and anthropogenic compounds. Over the last decade several studies have been conducted by different research groups, the efforts of which were mainly devoted to the establishing of a quantitative inventory of hydrocarbons and the relationships with other determined physical, chemical and biological parameters, at different depths and areas in the ocean. The identification of several classes of compounds by employing sophisticated techniques has been a further achievement.

From the inventories drawn up on natural and anthropogenic hydrocarbons in the marine environment, it is possible to arrive at an estimate of the bulk of natural hydrocarbons present in the oceans. From various estimates (Button, 1971; Farrington and Meyers, 1975), it would seem that the amount of natural hydrocarbons present in the marine environment exceed the estimated petroleum input.

With the exception of coastal or polluted areas, the majority of the values obtained for dissolved and particulate hydrocarbon concentrations lie within the range of 1 to 50 μg l^{-1} (Table I, section 2). The data show a general trend for the concentrations to increase near the surface and particularly in the surface microlayer; higher levels are found in inshore and productive coastal waters.

4.1. N-alkanes

Confirming the data of Blumer (1970), the isolation of n-alkanes ranging from about C_{14} to C_{36} without predominance of odd over even carbon number compounds was noticed by several authors; for example, Parker et al. (1972) for unfiltered surface samples from the Gulf of Mexico and the Carribean sea, Barbier et al. (1973) for filtered samples of coastal waters off the French coast and open waters from the eastern Atlantic Ocean, Iliffe and Calder, (1974) for filtered samples from the eastern Gulf of Mexico loop current and the Carribean Sea, Hardy et al. (1977b) for waters collected around the United Kingdom, and Saliot and Tissier (1977) for the Norwegian Sea.

Atlantic ocean waters collected from the surface to 4500 m show a similar composition in dissolved hydrocarbons. The hydrocarbons represent ca. 20% of the total lipid extract; n-alkanes occur to an extent of ca. 12% of the total hydrocarbons, ranging from C_{14} to C_{37}, with a maximum at n-C_{27} to n-C_{30} (Barbier et al., 1973). Iliffe and Calder (1974) reported that the dissolved non-polar hydrocarbon fraction was characterised by relatively large amounts of n-alkanes having between fifteen and twenty carbon atoms and small amounts of n-alkanes with more than twenty carbon atoms.

This n-alkane distribution pattern is in part similar to that of algae and phytoplankton (Clark and Blumer, 1967; Blumer et al., 1971; Halsall and Hills, 1971); with the exception of some productive coastal areas where the C_{15} and C_{17} predominate, the alkanes in the range C_{14}—C_{21} are present at low concentrations. This may be explained by weathering processes and to some extent by evaporation losses affecting the lower chain compounds. The hydrocarbons found in open sea waters at all depths, have probably a natural origin and are related to the marine biota, although Murray et al. (1977) using radiochemical techniques, found little evidence for the production of long-chain hydrocarbons (C_{22}—C_{33}), either by individual algae or by the mixed zooplankton population when fed on a *Phaeodactylum* diet. We have also to bear in mind that some residues of petroleum products such as tar balls (Morris et al., 1977) have similar n-alkane distributions. In this case, the differentiation between natural and anthropogenic compounds necessitates the analysis of all groups of hydrocarbons and the determination of the unresolved envelope of chromatograms.

4.2. N-alkenes

The olefins are more labile than saturated compounds and therefore the chance of isolation of this group from sea water is low. The $C_{21} : 6$ hexa-olefin present in algae serves some special biochemical function and is a very labile compound even in the dark. It was first isolated from sea water by Blumer (1970) and subsequently detected by Schultz and Quinn (1977) in phytoplankton culture and sea-water samples.

4.3. Isoprenoids

Many reports have been made on the presence of pristane and phytane. The values of the ratio pristane/phytane given by Ledet and Laseter (1974) for hydrocarbons from the air/sea interface, varying from 1.5 to 2.3, are close to those found for petroleum (1.5—2.5 after Blumer and Sass, 1972) and are consequently indicative of anthropogenic pollution.

4.4. Alicyclic hydrocarbons

Generally alicyclic hydrocarbons, present as complex mixtures, appear during the analysis as unresolved hydrocarbon components, part of the "hump" in the gas chromatograms. The composition of total hydrocarbons extracted from coastal water given by Barbier et al. (1973), by mass spectrometry according to the method of Hood and O'Neal (1959), is the following: a large predominance of normal and branched paraffins (51.5%) over mono-, bi- and tricyclic naphthenes (20.5%), monoaromatics (18%), four and more ringed naphthenes (4%), bi- and polycyclic aromatics (3.5 and 2.5%). The relative abundance of cyclic compounds with 1—4 rings was also observed by Brown et al. (1973) for ocean water surface samples.

4.5. Aromatic hydrocarbons

Data concerning aromatic hydrocarbons from sea water are available for samples collected along well-travelled tanker routes (Brown et al., 1973; Brown and Huffman, 1976). The figure of 3—5% aromatics, as observed by Brown and Huffman (1976) for hydrocarbons in the Mediterranean samples, is significantly lower than the aromatic content of the crude oil transported in the region. Biogenic sources do yield a few aromatic compounds which would probably serve to dilute the anthropogenic input. More recent data by Tissier and Dastillung (1978) for samples collected in the Amazon fan (Demerara plain, bottom water) indicate the predominance of the pyrene and fluoranthene over traces of phenanthrene, chrysene, triphenylene, benzo(a)anthracene, benzopyrenes and benzofluoranthenes, for the dissolved fraction. The particulate hydrocarbons are essentially composed of phenanthrene with minor contributions from fluoranthene, pyrene, benzofluoranthenes and benzopyrenes. The significance of phenanthrene together with perylene is also mentioned by Marty et al. (1978) for open sea surface samples collected in the eastern tropical Atlantic. These aromatic hydrocarbons are probably of anthropogenic origin characterised by the dominance of alkyl substituted polyaromatic components (Youngblood and Blumer, 1975), except for perylene, which is considered to be a terrestrial marker (Aizenshtat, 1973).

4.6. Correlation between hydrocarbons and biological activity

Some attempts have been made to correlate total hydrocarbons or alkanes with the biological activity of marine waters.

Zsolnay (1973b) reported the existence of a significant linear correlation ($r = 0.63$; $P \ll 0.001$) between the non-aromatic hydrocarbons and the chlorophyll a content in the euphotic zone of the water off West Africa during a short period (six days) of high biological activity in March 1972. It was suggested, therefore, that the non-aromatic hydrocarbons present resulted essentially from phytoplankton activity.

More recently, from the analyses of 23 water samples collected over a short period (two days) from the euphotic zone between the Gulf Stream and Nova Scotia, Zsolnay (1977b) developed a simple model relating hydrocarbon to chlorophyll contents. The equation proposed, ($H = 0.6 + 36.10 \, Chl - 55.327 \, Chl^2$), only explained 19% of the variance in the hydrocarbon distribution, with a correlation coefficient of 0.434.

From observations of the annual variations of dissolved, particulate and phytoplankton hydrocarbons and chlorophyll a content in Mediterranean "clean" coastal waters, Goutx and Saliot (1980) found a significant correlation ($r = 0.63$) between total hydrocarbons and chlorophyll a, and this supports the conclusions of Zsolnay for unpolluted areas (upwelling regions).

4.7. The sea surface microlayer

The process of surface-film formation is complex: rising bubbles, convection and diffusion bring the organic matter to the surface of the sea. The longer-chain surface active molecules such as fatty acids associated with other lipids and glycoproteins tend to concentrate at the surface. Analyses of hydrocarbons in films not associated with known pollution incidents have been reported (Garrett, 1967; Duce et al., 1972; Ledet and Laseter, 1974; Mackie et al., 1974; Morris, 1974; Wade and Quinn, 1975; Daumas et al., 1976; Marty and Saliot, 1976; Hardy et al., 1977b; and Marty et al., 1978, 1979). Hydrocarbons accumulate in the surface microlayer relative to the underlying water. The values of the enrichment factors for hydrocarbons depend on the sampling methods employed (e.g. metal screen or roller, Daumas et al., 1976).

Marty and Saliot (1976) gave the following values for the enrichment factor calculated for a 0.44 mm thick surface microlayer: about 20 for dissolved and particulate n-alkanes for the Mediterranean Sea and between 161 and 350 for coastal samples collected along the French coast. These results are comparable with those of Wade and Quinn (1975) who reported, for ultrasurface samples from the North Atlantic, values of the enrichment factor varying from 1.1 to 26. Analyses of sea surface samples from the Mediterranean by Morris (1974) indicated that the films collected, yielding a concentration of 40—230 mg of organic matter per m^2, were composed of both natural lipids in low amounts (<5% total extract) and a complex mixture of pollutant hydrocarbons.

The distribution pattern of n-alkanes in samples from the microlayer and underlying water is often similar, characterised by a regular distribution of n-alkanes centered around n-C_{26} to n-C_{30} (Marty and Saliot, 1976; Hardy et al., 1977b), with some exceptions. Samples collected around the United Kingdom commonly showed the presence of a small peak in the C_{16}—C_{19} region (Hardy et al., 1977b). The contribution of higher land based vegetation was more pronounced in the surface microlayer (probably an effect of aeolian contributions), and for surface layer samples collected directly above benthic algae a predominance of one compound, probably a C_{19} hydrocarbon, associated with the algal metabolism was found (Marty and Saliot, 1976).

Alkanes collected over a 12 months period, offshore from Louisiana and Florida, were characterised by Ledet and Laseter (1974); unexpectedly, methyl branched alkanes ranging in chain length from C_{15} to C_{35} and cycloalkanes were frequently the dominant components. Possible explanations for this enrichment may be: selective removal by autoxidation, preferential oxidation of n-alkanes by bacteria, adsorption onto particles, a contribution from some crude oils particularly rich in methylalkanes or a contribution from certain plants containing large quantities of 3-methyl branched alkanes (Weete et al., 1971).

A study of hydrocarbons collected at the air/sea interface in the eastern tropical Atlantic conducted by Marty et al. (1978, 1979) confirms the accumulation of aliphatic and polycyclic aromatic hydrocarbons in the surface film, a discontinuity in hydrocarbon composition between the underlying water and the microlayer and a similarity between the surface microlayer and the atmospheric aerosols. The underlying water is characterised by the predominance of phenanthrene and perylene (probably of continental origin) in the aromatic fraction. Phenanthrene together with alkyl substituted components predominates in the surface microlayer. The absence of perylene in the microlayer can be explained by the photo-oxidation of precursor perylene-quinone type pigments. A discontinuity is also discernable for the relative abundance of n-alkanes in samples, e.g. 9% of the total hydrocarbons in the underlying water, 6% for the sea surface microlayer and only 2% in aerosols.

5. HYDROCARBONS FROM SUPERFICIAL SEDIMENTS

Review papers on the subject of marine organic geochemistry have recently been published (Farrington and Meyers, 1975; Morris and Culkin, 1975; Yen, 1977; and Simoneit, 1978). In this section the sediment thickness considered will be limited to approximately 50 cm.

Marine sediments may be deposited under a wide variety of environmental conditions. A sediment includes material derived from the water column by transport and sedimentation processes, of both marine and terrestrial origin. Current work on superficial marine sediments has relied heavily on the analysis of lipids as an indicator of source of input and early diagenesis of organic material because lipids appear to be well preserved in the marine environment, even in ancient marine sediments (Simoneit and Burlingame, 1974). Transformations of organic matter occur at the sea water/sediment interface by way of chemical and biochemical reactions. This is illustrated by data obtained by Saliot and Tissier (1977) for the Norwegian Sea, and in the area of the Amazon River deep sea fan by Saliot and Tissier (1978), Tissier and Dastillung (1978) and Dastillung and Corbet (1978), which show no relationship between the hydrocarbon composition of the sediments and of the subsurface waters, particles and interstitial waters extracted from the sediments.

Concentrations of extractable organic material vary from 10 to about 500—700 μg per gram of dry sediment. Total hydrocarbons are in the range 1—100 μg g^{-1} of dry sediment. Higher concentrations are reported for inshore areas (Clark and Blumer, 1967) or polluted zones, for example the New York Bight (Farrington and Tripp, 1977).

5.1. N-alkanes

N-alkanes constitute generally the major fraction of saturated and unsaturated hydrocarbons in superficial sediments. The distribution pattern of n-alkanes is commonly characterised by the predominance of odd carbon number, high molecular weight compounds, with a maximum between n-C$_{25}$ and n-C$_{33}$, over a series of light n-alkanes in the range C$_{15}$—C$_{19}$. This latter distribution speaks in favour of a significant terrestrial contribution, transported over great distances (over 1000 km) into the ocean, to the organic matter of recent sediments.

High proportions of long-chain alkanes ($>$C$_{23}$) with often high carbon preference indices (CPI) have been reported for a number of near-shore sediments (Bray and Evans, 1961; Clark and Blumer, 1967; Mackie et al., 1974) and recent sediments of the eastern Atlantic (Simoneit et al., 1973a; Gaskell et al., 1975, for deeper layers, 30—73 cm), the Gulf of Mexico and western Atlantic (Aizenshtat et al., 1973; Farrington and Tripp, 1977; Dastillung and Corbet, 1978), the Dead Sea (Nissenbaum et al., 1972), the Norwegian Sea

(Dastillung et al., 1977), the Scotian shelf (Keizer et al., 1978), the Black Sea (Debyser et al., 1977; Simoneit, 1977b), the area around the United Kingdom (Hardy et al., 1977b), and the Baltic Sea (Debyser et al., 1977).

The presence of n-C_{15} or n-C_{17} hydrocarbons, which indicate a marine contribution by algae, has been documented for near-shore sediments (Clark and Blumer, 1967), but the residence time of lighter n-alkanes is certainly too short to allow their deposition without any significant loss during the slow sedimentation process from the water column to deep sediments.

A lack of odd/even predominance in the n-alkanes can be interpreted as being indicative of a predominantly marine origin for the lipid constituents (Koons, 1970). A low odd/even predominance may also be associated with microbiological activity (Johnson and Calder, 1973) or a terrestrial contribution from plants of low organisational level (Stransky et al., 1967). In strongly reducing environments, for example the Cariaco Trench, the sediment can be characterised by a large proportion of light, even carbon numbered n-alkanes (Simoneit, 1975; Dastillung and Corbet, 1978), probably derived by direct chemical reduction or biohydrogenation by micro-organisms (Eyssen et al., 1973) from even carbon numbered fatty acids with two maxima (between C_{14} and C_{18}, and around C_{26}–C_{28} which are typical components of plant waxes; Hitchcock and Nichols, 1971).

5.2. N-alkenes

There are a few reports of the presence of n-alkenes in surface sediments. Heneicosahexaene and heptadecene have been reported to be present in surface sediment by Ehrhardt and Blumer (1972). A series of mono-unsaturated n-alkenes have been characterised for sediments from the Norwegian Sea (Dastillung et al., 1977) and the Cariaco Trench (Dastillung and Corbet, 1978). The origin of these compounds may be related to natural sources such as algae (Gelpi et al., 1968), terrestrial plants (Sorm et al., 1964) and to early diagenesis from acids or alcohols.

5.3. Isoprenoids

The most common isoprenoids, pristane and phytane, are found in several marine organisms, in sea water and in recent or ancient sediments (Blumer and Snyder, 1965; Sever and Haug, 1971). They may also be derived from degradation products of the phytyl chain of chlorophyll. Reducing conditions are necessary to transform phytol to phytane, producing as intermediates dihydrophytol and phytene; oxidising conditions lead to the transformation of phytol into pristane (Ikan et al., 1975). Thus the ratio pristane/phytane is a useful indicator of the redox conditions in sediments (Brooks et al., 1969; Didyk et al., 1978). Values of the pristane/phytane ratio varying from 0.85 (0.4 m depth) to 1.43 are reported for anoxic Black Sea sediments

(Simoneit, 1977b). A higher value, 1.51, is given by Dastillung and Corbet (1978) for the reducing sediments from the Cariaco Trench.

Other branched alkanes derived by reduction of natural products include:

(a) Squalane, derived probably by reduction of squalene, identified in micro-organisms and zooplankton, from a core sampled in the Norwegian Sea (Dastillung et al., 1977). The maximum concentration of squalane occurs at a sediment depth where the bacteria population is highest (Bianchi et al., 1975).

(b) Lycopane reported for the sediment of the Cariaco Trench (Dastillung and Corbet, 1978), derived probably from the reduction of acyclic caro-tenoids found for example in Rhodobacteria (Van Niel, 1963) in higher amounts than β-carotene, the major carotenoid of photosynthetic organisms (Goodwin, 1970).

5.4. Alicyclic hydrocarbons

The cycloalkanes and cycloalkenes have not received as much attention as the n-alkanes from surface sediments.

Saturated cyclic hydrocarbons. Debyser et al. (1977) found for sediments from the Baltic, a predominant series of β,β hopanes over the thermody-namic stable series of α,β hopanes encountered generally in ancient sedi-ments (Ensminger et al., 1974; Van Dorsselaer et al., 1974). The hopane with 31 carbon atoms is present in both series as a uniform C_{22} isomer, which suggests that the sediment analysed was not polluted with petroleum products which contain equal quantities of the two 22 isomers of the α,β homohopane (Dastillung and Albrecht, 1976).

Hopanes (C_{27}–C_{35}) have been recently identified in different recent sedi-ments by Dastillung et al. (1977), Simoneit (1977b), and Dastillung and Corbet (1978). The correlation between the presence of active bacterial populations and the compounds of the hopane type seems to establish the value of these as biogeochemical markers.

Unsaturated cyclic hydrocarbons. Several hopenes, diploptene and related pentacyclic hydrocarbons have been identified in coastal and deep sediments (Baltic, Norwegian Sea) by Dastillung et al. (1977), Debyser et al. (1977); and Dastillung and Corbet (1978), and may be considered as tracers for micro-organism activity (Rohmer, 1975).

The identification of Δ^2-sterenes in various recent sediments by Simoneit (1977b) and Dastillung and Albrecht (1977) suggests that these compounds are degradation intermediates of the precursor sterols, via the corresponding stanols.

5.5. Aromatic hydrocarbons

Laflamme and Hites (1978), summarising the results of analyses of numerous sediment samples from various regions of the globe (New England, Walvis Bay, Cariaco Trench and Amazon River), suggest a dominance of non-alkylated polycyclic aromatic hydrocarbons (PAH) with three to five rings. The approximate relative composition is: phenanthrene 12%, fluoranthene 16%, pyrene 15%, $C_{18}H_{12}$ species 23%, and $C_{20}H_{12}$ species 35%.

Every sample shows not only the same parent PAH pattern, but also a complex mixture of alkylated PAH species, similar to that encountered in combustion products; this suggests a common source for these compounds: airborne particulates formed by the combustion of wood (forest fires) and petroleum products.

Analyses of the polycyclic aromatic hydrocarbons (PAH) in recent sediments (Giger and Blumer, 1974; Blumer and Youngblood, 1975; Hites and Biemann, 1975; Youngblood and Blumer, 1975; Tissier and Spyckerelle, 1977; Tissier and Dastillung, 1978; Laflamme and Hites, 1978) suggest this fraction to be highly complex, and the origin of these compounds in recent sediments is not completely resolved.

The hydrocarbon perylene, identified in surface sediments (Orr and Grady, 1967; Aizenshtat, 1973; Tissier and Oudin, 1973; Simoneit, 1977b; Tissier and Spyckerelle, 1977; Laflamme and Hites, 1978; Tissier and Dastillung, 1978), together with the identification of structures derived from higher plants triterpenes, suggests that these may be of terrestrial origin.

Another terrestrial marker, 1-methyl-7-isopropyl-phenanthrene (retene), from the hydrogenation of abietic acid, has been identified in recent sediments from the Black Sea (Simoneit, 1977b).

6. REFERENCES

Aizenshtat, Z., 1973. Perylene and its geochemical significance. Geochim. Cosmochim. Acta, 37: 559—567.

Aizenshtat, Z., Baedecker, M.J. and Kaplan, I.R., 1973. Distribution and diagenesis of organic compounds in JOIDES sediment from Gulf of Mexico and Western Atlantic. Geochim. Cosmochim. Acta, 37: 1881—1898.

Albrecht, P. and Ourisson, G., 1971. Biogene Substanzen in Sedimenten und Fossilien. Angew. Chem., 83: 221—260.

Andersen, N.R. (Editor), 1977. Concepts in Marine Organic Chemistry. Mar. Chem., 5: 303—640.

Avigan, J. and Blumer, M., 1968. On the origin of pristane in marine organisms. J. Lip. Res., 9: 350—352.

Barbier, M., Joly, D., Saliot, A. and Tourres, D., 1973. Hydrocarbons from sea water. Deep-Sea Res., 20: 305—314.

Bartha, R. and Atlas, R.M., 1977. The microbiology of aquatic oil spills. In: D. Perlman (Editor), Advances in Applied Microbiology, 22. Academic Press, London, pp. 225—266.

Bianchi, A., Jacq, V. and Bensoussan, M., 1975. La géochimie organique des sédiments marins profonds. Mission Orgon I. 1974 (Mer de Norvège). 1.4. Distribution des populations bactériennes hétérotrophes dans les sédiments et dans les eaux proches du fond en mer de Norvège. Rev. Inst. Fr. Pét., 30: 204—211.

Blumer, M., 1957. Removal of elemental sulphur from hydrocarbon fractions. Anal. Chem., 29: 1039—1041.

Blumer, M., 1965. Organic pigments: their long-term fate. Science, 149: 722—726.

Blumer, M., 1970. Dissolved organic compounds in sea water: saturated and olefinic hydrocarbons and singly branched fatty acids. In: D.W. Hood (Editor), Organic Matter in Natural Waters. Inst. Mar. Sci., Alaska, Occas. Publ., No. 1: 153—167.

Blumer, M. and Sass, J., 1972. Indigenous and petroleum-derived hydrocarbons in a polluted sediment. Mar. Pollut. Bull., 6: 92—94.

Blumer, M. and Snyder, W.D., 1965. Isoprenoid hydrocarbons in recent sediments: presence of pristane and probable absence of phytane. Science, 150: 1588—1589.

Blumer, M. and Thomas, D.W., 1965a. Phytadienes in zooplankton. Science, 147: 1148—1149.

Blumer, M. and Thomas, D.W., 1965b. "Zamene", isomeric C_{19} monoolefins from marine zooplankton, fishes and mammals. Science, 148: 370—371.

Blumer, M. and Youngblood, W.W., 1975. Polycyclic aromatic hydrocarbons in soils and recent sediments. Science, 188: 53—55.

Blumer, M., Mullin, M.M. and Thomas, D.W., 1963. Pristane in zooplankton. Science, 140: 974.

Blumer, M., Mullin, M.M. and Thomas, D.W., 1964. Pristane in the marine environment. Helgol. Wiss. Meeresunters., 10: 187—201.

Blumer, M., Robertson, J.C., Gordon, J.E. and Sass, J., 1969. Phytol-derived C_{19} di- and triolefinic hydrocarbons in marine zooplankton and fishes. Biochemistry, 8: 4067—4074.

Blumer, M., Mullin, M.M. and Guillard, R.R.L., 1970. A polyunsaturated hydrocarbon (3,6,9,12,15,18-heneicosahexaene) in the marine food web. Mar. Biol., 6: 226—235.

Blumer, M., Guillard, R.R.L. and Chase, T., 1971. Hydrocarbons of marine phytoplankton. Mar. Biol., 8: 183—189.

Blumer, M., Blokker, P.C., Cowell, E.B. and Duckworth, D.F., 1972. Petroleum. In: E.D. Goldberg (Editor), A Guide to Marine Pollution. Gordon and Breach, New York, N.Y., pp. 19—40.

Blumer, M., Ehrhardt, M. and Jones, J.H., 1973. The environmental fate of stranded crude oil. Deep-Sea Res., 20: 239—259.

Bodman, R.H., Slabaugh, L.V. and Bowen, V.T., 1961. A multipurpose large volume sea water sampler. J. Mar. Res., 19: 141—148.

Boehm, P.D. and Quinn, J.G., 1973. Solubilization of hydrocarbons by the dissolved organic matter in sea water. Geochim. Cosmochim. Acta, 37: 2459—2477.

Borneff, J., Selenka, F., Kunte, H. and Maximos, A., 1968. Experimental studies on the formation of polycyclic aromatic hydrocarbons in plants. Environ. Res., 2: 22—29.

Boutry, J.L., Bordes, M., Fevrier, A., Barbier, M. and Saliot, A., 1977. La diatomée marine *Chaetoceros simplex calcitrans* Paulsen et son environment. IV. Relations avec le milieu de culture: étude des hydrocarbures. J. Exp. Mar. Biol. Ecol., 28: 41—51.

Bray, E.E. and Evans, E.D., 1961. Distribution of *n*-paraffins as a clue to recognition of source beds. Geochim. Cosmochim. Acta, 22: 2—15.

Brooks, J.D., Gould, K. and Smith, J.W., 1969. Isoprenoid hydrocarbons in coal and petroleum. Nature, 222: 257—259.

Brooks, P.W., Cardoso, J.N., Didyk, B., Eglinton, G., Humbertson, M.J. and Maxwell, J.R., 1977. Analysis of lipid fractions from environmental and geological sources by computerised gas chromatography/mass spectrometry. In: R. Campos and J. Goni (Editors), Advances in Organic Geochemistry, 1975. Enadimsa, Madrid, pp. 433—453.

366

Brown, F.S., Baedecker, M.J., Nissenbaum, A. and Kaplan, I.R., 1972. Early diagenesis in a reducing fjord, Saanich Inlet, British Columbia. III. Changes in organic constituents of sediment. Geochim. Cosmochim. Acta, 36: 1185—1203.

Brown, R.A. and Huffman, H.L. Jr., 1976. Hydrocarbons in open ocean waters. Science, 191: 847—849.

Brown, R.A., Searl, T.D., Elliott, J.J., Phillips, B.G., Brandon, D.E. and Monaghan, P.H., 1973. Distribution of heavy hydrocarbons in some Atlantic ocean waters. Proc. Conf. on Prevention and Control of Oil Spills. American Petroleum Institute, Washington, D.C., pp. 505—519.

Brown, R.A., Elliott, J.J., Kelliher, J.M. and Searl, T.D., 1975. Sampling and analysis of nonvolatile hydrocarbons in ocean water. In: T.R.P. Gibb Jr. (Editor), Analytical Methods in Oceanography. American Chemical Society, Washington, D.C., pp. 172—187.

Burlingame, A.L. and Schnoes, H.K., 1969. Mass spectrometry in organic geochemistry. In: G. Eglinton and M.T.J. Murphy (Editors), Organic Geochemistry. Springer, Berlin, pp. 89—160.

Burns, K.A. and Teal, J.M., 1973. Hydrocarbons in the pelagic Sargassum community. Deep-Sea Res., 20: 207—211.

Butler, J.N., 1975. Evaporative weathering of petroleum residues: the age of pelagic tar. Mar. Chem., 3: 9—21.

Button, D.K., 1971. Petroleum — Biological effects in the marine environment. In: D.W. Hood (Editor), Impingement of Man on the Oceans. Wiley—Interscience, New York, N.Y., pp. 421—429.

Caldicott, A.B. and Eglinton, G., 1973. Surface waxes. In: L.P. Miller (Editor), Phytochemistry, III. Inorganic Elements and Special Groups of Chemicals. Van Nostrand—Reinhold, New York, N.Y., pp. 162—194.

Cardoso, J., Brooks, P.W., Eglinton, G., Goodfellow, R., Maxwell, J.R. and Philp, R.P., 1976. Lipids of recently deposited alga mats at Laguna Mormona, Baja California. In: J.O. Nriagu (Editor), Environmental Biogeochemistry. Ann Arbor Science, Mich., pp. 149—174.

Carlberg, S.R. and Skarstedt, C.B., 1972. Determination of small amounts of non-polar hydrocarbons (oil) in sea water. J. Cons. Int. Explor. Mer, 34: 506—515.

Clark, R.C. Jr. and Blumer, M., 1967. Distribution of n-paraffins in marine organisms and sediment. Limnol. Oceanogr., 12: 79—87.

Clark, R.C. Jr., Blumer, M. and Raymond, S.O., 1967. A large water sampler, rupture-disc triggered, for studies of dissolved organic compounds. Deep-Sea Res., 14: 125—128.

Conover, R.J., 1960. The feeding behavior and respiration of some marine planktonic crustacea. Biol. Bull. Woods Hole, 119: 339—415.

Cooper, J.E. and Bray, E.E., 1963. A postulated role of fatty acids in petroleum formation. Geochim. Cosmochim. Acta, 27: 1113—1127.

Copin, G. and Barbier, M., 1971. Substances organiques dissoutes dans l'eau de mer; premiers résultats de leur fractionnement. Cah. Océanogr. XXIII: 455—464.

Corner, E.D.S., Harris, R.P., Kilvington, C.C. and O'Hara, S.C.M., 1976. Petroleum compounds in the marine food web: short-term experiments on the fate of naphtalene in Calanus. J. Mar. Biol. Assoc. U.K., 56: 121—133.

Cranwell, P.A., 1975. Environmental organic chemistry of rivers and lakes, both water and sediment, In: G. Eglinton (Editor), Environmental Chemistry. The Chemical Society, London, pp. 22—54.

Cranwell, P.A., 1976. Decomposition of aquatic biota and sediment formation: lipid components of two blue-green algal species and of detritus resulting from microbial attack. Freshwater Biol., 6: 481—488.

Crow, S.A., Meyers, S.P. and Ahearn, D.G., 1974. Microbiological aspects of petroleum degradation in the aquatic environment. La Mer, 12: 95—112.

367

Dastillung, M. and Albrecht, P., 1976. Molecular test for oil pollution in surface sediments. Mar. Pollut. Bull., 7: 13—15.

Dastillung, M. and Albrecht, P., 1977. Δ²-sterenes as diagenetic intermediates in sediments. Nature, 269: 678—679.

Dastillung, M. and Corbet, B., 1978. Hydrocarbures saturés et insaturés des sédiments. In: A. Combaz and R. Pelet (Editors), Géochimie organique des sédiments marins profonds. ORGON II. Atlantique N—E Brésil. Editions du CNRS, Paris, pp. 293—323.

Dastillung, M., Albrecht, P. and Tissier, M.J., 1977. Hydrocarbures saturés et insaturés des sédiments. In: Géochimie organique des sédiments marins profonds. Orgon I. Mer de Norvège. Editions du CNRS, Paris, pp. 209—228.

Daumas, R.A., Laborde, P.L., Marty, J.C. and Saliot, A., 1976. Influence of sampling method on the chemical composition of water surface film. Limnol. Oceanogr., 21: 319—326.

Davis, J.B., 1968. Paraffinic hydrocarbons in the sulfate-reducing bacterium Desulfovibrio desulfuricans. Chem. Geol., 3: 155—160.

Dawson, R. and Ehrhardt, M., 1976. Determination of aromatic hydrocarbons in sea water. In: K. Grasshoff (Editor), Methods of Sea Water Analysis. Verlag Chemie, Weinheim, pp. 227—234.

Dawson, R., Riley, J.P. and Tennant, R.H., 1976. Two samplers for large-volume collection of chlorinated hydrocarbons. Mar. Chem., 4: 83—88.

Debyser, Y., 1975. Contamination des sédiments récents après leur prélèvement. Geochim. Cosmochim. Acta, 39: 531—534.

Debyser, Y., Pelet, R. and Dastillung, M., 1977. Géochimie organique de sédiments marins récents: Mer Noire, Baltique, Atlantique (Mauritanie). In: R. Campos and J. Goni (Editors), Advances in Organic Geochemistry, 1975. Enadimsa, Madrid, pp. 288—320.

Didyk, B.M., Simoneit, B.R.T., Brassell, S.C. and Eglinton, G., 1978. Organic geochemical indicators of paleoenvironmental conditions of sedimentation. Nature, 272: 216—222.

Dong, M., Locke, D.C. and Ferrand, E., 1976. High pressure liquid chromatographic method for routine analysis of major parent polycyclic aromatic hydrocarbons in suspended particulate matter. Anal. Chem., 48: 368—372.

Douglas, A.G., 1969. Gas chromatography. In: G. Eglinton and M.T.J. Murphy (Editors), Organic Geochemistry. Springer, Berlin, pp. 161—180.

Duce, R.A., Quinn, J.G., Olney, C.E., Piotrowicz, S.R., Ray, B.J. and Wade, T.L., 1972. Enrichment of heavy metals and organic compounds in the surface microlayer of Narragansett Bay, Rhode Island. Science, 176: 161—163.

Duursma, E.K. and Marchand, M., 1974. Aspects of organic marine pollution. Oceanogr. Mar. Biol. Annu. Rev., 12: 315—431.

Eganhouse, R.P. and Calder, J.A., 1976. The solubility of medium molecular weight aromatic hydrocarbons and the effects of hydrocarbon co-solutes and salinity. Geochim. Cosmochim. Acta, 40: 555—561.

Eglinton, G. and Hamilton, R.J., 1963. The distribution of alkanes. In: T. Swain (Editor), Chemical Plant Taxonomy. Academic Press, London, pp. 187—217.

Eglinton, G., Gonzalez, A.G., Hamilton, R.J. and Raphael, R.A., 1962. Hydrocarbon constituents of the wax coatings of plant leaves: a taxonomic survey. Phytochemistry, 1: 89—102.

Eglinton, G., Maxwell, J.R. and Philp, R.P., 1974. Organic geochemistry of sediments from contemporary aquatic environments. In: B. Tissot and F. Bienner (Editors), Advances in Organic Geochemistry, 1973. Editions Technip, Paris, pp. 941—961.

Ehrhardt, M., 1976. A versatile system for the accumulation of dissolved, non-polar compounds from seawater. "Meteor" Forschungsergeb., Reihe A, 18: 9—12.

Ehrhardt, M., 1977. Organic substances in sea water. Mar. Chem., 5: 307—316.

Ehrhardt, M. and Blumer, M., 1972. The source identification of marine hydrocarbons by gas chromatography. Environ. Pollut., 3: 179—194.

Ensminger, A., Van Dorsselaer, A., Spyckerelle, C., Albrecht, P. and Ourisson, G., 1974. Pentacyclic triterpenes of the hopane type as ubiquitous geochemical markers: origin and significance. In: B. Tissot and F. Bienner (Editors), Advances in Organic Geochemistry, 1973. Editions Technip, Paris, pp. 245—260.

Eyssen, H., De Paw, G. and De Somer, P., 1973. Biohydrogenation of long chain fatty acids by intestinal microorganisms. In: J.B. Heneghan (Editor), Germ Free Research. Academic Press, London, pp. 277—283.

Farrington, J.W., 1974. Some problems associated with the collection of marine samples and analysis of hydrocarbons. In: Proc. of Marine Environmental Implications of Offshore Drilling Eastern Gulf of Mexico, 1974 Conference/workshops. St. Petersburg, Fla., pp. 269—278.

Farrington, J.W. and Meyers, P.A., 1975. Hydrocarbons in the marine environment. In: G. Eglinton (Editor), Environmental Chemistry. The Chemical Society, London, pp. 109—136.

Farrington, J.W. and Quinn, J.G., 1973a. Petroleum hydrocarbons in Narragansett Bay I. Survey of hydrocarbons in sediments and clams (Mercenaria mercenaria). Estuarine Coastal Mar. Sci., 1: 71—79.

Farrington, J.W. and Quinn, J.G., 1973b. Biogeochemistry of fatty acids in recent sediments from Narragansett Bay, Rhode Island. Geochim. Cosmochim. Acta, 37: 259—268.

Farrington, J.W. and Tripp, B.W., 1975. A comparison of analysis methods for hydrocarbons in surface sediments. In: T.M. Church (Editor), Marine Chemistry in the Coastal Environment. American Chemical Society, Washington, D.C., pp. 267—284.

Farrington, J.W. and Tripp, B.W., 1977. Hydrocarbons in western north Atlantic surface sediments. Geochim. Cosmochim. Acta, 41: 1627—1641.

Farrington, J.W., Teal, J.M., Quinn, J.G., Wade, T. and Burns, K., 1973. Intercalibration of analyses of recently biosynthesized hydrocarbons and petroleum hydrocarbons in marine lipids. Bull. Environ. Contam. Toxicol., 10: 129—136.

Farrington, J.W., Teal, J.M., Medeiros, G.C., Burns, K.A., Robinson, E.A. Jr., Quinn, J.G. and Wade, T.L., 1976. Intercalibration of gas chromatographic analyses for hydrocarbons in tissues and extracts of marine organisms. Anal. Chem., 48: 1711—1716.

Finnerty, W.R., Kennedy, R.S., Lockwood, P., Spurlock, B.O. and Young, R.A., 1973. Microbes and petroleum: perspectives and implications. In: D.G. Ahearn and S.P. Meyers (Editors), The Microbial Degradation of Oil Pollutants. Center for Wetland Resources, Baton Rouge, La., LSU-SG-7301, pp. 105—125.

Gagosian, R.B. and Stuermer, D.H., 1977. The cycling of biogenic compounds and their diagenetically transformed products in sea water. Mar. Chem., 5: 605—632.

Garrett, W.D., 1965. Collection of slick-forming materials from the sea surface. Limnol. Oceanogr., 10: 602—605.

Garrett, W.D., 1967. The organic chemical composition of the ocean surface. Deep-Sea Res., 14: 221—227.

Gaskell, S.J., Morris, R.J., Eglinton, G. and Calvert, S.E., 1975. The geochemistry of a recent marine sediment off northwest Africa. An assessment of source of input and early diagenesis. Deep-Sea Res., 22: 777—789.

Gastaud, J.M., 1977. Biochimie des lipides du zooplancton de mers froides et tempérées. Rev. Int. Océanogr. Méd., XLV, XLVI: 99—123.

Gelpi, E., Oro, J., Schneider, H.J. and Bennett, E.O., 1968. Olefins of high molecular weight in two microscopic algae. Science, 161: 700—702.

Gelpi, E., Schneider, H., Mann, J. and Oro, J., 1970. Hydrocarbons of geochemical significance in microscopic algae. Phytochemistry, 9: 603—612.

Giger, W. and Blumer, M., 1974. Polycyclic aromatic hydrocarbons in the environment: isolation and characterization by chromatography, visible, ultraviolet and mass spectrometry. Anal. Chem., 46: 1663—1671.

Giger, W. and Schaffner, C., 1977. Aliphatic, olefinic, and aromatic hydrocarbons in recent sediments of a highly eutrophic lake. In: R. Campos and J. Goni (Editors), Advances in Organic Geochemistry, 1975. Enadimsa, Madrid, pp. 375—390.

Giger, W., Reinhard, M., Schaffner, C. and Zürcher, F., 1976. Analyses of organic constituents in water by high-resolution gas chromatography in combination with specific detection and computer-assisted mass spectrometry. In: L.H. Keith (Editor), Identification and Analysis of Organic Pollutants in Water. Ann Arbor Science, Ann Arbor, Mich., pp. 433—452.

Goodwin, T.W., 1970. Algal carotenoids. In: T.W. Goodwin (Editor), Aspects of Terpenoid Chemistry and Biochemistry. Academic Press, London, pp. 314—356.

Gordon, D.C. Jr., Keizer, P.D. and Dale, J., 1974. Estimates using fluorescence spectroscopy of the present state of petroleum hydrocarbon contamination in the water column of the northwest Atlantic ocean. Mar. Chem., 2: 251—261.

Goutx, M. and Saliot, A., 1980. Relationship between dissolved and particulate fatty acids and hydrocarbons, chlorophyll a and zooplankton biomass in Villefranche Bay, Mediterranean Sea. Mar. Chem., 8: 299—318.

Grice, G.D., Harvey, G.R., Bowen, V.T. and Backus, R.H., 1972. The collection and preservation of open ocean marine organisms for pollutant analysis. Bull. Environ. Contam. Toxicol., 7: 125—132.

Grimes, D.J. and Morrison, S.M., 1975. Bacterial bioconcentration of chlorinated hydrocarbon insecticides from aqueous systems. Microb. Ecol., 2: 43—59.

Grimmer, G. and Duevel, D., 1970. Biosynthetic formation of polycyclic hydrocarbons in higher plants VIII. Carcinogenic hydrocarbons in human environment. Z. Naturforsch. Wiss., 25 B: 1171—1175.

Halsall, T.G. and Hills, I.R., 1971. Isolation of heneicosa-1,6,9,12,15,18-hexaene and -1,6,9,12,15-pentaene from the alga *Fucus vesiculosus*. Chem. Comm., 448: 449.

Han, J. and Calvin, M., 1969. Hydrocarbon distribution of algae and bacteria and microbiological activity in sediments. Proc. Natl. Acad. Sci. U.S.A., 64: 436—443.

Hancock, J.L., Applegate, H.G. and Dodd, J.D., 1970. Polynuclear aromatic hydrocarbons on leaves. Atmos. Environ., 4: 363—370.

Hansen, H.P., 1977. Photodegradation of hydrocarbon surface films. Rapp. P.-V. Réun. Cons. Int. Explor. Mer, 171: 101—106.

Hardy, R., Mackie, P.R. and Whittle, K.J., 1977a. Hydrocarbons and petroleum in the marine ecosystem — a review. Rapp. P.-V. Réun. Cons. Int. Explor. Mer, 171: 17—26.

Hardy, R., Mackie, P.R., Whittle, K.J., McIntyre, A.D. and Blackman, R.A.A., 1977b. Occurrence of hydrocarbons in the surface film, sub-surface water and sediment in the waters around the United Kingdom. Rapp. P.-V. Réun. Cons. Int. Explor. Mer, 171: 61—65.

Hase, A. and Hites, R.A., 1976. On the origin of polycyclic aromatic hydrocarbons in recent sediments: biosynthesis by anaerobic bacteria. Geochim. Cosmochim. Acta, 40: 1141—1143.

Hitchcock, C. and Nichols, B.W., 1971. Plant Lipid Biochemistry. Academic Press, London, 387 pp.

Hites, R.A. and Biemann, W.G., 1972. Water pollution: organic compounds in the Charles River, Boston. Science, 178: 158—160.

Hites, R.A. and Biemann, W.G., 1975. Identification of specific organic compounds in a highly anoxic sediment by GC/MS and HRMS. Adv. Chem. Ser., 147: 188—201.

Hood, A. and O'Neal, M.J., 1959. Status of application of mass spectrometry to heavy oil analysis. In: J.D. Waldron (Editor), Advances in Mass Spectrometry. Pergamon Press, New York, N.Y., pp. 175—191.

Hunt, D.C., Wild, P.J. and Crosby, N.T., 1977. Application of high pressure liquid chromatography to the analysis of residual levels of polycyclic aromatic hydrocarbons. Rapp. P.-V. Réun. Cons. Int. Explor. Mer, 171: 41—48.

Iliffe, T.M. and Calder, J.A., 1974. Dissolved hydrocarbons in the eastern Gulf of Mexico loop current and the Caribbean Sea. Deep-Sea Res., 21: 481—488.

Ikan, R., Baedecker, M.J. and Kaplan, I.R., 1975. Thermal alteration experiments on organic matter in recent marine sediments. II. Isoprenoids. Geochim. Cosmochim. Acta, 39: 187—194.

Jaenicke, L., 1977. Sex and sex-attraction in seaweed. Trends Biol. Sci., 152—155.

Jeffrey, L.M., Pasby, B.F., Stevenson, B. and Hood, D.W., 1964. Lipids of ocean water. In: U. Colombo and G.D. Hobson (Editors), Advances in Organic Geochemistry. Pergamon Press, Oxford, pp. 175—197.

Johnson, R.W. and Calder, J.A., 1973. Early diagenesis of fatty acids and hydrocarbons in a salt marsh environment. Geochim. Cosmochim. Acta, 37: 1943—1955.

Kawahara, F.K., 1969. Identification and differentiation of heavy residual oil and asphalt pollutants in surface waters by comparative ratios of infrared absorbances. Environ. Sci. Technol., 3: 150—153.

Keizer, P.D. and Gordon, D.C. Jr., 1973. Detection of trace amounts of oil in sea water by fluorescence spectroscopy. J.Fish. Res. Board Can., 30: 1039—1046.

Keizer, P.D., Gordon, D.C. Jr. and Dale, J., 1977. Hydrocarbons in Eastern Canadian marine waters determined by fluorescence spectroscopy and gas-liquid chromatography. J. Fish. Res. Board Can., 34: 347—353.

Keizer, P.D., Dale, J. and Gordon, D.C. Jr., 1978. Hydrocarbons in surficial sediments from the Scotian shelf. Geochim. Cosmochim. Acta, 42: 165—172.

Khan, S.U. and Schnitzer, M., 1972. The retention of hydrophobic organic compounds by humic acid. Geochim. Cosmochim. Acta, 36: 745—754.

Klimisch, H.J., 1973. Separation of polycyclic aromatic hydrocarbons by high-pressure liquid chromatography. Selective separation system for the quantitative determination of isomeric benzpyrenes and of coronene. J. Chromatogr., 83: 11—14.

Kolattukudy, P.E., 1968. Species specificity in the biosynthesis of branched paraffins in leaves. Plant. Physiol., 43: 1423—1429.

Koons, C.B., 1970. Joides cores: organic geochemical analyses of four Gulf of Mexico and western Atlantic sediment samples. Geochim. Cosmochim. Acta, 34: 1353—1356.

Kovats, E., 1958. Gas chromatographische Charakterisierung organischer Verbindungen. Teil 1: Retentions indices aliphatischer Halogenide, Alkohole, Aldehyde und Ketone. Helv. Chim. Acta, 41: 1915—1932.

Laflamme, R.E. and Hites, R.A., 1978. The global distribution of polycyclic aromatic hydrocarbons in recent sediments. Geochim. Cosmochim. Acta, 42: 289—303.

Ledet, E.J. and Laseter, J.L., 1974. Alkanes at the air/sea interface from offshore Louisiana and Florida. Science, 186: 261—263.

Lee, R.F. and Loeblich, A.R., III., 1971. Distribution of 21 : 6 hydrocarbon and its relationship to 22 : 6 fatty acid in algae. Phytochemistry, 10: 593—602.

Lee, R.F., Nevenzel, J.C., Paffenhöfer, G.A., Benson, A.A., Patton, S. and Kavanagh, T.E., 1970. A unique hexaene hydrocarbon from a diatom Skeletonema costatum. Biochim. Biophys. Acta, 202: 386—388.

Levy, E.M. and Walton, A., 1973. Dispersed and particulate petroleum residues in the gulf of St. Lawrence. J. Fish. Res. Board Can., 30: 261—267.

Lima-Zanghi, C., 1968. Bilan des acides gras du plancton marin et pollution par le benzo-3,4 pyrène. Cah. Océanogr., 20: 203—216.

Loheac, J., Martin, M. and Guiochon, G., 1973. Analyse d'hydrocarbures aromatiques polynucléaires par chromatographie en phase liquide sous pression. Analusis, 2: 168—175.

Mackie, P.R., Whittle, K.J. and Hardy, R., 1974. Hydrocarbons in the marine environment. I. n-alkanes in the firth of Clyde. Estuarine Coastal Mar. Sci., 2: 359—374.

Malins, D.C. (Editor), 1977. Effects of Petroleum on Arctic and Subarctic Marine Environments and Organisms, 2. Biological Effects. Academic Press, New York, N.Y., 500 pp.

Marty, J.C. and Saliot, A., 1976. Hydrocarbons (normal alkanes) in the surface microlayer of sea water. Deep-Sea Res., 23: 863—873.

Marty, J.C., Saliot, A. and Tissier, M.J., 1978. Inventaire, répartition et origine des hydrocarbures aliphatiques et polyaromatiques dans l'eau de mer, la microcouche de surface
et les aérosols marins en Atlantique tropical Est. C.R. Acad. Sc. Paris, 286 D: 833—
836.

Marty, J.C., Saliot, A., Buat-Menard, P., Chesselet, R. and Hunter, K.A., 1979. Relationship between the lipid compositions of marine aerosols, the sea surface microlayer and
subsurface water. J. Geophys. Res., 84: 5707—5716.

May, W.E., Chesler, S.N., Cram, S.P., Gump, B.H., Hertz, H.S., Enagonio, D.P. and
Dyszel, S.M., 1975. Chromatographic analysis of hydrocarbons in marine sediments
and sea water. J. Chromatogr. Sci., 13: 535—540.

Mazliak, P., 1968. Chemistry of plant cuticles. In: L. Reinhold and Y. Liwschitz (Editors),
Progress in Phytochemistry. Wiley, New York, N.Y., pp. 49—111.

McAuliffe, C., 1966. Solubility in water of paraffin, cycloparaffin, olefin, acetylene,
cycloolefin, and aromatic hydrocarbons. J. Phys. Chem. Wash., 70: 1267—1275.

McIntyre, A.D. and Whittle, K.J. (Editors), 1977. Petroleum hydrocarbons in the marine
environment. Rapp. P.-V. Réun. Cons. Int. Explor. Mer, 171: 230 pp.

Meyers, P.A. and Quinn, J.G., 1973. Association of hydrocarbons and mineral particles
in saline solution. Nature, 244: 23—24.

Morris, R.J., 1974. Lipid composition of surface films and zooplankton from the eastern
Mediterranean. Mar. Pollut. Bull., 5: 105—109.

Morris, B.F., Butler, J.N., Sleeter, T.D. and Cadwallader, J., 1976. Transfer of particulate
hydrocarbon material from the ocean surface to the water column. In: H.L. Windom
and R.A. Duce (Editors), Marine Pollutant Transfer. Lexington Books, Lexington, Ky.,
pp. 213—234.

Morris, B.F., Butler, J.N., Sleeter, T.D. and Cadwallader, J., 1977. Particulate hydrocar
bon material in ocean waters. Rapp. P.-V. Réun. Cons. Int. Explor. Mer, 171: 107—
116.

Morris, R.J. and Culkin, F., 1975. Environmental organic chemistry of oceans, fjords and
anoxic basins. In: G. Eglinton (Editor), Environmental Chemistry. The Chemical
Society, London, pp. 81—108.

Mortimer, J.V. and Luke, L.A., 1967. The determination of normal paraffins in petroleum products. Anal. Chim. Acta, 38: 119—126.

Murphy, M.T.J., 1969. Analytical methods. In: G. Eglinton and M.T.J. Murphy (Editors),
Organic Geochemistry. Springer, Berlin, pp. 74—88.

Murray, J., Thompson, A.B., Stagg, A., Hardy, R., Whittle, K.J. and Mackie, P.R., 1977.
On the origin of hydrocarbons in marine organisms. Rapp. P.-V. Réun. Cons. Int.
Explor. Mer, 171: 84—90.

Neff, J.M., Cox, B.A., Dixit, D. and Anderson, J.W., 1976. Accumulation and release of
petroleum-derived aromatic hydrocarbons by four species of marine animals. Mar.
Biol., 38: 279—289.

Niaussat, P., Mallet, L. and Ottenwaelder, J., 1969. Apparition de benzo-3,4-pyrène dans
diverses souches de phytoplancton marin cultivées in vitro. Rôle éventuel des bactéries
associées. C.R. Acad. Sci. Paris, 268 D: 1109—1112.

Niaussat, P., Auger, C. and Mallet, L., 1970. Appearance of carcinogenic hydrocarbons in
pure Bacillus badius cultures relative to the presence of certain compounds in the
medium. C.R. Acad. Sci. Paris, 270 D: 1042—1045.

Nissenbaum, A., Baedecker, M.J. and Kaplan, I.R., 1972. Organic geochemistry of Dead
Sea sediments. Geochim. Cosmochim. Acta, 36: 709—728.

Oro, J., Tornabene, T.G., Nooner, D.W. and Gelpi, E., 1967. Aliphatic hydrocarbons and
fatty acids of some marine and freshwater microorganisms. J. Bacteriol., 93: 1811—
1818.

372

Orr, W.L. and Grady, J.R., 1967. Perylene in sediments off southern California. Geochim. Cosmochim. Acta, 31: 1201—1209.

Paoletti, C., Pushparaj, B., Florenzano, G., Capella, P. and Lercker, G., 1976. Unsaponifiable matter of green and blue-green algal lipids as a factor of biochemical differentiation of their biomasses. I. Total unsaponifiable and hydrocarbon fraction. Lipids, 11: 258—265.

Paradis, M. and Ackman, R.G., 1977. Influence of ice cover and man on the odd-chain hydrocarbons and fatty acids in the waters of Jeddore Harbour, Nova Scotia. J. Fish. Res. Board Can., 34: 2156—2163.

Parker, P.L., Winters, J.K. and Morgan, J., 1972. A base-line study of petroleum in the Gulf of Mexico. In: Baseline Studies of Pollutants in the Marine Environment and Research Recommendations. Proc. of IDOE Baseline Conf., New York, N.Y., pp. 555—582.

Pelet, R., 1977. Géochimie organique des sédiments marins profonds de la mer de Norvège. Vue d'ensemble, In: Géochimie organique des sédiments marins profonds, Orgon I. Mer de Norvège. Editions du CNRS, Paris, pp. 281—296.

Quinn, J.G. and Meyers, P.A., 1971. Retention of dissolved organic acids in sea water by various filters. Limnol. Oceanogr., 16: 129—131.

Rohmer, M., 1975. Triterpénoïdes de procaryotes. Thesis, University of Strasbourg, 101 pp.

Rohmer, M. and Ourisson, G., 1976. Dérivés du bactériohopane: variations structurales et répartition. Tetrahedron Lett., 40: 3637—3640.

Sackett, W.M. and Brooks, J.M., 1975. Origin and distributions of low molecular weight hydrocarbons in Gulf of Mexico coastal waters. In: T.M. Church (Editor), Marine Chemistry in the Coastal Environment. American Chemical Society, Washington, D.C., pp. 211—230.

Saliot, A. and Tissier, M.J., 1977. Interface eau—sédiment: acides gras et hydrocarbures dissous et particulaires dans l'eau de mer. In: Géochimie organique des sédiments marins profonds. Orgon I. Mer de Norvège. Editions de CNRS, Paris, pp. 197—208.

Saliot, A. and Tissier, M.J., 1978. Inventaire et dynamique des lipides à l'interface eau de mer/sédiment. IV Hydrocarbures aliphatiques et alicycliques de l'eau de mer et de l'eau interstitielle. In: A. Combaz and R. Pelet (Editors), Géochimie organique des sédiments marins profonds. Orgon II. Atlantique N—E Brésil. Editions du CNRS, Paris, pp. 263—273.

Schultz, D.M. and Quinn, J.G., 1977. Suspended material in Narragansett Bay: fatty acid and hydrocarbon composition. Org. Geochem., 1: 27—36.

Scranton, M.I. and Brewer, P.G., 1977. Occurrence of methane in the near-surface waters of the western subtropical north Atlantic. Deep-Sea Res., 24: 127—138.

Sever, J.R. and Haug, P., 1971. Fatty acids and hydrocarbons in Surtsey sediment. Nature, 234: 447—450.

Shimoyama, A. and Johns, W.D., 1972. Formation of alkanes from fatty acids in the presence of $CaCO_3$. Geochim. Cosmochim. Acta, 36: 87—91.

Simoneit, B.R.T., 1975. Sources of Organic Matter in Oceanic Sediments. Thesis, University of Bristol, Bristol, 300 pp.

Simoneit, B.R.T., 1977a. Organic matter in eolian dusts over the Atlantic Ocean. Mar. Chem., 5: 443—464.

Simoneit, B.R.T., 1977b. The Black Sea, a sink for terrigenous lipids. Deep-Sea Res., 24: 813—830.

Simoneit, B.R.T., 1977c. Diterpenoid compounds and other lipids in deep-sea sediments and their geochemical significance. Geochim. Cosmochim. Acta, 41: 463—476.

Simoneit, B.R.T., 1978. The organic chemistry of marine sediments. In: J.P. Riley and R. Chester (Editors), Chemical Oceanography, 7. Academic Press, London, 2nd ed., pp. 233—311.

Simoneit, B.R.T. and Burlingame, A.L., 1974. Study of organic Matter in DSDP (JOIDES) cores, legs 10—15. In: B. Tissot and F. Bienner (Editors), Advances in Organic Geochemistry, 1973. Technip, Paris, pp. 629—648.

Simoneit, B.R.T. and Eglinton, G., 1977. Organic matter of eolian dust and its input to marine sediments. In: R. Campos and J. Goni (Editors), Advances in Organic Geochemistry, 1975. Enadimsa, Madrid, pp. 415—430.

Simoneit, B.R.T., Scott, E.S. and Burlingame, A.L., 1973a. Preliminary organic analyses of DSDP cores, Leg 14, Atlantic ocean. In: T.H. van Andel, G.R. Heath et al., Initial Reports of the Deep Sea Drilling Project, XVI. U.S. Government Printing Office, Washington, D.C., pp. 575—600.

Simoneit, B.R.T., Smith, D.H., Eglinton, G. and Burlingame, A.L., 1973b. Application of real-time mass spectrometric techniques to environmental organic geochemistry. II. Organic matter in San Francisco Bay area water. Arch. Environ. Contamin. Toxicol., 1: 193—208.

Skopintsev, B.A., 1971. Recent advances in the study of organic matter in the oceans. Oceanology, 6: 775—789.

Sorm, F., Wollrab, V., Jarolimek, P. and Streibl, M., 1964. Olefins in plant waxes. Chem. Ind., 1833—1834.

Stegeman, J.J. and Teal, J.M., 1973. Accumulation, release and retention of petroleum hydrocarbons by the oyster *Crassostrea virginica*. Mar. Biol., 22: 37—44.

Streibl, M. and Herout, V., 1969. Terpenoids — especially oxygenated mono-, sesqui-, di-, and triterpenes. In: G. Eglinton and M.T.J. Murphy (Editors), Organic Geochemistry. Springer, Berlin, 401—424.

Stransky, K., Streibl, M. and Herout, V., 1967. On natural waxes VI. Distribution of wax hydrocarbons in plants at different evolutionary levels. Czechoslov. Chem. Commun., 32: 3213—3220.

Sutton, C. and Calder, J.A., 1974. Solubility of higher-molecular-weight n-paraffins in distilled water and sea water. Environ. Sci. Technol., 8: 654—657.

Sutton, C. and Calder, J.A., 1975. Solubility of alkylbenzenes in distilled water and sea water at 25.0°C. J. Chem. Eng. Data, 20: 320—322.

Swetland, P.J. and Wehmiller, J.F., 1975. Lipid geochemistry of recent sediments from the Great marsh, Lewes, Delaware. In: T.M. Church (Editor), Marine Chemistry in the Coastal Environment. American Chemical Society, Washington, D.C., pp. 285—303.

Swinnerton, J.W. and Lamontagne, R.A., 1974. Oceanic distribution of low-molecular-weight hydrocarbons; baseline measurements. Environ. Sci. Technol., 8: 657—663.

Tissier, M.J. and Dastillung, M., 1978. Inventaire et dynamique des lipides à l'interface eau de mer/sédiment. V. Hydrocarbures polyaromatiques des sédiments, de l'eau de mer et de l'eau interstitielle. In: A. Combaz and R. Pelet (Editors), Géochimie organique des sédiments marins profonds. Orgon II. Atlantique N—E, Brésil. Editions du CNRS, Paris, pp. 275—283.

Tissier, M.J. and Oudin, J.L., 1973. Characteristics of naturally occurring and pollutant hydrocarbons in marine sediments. Proc. Conf. on Prevention and Control of Oil Spills. American Petroleum Institute, Washington, D.C., pp. 205—214.

Tissier, M.J. and Oudin, J.L., 1974. Influence de la pollution pétrolière sur la répartition des hydrocarbures de vases marines. In: B. Tissot and F. Bienner (Editors), Advances in Organic Geochemistry, 1973. Technip, Paris, pp. 1029—1041.

Tissier, M.J. and Spyckerelle, C., 1977. Hydrocarbures polyaromatiques des sédiments. In: Géochimie organique des sédiments marins profonds. Orgon I. Mer de Norvège. Editions du CNRS, Paris, pp. 229—236.

Tornabene, T.G., Kates, M. and Volcani, B.E., 1974. Sterols, aliphatic hydrocarbons, and fatty acids of a nonphotosynthetic diatom, *Nitzschia alba*. Lipids, 9: 279—284.

Van Dorsselaer, A., Ensminger, A., Spyckerelle, C., Dastillung, M., Sieskind, O., Arpino,

374

P., Albrecht, P., Ourisson, G., Brooks, P.W., Gaskell, S.J., Kimble, B.J., Philp, R.P., Maxwell, J.R. and Eglinton, G., 1974. Degraded and extended hopane derivatives (C_{27} to C_{35}) as ubiquitous geochemical markers. Tetrahedron Lett., 14: 1349—1352.

Van Niel, C.B., 1963. A survey of the photosynthetic bacteria. In: H. Guest, A. San Pietro and L.P. Vernon (Editors), Bacterial Photosynthesis. Antioch. Press, Yellow Springs, Ohio, pp. 459—467.

Wade, T.L. and Quinn, J.G., 1975. Hydrocarbons in the Sargasso Sea surface microlayer. Mar. Pollut. Bull., 6: 54—57.

Weete, J.D., Venketeswaran, S. and Laseter, J.L., 1971. Two populations of aliphatic hydrocarbons of teratoma and habituated tissue cultures of tobacco. Phytochemistry, 10: 939—943.

Whittle, K.J., 1977. Marine organisms and their contribution to organic matter in the ocean. Mar. Chem., 5: 381—411.

Whittle, K.J., Mackie, P.R., Hardy, R., McIntyre, A.D. and Blackman, R.A.A., 1977. The alkanes of marine organisms from the United Kingdom and surrounding waters. Rapp. P.-V. Réun. Cons. Int. Explor. Mer, 171: 72—78.

Williams, P.J. Le B., 1975. Biological and chemical aspects of dissolved organic material in sea water. In: J.P. Riley and G. Skirrow (Editors), Chemical Oceanography, 2. Academic Press, London, 2nd ed., pp. 301—363.

Williams, P.M., 1971. The distribution and cycling organic matter in the ocean. In: S.J. Faust and J.V. Hunter (Editors), Organic Compounds in Aquatic Environments. Marcel Dekker, New York, N.Y., pp. 145—163.

Winters, K., Parker, P.L. and Van Baalen, C., 1969. Hydrocarbons of blue-green algae: geochemical significance. Science, 163: 467—468.

Wolfe, D.A. (Editor), 1977. Fate and Effects of Petroleum Hydrocarbons in Marine Organisms and Ecosystems. Pergamon Press, New York, N.Y., 478 pp.

Yen, T.F., 1975. Genesis and degradation of petroleum hydrocarbons in marine environments. In: T.M. Church (Editor), Marine Chemistry in the Coastal Environment. American Chemical Society, Washington, D.C., pp. 231—266.

Yen, T.F. (Editor), 1977. Chemistry of Marine Sediments. Ann Arbor Sci., Ann Arbor, Mich., 265 pp.

Youngblood, W.W. and Blumer, M., 1975. Polycyclic aromatic hydrocarbons in the environment: homologous series in soils and recent marine sediments. Geochim. Cosmochim. Acta, 39: 1303—1314.

Youngblood, W.W., Blumer, M., Guillard, R.L. and Fiore, F., 1971. Saturated and unsaturated hydrocarbons in marine benthic algae. Mar. Biol., 8: 190—201.

Zafiriou, O.C., 1973. Petroleum hydrocarbons in Narragansett Bay. II. Chemical and isotopic analysis. Estuarine Coastal Mar. Sci., 1: 81—87.

Zitko, V., 1975. Aromatic hydrocarbons in aquatic fauna. Bull. Environ. Contam. Toxicol., 14: 621—631.

Zobell, C.E., 1969. Microbial modification of crude oil in the sea. In: Proc. API/FWPCA Conf. on Prevention and Control of Oil Spills. American Petroleum Institute, Washington, D.C., pp. 317—326.

Zobell, C.E., 1971. Sources and biodegradation of carcinogenic hydrocarbons. In: Proc. API/EPA/USCG Conf. on Prevention and Control of Oil Spills. American Petroleum Institute, Washington, D.C., pp. 441—451.

Zsolnay, A., 1973a. The relative distribution of non-aromatic hydrocarbons in the Baltic in September 1971. Mar. Chem., 1: 127—136.

Zsolnay, A., 1973b. Hydrocarbon and chlorophyll: a correlation in the upwelling region off West Africa. Deep-Sea Res., 20: 923—925.

Zsolnay, A., 1977a. Sorption of benzene on particulate material in sea water. Rapp. P.-V. Réun. Cons. Int. Explor. Mer, 171: 117—119.

Zsolnay, A., 1977b. Hydrocarbon content and chlorophyll correlation in the waters between Nova Scotia and the Gulf Stream. Deep-Sea Res., 24: 199—207.

Chapter 12

NATURAL HALOGENATED ORGANICS

WILLIAM FENICAL

1. INTRODUCTION

Within the past decade it has been recognized that various groups of marine organisms have the ability to produce halogen-containing organic compounds. This process is recognized for some terrestrial organisms, particularly for several groups of fungi (Siuda and DeBernardis, 1973), which produce various medically utilized chlorinated antibiotics such as chlortetracycline (aureomycin) from *Streptomyces aureofaciens* and griseofulvin from cultures of *Penicillium griseofulvum*. In contrast to these chlorinated substances from terrestrial microorganisms, the marine biota produce a variety of halogen-containing compounds possessing from one (halomethanes) to thirty carbon atoms. These compounds are of diverse biosynthetic origins, and bromine, rather than chlorine, is the most prevalent halogen found in these marine-derived molecules. Halogenation is not ubiquitous in the marine biota, but species of halogenating organisms have been found among selected species of marine bacteria (Schizophyta), blue-green algae (Cyanophyta), green algae (Chlorophyta), red algae (Rhodophyta), as well as several classes of marine invertebrates such as the sponges (Porifera), the molluscs (Mollusca), several coelenterates (Cnidaria), and several marine worms (Annelida). Halogenation (involving bromine and chlorine) has not been described for the higher marine vertebrates, nor is this process recognized for analogous fresh-water algae and invertebrates.

It seems clear that the sources for the halogens are the rather large amounts of these elements which exist as their anions in seawater, the concentrations of which average 19,000 mg l^{-1} Cl^-, 65 mg l^{-1} Br^- and 0.06 mg l^{-1} I^- (IO_3^-) (Goldberg, 1963). It is certainly surprising that bromine is the major covalently bound halogen, considering that in seawater chloride ion is 300 times as concentrated. It is also curious, that although iodide (iodate) concentrations are low, iodine-containing compounds are found in modest amounts in several unique red algal species.

Among the halogens, iodine must be considered, in part, separately owing to its clear-cut involvement in primary metabolism. In mammals, iodine is intimately involved in thyroid function (Roche and Michel, 1955), being condensed with the amino acid tyrosine to produce the iodinated hormone thyroxine. In the marine environment, related iodinated amino acids have been found in red algae (Scott, 1954), but the major biological sinks for iodine appear to be the large brown seaweeds or kelps, which is a topic earlier reviewed by Fritsch (1945). It appears that large amounts of iodide are sequestered by brown algae such as *Laminaria*, as a necessity for their proper growth (Pedersen, 1969). Although high concentrations of iodide become available in these algae, little appears to be converted into covalent organic form. Some iodide may, however, be oxidized to elemental iodine (Kylin, 1930; Rönnerstrand, 1968).

It is, then, the bromine and chlorine-containing compounds which are likely to be excreted into sea water and recognized as a component of the dissolved organics, and it should be pointed out that most halogen-containing compounds are produced and stored at levels of approximately 5—6% of the dry weight of the plant or animal. These massive amounts clearly serve to distinguish these natural compounds from the ppb or ppm quantities of halogenated synthetics encountered in pollution analyses.

In the following sections of this chapter, I have attempted to briefly summarize the structural organic chemistry of these compounds, their sources in the biota and their conceived biological functions.

2. NATURAL HALOGEN-CONTAINING ORGANICS

It is clearly beyond the context of this chapter to rigorously review the structural organic chemistry and biological sources for the approximately 400 known marine-derived halogenated compounds. It is important, however, that the general sources and prominent chemical features of this phenomenon be discussed. The subsections below are, therefore, organized as to biological sources, and in each case the prominent aspects of halogen chemistry are discussed.

2.1. Marine microorganisms

As a consequence of the difficulty in obtaining pure cultures, or natural collections of homogeneous species, the unicellular marine biota have received little attention. Lunde (1967) has, however, provided ample evidence that non-described species of both diatoms (Bacillariophyta) and dinoflagellates (Pyrophyta) (collected as mixtures) produce halogenated organics. This observation has not subsequently led to the identification of a halogenating diatom or dinoflagellate, but recent work in the author's laboratory (Fenical and Look, work in progress) has shown that pure cultures of the estuarine blue-green alga *Anacystis marina* produce the chlorinated diphenyl methane derivative *1*. This phenol is strongly antibiotic and its production may explain, in part, the enhanced survival of the blue-green algae in estuarian ecosystems.

Other filamentous blue-green algae, such as *Lyngbya*, *Oscillatoria*, and

Schizothrix, abound in estuarine ecosystems, and these algae are known to produce various modifications of the toxic bromophenol aplysiatoxin (2) (Mynderse et al., 1977; Mynderse and Moore, 1978). Compound 2 derives its name since it was isolated initially as the toxic component of the sea hare *Stylocheilus longicauda* (Aplysidae) (Kato and Scheuer, 1974).

Several species of marine bacteria are also known to produce bromine-containing compounds. *Pseudomonas bromoutilis* (Burkholder et al., 1960) and an undescribed *Chromobacterium* sp. (Anderson et al., 1974) have been shown to produce, among other compounds, the pyrrole derivative *3*, which contains 72% bromine by weight. The list of halogenating microorganisms will no doubt grow as more is learned about these organisms. The major

phytoplanktonic species are not likely to contribute to this picture, however, as species of *Skeletonema*, *Lauderia* (diatoms), *Gonyaulax*, *Amphidinium*, *Glenodinium*, *Peridinium* (dinoflagellates), inter alia, have been found to be void of halogenated organics (Fenical, work in progress).

2.2. Brown and green algae

The brown algae (Phaeophyta) and green algae (Chlorophyta) have been extensively investigated and found to be essentially void of halogenated organics, with only two exceptions. As earlier mentioned, the role of iodine in the metabolism of the brown algae has yet to be adequately described. Iodinated organics may be produced in low levels in these algae, as increased levels of methyl iodide (4) have been detected near beds of the alga *Laminaria digitata* (Lovelock, 1975). The green algae are essentially void of halogen chemistry, but a single tropical species, *Cymopolia barbata*, has been shown to produce a series of brominated compounds, the major of which is the hydroquinone 5 (Högberg et al., 1976).

CH_3I

378

2.3. Red algae

The red algae (Rhodophyta) are the most prolific sources of halogenated organics in the marine environment, and this topic was recently reviewed (Fenical, 1975). At least six orders, representing some ten families of red algae are now known to produce a wide variety of structure types from halomethanes (C_1) to halogenated products derived from squalene (C_{30}). Within this group are aromatic and acyclic compounds produced from acetate (polyketide) biosynthesis and monoterpenes (C_{10}), sesqui- (C_{15}) and diterpenoids (C_{20}). The structures of well over 200 compounds have now been firmly established.

The halogenated compounds produced from acetate biosynthesis can essentially be discussed as three independent groups. The most wide-spread group consists of the bromophenols typified by the aldehyde 6 isolated first from *Polysiphonia lanosa* (Augier and Henry, 1950). Another example of this structure type, now generally shown to be derived from tyrosine (Manley, 1977), is the phenylacetic acid derivative 7 obtained from the European alga *Halopytis incurvus* (Chantraine et al., 1973). These compounds, of which there have been approximately twenty described, are

found in unusually large amounts in red seaweeds of the family Rhodomelaceae, and particularly in the genera *Polysiphonia, Rhodomela* and *Odonthalia* (Fenical, 1975). A more recent report describes several bromophenols as constituents of the blue-green alga *Calothrix brevissima* (Pedersen and DaSilva, 1973), which probably indicates these compounds to be more widely distributed then expected.

The red alga *Laurencia* produces an unusual group of bromine and chlorine-bearing ethers based upon an *n*-pentadecane skeleton, of probable fatty acid origin. Specific examples of the approximately 25 described substances are the halogenated ethers 8 and 9, isolated from *L. glandulifera* in Japan (Irie et al., 1965) and an undescribed *Laurencia* species from the Gulf of California, Mexico (Fenical et al., 1974).

Perhaps the most exciting of the acetate-derived halogenated organics are

Compound	Formula	
10	$CHBr_3$	descending order of abundance in *Asparagopsis* spp.
11	CH_2Br_2	
12	$CHBr_2Cl$	
13	CBr_4	
14	$CHBr_2I$	
15	CH_2ClI	
16	$CHBrCl_2$	
17	CCl_4	
18	$CHCl_3$	
3	CH_3I	

the naturally-occurring haloforms produced by the widely distributed red alga *Asparagopsis* (Burreson et al., 1975; McConnell and Fenical, 1976). Recent studies of *A. taxiformis* from Mexico and Hawaii, and *A. armata* from Europe have shown that both species produce exceptional quantities (2—3% dry wt.) of bromoform (*10*) accompanied by lesser amounts of a variety of halomethanes, *11—18*. It should be pointed out that carbon tetrachloride (*17*), chloroform (*18*) and methyl iodide (*3*), are among the halomethanes produced by *Asparagopsis*, and that these compounds were predicted to be of a totally anthropogenic origin (Galbally, 1976). Careful analyses of the minor compounds from *Asparagopsis* have illustrated the probable origins of these halomethanes in a classical haloform reaction of haloacetone. The lesser volatile components of *A. armata*, for example, consist of mixtures of chlorinated and brominated acetones as well as halogenated acrylic and acetic acids (McConnell and Fenical, 1976). Based on these components, it appears that polyhaloacetones undergo haloform reactions to yield halo-acetic acids and halomethanes, as well as Favorsky rearrangements to yield haloacrylic acids:

It should be pointed out that the carbon tetrahalides are difficult to rational-ize via this mechanistic proposal. A conceivable but unprecedented possibil-ity would be the reaction of the haloform-produced trihalomethyl anion an enzymatically produced halonium ion (X^+) or its equivalent.

The unique genus *Asparagopsis* is classified within the family Bonne-maisoniaceae, and several related genera have been subsequently investigated chemically. None have been found to produce the volatile compounds of *Asparagopsis*, but instead these relatives have been recognized to produce C_7—C_9 halogenated compounds based again upon simple acetate (poly-ketide) metabolism. Various species of the genus *Bonnemaisonia*, for example, have been shown to yield the C_7 compounds below (*19—22*), the C_9 analogs of these compounds, and mixtures of polyhalogenated 1-octen-

3-ones (*23*) (Siuda et al., 1975; McConnell and Fenical, 1977a, b). Two related algae, *Ptilonia australasica* and *Delisea fimbriata* are also known to produce C$_8$ ketones as in *23*, and *D. fimbriata* from both Antarctic and Australian waters is also reported to produce a unique group of polyhalolactones such as *24* (Pettus et al., 1977; Kazlauskas, et al., 1976, 1977a).

The halogenated and non-halogenated acetate-derived algal metabolites have recently been reviewed by Moore (1978).

A major aspect of the halogenation process in red seaweeds is the production of halogenated terpenoids. Monoterpenoids (C$_{10}$), sesquiterpenoids (C$_{15}$), diterpenoids (C$_{20}$) and a single triterpenoid (C$_{30}$), are known to be produced, most of which contain bromine and chlorine and not iodine. Halogenated monoterpenes, such as the halo-hydrocarbons *25—28*, are known to be produced by algae of the genera *Plocamium*, *Chondrococcus* and *Ochtodes* (Ichikawa et al., 1974; Mynderse and Faulkner, 1975a, b; McConnell and Fenical, work in progress).

The halogen-containing sesquiterpenoids from red algae have recently been reviewed by Martín and Darias (1978). Over only a few years some 60 new halogenated sesquiterpenoids, representing 18 various carbon skeletons, have been isolated. A large majority of these compounds are products of the cosmopolitan intertidal algal genus *Laurencia* (Fenical, 1975). Four examples of the more common structure types are compounds *29—32*, isolated from Japanese, Californian and Spanish collections of various *Laurencia* spp.

In contrast to the abundance of the sesquiterpenoids, the halogenated diterpenoids are rare compounds found only in some *Laurencia* species and in the Mediterranean alga *Sphaerococcus coronopifolius* (Fenical et al., 1976; Fattorusso et al., 1976). The diterpenoids concinndiol (*33*) (Sims et al., 1973) and iriediol (*34*) (Fenical et al., 1975) represent most diterpenoid skeletons in *Laurencia*, while the uncommon primary halide functionality of sphaerococcenol A (*35*) is unique to *S. coronopifolius*. The sole example of a halogenated triterpenoid is the squalene-derived compound *36* from the

New Zealand alga *Laurencia thysifera* (Blunt et al., 1978).

Recently several halogenated compounds of possible amino-acid origin have been isolated from red algae. In a survey of the New Zealand seaweeds, Brennan and Erickson (1978) discovered the red alga *Rhodophyllis*

382

membranacea to produce a complex mixture of halogenated indoles (*37*). Compounds containing chlorine and bromine were isolated as mixtures which could not be separated.

$n + n' = 3,4,5,6$

28 compounds with various ratios
of Cl and Br were obtained

37

2.4. Marine sponges

The secondary metabolite chemistry of the marine sponges has recently been extensively reviewed (Minale, 1976, 1978; Minale et al., 1976). A significant, but not major, portion of the sponge-derived compounds do contain halogens. Particularly abundant world-wide are sponges of the family Verongidae, which are known to yield large amounts of brominated derivatives, such as *38* and *39*, apparently derived from 3,5-dibromotyrosine. The dibromocyclohexadienone *38* was isolated from both *Verongia fistularis* and

38

39

V. cauliformis (Sharma and Burkholder, 1967), while the related compound *39* was isolated from *V. lacunosa* (Borders et al., 1974). From another sponge, *Agelas oroides*, the dibromopyrrole derivative *40* was isolated (Forenza et al., 1971), and from *Disidea herbacea* the pentabromophenol *41* was obtained (Sharma and Vig, 1972). Only two examples of chlorine-

40

41

containing compounds have been described from sponges. In a study of Australian *Disidea herbacea*, Kazlauskas et al. (1977b), reported finding mixtures of *41* and hexabromo compounds, but in addition isolated and described the unique bis-trichloromethyl compound *42*. In another recent report, the sponge *Pseudaxinyssa pitys* was shown to contain several sesquiterpenoid carbonimidic dichlorides, the major compound being the tetrachloride *43* (Wratten and Faulkner, 1977).

42

43

2.5. Miscellaneous marine invertebrates

In contrast to the sponges, other marine invertebrates must be considered casual producers of halogenated metabolites. One group, the marine worms, which are broadly classified within the Phyla Hemichordata, Annelida, and Phoronida, share a similar synthetic capability in the production of both simple and complex bromophenolics. In the most simple case, the mud-dwelling tube worm *Phoronopsis viridis* has been shown to produce 2,6-dibromophenol (*44*) (Sheikh and Djerassi, 1975). Curiously, the same compound has been isolated as the major odorous component of the acorn worm *Balanoglossus biminiensis* (Hemichordata) (Ashworth and Cormier, 1967). The Hawaiian acorn worm *Ptychodera flava laysanica* also contains bromophenols but the major metabolite was found to be tetrabromohydroquinone (*45*) (Higa and Scheuer, 1977). The major odorous component of this acorn worm was found to be 3-chloroindole (Higa and Scheuer, 1975a). The

44   45

annelid worm *Thelepus setosus* was found to contain several bromophenols, the major being 3,5-dibromo-4-hydroxybenzyl alcohol (*46*) and the more complex dimeric product *47* (Higa and Scheuer, 1975b). It may be that the bromo-phenolics from these mud-dwelling worms are obtained by dietary

46   47

concentration, as it seems unusual that similar and identical metabolites should be found in these taxonomically diverse organisms.

Only three reports describe halogen-containing compounds in marine molluscs. The sea hare *Stylocheilus longicauda*, an opisthobranch, contains the chlorinated amide *48* (Rose and Scheuer, 1975), but, based upon the feeding habits of sea hares, this compound is likely to have an algal origin. Various gastropod molluscs related to *Murex* produce the ancient blue-purple dye Tyrian purple (*49*), which is the dibromo analog of indigotin (Baker and Duke, 1974). Tyrian purple was, perhaps, the first useful marine

48   49

compound as it was extensively used by the Phoenicians to dye ceremonial clothing.

When the edible Japanese gastropod *Babylonia japonica* suddenly became toxic, the active toxin, isolated from the midgut of the animal, was shown to be the complex structure *50* containing the bromoindole moiety (Kosuge et al., 1972). Recent evidence suggests that the toxin was produced by a marine bacterium which sporadically proliferated as the result of heavy pulp mill pollution (Y. Shimizu, pers. comm., 1978).

The coelenterates, while producing considerable unique secondary metabolites, do not generally engage in halogenation. A small group of alcyonarians

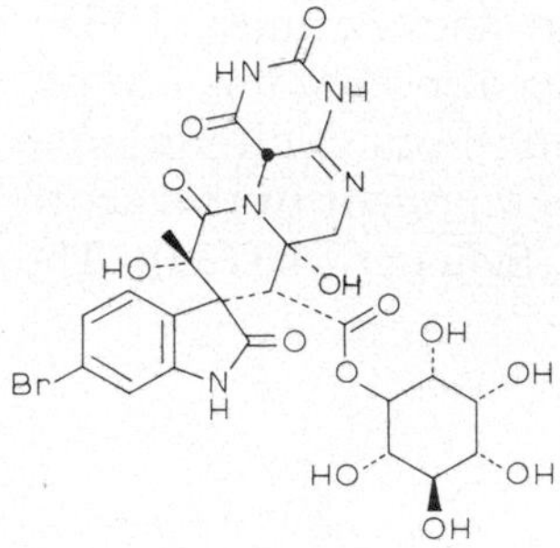

50

(soft corals) do, however, produce unique chlorine-containing diterpenoids. The Caribbean gorgonian *Briareum asbestinum*, for example, produces the pentaacetate *51* (Burks et al., 1977) and the related sea pen *Ptilosarcus gurneyi* produces the toxic ketone *52* (Wratten et al., 1977).

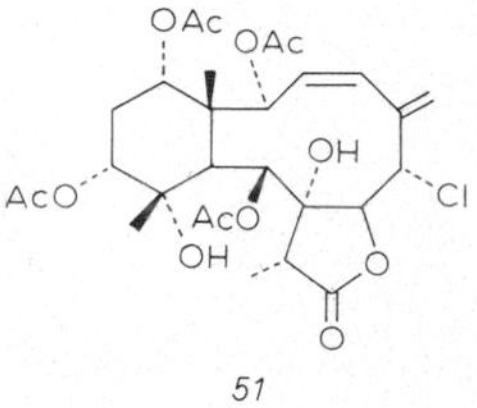

51

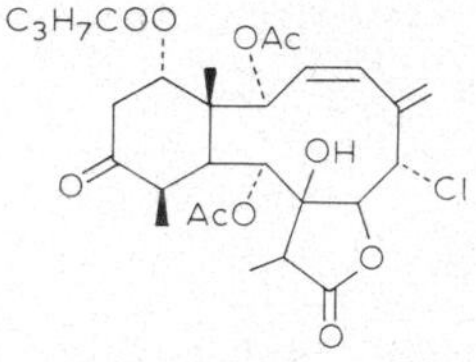

52

3. HALOGEN BIOSYNTHESIS

The enzymatic incorporation of iodine into the hormone thyroxine is a well-known process in man (Roche and Michel, 1955). In the same regard, the enzymatic chlorination reactions of several microorganisms have been extensively investigated. In a now classic paper, Morris and Hager (1966) reported the isolation and complete purification of the first halogenating enzyme "chloroperoxidase", from the fungus *Caldariomyces fumago*. *C. fumago* produces the dichloro-antibiotic caldariomycin (*53*), presumably via the chlorination of an intermediate such as compound *54*. *In vitro* experiments with the enzyme have shown that bromide could be substituted for chloride ion and, hence, incorporated into organic substrates.

54   53

A completely purified halogenating enzyme has not been isolated from marine sources, but the existence of haloperoxidases has been securely established. Pedersen (1976) succeeded in demonstrating the existence of a brominating and hydroxylating enzyme in the red alga *Cystoclonium purpureum*, and Manley (1977) has recently illustrated the *in vitro* bromination of precursor molecules in cell-free extracts of the red alga *Odonthalia floccosa*. Both enzymes were found to catalyse the bromination of phenolic precursors such as *p*-hydroxybenzyl alcohol (55) to yield the halo derivatives as in 56.

55   56

More recently, the Hager group (Theiler et al., 1978), has reported the partial purification of a "bromoperoxidase" enzyme from the red alga *Bonnemaisonia hamifera*. An *in vitro* bromination assay utilizing 3-keto-octanoic acid (57) as a substrate yielded, among other products, 1-bromo-2-heptanone (58), which is known as a natural component of *B. hamifera*. In

57   58

contrast to the behavior of chloroperoxidase, the brominating enzyme from *B. hamifera* is incapable of catalysing chlorination reactions.

4. NATURAL HALOGENS IN THE ENVIRONMENT

4.1. Biological functions

While strong evidence is lacking and difficult to obtain, many if not all of the aforementioned halogenated metabolites are likely produced as defensive environmental adaptations. Just as chlorine substitution increases the potency of commercial pesticides, the toxicity of simple terpenes is greatly increased by virtue of added bromine substituents. Those marine organisms that produce substances of this nature clearly benefit by enhanced survival against potential predators.

It is still unclear whether halogen-containing compounds are produced and simply stored, or whether they are products of excretion in dynamic processes. In our studies of the red seaweeds (Howard and Fenical, in prep.; McConnell and Fenical, work in progress), we find both cases to prevail. In *Laurencia*, few metabolites are excreted in culture, but rather they are stored in membrane-bound lipid bodies referred to as the "corps en cerise". Hence, bromine metabolism is slow and directly proportional to plant growth. With seaweeds of the family Bonnemaisoniaceae the situation is clearly different. In life, *Asparagopsis* species exude large amounts of bromoform (*10*) and several *Bonnemaisonia* species exude tetrabromo-2-heptanone (*19*). The very striking odors of these algae can clearly be attributed to the exudation of these halogenated metabolites. In culture these algae have impressive bromide requirements and rapidly metabolize acetate into halogenated compounds (McConnell and Fenical, work in progress).

The nature of organic halogen synthesis in the invertebrates has received less study. In general, the large amounts of the compounds (averaging 1—5% dry weight), and their toxicities, would suggest that they are carefully stored in metabolic isolation or efficiently excreted. In either case it stands by definition that these metabolites are ultimately liberated upon death or excretion into sea water, and, hence, form part of the dissolved organics.

4.2. Liberation and transport in sea water

Very little is known concerning the rate of input of natural halogenated organics into sea water. In an attempt to provide some insight into this question for coastal waters, Sleeper and Fenical (in progress) have investigated the dynamic liberation of dissolved halo-organics into a large tide pool, mainly inhabited by red seaweeds, in La Jolla, California. During a 3-h period of low tide isolation, 20-l water samples were removed from the pool every 1.5 h. The water samples were filtered (0.45 μm) and extracted with purified hexane following techniques described for pesticide residue analysis. The hexane residues were then quantitatively analysed in consistent fashion, by electron capture gas chromatography, a method highly sensitive and essentially selective for volatile organohalogen compounds. The results of these analyses, illustrated in Fig. 1, provide evidence for the accumulation of organohalogen compounds. The identities of the substances in the figure were not determined, but the main algal species within the tide pool were of the family Corallinaceae, and a prominent member of this group, *Corallina officinalis*, has been shown to contain the bromophenol lanosol (*59*) (Pedersen et al., 1974). The main peak in trace C (retention time ca. 12.5 min) corresponds roughly to compounds such as *59*.

Halogenated compounds have been detected in sea water. Using gas chromatography—mass spectrometry (GC—MS), Pedersen et al. (1974) were able to illustrate that lanosol (*59*) was a component of the dissolved organics

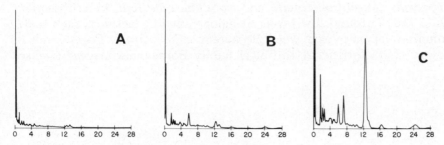

Fig. 1. Electron capture—gas chromatographic analyses of hexane extracts of a La Jolla, California tidepool at low tide. Trace A is at zero time, the time at which the pool became totally isolated. Trace B is the analysis after 1.5 h and trace C after 3.0 h. The GC conditions were: column, 6' × 1/4" 10% DC-200 on Chromosorb Q 80/100 mesh; carrier gas 95% argon, 5% methane; detector [63]Ni pulse; conditions, isothermal at 220°C.

59

of near-shore water proximate to beds of the red alga *Polysiphonia brodiaei*. In addition, circumstantial evidence has been provided that halogenated algal metabolites are transmitted through sea water being adsorbed onto lipophilic surfaces. Crews et al. (1976) have reported that *Microcladia coulteri* contains small amounts of halogenated monoterpenes characteristic of those found only in the unrelated genus *Plocamium* (see section 2.3). In that study *Plocamium* spp. were abundantly growing in close proximity to *M. coulteri*. Subsequent investigations of *Microcladia* species have not yielded halo-monoterpenes, nor has *M. coulteri* yielded these compounds when collected in areas void of *Plocamium*.

Rinehart et al. (1975) have reported a similar wide-spread occurrence of the bromo-terpene laurinterol (*60*), at undescribed levels, in a variety of marine organisms including several unrelated red algae, sponges, echinoderms, and bryozoans. Laurinterol is a well-known component of several species of

60

the red alga *Laurencia*, and the production of laurinterol by unrelated plants and diverse invertebrate groups is unlikely. A similar situation recently occurred in Japan. Ohta and Takagi (1977) found very low levels of laurinterol, and several of its derivatives, in three coralline algae *Margini-*

sporum aberrans, Amphiroa zonata and *Corallina pilulifera.* In a subsequent study, Ohta (1977) found low levels of halogenated C_9 lactones, such as *61*, in the common Japanese agar weed *Beckerella subcostatum.* Compounds of this type are characteristic of the algal family Bonnemaisoniaceae (*Delisea*

61

spp.), and isolation of these compounds from *B. subcostatum* is not expected based upon taxonomic grounds. It should be pointed out that these latter compounds were isolated in very minor amounts (~5 ppm), and that the *in vivo* concentrations of halogenated metabolites in most clear-cut sources is as high as 6% of the dry weight.

The logical conclusion, based upon the known dynamic production of halogenated organics, and their detection in sea water, is that dissolved organics liberated by one organism are transported short distances through sea water and ultimately become adsorbed onto various lipophilic surfaces. An investigation of the synthetic activities of *Beckerella* and *Marginisporum* in pure culture, and in mixed culture with *Laurencia* spp., should solve this problem.

4.3. Biodegradation

Very little is known concerning the ultimate fate of natural halogenated organics in the sea. Since these compounds are natural products, pre-established methods for their degradation must exist. In an attempt to establish a rough rate of degradation, my collaborators and I (Fenical et al., 1974) measured the rate of loss of the bromo-terpene laurinterol (*60*) in mixed anerobic bacterial cultures from various local isolates. Undefined cultures from near-shore sediments degraded laurinterol with a half-life of ca. 5—8 days at 20°C. The mechanism of degradation and the products produced were not explored, but the facile rate observed here would appear to contrast the rather poor degradability of the halogenated pollutant hydrocarbons such as DDT, PCB, etc.

4.4. Natural versus unnatural halogenated organics

There has been considerable recent concern over the accumulation of commercially produced halogenated organics in the atmosphere, in drinking water, and in the sea. In this regard, it is curious to compare the natural compounds described here with the synthetic pollutants which have surprising structural similarities. Considering the massive production of organohalogens by many species, the assessment of halogenated pollutant levels (DDT, PCB, etc.) may present difficulties. To assess the volatilities and chromatographic behaviors of several natural halogen-containing compounds, and hence the degree to which they might interfere with DDT and PCB analysis, extracts of several known halogenating organisms were compared, by electron capture—gas chromatography, with several pesticide standards. The results of this comparison are presented in Fig. 2. Traces A and B are pesticide standards of

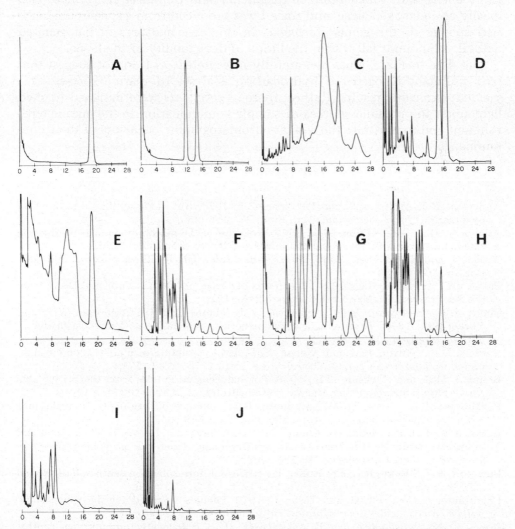

Fig. 2. Electron capture—gas chromatographic analyses of standard pollutants and extracts of several marine organisms: A, DDT; B, DDE (first peak), DDD; C, *Laurencia pacifica*; D, *Laurencia subopposita*; E, *Chromobacterium* sp.; F, Arochlor 1242; G, Arochlor 1254; H, *Plocamium cartilagineum*; I, *Verongia* sp.; J, *Chondria californica*. The GC conditions are identical to those used to produce Fig. 1.

DDT (trace A) and its metabolites DDE and DDD (trace B), and traces F and G are the commercial polychlorinated biphenyl (PCB) mixtures Arochlor 1242 (trace F) and Arochlor 1254 (trace G). As is illustrated, extracts of local *Laurencia* species (traces C and D), and the extract of a halogenating bacterium *Chromobacterium* sp. (trace E), contain compounds of similar retention time to DDT and its metabolites. Traces H and J were produced from extracts of *Plocamium cartilagineum* and *Chondria californica*, two locally abundant red algae, and trace I was produced from a species unidentified sponge of the genus *Verongia*. In this case mixtures of halogenated natural compounds fall within the limits of detectability of the PCBs.

The data in Fig. 2 must be carefully interpreted, as I do not suggest that past pollutant halogenated hydrocarbon analyses of marine species have yielded erroneous results. Rather, these experiments were designed to shed light upon the possible sources of sample contamination in the marine environment and to illustrate the similar chromatographic behaviors of these contaminants.

5. REFERENCES

Anderson, R.J., Wolfe, M.S. and Faulkner, D.J., 1974. Autotoxic antibiotic production by a marine *Chromobacterium*. Mar. Biol., 27: 281—285.

Ashworth, R.B. and Cormier, M.J., 1967. Isolation of 2,6-dibromophenol from the marine hemichordate *Balanoglossus biminiensis*. Science, 155: 1558—1559.

Augier, J. and Henry, M-H., 1950. Au sujet du bromé dans les Rhodophycées. Bull. Soc. Bot. Fr., 97: 29—30.

Baker, J.T. and Duke, C.C., 1974. Precursors of tyrian purple. In: Food and Drugs from the Sea Proceedings. Mar. Technol., soc., 345—353.

Blunt, J.W., Hartshorn, M.P., McLennan, T.J., Munro, M.H.G., Robinson, W.T. and Yorke, S.C., 1978. Thysiferol, a squalene-derived metabolite of *Laurencia thysifera*. Tetrahedron Lett., 1978: 69—72.

Borders, D.B., Morton, G.O. and Wetzel, E.R., 1974. Structure of a novel bromine compound isolated from a sponge. Tetrahedron Lett., 1974: 2709—2712.

Brennan, M.R. and Erickson, K.L., 1978. Polyhalogenated indoles from the marine alga *Rhodophyllis membranaceae* Harvey. Tetrahedron Lett., 1978: 1637—1640.

Burkholder, P.R., Pfister, R.M. and Leitz, R.M., 1960. Production of a pyrrole antibiotic by a marine bacterium. Appl. Microbiol., 14: 649—653.

Burks, J.E., Van der Helm, D., Chang, C.Y. and Ciereszko, L., 1977. The crystal and molecular structure of briarein A, a diterpenoid from the gorgonian *Briareum asbestinum*. Acta Crystallogr., 33: 704—709.

Burreson, B.J., Moore, R.E. and Roller, P., 1975. Haloforms in the essential oil of the red alga *Asparagopsis taxiformis*. Tetrahedron Lett., 1975: 473—476.

Chantraine, J., Combaut, G. and Teste, J., 1973. Phenols bromés d'une alga rouge, *Halopytis incurvus*: acides carboxyliques. Phytochemistry, 12: 1793—1795.

Crews, P., Ng, P., Kho-Wiseman, E. and Pace, C., 1976. Halogenated monoterpene synthesis by the red alga *Microcladia*. Phytochemistry, 15: 1707—1710.

Fattorusso, E., Magno, S., Santacroce, C., Sica, D., DiBlasio, B. and Pedone, C., 1976. Bromosphaerol, a new bromine-containing diterpenoid from the red alga *Sphaerococcus coronopifolius*. Gazz. Chim. Ital., 106: 779—780.

Fenical, W., 1975. Halogenation in the Rhodophyta; a review. J. Phycol., 11: 245—259.

Fenical, W., Gilkins, K.B. and Clardy, J., 1974. X-ray determination of chondriol; a reassignment of structure. Tetrahedron Lett., 1974: 1507—1510.

Fenical, W., Howard, B.M., Gilkins, K.B. and Clardy, J., 1975. Irieol A and iriediol, dibromoditerpenes of a new skeletal class from *Laurencia*. Tetrahedron Lett., 1975: 69—72.

Fenical, W., Finer, J. and Clardy, J., 1976. Sphaerococcenol A; a new rearranged bromo-diterpene from the red alga *Sphaerococcus coronopifolius*. Tetrahedron Lett., 1976: 731—734.

Forenza, S.L., Minale, L., Riccio, R. and Fattorusso, E., 1971. New bromopyrrole deriva-tives from the sponge *Agelas oroides*. J. Chem. Soc., Chem. Commun., 1971: 1129.

Fritsch, F.E., 1945. The Structure and Reproduction of the Algae, II. Cambridge Univer-sity Press, London, 939 pp.

Galbally, J.E., 1976. Man-made carbon tetrachloride in the atmosphere. Science, 193: 573—576.

Goldberg, E.D., 1963. The oceans as a chemical system. In: M.N. Hill (Editor), The Sea, 2. Wiley—Interscience, New York, N.Y., pp. 3—25.

Higa, T. and Scheuer, P.J., 1975a. 3-chloroindole, principal odorous constituent of the hemichordate *Ptychodera flava laysanica*. Naturwissenschaften, 62: 395—396.

Higa, T. and Scheuer, P.J., 1975b. Constituents of the marine annelid *Thelepus setosus*. Tetrahedron, 31: 2379—2381.

Higa, T. and Scheuer, P.J., 1977. Constituents of the hemichordate *Ptychodera flava laysanica*. In: D.J. Faulkner and W.H. Fenical (Editors), Marine Natural Products Chemistry. Plenum Press, London, pp. 34—43.

Högberg, H-E., Thomson, R.H. and King, T.J., 1976. The cymopols, a group of prenylated bromohydroquinones from the green calcareous alga *Cymopolia barbata*. J. Chem. Soc., Chem. Commun., 11: 245—259.

Ichikawa, N., Naya, Y. and Enomoto, S., 1974. New halogenated monoterpenes from *Desmia (Chondrococcus) hornemanni*. Chem. Lett., 1974: 1333—1336.

Irie, T., Suzuki, M. and Masamune, T., 1965. Laurencin, a constituent from *Laurencia* sp., Tetrahedron Lett., 1965: 1091—1094.

Kato, Y. and Scheuer, P.J., 1974. Aplysiatoxin and dibromoaplysiatoxin, constituents of the marine mollusk *Stylocheilus longicauda*. J. Am. Chem. Soc., 96(7): 2245—2246.

Kazlauskas, R., Murphy, P.T., Quinn, R.J. and Wells, R.J., 1976. Abstracts Int. Symp. Chem. Nat. Prod. 10th No. Cb. Dunedin, New Zealand, August 1976.

Kazlauskas, R., Murphy, P.T., Quinn, R.J. and Wells, R.J., 1977a. A new class of halogenated lactones from the red alga *Delisea fimbriata*. Tetrahedron Lett., 1977: 37—40.

Kazlauskas, R., Lidgard, R.O., Wells, R.J. and Vetter, W., 1977b. A novel hexachloro-metabolite from the sponge *Dysidea herbacea*. Tetrahedron Lett., 1977: 3183—3186.

Kosuge, T., Zenda, H., Ochiae, A., Masaki, N., Noguchi, M., Kimura, S. and Narita, H., 1972. Isolation and structure determination of a new marine toxin, surugatoxin from the Japanese ivory shell, *Babylonia japonica*. Tetrahedron Lett., 1972: 2545—2548.

Kylin, H., 1930. Über die jodidspaltende Fähigkeit der Phäeophyceen. Hoppe-Seyler's Z. Physiol. Chem., 191: 200—210.

Lovelock, J.E., 1975. Natural halocarbons in the air and in the sea. Nature, 256: 193—194.

Lunde, G., 1967. Activation analysis of bromine, iodine and arsenic in oils from fishes, whales, phyto- and zooplankton of marine and limnetic biotopes. Int. Rev. Gesamten Hydrobiol., 52(2): 265—279.

Manley, S.L., 1977. Metabolism of tyrosine by cell free fractions of the red alga *Odonthalia floccosa*: a proposed biosynthetic pathway for the bromophenols. Abstracts IX Int. Seaweed Symp., Santa Barbara, Calif., August 1977.

Martín, J.D. and Darias, J., 1978. Algal sesquiterpenoids. In: P.J. Scheuer (Editor),

Marine Natural Products; Chemical and Biological Perspectives. Academic Press, London, pp. 125—171.

McConnell, O.J. and Fenical, W., 1976. Halogen chemistry of the red alga *Asparagopsis*. Phytochemistry, 16: 367—374.

McConnell, O.J. and Fenical, W., 1977a. Polyhalogenated 1-octene-3-ones, antibacterial metabolites from the red seaweed *Bonnemaisonia asparagoides*. Tetrahedron Lett., 1977: 1851—1854.

McConnell, O.J. and Fenical, W., 1977b. Halogenated metabolites — including Favorsky rearrangement products — from the red seaweed *Bonnemaisonia nootkana*. Tetrahedron Lett., 1977: 4159—4162.

McConnell, O.J. and Fenical, W., 1978. Ochtodene and ochtodiol, novel polyhalogenated monoterpenes from the red alga *Ochtodes secundiramea*. J. Org. Chem., 43: 4328—4241.

Minale, L., 1976. Natural product chemistry of the marine sponges. Pure Appl. Chem., 48: 7—23.

Minale, L., 1978. Terpenoids from marine sponges. in: P.J. Scheuer (Editor), Marine Natural Products; Chemical and Biological Perspectives. Academic Press, London, pp. 175—238.

Minale, L., Cimino, G., DeStefano, S. and Sodano, G., 1976. Natural products from porifera. Progr. Org. Chem. Nat. Prod., 33: 1—72.

Moore, R.E., 1978. Algal nonisoprenoids. In: P.J.Scheuer (Editor), Marine Natural Products; Chemical and Biological Perspectives. Academic Press, London, pp. 44—121.

Morris, D.R. and Hager, L.P., 1966. Chloroperoxidase: isolation and properties of the crystalline glycoprotein. J. Biol. Chem., 421: 1763.

Mynderse, J.S. and Faulkner, D.J., 1975a. Polyhalogenated monoterpenes from the red alga *Plocamium cartilagineum*. Tetrahedron, 31: 1963—1967.

Mynderse, J.S. and Faulkner, D.J., 1975b. (1R,2S,4S,5R)-1-bromo-*trans*-2-chlorovinyl-4,5-dichloro-1,5-dimethylcyclohexane, a new monoterpene skeletal type from the red alga *Plocamium violaceum*. Tetrahedron Lett., 1975: 2175—2178.

Mynderse, J.S., Moore, R.E., Kashiwagi, M. and Norton, T.R., 1977. Antileukemia activity in the Oscillatoriaceae; isolation of dibromoaplysiatoxin from *Lyngbya*. Science, 196: 538—540.

Mynderse, J.S. and Moore, R.E., 1978. Toxins from blue-green algae: structures of oscillatoxin A and three related bromine-containing toxins. J. Org. Chem., 43(11): 2301—2303.

Ohta, K., 1977. Antimicrobial compounds in the marine red alga *Bekerella subscostatum*. Agric. Biol. Chem., 41: 2105—2106.

Ohta, K. and Takagi, M., 1977. Halogenated sesquiterpenes from the marine red alga *Marginisporum aberrans*. Phytochemistry, 16: 1062—1063.

Pedersen, M., 1969. The demand for iodine and bromine of three marine brown algae grown in bacteria-free cultures. Physiol. Plant., 22: 680—685.

Pedersen, M., 1976. A brominating and hydroxylating peroxidase from the red alga *Cystoclonium purpureum*. Physiol. Plant., 37: 6—11.

Pedersen, M. and DaSilva, E.J., 1973. Simple brominated phenols in the bluegreen alga *Calothrix brevissima* West. Planta, 115: 83—86.

Pedersen, M., Saenger, P. and Fries, L., 1974. Simple bromophenols in red algae. Phytochemistry, 13: 2273—2279.

Pettus, J.A. Jr., Wing, R.M. and Sims, J.J., 1977. Marine natural products XII, isolation of a family of multihalogenated gamma-methylene lactones form the red seaweed *Delisea fimbriata*. Tetrahedron Lett., 1977: 41—44.

Rinehart, K.L. Jr., Johnson, R.D., Suida, J.F., Krejcarek, G.E., Shaw, P.D., McMillan, J.A. and Paul, I.C., 1975. Structures of halogenated and antimicrobial organic compounds from marine sources. In: E.D. Goldberg (Editor), Dahlem Conference on the Nature of Sea Water. Abakon Verlags Gesellschaft, Berlin, pp. 651—666.

Roche, J. and Michel, R., 1955. Nature, biosynthesis and metabolism of thyroid hormones. Physiol. Rev., 35: 583—614.

Rönnerstrand, S., 1968. Investigations into polyphenols of the oxidase systems of some algae. Bot. Mar., 11: 107—114.

Rose, A.F. and Scheuer, P.J., 1975. 30th Northwest Regional Meet. Am. Chem. Soc., Honolulu, Hawaii, Abstract 186.

Scott, R., 1954. Observations on the iodoamino acids of marine algae using iodine-131. Nature, 173: 1098—1099.

Sharma, G.M. and Burkholder, P.R., 1967. Studies on the antimicrobial substances of sponges II. Structure and synthesis of a bromine-containing antibacterial compound from a marine sponge. Tetrahedron Lett., 1967: 4147—4150.

Sharma, G.M. and Vig, B., 1972. Studies on the antimicrobial substances of sponges VI. Structures of two antibacterial substances isolated from the marine sponge *Disidea herbacea*. Tetrahedron Lett., 1972: 1715—1718.

Sheikh, Y.M. and Djerassi, C., 1975. 2,6-dibromophenol and 2,4,6-tribromophenols — antiseptic secondary metabolites of *Phoronopsis viridis*. Experientia, 31: 265—266.

Sims, J.J., Lin, G.H.Y., Wing, R.M. and Fenical, W., 1973. Marine natural products. Concinndiol, a bromo-diterpene alcohol from the red alga *Laurencia concinna*. J. Chem. Soc., Chem. Commun., 1973: 470—471.

Siuda, J.F. and DeBernardis, J.F., 1973. Naturally occurring halogenated organic compounds. Lloydia, 36: 107—143.

Siuda, J.F., Van Blaricom, G.R., Shaw, P.D., Johnson, R.D., White, R.H., Hager, L.P. and Rinehart, K.L. Jr., 1975. 1-Iodo-3,3-dibromo-2-heptanone, 1,1,3,3-tetrabromo-2-heptanone and related compounds from the red alga *Bonnemaisonia hamifera*. J. Am. Chem. Soc., 197: 937—938.

Theiler, R.F., Siuda, J.F. and Hager, L.P., 1978. Bromoperoxidase from the Red Alga *Bonnemaisonia hamifera*. In: P.N. Kaul and C.J. Sindermann (Editors), Food and Drugs from the Sea. The University of Oklahoma Press, Norman, Okla., pp. 153—169.

Wratten, S.J. and Faulkner, D.J., 1977. Carbonimidic dichlorides from the marine sponge *Pseudaxinyssa pitys*. J. Am. Chem. Soc., 99: 7367—7368.

Wratten, S.J., Fenical, W., Faulkner, D.J. and Wekell, J.C., 1977. Ptilosarcone, the toxin from the sea pen *Ptilosarcus gurneyi*. Tetrahedron Lett., 1977: 1559—1562.

Chapter 13

ORGANIC SULPHUR IN THE MARINE ENVIRONMENT [1]

WOLFGANG BALZER

1. INTRODUCTION

Organically combined sulphur is synthesized by direct assimilation of sulphate by living plants and microbiological processes. Living organisms contain highly variable amounts (0.01—5%) of sulphur; Orr (1974) reports mean values of 0.5% for plants and decomposers and 1.3% for animals. Average values of 0.9 and 1.1% sulphur content in dry biological residue have been given for marine algae and animals respectively (Kaplan et al., 1963), and according to Postgate (1968), sulphur usually accounts for 0.4—0.8% of the microbial dry weight.

Organosulphur compounds do not only constitute a significant part of proteinaceous or structural materials, but have also been found, e.g. in the form of a polynucleotide—peptide complex, to exert a stimulating effect on growth and cell division of certain blue-green algae (Volodin, 1975). Sulphhydryl groups are further required for algal enzyme activity (Desai et al., 1972).

A few inorganic processes in seawater are also influenced by organic sulphur compounds since these are able to form relatively strong complexes with certain transition metals (Gardner, 1974; Boulegue, 1979). In organic geochemical reactions sulphur is involved in dehydrogenation, oxidation, and ring closure (De Roo and Hodgson, 1978).

In recent years some comprehensive books have appeared covering most aspects of the environmental chemistry of sulphur with special emphasis on biological, chemical, geological, and clinical studies of pollutant sulphur (Nriagu, 1976; Meyer, 1977; Nriagu, 1978). Nevertheless the impact of organic sulphur in the marine cycle of organic matter is not matched by an adequate amount of knowledge about this class of compounds.

This survey opens with a short compilation of analytical methods employed in the marine organic chemistry of sulphur and the individual (dissolved) compounds known so far to occur in seawater. Thereafter it proceeds to sulphur-releasing processes in living and decomposing organic matter in

[1] Contribution No. 233 from the Sonderforschungsbereich 95 of the University of Kiel, Olshausenstr. 40, D-2300 Kiel (Federal Republic of Germany).

seawater and sediments, and finally describes possible impacts from petroleum and anthropogenic sources.

2. ANALYTICAL CHEMISTRY OF ORGANIC SULPHUR

An excellent review of modern methods for the determination of different forms of sulphur with particular emphasis on the use of highly sensitive and sophisticated equipment is given by Tanner et al. (1978); although intended to be used with atmospheric samples, a detailed outline of methods is given (including coulometry and gas chromatography using flame ionisation and flame photometric detection) which enables adaptation to analysis of marine samples. A microcoulometric method independent of the organo-sulphur linkage was developed for the determination of 10—500 ng of organic sulphur and was applied to analysis of surface waters (De Groot et al., 1975).

A good deal of progress in the analysis of marine samples (e.g., Lovelock et al., 1972; Warner, 1975; Bates and Carpenter, 1979a, b) has been made by the application of the sulphur-sensitive flame photometric detector (FPD, Fig. 1) in gas chromatography (see also: Maruyama and Kakemoto, 1978). Systematic use of UV-spectra were found by Boulegue (1978) to be effective in distinguishing organic and inorganic polysulphides in brines.

A potentiometric method for the determination of different dissolved sulphur species, including organosulphides, was proposed by Boulegue and Popoff (1979) employing titration with $HgCl_2$ and end-point determinations with a sulphide-sensitive electrode. GLC analysis using pentafluorobenzyl

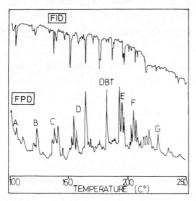

Fig. 1. Gas chromatogram of an environmental sample using FID and FPD detection. Effluent entering Puget Sound from a sewage treatment plant. The lettered peaks are postulated to be: A = benzothiophene; B = methylbenzothiophenes; C = dimethylbenzothiophenes; D = trimethylbenzothiophenes; DBT = dibenzothiophene; E = methyldibenzothiophenes; F = dimethyldibenzothiophenes; G = naphthobenzothiophene. (After Bates and Carpenter, 1979b.)

derivatives was used for the determination of mercaptans in surface waters (Kawahara, 1971). Bates and Carpenter (1979a) discuss extraction procedures for marine sediments in the presence of elemental sulphur: removal of the sulphur interference (prior to GLC—FPD analysis) by an activated Cu column was found to be the only method avoiding artefacts. This method, however, removed also mercaptans and most disulphides. For the identification of individual organosulphur compounds in marine samples GLC—mass spectrometry has been employed successfully (Yamaoka and Tanimoto, 1976; Bouchertall, 1979).

3. DISSOLVED ORGANOSULPHUR COMPOUNDS

The chemistry of organic sulphur species dissolved in oceanic waters started with the detection of sulphhydryl group containing polar compounds and "hydrocarbon-like" mercaptans in an anoxic fjord by Adams and Richards (1968). Dimethyl sulphide (DMS) and carbon disulphide were

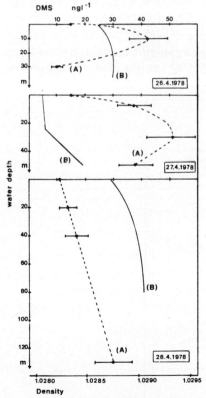

Fig. 2. Dimethyl sulphide profiles (*A*) in Mediterranean waters and the respective density (*B*) distribution. (After Nguyen et al., 1978.)

398

found in different oceanic environments (Lovelock et al., 1972; Lovelock, 1974; Nguyen et al., 1978); these compounds are believed to originate from decomposing plant material (Fig. 2). Other volatile organosulphur compounds are also derived from natural sources: methyl mercaptan (MeSH) and dimethyl disulphide (DMDS), which were detected in Hiro Bay seawater by Yamaoka and Tanimoto (1976). Dibenzothiophene (DBT) and napthobenzothiophene, and benzothiazol were found in oxic surface waters of the Baltic proper (Bouchertall, 1979); DBT in the marine sediment of Puget Sound was believed to originate from natural sources (Bates and Carpenter, 1979b). The presence of thiamine and biotin in seawater is discussed by Benzhitskiy (1974).

Sulphur-containing petroleum components such as benzothiophene, DBT, and naphthobenzothiophene were obtained from seawater extracts (Warner, 1975) and were related to an oil spill. The detection of diphenyl sulphone and bis-(*p*-chlorphenyl) sulphone in ng amounts in the Mediterranean, as well as in oxic and anoxic Baltic waters, suggests anthropogenic input (Bouchertall, 1979) — the former compound is used as wood preserving agent and ovicide.

4. ORGANOSULPHUR IN ORGANISMS

The synthesis of organic sulphur by marine organisms leads to a variety of compounds comprising mainly sulphur-containing amino acids, sulphonates

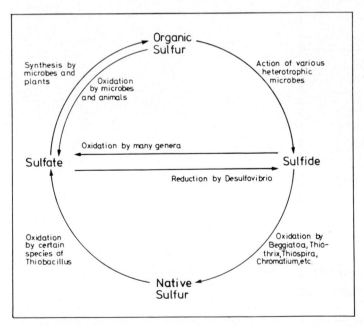

Fig. 3. Salient features of the sulphur cycle in the biosphere. (After Zobell, 1963.)

in which sulphur is directly bonded to carbon, and organic esters of sulphuric acid which contain a C—O—SO$_3$H linkage. The gross cycle of organic sulphur is depicted in Fig. 3.

4.1. Production and requirements of S-containing amino acids

Most microorganisms are capable of using sulphate as the sole source of sulphur for growth; in doing so, they reduce it intracellularly to hydrogen sulphide and then replace the hydroxyl groups on serine and homoserine with sulphhydryl groups (Roy and Trudinger, 1970). A few microorganisms reduce sulphur of lower oxidation states, while others require hydrogen sulphide or preformed sulphur-containing amino acids.

The more complex sulphur requirements of the marine animals are met largely by cysteine, cystine, methionine, biotin, and thiamine (Young and Maw, 1958) (Fig. 4). Cysteine is a component of the tripeptide glutathione and a precursor of taurine. Methionine is as an essential amino acid involved in biosynthesis of proteins, creatine and adrenaline. Adenosylmethionine is considered to be the active part in transmethylation, e.g. of choline. Methionine is part of the pathways to homocysteine, cystathionine and methylthioadenosine (Young and Maw, 1958). Various organisms convert cysteine and/or cystine into mercapturic acids, cysteine sulphinic acid, and thiazolidine derivatives (Zobell, 1963).

From the extensive literature on amino acids as intermediate products of protein decomposition only a few papers, which include S-containing amino

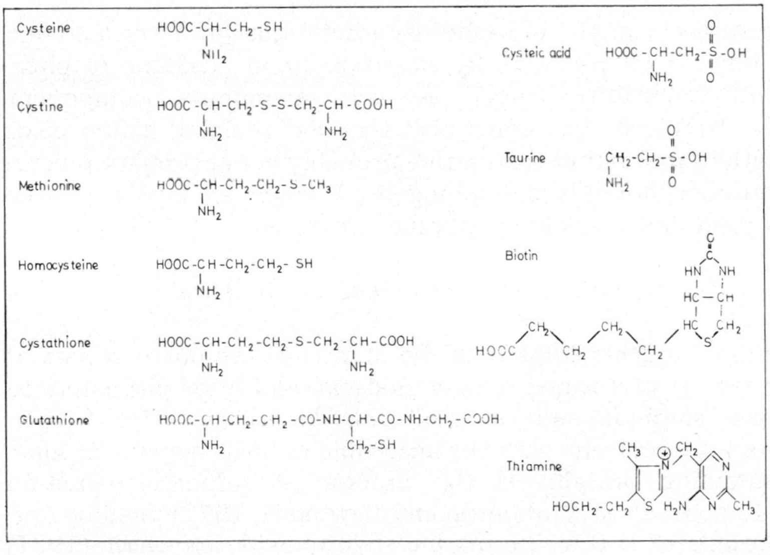

Fig. 4. Bioactive organosulphur compounds occurring in marine organisms.

acids, can be mentioned here within the limited space. In an early study Degens et al. (1964) found plankton samples offshore California to contain 2.1 mole% of cysteine and 0.4 mole% of methionine among the eighteen amino acids determined. Except for methionine in the combined fraction (CAA), the S-containing amino acids could not be detected in seawater, nor were they present in the sediments in quantifiable amounts. They have been estimated to be present in concentrations near the detection limit in the Irish Sea and North Atlantic Ocean by Riley and Segar (1970) and Pocklington (1971), respectively. Dawson and Gocke (1978) and Dawson and Pritchard (1978) distinguished between cysteic acid, taurine, methionine and cysteine in the pool of free amino acids (FAA) collected from the Baltic and found only the first amino acid (5 mole%) to be present in more than traces. Relatively high concentrations, found in earlier investigations, may have been overestimated due to systematic analytical errors. An example is taurine, which is suggested to play an important role in the regulation of osmotic pressure in marine life (Stuart et al., 1979). The detected levels of this substance were found to arise solely from contamination (Garrasi et al., 1979) when using a commonly employed desalting technique (cation resin). Taurine, however, was present in samples of seawater analysed directly by HPLC techniques (Mopper, pers. comm., 1979). Studying offshore water from the North Sea, the Scandinavian Shelf, the Norwegian Sea and the northeast Atlantic Ocean by a direct injection technique, Garrasi et al. (1979) stated that S-containing amino acids, while non-detectable in DFAA, were often present in low concentrations in the combined fraction. The conclusions of Dawson and Liebezeit (this volume, Chapter 15) point in the same direction.

In certain organisms, e.g. the non-photosynthetic diatom *Nitzschia alba*, lecithine was found to be replaced by its sulphonium analogue (a phosphatidyl-S,S-dimethylmercaptoethanol) as cell membrane component (Anderson et al., 1976); it was concluded that S-containing amino acids (e.g. cysteine, methionine) rather than serine probably act as primary precursors of the glycerophospholipids and sphingolipids which are involved in the formation of hydrophobic layers in membrane structures.

4.2. Degradation and remineralisation of proteinaceous material

The degradation of organic sulphur in the marine environment is part of the gross breakdown of preformed organic material and leads ultimately to sulphate, hydrogen sulphide, and/or small volatile organic sulphur compounds (see below). Concurrent with the anaerobic reduced sulphur production from S-containing proteins is the process of sulphate reduction (Goldhaber and Kaplan, 1975; Middleton and Lawrence, 1977) yielding normally greater amounts of H_2S in the marine environment. Jørgensen (1977) found that only 3% of the observed sulphide in a coastal marine sediment was derived from organic sulphur. Investigating temperature effects on sulphide production from a marine sediment, Nedwell and Floodgate (1972) found in addition to a decrease in the rate with temperature drop that there was a change in the origin of sulphur: at low temperatures (5 and 10°C) the majority of sulphur originated from organic sulphur, while sulphate contributed a greater proportion at 25° and 30°C. These changes may reflect those in the natural environment during winter and summer.

4.3. Dimethylpropiothetin and dimethyl sulphide

Challenger and Simpson (1948) first described the occurrence of a sulphonium compound in marine plants. From the constituent dimethyl-2-carboxethyl sulphonium chloride of the marine algae *Polysiphonia fastigiata* and *Polysiphonia nigrescens* dimethyl sulphide (DMS) is evolved by enzymatic decomposition on exposure to air. Some years later it was concluded that marine algae are prominent among different living systems in producing DMS (Challenger, 1951; see for a review also Challenger, 1959). Dimethyl-β-propiothetin (β-DMPT), a precursor of DMS, may pass through the food chain from phytoplankton through zooplankton to accumulate in fish where it may decompose spontaneously to DMS which has been blamed for the "petroleum"-like odour of fish (Ackman and Dale, 1965; Ackman et al., 1966). In plankton samples from brackish waters Granroth and Hattula (1976) found 310—515 μg g^{-1} dry weight β-DMPT bromide and 50—83 μg g^{-1} dry weight DMS, and concluded that DMS is generally present as a minor component of the aroma of Baltic herring where DMS concentrations less than 1 ppm were found in muscle tissue, the content increasing with age.

In different specimens of dried laver from various culture grounds in Japan and Korea (Noda, 1975), DMS concentration varies considerably in the range 168—292 ppb reflecting the strength of odour. Variations in β-DMPT concentration both among different shrimps and within a single organism Tokunaga (1977) showed to occur in the Antarctic krill where β-DMPT in the head part almost exceeded the concentrations in the tail muscle; considerably increasing concentrations of DMS with time in frozen raw krill indicated uninterrupted enzymatic production even at −10 to −20°C.

The occurrence of β-DMPT in certain unicellular algae was used by Craigie et al. (1967) to differentiate between the two similar classes of Chlorophyceae and Prasinophyceae, the latter probably contains β-DMPT generally. The metabolic production of DMS from methionine may begin with a methylation step to S-methyl-methionine sulphonium ion occurring mainly in higher plants (Hattula and Granroth, 1974); enzymatic cleavage leads to DMS and homoserine. Alternatively it includes the primary biosynthesis of β-DMPT from methionine followed by the enzymatic decomposition to DMS and acrylic acid (Granroth and Hattula, 1976) (Fig. 5). DMS is not neces-

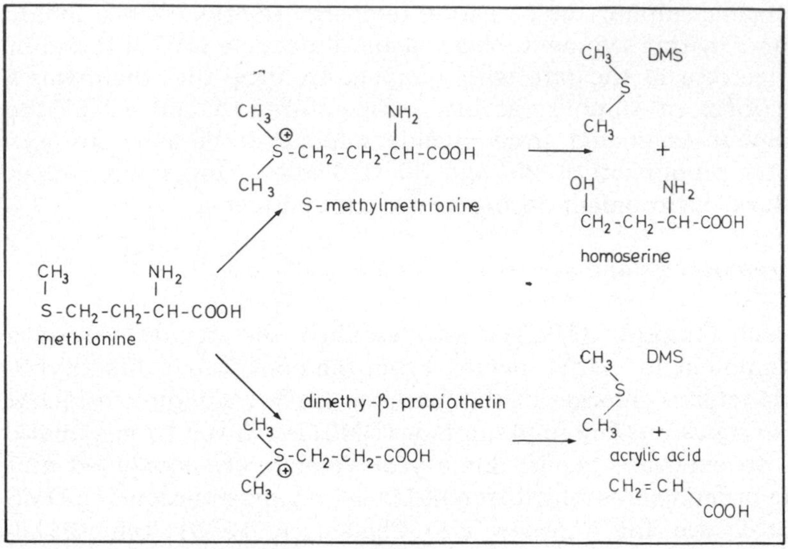

Fig. 5. Possible pathways for the formation of dimethylsulphide (DMS) from methionine. (After Hattula and Granroth, 1974; Granroth and Hattula, 1976.)

sarily a stable endproduct: Zinder and Brock (1978a) identified anaerobic habitats as possible sinks for MeSH, dimethyl disulphide (DMDS), and DMS when analyzing microbial populations in freshwater sediments and anaerobic sewage sludge capable of metabolizing carbon in these compounds to methane and carbon dioxide.

Direct determinations of DMS in oceanic waters are scarce. In all seawater samples of a cruise from Montevidea to the United Kingdom analyzed by Lovelock et al. (1972) DMS was present with an average oceanic concentration of 12 ng l^{-1}. The possibility of large variations in the oceanic concentration of DMS was predicted by Lovelock (1974): seawater in "equilibrium" with *Polisiphonia fastigiata* contained five orders of magnitude more DMS than the open ocean. Based on improved methods applied to numerous oceanic samples, Nguyen et al. (1978) report an average concentration of 30 ng l^{-1} along with much higher values for Mediterranean waters (580 ng l^{-1}) and even 1500 ng l^{-1} in its surface microlayer (400 μm). DMS concentrations in surface waters of Hiro Bay, Japan ($Cl^- = 12-15\%_{00}$; T ca. 14°C), varied from 88 to 624 ppb with a mean of 341 ppb in ten samples (Yamaoka and Tanimoto, 1976).

Based on their DMS determinations in air and oceanic waters, of the emission of DMS from marine algae and soils, Lovelock et al. (1972) concluded that DMS may fulfill the role in the global sulphur cycle (providing a vehicle for the transfer of sulphur from the sea through air to the land surface) that had previously been assigned to H$_2$S. Re-evaluating this role of DMS in connection with new determinations of the DMS concentration gradient in the sea surface microlayer, and using exchange coefficients provided by Liss and Slater (1974), Nguyen et al. (1978) calculated that 30% of the sulphur required by the global budget (Kellogg et al., 1972) could originate from the oceans. In a diurnal kinetic photochemical model of the natural marine atmosphere, Graedel (1979) included the reduced sulphur compounds H$_2$S and DMS and found that conversion of reduced sulphur gases to SO$_2$ did not result in SO$_2$ concentrations as high as those observed over the oceans. He concluded that either a natural or "transport-source" of SO$_2$ is present or additional reduced sulphur compounds are involved in oceanic SO$_2$ production.

4.4. Mercaptans

Along with other volatile sulphur compounds methyl mercaptan (MeSH) is produced from methionine by various bacteria and fungi (Segal and Starkey, 1969). Blue-green algal mats incubated anaerobically rapidly produce large amounts of volatile sulphur compounds among which MeSH constitutes the major product, in contrast to previous results with marine eucaryotic algae (Zinder et al., 1977). Free MeSH has been observed in oxic sea water samples from Hiro Bay (Yamaoka and Tanimoto, 1976) and "non-polar" as well as "polar" mercaptans have been detected in anoxic waters (determined as sulphhydryl groups) by Adams and Richards (1968), predicted by Duursma (1965) to result from degradation of organic material under anoxic conditions.

4.5. Dimethyl disulphide, sulphoxides, and sulphones

Besides MeSH and DMS, dimethyl disulphide (DMDS) is an additional S-compound produced from the methiol group of methionine by both aerobic and anaerobic organisms (Kadota and Ishida, 1972). During their investigation of volatile sulphur compounds in algal mats, Zinder et al. (1977) showed in experiment that large amounts of DMDS were evolved when methionine was added. Yamaoka and Tanimoto (1976) found concentrations of DMDS ranging from 99 to 780 ppb with a mean of 452 ppb in ten samples of Hiro Bay seawater. Natural DMDS is a possible oxidation product of MeSH (Zinder et al., 1977) or may be derived from cystathionine via S-methylcysteinesulphoxide, which hydrolyzes to DMDS and methyl-methanethiosulphonate (Ostermaier and Tarbell, 1959, as quoted by Lewis, 1976). The organic disulphides are the most stable among the organic polysulphides in aqueous solution (Boulegue, 1978) and DMDS together with MeSH and DMS is more stable towards oxidation than sulphur and inorganic polysulphides (Boulegue, 1979).

As a photochemical oxidation product of DMS (Bentley et al., 1971) dimethyl sulphoxide may be detected in seawater which has been reported to inhibit photosynthesis in algae (Cheng et al., 1972). Methionine and biotin sulphoxides are oxidation products of methionine and biotin respectively, while microorganisms have been reported which reduce these compounds as well as DMDS (Zinder and Brock, 1978b). These authors also comment on the chemical stability of the sulphones. The methyl sulphones which have been identified in Baltic seal by Jensen and Jansson (1976) are, however, metabolically produced from PCB and DDE within the animal.

4.6. Carbon disulphide

Besides being a widely used industrial product (Meyer, 1977) and therefore probably delivered by air to the oceans, CS_2 can also emanate from the ocean itself. Lovelock (1974) found CS_2 to be present in every sampled coastal and open ocean locality with relatively high amounts in stagnant bays. Far higher concentration in anaerobic muds (2.95×10^{-11} g ml^{-1}) suggested the origin of CS_2 to be a product of the anaerobic conditions at the sea floor. CS_2 is stable for at least ten days in oxygen-containing seawaters and may therefore participate in the transfer of sulphur from the sea floor back to surface waters. CS_2 may also be a product of the decomposition of cysteine, as found in soils (Banwart and Bremner, 1974).

4.7. Sulphate esters, sulphonic acids, and sulpholipids

In soils nearly 50% of the total sulphur is found in the "hydroiodic acid reducible fraction", consisting of sulphate esters, which are probably in part polysaccharide sulphates (e.g. condroitin sulphate), keratin sulphates, choline sulphate, and arylsulphates (Zinder and Brock, 1978b). The heteropolysaccharide sulphates, which are also reported from marine species (e.g. Torres-Pombo et al., 1969; Abdel-Fattah et al., 1973; Batey and Turvey, 1975), serve mainly as structural materials reaccessible to the marine sulphur cycle as SO_4^{2-}, depending on the ability of microorganisms, plants, and mammals to produce enzymes (sulphhydrolases) to hydrolyze these esters (Fitzgerald, 1978).

Sulphonic acids are widely distributed among the algae. In addition to a sulphoglycolipid which appears to play a functional role in all plant chloroplast membranes (Benson, 1964), free aminosulphonic acids are present in certain macroscopic red, green, and brown algae. Examples of these compounds include taurine, its N-substituted derivatives, and D-cysteinolic acid plus its isomers and derivatives (c.f. Busby and Benson, 1973). Metabolism of cysteinolic acid, sulphopropanediol and the plant sulpholipid have been studied in the diatom *Navicula pelliculosa* during silicate starvation synchrony, sulphur deficiency and sulphonic acid turnover experiments (Busby and Benson, 1973). These authors suggest a possible metabolic pathway from sulphoquinovose precursors over cysteinolic acid to sulphopropanediol.

Sulpholipids of various chemical composition have been isolated from diatoms (e.g. Anderson et al., 1975). Mono- and polyester glycosyl sulphates or phosphate diglycerides account for a group of polar lipids which are found in large amounts in certain Fucaceae (Liem and Laur, 1976).

4.8. Other organosulphur compounds

In Scheuer (1973), investigations concerning sulphur-containing compounds in lipids of brown algae are reviewed which are biogenetically related to the dictyopterenes (Moore et al., 1972). The spectroscopically identified compounds belong to the classes of alkyl thioacetates, open-chain disulphides, and 1,2-dithiepin-5-ones as well as organic tri- and tetra-sulphides.

5. ORGANOSULPHUR IN SEDIMENTS

Two very thorough studies deal with organic sulphur compounds (OSC) in lipophilic extracts of Puget Sound sediments; for operational reasons, however, only OSC other than mercaptans and most disulphides could be determined with an efficiency of 80—90% (Bates and Carpenter, 1979a, b). Normalizing data to the organic carbon content, the authors found a baseline concentration of 10 mg g^{-1} organic carbon for total OSC and 0.5 μg g^{-1} for dibenzothiophene (Fig. 6). A flux rate of 51—140 ng cm^{-2} yr^{-1} to surface sediments was calculated. In contrast to total aliphatic hydrocarbons analyzed simultaneously, DBT was evenly distributed with depth thus suggesting

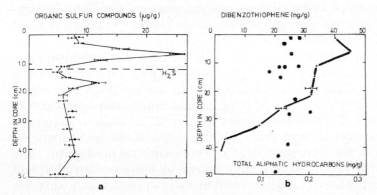

Fig. 6. Concentrations per gram dry weight of sediment of (a) total organosulphur compounds, and (b) of dibenzothiophene ($\bullet$) and total aliphatic hydrocarbons (o———o) in a Puget Sound sediment core. (After Bates and Carpenter, 1979b.)

that DBT is neither created nor destroyed within the sedimentary column (Fig. 6). Only one subsurface maximum was believed to be produced *in situ* because of the unusual spectrum of compounds and was attributed to microbial activity or redox reactions. No anthropogenic source, such as petroleum, terrestrial run-off, or sewage-plant effluent could be identified as a source for DBT. Thus natural inputs from forest and prairie fires via atmospheric transport were proposed to account for the uniform distribution found in sediments, soil, and the open sea. The relatively constant organic sulphur distribution with depth found in Delaware estuary sediments was according to Boulegue (1979) the result of a steady state of production and consumption or a consequence of the extremely low reactivity of slowly accumulating OSC. Out of the class of organic polysulphides two were detected by Ciereszko and Youngblood (1971) in marine sediments: *n*-hexadecyl and *n*-octodecyl disulphides.

Although organic sulphur accounts for less than 10% of the total sedimentary sulphur (Goldhaber, 1974), there is probably a far greater number of as yet undetected OSC, especially in combined or condensed form, which tend to accumulate in the humic fraction of the marine sediments (Nissenbaum and Kaplan, 1972). Effective methods of differentiating between classes of sedimentary organic sulphur are provided by the fractionation schemata of soil chemistry (see for a review Fitzgerald, 1978). Three fractions of organic sulphur can be distinguished (Williams, 1975): a fraction that is reduced by hydroiodic acid to H_2S consisting mainly of ester sulphates and sulphamates, a fraction containing the carbon-bonded sulphur, and finally a fraction which is characterized as Raney-nickel-reducible sulphur. The latter one is also subfraction of carbon-bonded sulphur representing mainly the cystine and methionine in the soil organic matter. In a long-term experiment, Freney et al. (1971) studied the fate of ^{35}S-sulphate added to soil: about 18—25% of the sulphur added became insoluble carbon-bonded sulphur (assumed to be amino acids and perhaps some humic material), 10—20% was incorporated as soluble carbon-bonded sulphur and most of the rest were sulphate esters.

This concept of sulphur fractionation in soils has been successfully applied to sewage sludge (Sommers et al., 1977) and to freshwater and marine derived peat-forming systems (Casagrande et al., 1977). In the marine peat, carbon-bonded sulphur accounted for 50% of total sulphur while ester sulphate constituted only ca. 25%. These authors noted an overall increase in sulphur in going from plant samples to peat in the marine environment, and concluded that plants were not the dominant sulphur-concentrating mechanism. The sulphur was probably delivered to a large extent by sulphate diffusion and microbial reduction, whereby carbon-bonded sulphur acted as a sink for sulphur in the peat.

6. ORGANOSULPHUR FROM PETROLEUM SOURCES

Significant amounts of organic sulphur compounds are not directly synthesized in the natural S-cycle and are introduced into the oceans by the discharge of petroleum and heavy fuel oils. Crude oils contain from 0.05 to 14% sulphur; the great majority of which have been 0.1 and 3% (Meyer, 1977; Gransch and Postuma, 1973).

S-containing petroleum compounds have been identified in marine molluscs (Lake and Hershner, 1977). In an investigation of seawater, sediment, and organisms near an oil spill after a collision, Warner (1975) found that the S-components of petroleum can be preferentially concentrated in the marine environment; e.g., in the isopod *Lygia* sp. which had crawled over

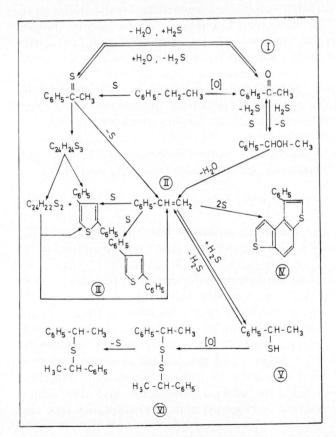

Fig. 7. Reaction scheme for the generation of S-compounds from ethylbenzene and sulphur: *I* = acetophenone; *II* = styrene; *III* = 2,4-diphenylthiophene; *IV* = 1-phenylbenzo (1,2-b, 4,3-b) dithiophene; *V* = 1-phenylethylthiol; *VI* = di(1-phenylethyl)disulphide. (After De Roo and Hodgson, 1978.)

an oil spill; the S-components were enriched by a factor of 5 over other non-sulphur petroleum components.

It is generally accepted that most sulphur compounds in petroleum are synthesized by secondary reactions with elemental sulphur as shown by Martin and Hodgson (1973, 1977) in laboratory experiments. Especially the high concentration of sulphur in crudes suggests it to be unlikely that petroleum-type S-compounds are formed from sedimentary organic sources: most sulphur is probably provided by sulphate reduction and secondary reactions of the products (H_2S and S) during early diagenesis (Gransch and Postuma, 1973).

In an attempt to simulate petroleum formation under mild conditions (T < 150°C) De Roo and Hodgson (1978) reacted ethylbenzene and sulphur in the presence of oxygen and water and identified a number of S-compounds, among which substituted thiophenes seemed to be the major products (Fig. 7). Baker (1973), however, found only three poorly defined organosulphur compounds on reaction of Beaufort Sea sediments with elemental sulphur, when studying whether hydrocarbons could be generated from the humic material present in the area.

A detailed analysis of individual sulphur compounds in oil is provided by Rall et al. (1972). In marine samples S-containing petroleum components such as benzothiophene, dibenzothiophene, naphthobenzothiophene, and their alkyl derivatives were obtained by Warner (1975) together with aromatic hydrocarbons. Alterations in the composition of S-compounds in crude oils exposed to the sea surface were analyzed by Utashiro and Hatsuo (1977).

7. ANTHROPOGENIC ORGANOSULPHUR COMPOUNDS

The main channels of organic sulphur pollution into the marine environment are aerosol transport of land-derived materials and precipitation, as well as terrestrial run-off.

In the list of gaseous sulphur-containing compounds of practical significance occurring in the atmosphere, Graedel (1977) included paraffinic mercaptans and dialkylsulphides which are partly transferred to the oceans by precipitation. In a review of organosulphur emissions from industrial sources, Bhatia (1978) mentions carbonyl sulphide, carbon disulphide, mercaptans of lower molecular weight, and thiophenes plus derivatives as being present in coal and oil gas streams. Organosulphur in synthesis gases appears mainly as carbonyl sulphide, carbon disulphide, and thiophenes, while petroleum distillates contain mercaptans and alkyl sulphides (see also Peyton et al., 1976).

Paper mills using Kraft's process, in which certain types of wood are pulped in digesters with a strong solution of sodium hydroxide, sodium sulphate and sodium sulphide, contribute MeSH, DMS, and DMDS (Sivelä and Sundman, 1975) to environmental pollution.

TABLE I

Annual global organic sulphur emissions (after Hitchcock, 1976)

Material	Average standing crop (g)	Reported DMS-S emissions (10^{-12} g gm^{-1} h^{-1})	Assumed mean S emissions (10^{-12} g gm^{-1} h^{-1})	Calculated global S emissions (10^{12} g S yr^{-1})
Marine algae	5×10^{15}	1060	1060	0.05
Intact leaves	1×10^{17}	1–23	12	0.011
Senescent leaves	5×10^{16}	"10 to 100 times greater"	1200	0.53
Soils	6×10^{18} to 2×10^{19}	11–45	28	1.5–4.9
			Total	2.1–5.5

Contamination of coastal waters results from the discharge of anionic surf-actants used in detergents such as alkyl-arylsulphonates and some alkyl sulphates. Of more localized significance, too, is terrestrial run-off of organo-sulphur containing fertilizers, fungicides, and insecticides (see Meyer, 1977; Nyborg, 1978).

It is difficult to quantify anthropogenic organosulphur contributions to the marine environment in relation to biogenic sources because no data are available at present. As far as this relationship is reflected in the atmosphere, the studies of Hitchcock (1975, 1976) give interesting hints considering, however, only total sulphur budgets. In contrast to 70×10^6 tons of pollutant sulphur per year, biogenic sulphur emission accounts for 100 to 230×10^6 tons per year (estimated from the literature) of which only 2 to 5×10^6 tons are contributed by the organic sulphur sources listed in Table I. In addition to these, on a station remote from anthropogenic sources (Oahu, Hawaii), Hitchcock found sulphate reduction to be a probable source for biogenic sulphur of marine origin, which contributes significant amounts of gaseous sulphur in the form of H_2S.

8. ACKNOWLEDGEMENTS

Numerous people have assisted me in search for literature and gave valuable comments while preparing this chapter. I particularly wish to thank Drs. D. Adams, J. Boulegue, and R. Carpenter, and Prof. Dr. W. Walter. The brush-up of the scientific English by Dr. R. Dawson and Barbara Heywood is gratefully acknowledged. I am indebted to Ursula Seiffert for drafting figures and her delightful presence.

9. REFERENCES

Abdel-Fattah, A.F., Hussein, M.M.-E. and Salem, H.M., 1973. Sargassan: A sulfated heteropolysaccharide from Sargassum linifolium. Phytochemistry, 12: 1995—1998.

Ackman, R.G. and Dale, J., 1965. Reactor for determination of dimethyl-β-propiothetin in tissue of marine origin by gas—liquid chromatography. J. Fish. Res. Board Can., 22: 875—883.

Ackman, R.G., Tocher, C.S. and McLachlan, J., 1966. Occurrence of dimethyl-β-propio-thetin in marine phytoplankton. J. Fish. Res. Board Can., 23: 357—364.

Adams, D.D. and Richards, F.A., 1968. Dissolved organic matter in an anoxic fjord, with special reference to the presence of mercaptans. Deep-Sea Res., 15: 471—481.

Anderson, R., Livermore, B.P., Volcani, B.E. and Kates, M., 1975. A novel sulfonolipid in diatoms. Biochim. Biophys. Acta, 409: 259—263.

Anderson, R., Kates, M. and Volcani, B.E., 1976. Sulfonium analogue of lecithin in diatoms. Nature, 263: 51—53.

Baker, B.L., 1973. Generation of alkane and aromatic hydrocarbons from humic materials in arctic marine sediments. In: B. Tissot and F. Bienner (Editors), Advances in Organic Geochemistry. Pergamon Press, London, pp. 137—153.

Banwart, W.L. and Bremner, J.M., 1974. Gas chromatographic identication of sulfur gases in soil atmospheres. Soil Biol. Biochem., 6: 113—115.

411

Bates, T.S. and Carpenter, R., 1979a. Determination of organosulfur compounds extracted from marine sediments. Anal. Chem., 51: 551—554.
Bates, T.S. and Carpenter, R., 1979b. Organo-sulfur compounds in sediments of the Puget Sound basin. Geochim. Cosmochim. Acta, 43: 1209—1221.
Batey, J.F. and Turvey, J.R., 1975. The galactan sulphate of the red alga *Polysiphonia lanosa*. Carbohydrate Res., 43: 133—143.
Benson, A.A., 1964. Plant membrane lipids. Annu. Rev. Plant Physiol., 15: 1—16.
Bentley, M.D., Douglas, I.B., Lacadie, J.A. and Whittier, D.R., 1971. The photolysis of dimethyl sulfide in air. Paper presented at the APCA meeting, Atlantic City, N.J., June 27, 1971.
Benzhitskiy, A.G., 1974. The content, distribution and ecological significance of the thiamine and biotin in seawater. Hydrobiol. J., 10: 91—96 (a translation of Gidrobiol. Zh.).
Bhatia, S.P., 1978. Organosulfur emissions from industrial sources. In: J.O. Nriagu (Editor), Sulfur in the Environment, 1. Wiley, New York, N.Y., pp. 51—84.
Bouchertall, F., 1979. Fluoreszierende Verbindungen im Seewasser (unpublished manuscript).
Boulegue, J., 1978. Metastable sulfur species and trace metals (Mn, Fe, Cu, Zn, Cd, Pb) in hot brines from the French Dogger. Am. J. Sci., 278: 1394—1411.
Boulegue, J., 1979. Evolution des composés soufrés dans la sédimentation récente (unpublished manuscript).
Boulegue, J. and Popoff, G., 1979. Nouvelles méthodes de détermination des principales espèces ioniques du soufre dans les eaux naturelles. J. Fr. Hydrol., in press.
Busby, W.F. and Benson, A.A., 1973. Sulfonic acid metabolism in the diatom *Navicula pelliculosa*. Plant Cell Physiol., 14: 1123—1132.
Casagrande, D.J., Siefert, K., Berschinski, C. and Sutton, N., 1977. Sulfur in peat-forming systems of the Okefenokee Swamp and Florida Everglades: origins of sulfur in coal. Geochim. Cosmochim. Acta, 41: 161—167.
Challenger, F., 1951. Biological Methylation. Adv. Enzymol., 12: 429—451.
Challenger, F., 1959. Aspects of the Organic Chemistry of Sulfur. Butterworth, London, 253 pp.
Challenger, F. and Simpson, M.I., 1948. Studies on biological methylation. Part XII. A precursor of the dimethylsulfide evolved by *Polysiphonia fastigiata*. Dimethyl-2-carboxyethyl sulphonium hydroxide and its salts. J. Chem. Soc., 1948: 1591—1597.
Cheng, K.H., Grodzinski, B. and Colman, B., 1972. Inhibition of photosynthesis in algae by dimethyl sulfoxide. J. Phycol., 8: 399—400.
Ciereszko, L.S. and Youngblood, W.W., 1971. n-Hexadecyl and n-octadecyl disulfides in "sediment" derived from the Gorgonian *Pseudoplexaura porosa*. Geochim. Cosmochim. Acta, 35: 851—853.
Craigie, J.S., McLachlan, J., Ackman, R.G. and Tocher, C.S., 1967. Photosynthesis in algae. III. Distribution of soluble carbohydrates and dimethyl-β-propiothetin in marine unicellular chlorophyceae and prasinophyceae. Can. J. Bot., 45: 1327—1334.
Dawson, R. and Gocke, K., 1978. Heterotrophic activity in comparison to the free amino acid concentrations in Baltic sea water samples. Oceanol. Acta, 1: 45—54.
Dawson, R. and Pritchard, R.G., 1978. The determination of α-amino acids in seawater using a fluorimetric analyser. Mar. Chem., 6: 27—40.
Degens, E.T., Reuter, J.H. and Shaw, K.N.F., 1964. Biochemical compounds in offshore California sediments and seawater. Geochim. Cosmochim. Acta, 28: 45—66.
De Groot, G., Greve, P.A. and Maes, R.A.A., 1975. A microcoulometric method for the determination of nanogram amounts of sulfur in organic compounds. Anal. Chim. Acta, 79: 279—284.
De Roo, J. and Hodgson, G.W., 1978. Geochemical origin of organic sulfur compounds: thiophene derivatives from ethylbenzene and sulfur. Chem. Geol., 22: 71—78.

Desai, I.D., Laub, D. and Antia, N.J., 1972. Comparative characterization of L-threonine dehydratase in seven species of unicellular marine algae. Phytochemistry, 11: 277—287.

Duursma, E.K., 1965. The dissolved organic constituents of seawater. In: J.P. Riley and G. Skirrow (Editors), Chemical Oceanography, 1. Academic Press, London, pp. 433—475.

Fitzgerald, J.W., 1978. Naturally occurring organosulfur compounds in soils. In: J.O. Nriagu (Editor), Sulfur in the Environment, II. Wiley, New York, N.Y., pp. 391—444.

Freney, J.R., Melville, G.E. and Williams, C.H., 1971. Organic sulphur fractions labelled by addition of ^{35}S-sulphate to soil. Soil Biol. Biochem., 3: 133—141.

Gardner, L.R., 1974. Organic versus inorganic metal complexes in sulfidic marine waters — some speculative calculations based on available stability constants. Geochim. Cosmochim. Acta, 38: 1297—1302.

Garrasi, C., Degens, E.T. and Mopper, K., 1979. The free amino acid composition of seawater obtained without desalting and preconcentration. Mar. Chem., 8: 71—85.

Goldhaber, M.B., 1974. Equilibrium and Dynamic Aspects of the Marine Geochemistry of Sulfur. Thesis, University of California, Los Angeles, Calif., 399 pp.

Goldhaber, M.B. and Kaplan, I.R., 1975. Controls and consequences of sulfate reduction rates in recent marine sediments. Soil Sci., 119: 42—55.

Graedel, T.E., 1977. The homogeneous chemistry of atmospheric sulfur. Rev. Geophys. Space Phys., 15: 421.

Graedel, T.E., 1979. Reduced sulfur emission from the open oceans. Geophys. Res. Lett., 6: 329—331.

Granroth, B. and Hattula, T., 1976. Formation of dimethyl sulfide by brackish water algae and its possible implication for the flavor of Baltic herring. Finn. Chem. Lett., 6: 148—150.

Gransch, J.A. and Postuma, J., 1973. On the origin of sulfur in crudes. Advances in organic geochemistry, Actes du 6e Congrès international de Géochimie organique, 18—21 Septembre 1973, Rueil—Malmaison, France, pp. 727—738.

Hattula, T. and Granroth, B., 1974. Formation of dimethyl sulfide from S-methyl-methionine in onion seedlings (*Allium cepa*). J. Sci. Food Agric., 25: 1517—1521.

Hitchock, D.R., 1975. Biogenic contributions to atmospheric sulfate levels. Paper presented at 2nd Annual Conference on Water Reuse. Chicago, Ill., May 1975.

Hitchcock, D.R., 1976. Microbiological contributions to the atmospheric load of particulate sulfate. In: J.O. Nriagu (Editor), Environmental Biogeochemistry, 1. Ann Arbor Science, Ann Arbor, Mich., pp. 351—367.

Jensen, S. and Jansson, B., 1976. Methyl sulfone metabolites of PCB and DDE. Ambio, 5: 257—260.

Jørgensen, B.B., 1977. The sulfur cycle of a coastal marine sediment (Limfjorden, Denmark). Limnol. Oceanogr., 22: 814—832.

Kadota, H. and Ishida, Y., 1972. Production of volatile sulfur compounds by micro-organisms. Annu. Rev. Microbiol., 26: 127—138.

Kaplan, I.R., Emery, K.O. and Rittenberg, S.C., 1963. The distribution and isotopic abundance of sulfur in recent marine sediments off Southern California. Geochim. Cosmochim. Acta, 27: 297—331.

Kawahara, F.K., 1971. Gas chromatographic analysis of mercaptans, phenols and organic acids in surface waters with use of pentafluorobenzyl derivatives. Environ. Sci. Technol., 5: 235—239.

Kellogg, W.W., Cadle, R.D., Allen, E.R., Lazrus, A.L. and Martell, E.A., 1972. The sulfur cycle. Science, 175: 587—596.

Lake, J.L. and Hershner, C., 1977. Petroleum sulfur-containing compounds and aromatic hydrocarbons in the marine mollusks *Modiolus demissus* and *Crassostrea virginica*. In:

1977 Oil Spill Conference. American Petroleum Institute, Washington, D.C., 4284: p. 640.

Lewis, B.A.G., 1976. Selenium in biological systems, and pathways for its volatilization in higher plants. In: J.O. Nriagu (Editor), Environmental Biogeochemistry, 1. Ann Arbor Science, Ann Arbor, Mich., pp. 389—409.

Liem, P.Q. and Laur, M.-H., 1976. Structures, teneurs et compositions des esters sulfuriques, sulfoniques, phosphoriques des glycosyldiglycérides de trois fucacées. Biochimie, 58: 1367—1380.

Liss, P.S. and Slater, P.G., 1974. Flux of gases across the air—sea interface. Nature, 247: 181—184.

Lovelock, J.E., 1974. CS_2 and the natural sulfur cycle. Nature, 248: 625—626.

Lovelock, J.E., Maggs, R.J. and Rasmussen, R.A., 1972. Atmospheric dimethyl sulfide and the natural sulfur cycle. Nature, 237: 452—453.

Martin, T.H. and Hodgson, G.W., 1973. Geochemical origin of organic sulfur compounds: reaction of phenylalanine with elemental sulfur. Chem. Geol., 12: 189—208.

Martin, T.H. and Hodgson, G.W., 1977. Geochemical origin of organic sulfur compounds: precursor products in the reactions of phenylalanine and benzylamine with elemental sulfur. Chem. Geol., 20: 9—25.

Maruyama, M. and Kakemoto, M., 1978. Behaviour of organic sulfur compounds in flame photometric detectors. J. Chromatogr. Sci., 16: 1—7.

Meyer, B., 1977. Sulfur, Energy and the Environment. Elsevier, Amsterdam, 448 pp.

Middleton, A.C. and Lawrence, A.W., 1977. Kinetics of microbial sulfate reduction. J. Water Pollut. Control Fed., 49: 1659—1670.

Moore, R.E., Mistysyn, J. and Pettus jr., J.A., 1972. (-)-Bis-(3-acetoxyundec-5-enyl) disulphide and S-(-)-3-acetoxyundec-5-enyl thioacetate, possible precursors to undeca-1,3,5-trienes in *Dictyopteris*. J. Chem. Soc. Chem. Comm., 1972: 326—327.

Nedwell, D.B. and Floodgate, G.D., 1972. Temperature-induced changes in the formation of sulphide in a marine sediment. Mar. Biol., 14: 18—24.

Nguyen, B.C., Gaudry, A., Bonsang, B. and Lambert, G., 1978. Reevaluation of the role of dimethyl sulphide in the sulfur budget. Nature, 275: 637—639.

Nissenbaum, A. and Kaplan, I.R., 1972. Chemical and isotopic evidence for the in situ origin of marine humic substances. Limnol. Oceanogr., 17: 570—582.

Noda, H., 1975. Studies on the flavor substances of Nori, the dried laver *Porphyra tenera* — I: Dimethylsulfide and dimethyl-β-propiothetin. Nikon suisan-gakkai shi, 41: 481—486 (as quoted in Oceanic Abstr., 13(1): 1976).

Nriagu, J.O. (Editor), 1976. Environmental Biogeochemistry, 1. Ann Arbor Science, Ann Arbor, Mich., 423 pp.

Nriagu, J.O. (Editor), 1978. Sulfur in the Environment, 1 and 2. Wiley, New York, N.Y.

Nyborg, M., 1978. Sulfur pollution and soils. In: J.O. Nriagu (Editor), Sulfur in the Environment, 2. Wiley, New York, N.Y., pp. 359—390.

Orr, W., 1974. Biogeochemistry of sulfur, II/4. Springer, New York, N.Y.

Ostermayer, F. and Tarbell, D., 1959. Products of acidic hydrolysis of S-methyl-L-cysteine sulfoxide; the isolation of methylmethane-thiosulfonate and mechanism of hydrolysis. J. Am. Chem. Soc., 82: 3752—3755.

Peyton, T.O., Steele, R.V. and Mabey, W.R., 1976. Carbon Disulfide, Carbonyl Sulfide: Literature Review and Environmental Assessment. Stanford Res. Inst., Menlo Park, Calif., U.S. NTIS., PB Rep., PB-257947, 64 pp.

Pocklington, R., 1971. Free amino acids dissolved in North Atlantic Ocean waters. Nature, 230: 374—375.

Postgate, J.R., 1968. The sulfur cycle. In: G. Nickless (Editor), Inorganic Sulfur Chemistry. Elsevier, New York, N.Y.

Rall, H.T., Thompson, C.J., Coleman, H.J. and Hopkins, R.L., 1972. Sulfur compounds in crude oil. U.S. Dep. Interior, Bur. Mines Bull., 659.

414

Riley, J.P. and Segar, D.A., 1970. The seasonal variation of the free and combined dissolved amino acids in the Irish Sea. J. Mar. Biol. Assoc. U.K., 50: 713—720.

Roy, A.B. and Trudinger, P.A., 1970. The Biochemistry of Inorganic Compounds of Sulfur. Cambridge University Press, London, 400 pp.

Scheuer, P.J., 1973. Chemistry of Marine Natural Products. Academic Press, New York, N.Y., 201 pp.

Segal, W. and Starkey, R.L., 1969. Microbial decomposition of methionine and identity of the resulting sulfur products. J. Bacteriol., 98: 908—913.

Sivelä, S. and Sundman, V., 1975. Demonstration of *Thiobacillus*-type bacteria, which utilize methyl sulfides. Arch. Microbiol., 103: 303—304.

Sommers, L.E., Tabatabai, M.A. and Nelson, D.W., 1977. Forms of sulphur in sewage sludge. J. Environ. Qual., 6: 42—46.

Stuart, J.D., Hill, D.W., Wilson, T.D., Walters, F.H. and Feng, S.Y., 1979. Liquid chromatographic separation and fluorescence measurement of taurine — a key amino acid. Abstracts of the Pittsburgh Conference on Analytical Chemistry and Applied Spectroscopy, March 5—9, 1979, Cleveland Convention Center, Cleveland, Ohio, p. 379.

Tanner, R.L., Forrest, J. and Newman, L., 1978. Determination of atmospheric gaseous and particulate sulfur compounds. In: J.O. Nriagu (Editor), Sulfur in the Environment, 1. Wiley, New York, N.Y., pp. 371—452.

Tokunaga, T., 1977. Formation of dimethyl sulfide in Antarctic krill, *Euphausia superba*. Nihon suisan-gakkai shi, 43: 1209—1217.

Torres-Pombo, J., Seoane-Camba, J. and Ribas, I., 1969. A sulfuric ester of choline in *Gelidium sesquipedale*. In: Margaleff (Editor), Proceedings of the Sixth International Seaweed Symposium. Subsecretaria de la marine mercante direccion general de pesca maritima, Madrid, Spain, pp. 611—614.

Utashiro, S. and Hatsuo, H., 1977. Studies on the sulfur compounds in crude oils and heavy fuel oils exposed on the sea surface by the FPD gas chromatography. Kaijo Hoan Daigakko Kenkyu Hokoku, Dai-2-Bu, 23(1-2, Sekt. 2): 37—54.

Volodin, B.B., 1975. Participation of sulfur in processes of multiplication of certain blue-green algae. Sov. Plant Physiol., 22: 255—258.

Warner, J.S., 1975. Determination of sulfur-containing petroleum components in marine samples. Proc. Conf. on Prevention and Control of Oil Pollution Washington, D.C., pp. 97—101.

Williams, C.H., 1975. The chemical nature of sulphur compounds in soils. In: K.D. McLachlan (Editor), Sulfur in Australia Agriculture, pp. 21—30.

Yamaoka, Y. and Tanimoto, T., 1976. On the distribution of dimethyl sulfide and dimethyl disulfide in sea water in Hiro Bay. J. Agric. Chem. Soc. Jap., 50: 147—148.

Young, L. and Maw, G.A., 1958. The Metabolism of Sulfur Compounds. Wiley, New York, N.Y., 180 pp.

Zinder, S.H. and Brock, T.D., 1978a. Production of methane and carbon dioxide from methane thiol and dimethyl sulfide by anaerobic lake sediments. Nature, 273: 226—228.

Zinder, S.H. and Brock, T.D., 1978b. Microbial transformations of sulfur in the environment. In: J.O. Nriagu (Editor), Sulfur in the Environment, 2. Wiley, New York, N.Y., pp. 445—466.

Zinder, S.H., Doemel, W.N. and Brock, T.D., 1977. Production of volatile sulfur compounds during the decomposition of algal mats. Appl. Environ. Microbiol., 34: 859—860.

Zogell, C.E., 1963. Organic geochemistry of sulfur. In: I.A. Breger (Editor), Organic Chemistry. Pergamon Press, Oxford, pp. 543—578.

Chapter 14

THE MEASUREMENT OF ORGANIC CARBON IN SEA WATER [1]

M.D. MACKINNON

1. INTRODUCTION

Organic matter is a minor but important component in sea water. The effects of organic matter on the biological (productivity, heterotrophy), chemical (chelation, photochemistry, adsorption), geological (sedimentation, remobilization), and physical (wave damping, non-conservative distributions) processes in the ocean have been reported (Wangersky, 1965, 1972, 1976a, 1978; Hood, 1970; Riley, 1971; Parsons, 1975; Williams, 1975; Andersen, 1977). In most oceanic areas, the concentration of total organic matter is low (0.5—3.0 ppm) and individual components may be present at the parts per billion (ppb) level (Williams, 1975). Only about 10—20% of the total organic matter in sea water has been characterized into specific components (Ehrhardt, 1977). Many of these analyses are based on class reactions which are often subject to interferences and difficulties in interpretation (Wangersky and Zika, 1978). The molecular weight distribution of the organic components reported in sea water covers a wide range (Sharp, 1973a; Ogura, 1974; Brown, 1975; Baturina et al., 1975; Smith, 1976; Starikova et al., 1976). Analytical methods for many of the natural organic components in sea water have been reviewed (Vallentyne, 1957; Wagner, 1969; Strickland and Parsons, 1972; Williams, 1975; Grasshoff, 1976; Gagosian and Stuermer, 1977; Wangersky and Zika, 1978). Much effort has been directed to the determination of compounds of environmental concern (Duursma and Marchand, 1974; Goldberg, 1976), and a protocol has been issued by the U.S. Environmental Protection Agency (1977). With improvement of the analytical procedures for concentration (purging, extractions, adsorbents) and analysis (ultraviolet, fluorescence, infrared spectroscopy; thin-layer, gas, or liquid chromatography; chromatographic methods interfaced directly with mass spectrometry) of organic compounds in water, the level of detection has been lowered. The need for and justification of the effort required for complete characterization of natural organic matter has been questioned (Blumer, 1975). Most such highly specific methods are not suited to routine analysis because of time, expertise, and equipment required. Therefore,

---

[1] NRCC No. 16920.

416

much of the work done on the distribution, cycling, and role of the organic matter in sea water has been based on non-specific methods.

These non-specific methods would appear to be relatively simple but analytical problems (accuracy, precision, sampling, and storage) have been encountered and conflicting results of the organic matter concentrations in natural waters have been reported. If the sources of these discrepancies in results from different analytical methods used to measure the various fractions of organic matter in sea water could be understood, then a comparison of past and future studies may be possible.

Total organic matter is classed into broad arbitrary divisions based on physical, operational, or working definitions. With most analytical methods, analysis of organic matter is based on the measurement of the oxidation (CO_2) or reduction (CH_4) products. The concentration of the organic matter is reported in terms of organic carbon which makes up about 40—50% of most natural organic compounds. The total organic carbon (TOC) has been categorized into dissolved (DOC), particulate (POC), and volatile (VOC) organic carbon fractions (Fig. 1).

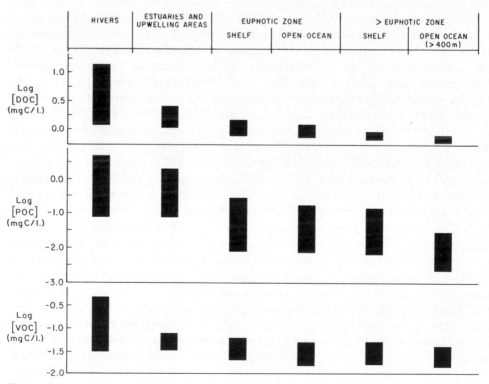

Fig. 1. Range of concentrations of dissolved, particulate and volatile fractions of the total organic carbon in various marine environments.

The results from analyses of these various fractions of the TOC have been used in the examination of the cycle (source, pathway, fate), role, and distribution of the organic matter in sea water. The use of organic carbon analyses for understanding the role of organic matter in sea-water systems has been discussed in general reviews by Duursma (1961, 1965), Provasoli (1963), Wangersky (1965, 1972, 1978), Wagner (1969), Riley (1970), Hood (1970, 1971), Riley (1971), Faust and Hunter (1971), Woodwell and Pecan (1973), Menzel (1974), Williams (1975), and in the Symposium on Concepts in Marine Organic Chemistry (Andersen, 1977). The measurement of TOC concentrations can be used as a general parameter to predict the quality of waters (Malcolm and Leenheer, 1973) and to identify areas where organic pollution may be increasing (Carlberg, 1976).

The principal source of organic matter in sea water is *in situ* biological activity (photosynthesis, secretion, excretion, decomposition) (Whittle, 1977, Wangersky, 1978) with external input coming from land masses (fluvial, atmospheric) (Handa, 1977; Hunter and Liss, 1977) and the activities of man (agricultural, industrial, and urban wastes) (Duce and Duursma, 1977). The distribution of organic matter is affected by these sources, and can be generalized (Menzel and Ryther, 1970; Starikova, 1970; Williams, 1971; Menzel, 1974; Wangersky, 1976b; Morris and Eglington, 1977; MacKinnon, 1978) (Fig. 1). The highest concentrations of organic carbon (DOC, POC, and VOC) will usually be found in areas of high productivity or in coastal areas under the influence of terrestrial input. Average concentrations of organic matter decrease as distance from coastal input and as depth of the water increase (Table I and Fig. 2). At depths greater than 500 m, organic carbon values are low and relatively constant unless the otherwise homogeneous system is affected by microstructure (e.g. density layers) (Wangersky, 1974). Since little geographic or seasonal variation in the concentrations of organic matter are observed, the organic matter in deep water has been considered to be refractory or resistant to biological or chemical degradation (Menzel and Ryther, 1970; Menzel, 1974; Bada and Lee, 1977). To explain the observed distribution of TOC values in natural waters, a steady-state relationship between the rate of production of the organic matter and its rate of utilization or remineralization has been postulated (Wangersky, 1972). Organic material, produced in or added to the water column, may be lost to the sediments but no increase in organic carbon content in the sediments has been noted (Eadie and Jeffrey, 1973). In deep water, the age of dissolved organic matter has been estimated at 1000—3000 years (Williams et al., 1969; Skopintsev, 1972a) which has been used to support arguments that organic matter in deep water (>500 m) is either refractory (Menzel, 1974) or slowly utilized (Wangersky, 1978). It has been estimated that only about 0.3—1% of the annual production of organic carbon is added to the organic pool each year. To explain steady-state relationship of organic matter in the oceans, mechanisms of removal including heterotrophic utilization (Williams,

TABLE I

Variability of the organic carbon concentrations in an area of the N.W. Atlantic

Station (see Fig. 2)	Depth zone (m)	Averaged POC concentration (μg C l^{-1}) (Gordon, 1977) [1]	Averaged TOC concentration (mg C l^{-1}) (MacKinnon, 1977)	Averaged VOC concentration (μg C l^{-1}) (MacKinnon, 1977)
A				
(a) Pre-bloom	0—40	40.74 [2]	1.22 ± 0.14	29.92 ± 6.4
(b) Spring bloom	0—40	217.33 [2]	1.67 ± 0.31	33.01 ± 6.2
B	0—70	41.78	1.08 ± 0.13	41.15 ± 16.0
C	0—100	30.28	1.04 ± 0.13	31.88 ± 9.8
	>100	5.19	0.74 ± 0.05	28.70 ± 4.8
D	0—100	16.59	0.90 ± 0.05	28.39 ± 5.3
	>100	4.63	0.79 ± 0.10	25.99 ± 3.3
E	0—100	17.00	0.99 ± 0.14	27.81 ± 7.2
	>100	4.51	0.73 ± 0.09	24.80 ± 4.5
F	0—100	14.38	0.97 ± 0.05	37.83 ± 12.8
	>100	3.46	0.75 ± 0.06	23.19 ± 2.4

[1] Geometric mean.
[2] MacKinnon (1977).

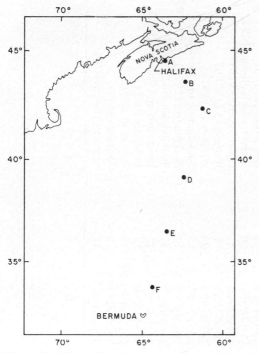

Fig. 2. Position of stations described in Table I.

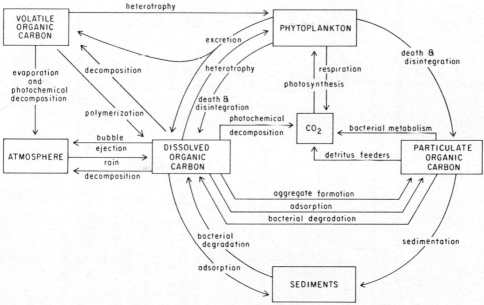

Fig. 3. Cycle of organic matter in the marine system. This is based on that proposed by Wangersky (1972).

TABLE II

Representative values of particulate organic carbon in the surface mixed layer of different areas of the ocean (Pocklington and MacKinnon, in press)

Location (period)	Depth range (m)	Number of samples	POC (μg C l^{-1}) min.–mean[1]–max.	Coefficient of variation (%)	Reference
Labrador Sea (Feb.)	20	28	4–6–8	22	Wangersky and Hincks (1978)
North of Oahu, Pacific (all)	0–100	30	8–17–62	34	Gordon (1971)
NW Atlantic (all)	0–100	100	5–19–145	23	Gordon (1977)
Nova Scotian Shelf (all except fall)	0–50	40	17–29–124	16	Pocklington (1973)
Pacific, 150°W (Apr.–June)	0–100	116	13–30–88	14	Wangersky (1976b)
West Mediterranean	0–90	18	18–36–57	26	Banoub and Williams (1972)
Baffin Bay (Aug.–Sept.)	0–30	11	27–39–59	31	Pocklington (1978)
S. Atlantic (Dec.–Jan.)	0–100	125	31–54–134	9	Wangersky (1976b)
Guinea Dome (Sept.)	0–60	80	30–60–90	25	Minas et al. (1978)
Off Senegal and The Gambia (Feb.–Mar.)	0–50	87	31–68–231	9	Pocklington and MacKinnon (in press)
Gulf of St. Lawrence (all except winter)	0–30	260	16–101–942	17	Pocklington (1973)
Off Peru (Nov.)	0–100	30	15–107–810	27	Pocklington et al. (1978)
Off NW Africa (Feb.–May)	0–50	245	23–133–390	14	Huntsman and Barber (1977)
Off SW Africa (Apr.–May)	0–50	18	68–152–864	17	Hobson (1971)
Off South California (all)	0–60	433	50–168–1000	15	Eppley et al. (1977)
Baltic Sea (Sept.)	0–30	22	68–180–289	6	Zsolnay (1973)
Kaneohe Bay, Oahu (spring through fall)	0–10	244	47–215–579	12	Caperon et al. (1976)

[1] Geometric mean.

1971; Ogura, 1972, Khaylov and Ayzatullin, 1972; Skopintsev, 1972b; Wangersky, 1978), chemical processes (Zika, 1977; Zafiriou, 1977), and physical processes such as particle formation (Johnson, 1976; Sholkovitz, 1976; Wangersky, 1977; Cauwet, 1978) and evaporization (Liss, 1975; Mac-Kinnon, 1977) have been postulated. This cycle for production and utilization of organic matter in sea water systems is summarized in Fig. 3.

The distributions of various fractions of the TOC in many marine areas have been reported. In Table II, a summary of some of the reported particulate organic carbon (POC) results from various oceanic environments is presented. A survey of some of the geographic areas where analyses of dissolved (DOC) or total (TOC) concentrations have been reported is shown in Fig. 4. The depth and geographic relationships of TOC results which I have analyzed are summarized in Table III. In Fig. 5, depth profiles of averaged

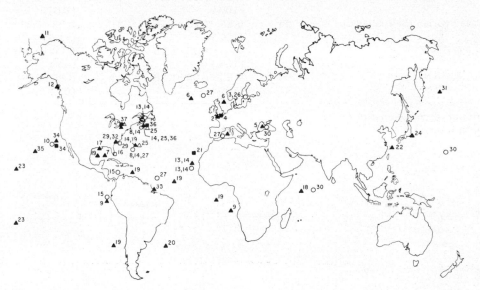

Fig. 4. Locations where DOC concentrations have been measured using: ▲ = wet oxidation; ○ = dry oxidation; ■ = photo-oxidation; △ = direct injection; □ = UV absorbance.
References: 1 = Banoub and Williams (1972); 2 = Banoub and Williams (1973); 3 = Carlberg (1976); 4 = Collins and Williams (1977); 5 = Deuser (1971); 6 = Duursma (1961); 7 = Fredericks and Sackett (1970); 8 = Gordon and Sutcliffe (1973) 9 = Hobson et al. (1973); 10 = Holm-Hansen et al. (1966); 11 = Kinney et al. (1971); 12 = Loder and Hood (1972); 13 = MacKinnon (1977); 14 = MacKinnon (1978); 15 = MacKinnon (this paper); 16 = Mattson et al. (1974); 17 = Maurer and Parker (1972); 18 = Menzel (1964); 19 = Menzel (1970); 20 = Menzel and Ryther (1968); 21 = Moore (1977); 22 = Ogura (1970a); 23 = Ogura (1970b); 24 = Ogura (1974); 25 = Sharp (1973b); 26 = Skopintsev et al. (1962); 27 = Skopintsev et al. (1966); 28 = Skopintsev et al. (1968); 29 = Smith (1976); 30 = Starikova (1970); 31 = Starikova and Yablokova (1974); 32 = Wheeler (1977); 33 = Williams (1968); 34 = P.J. Williams (1969); 35 = Williams (1971); 36 = Gershey et al. (1979); 37 = Goulden and Brooksbank (1975).

TABLE III

TOC measured with the dry oxidation method of MacKinnon (1978) in various ocean areas

TOC concentrations (mg C l^{-1}): minimum—mean—maximum (coefficient of variation)

Area	n [1]	0—10 m [2]	n	10—50 m
1. Coastal				
a. pre-bloom	23	1.15—1.32—1.57 (8%)	12	0.92—1.14—1.36 (10%)
b. spring bloom	31	1.41—1.84—2.40 (14%)	6	1.13—1.30—1.55 (12%)
2. Peru, upwelling area 11/77	15	0.86—1.75—3.95 (46%)	22	0.67—1.02—2.80 (44%)
3. Gulf of St. Lawrence	38	1.04—1.40—2.28 (20%)	15	1.05—1.21—1.42 (9%)
4. Senegal, upwelling area 2/76	13	1.22—1.40—1.67 (10%)	19	0.85—1.18—1.45 (14%)
5. Nova Scotian shelf and slope	24	0.85—1.25—1.45 (10%)	31	0.93—1.11—1.38 (11%)
6. Sargasso Sea	33	0.83—1.02—1.18 (10%)	37	0.80—0.99—1.21 (9%)
7. Caribbean Sea	6	0.92—1.03—1.18 (10%)	10	0.86—0.96—1.04 (6%)

[1] Number of determinations.
[2] Depth interval.

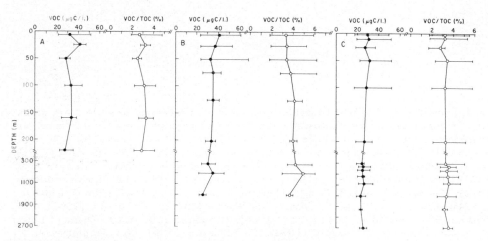

Fig. 5. Depth-averaged profiles of volatile organic carbon (VOC) concentrations and its fraction of total organic carbon (VOC/TOC) measured in various geographic areas. Bars represent range of values measured; ● = VOC; ○ = VOC/TOC. A. Gulf of St. Lawrence. B. Nova Scotian shelf and slope. C. Sargasso Sea.

n	50—100 m	n	100—200 m	n	200—500 m	n	>500 m
11	0.57—0.77—0.92 (14%)						
17	0.96—1.11—1.35 (10%)	21	0.79—1.01—1.28 (12%)	10	0.76—0.89—1.03 (11%)		
10	0.74—0.94—1.22 (15%)	11	0.76—0.84—0.97 (8%)	27	0.65—0.81—0.93 (7%)	26	0.66—0.77—0.89 (6%)
14	0.79—0.93—1.09 (10%)	12	0.69—0.83—0.96 (10%)	15	0.60—0.74—0.87 (8%)	21	0.68—0.74—0.84 (6%)
23	0.69—0.90—1.12 (11%)	32	0.67— 0.83—1.07 (11%)	33	0.62—0.76—0.96 (11%)	88	0.58—0.71—0.89 (8%)
9	0.76—0.93—1.03 (10%)	11	0.66—0.80—0.91 (10%)	11	0.56—0.67—0.84 (12%)	4	0.55—0.60—0.66 (8%)

volatile organic carbon (VOC) concentrations from various marine areas are presented (MacKinnon, 1977). However, there are discrepancies in reported results of TOC determinations from similar areas which can not be correlated with, or explained completely by, variations in the time of sampling and the movement of water masses (Fig. 6). While precision of each of the analytical methods is usually high, accuracy is difficult to assess. Until interlaboratory and method intercomparisons using certified reference materials are routinely undertaken, the source of these discrepancies in reported values will remain uncertain and the results of measurements made by different workers can not be compared with sufficient confidence.

2. MEASUREMENT OF ORGANIC MATTER IN SEA WATER

While total organic matter is defined as arbitrary physical fractions (dissolved, particulate, and volatile), the recommended methods for their measurement are similar. An outline of the methods used for analysis of organic carbon in sea-water samples is shown in Fig. 7. Most of these methods are based on oxidation of organic matter to CO_2 which is determined by volumetric, gravimetric, conductometric, titrimetric, or coulometric methods

424

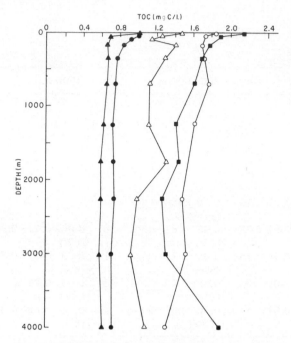

Fig. 6. Indirect comparison of depth-averaged TOC concentrations measured in the north-western Atlantic with different oxidation methods (MacKinnon, 1978). ▲, Menzel (1970), persulfate oxidation; ●, MacKinnon (1978), dry oxidation; △, Sharp (1973b), direct injection; ○, Skopintsev et al. (1966), dry oxidation; ■, Gordon and Sutcliffe (1973), dry oxidation.

(Duursma, 1961), or by non-dispersive infrared spectrometry (Menzel and Vaccaro, 1964), gas chromatography (West, 1964), mass spectrometry (Games and Hayes, 1976) and plasma-emission spectrometry (Mitchell et al., 1977). The concentration of organic matter is expressed in terms of organic carbon (mg Cl^{-1} or ppm C). Oxidation of organic matter would appear to be straightforward, but in sea water the accuracy of the analytical methods has been dependent upon some of the following considerations: (1) low concentration of organic matter (DOC = 0.5—2.0 ppm; POC = 0.005—1.0 ppm; VOC = 0.01—0.1 ppm); (2) high concentration of inorganic carbon (20—30 ppm) as carbonates; (3) contamination during sampling, storage, preparation, and analysis; (4) difficulty in accurately determining reagent or method blanks; (5) incompleteness of oxidation; (6) lack of certified standard or well-characterized reference materials for interlaboratory of method comparisons; (7) length of time required for analysis.

The influence of these factors on accuracy and precision of the measurement of organic carbon will depend on the specific oxidation method, the sampling, storage and preparation procedures, and the type of detection sys-

Scheme for Analysis of Organic Carbon in Sea Water

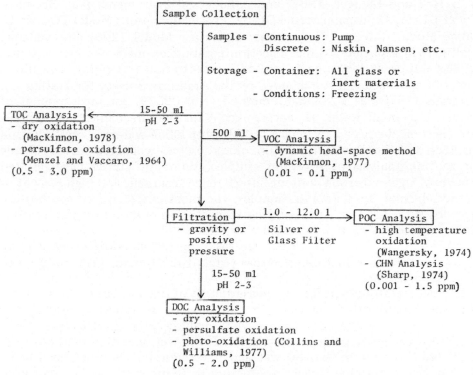

Fig. 7. Outline of the sampling, handling, and analytical procedures recommended for analysis of the fractions (DOC, POC, VOC) of total organic carbon (TOC) in sea-water samples.

tem. For analysis of organic matter in sea water, the oxidation methods can be classed broadly into either wet oxidation or dry combustion methods. In both approaches, organic matter is oxidized and then measured as CO_2 or as CH_4 after a subsequent reduction step. In wet oxidation procedures, oxidation is performed in the aqueous sample by addition of a chemical oxidant such as potassium permanganate ($KMnO_4$), potassium dichromate ($K_2Cr_2O_7$) (Kay, 1954), silver-catalyzed dichromate (Duursma, 1961; Starikova and Yablokova, 1974; Ljutsarev et al., 1975), potassium persulfate ($K_2S_2O_8$) (Wilson, 1961; Menzel and Vaccaro, 1964), or silver-catalyzed persulfate (Goulden and Brooksbank, 1975; Skopintsev, 1976), or by photo-oxidation with ultraviolet irradiation (Armstrong and Tibbitts, 1968; Ehrhardt, 1969; Collins and Williams, 1977). In dry oxidation methods, the sample is dried and the resulting residue is oxidized. In freshwater systems of high organic carbon concentrations (>2 ppm), reliable methods using direct injection of the sample into a high-temperature oxidation system have been developed (Van Hall and Stenger, 1967) and commerical instruments (e.g. Beckman Model 915 A, Dohrmann Model DC 52, Phase Separations Model TOC sin II, Ionics Model 445, and C. Matheson Scientific Model 1200) are available. Croll (1972) reported a direct injection combustion method which was useful for waters of low organic carbon concentration (<1 ppm). The direct injection approach is recommended by the American Society for Testing and Materials (1976) as a standard method (No. 02579) for measurement of organic carbon in water. However, except for the direct injection combustion method described by Sharp (1973b) for saline waters, direct injection analyzers based on high temperature oxidation have yet to be proved reliable for routine analysis of sea-water samples. Major difficulties have resulted from low organic carbon content, interference from salt, and high reagent or method blanks. In sea-water samples, the most common dry combustion methods have used an initial drying step such as evaporation (Skopintsev, 1960), freeze-drying (Gordon and Sutcliffe, 1973), or vacuum evaporation (MacKinnon, 1978). When dried, the resulting salt is oxidized either in an O_2 atmosphere in a high temperature combustion furnace (700–900°C) or in a CHN analyzer.

The choice of method for the measurement of organic carbon in sea water has been complicated by the wide discrepancy in reported results. For example, dry oxidation results for DOC analyses from similar areas are up to three times greater than those obtained with wet oxidation methods (Fig. 6). These differences between wet and dry oxidation methods need to be examined to determine if they result from differing extents of oxidation or from other analytical errors (blank calculation, contamination, interference with detector). As oxidants used in chemical wet oxidation methods have become more effective through the years, an increase in measured organic carbon values have been observed (Wangersky, 1975). An incompleteness of the oxidation of specific compounds with a chemical wet oxidation procedure was noted by Martinie and Schilt (1976). A similar problem of incompleteness of oxidation of certain organic compounds has been reported with photo-oxidation methods. However, except for a few reported organic compounds (those containing sulfur) this incompleteness of ultraviolet-assisted photo-oxidation systems may have been due to the method of determining completeness rather than the actual oxidation procedure (Gershey et al., 1979). Nitrogen-containing organic compounds have been reported to be incompletely recovered if measurement was based on the oxidation products of nitrogen (Armstrong et al., 1966), whereas complete recovery has been observed if analysis is based on the oxidation products of carbon (Gershey et al., 1979). Problems of contamination or blank correction in dry oxidation methods (Skopintsev, 1960; Gordon and Sutcliffe, 1973) may be the main reasons for their higher reported values although this is disputed

(Skopintsev and Timofeyeva, 1962; Skopintsev, 1976). However, when the TOC results from wet and dry oxidation were compared directly, the procedures agreed within 15% (MacKinnon, 1978) and were comparable to results with the photo-oxidation method (Gershey et al., 1979) when contamination was controlled by handling the samples in an organic-free atmosphere. However, it must be remembered that the absolute accuracy of all of these oxidation methods is still uncertain.

The methods which have been described for measurement of the organic content of water are based on oxidation of organic matter and measurement of the products. Another direct method for determining the organic content of natural waters based on measurement of ultraviolet absorbance of water (Mattson et al., 1974). This method can only be successfully employed when the composition of organic matter remains similar over the area sampled and as such may find only limited application. Correlation of results of the ultraviolet absorbance method with those of the persulfate oxidation method for samples of coastal and marsh waters was very poor (Wheeler, 1977).

Besides these direct methods, another approach, based on determination of oxidizability, has been used to estimate the quality of waters. With these methods, which include permanganate oxidation (Skopintsev et al., 1966), chemical oxygen demand (COD) (Method No. 01252 in A.S.T.M., 1976) and total oxygen demand (TOD) (Method No. 03250 in A.S.T.M., 1976), an estimate of organic content of water is obtained by measuring the amount of oxidant consumed. These methods are used mainly for the analysis of waste and fresh waters with high organic loads, and are of little use for most unpolluted marine samples. The direct methods in which organic matter is measured as the organic carbon (TOC) are recommended because they are faster, more independent of biological and inorganic interferences, and easier to interpret (Malcolm and Leenheer, 1973; Poirier and Wood, 1978). Results from oxidizability determinations with permanganate correlate poorly with those determined by direct oxidation (Tarasova, 1972; Carlberg, 1976). While not recommended for measurement of organic matter in sea water, the oxidizability parameter may provide information on biological or chemical lability and on the ease with which organic matter will be oxidized, although interpretation of these results may be difficult.

In seawater systems, the oxidation method chosen for analysis of the organic carbon depends on the fraction (DOC, POC or VOC) of interest, the origin of the sample, and the type of information required. However, analytical results from different methods cannot be compared readily. By examining these various analytical methods for the measurement of the dissolved, particulate, and volatile fractions of the organic matter, the sources of the differences in the reported results may be better understood.

2.1. Particulate organic carbon

The particulate organic fraction is usually a small component (<10%) of the total organic matter except in areas of high productivity or in coastal areas of high turbulence or high terrestrial input (Sharp, 1973b; Gordon and Sutcliffe, 1973) (Table II). The contribution of plankton to the calculated POC concentrations varies from 20—50% in oligotrophic waters up to 70—90% in areas of high productivity (Hobson et al., 1973; Wangersky, 1977; Cauwet, 1978). Of the total suspended matter (seston) filtered from natural waters, the proportion of this filter-retained material, which is organic, will be dependent on the source of samples (Gordon, 1970; Copin-Montegut and Copin-Montegut, 1972a; Chester and Stoner, 1974).

The particulate organic matter is classed as that material retained by a 0.45-μm filter. This is an operational definition based on pore size of the membrane filters first available. However, in natural waters, particle size distribution is continuous (Sharp, 1973a), so that division between dissolved and particulate classes is not rigid. Further, the nominal pore size of a filter does not define accurately the size of particles retained (Sheldon, 1972), and the efficiency of filtration depends on the type of filter used (Wangersky and Hincks, 1978). Most of the uncertainties in determination of the POC fraction probably arise not from the actual oxidation procedure, but from such other aspects of methods as choice of filters, blank correction, carbonate interference, and contamination in sampling and analysis. Results of different high-temperature oxidation methods for the filter-retained material are in good agreement. However, with the wet oxidation procedure (Kinney et al., 1971) for measuring POC concentrations, lower POC results are obtained than is the case with high-temperature oxidation methods (Copin-Montegut and Copin-Montegut, 1972b).

While centrifugation has been used to separate particulate material from sea-water samples (Chester and Stoner, 1974), filtration is the most common separation method. Gravity or positive-pressure filtration are recommended over vacuum filtration because of the possibility of particle formation by breaking of bubbles on the underside of the filter during vacuum filtration (Sharp, 1974; Gordon and Sutcliffe, 1974). Obviously, filters made from organic materials such as Millipore (cellulose acetate) and Nucleopore (polycarbonate) type filters, are not suitable for measurement of particulate organic carbon with oxidation methods. Most studies have used silver (Selas Flotronics) or glass (Whatman GF/C, or GF/F, or Gelman AEE) filters. The retention characteristics of different types of filters with varying nominal pore sizes have been studied using a Coulter-counter to examine the particles not retained (Sheldon, 1972) and by using particles of defined sizes to determine the efficiency of filtration (Cranston and Buckley, 1972). The ranges of particle sizes separated overlapped. The carbon blanks of filters are variable and high when compared with the amount of particulate organic material expected to be found in open ocean samples. A combustion clean-up (400—450°C for several hours) is recommended but can cause distortion of pore sizes of silver filters (Wangersky, 1975). Glass filters are of a matted structure so that the filtering process is less well-defined than with silver filters of controlled pore size. While nominal pore sizes of 0.7 μm (Whatman GF/F) to 1.32 μm (Whatman GF/C) are stated for the commonly used glass-fiber filters, a clearly defined separation can not be expected, and retention of smaller particles is observed. In a direct comparison of glass and silver filters for retention of POC in natural samples, Wangersky and Hincks (1978) reported that the amount of organic material retained by glass filters (Whatman GF/C and Gelman AEE) was significantly higher (up to three times) than with silver (0.45-μm and 0.8-μm) filters. However, whether this difference in measured organic carbon is the result of efficiency of particle retention or adsorption of dissolved organic species is uncertain.

Both glass and silver filters can be used either in high-temperature oxidation systems (Armstrong et al., 1971; Wangersky, 1974) or in commercial CHN analyzers (Sharp, 1974; Ehrhardt, 1976; Gordon, 1977). However, many workers using a CHN analyzer prefer to use the more expensive silver filters because they are easier to handle. Adsorption of dissolved organic material and retention of water are reported to be higher with glass-fiber filters (Quinn and Meyers, 1971; Sharp, 1974) than with silver filters (Gordon and Sutcliffe, 1974). Attempts to correct for this adsorbed dissolved material have included the use of several filters in series (Kinney et al., 1971; Banoub and Williams, 1972) for the filtration. The organic carbon measured on the lower or bottom filters is subtracted from that measured on the top filter. Gordon and Sutcliffe (1974) showed that the use of filter stacks (two or more filters) may affect accuracy of results by overcompensating the blank because of possible particle formation on the underside of the bottom filter. Other errors can arise from the use of filter stacks because of a longer filtration time required and by an increase in the possibility of sample contamination. Overestimated blanks obtained by this approach would result in POC concentrations that are too low (up to 50% in oligotrophic areas). Other than in surface waters, the amount of adsorbed organic carbon on glass-fiber filters, measured as the blank, should be small and negligible so that use of the bottom filter as a blank is not recommended (Wangersky and Hincks, 1978).

In areas of low POC concentrations (1—10 μg Cl^{-1}), accuracy of the analysis will depend on a careful determination of the filter blank. Even after cleaning the filter by precombustion, the carbon blank of the filter (glass or silver) may be significant and variable (Wangersky, 1974). Accuracy can be affected by the adsorption of dissolved organic carbon during filtration. Inorganic carbonates may also interfere with organic carbon determinations when samples are collected from areas where there are significantly high concentrations of calcareous suspended matter. Removal of carbonates by drying the filter after it has been acidified with HCl (Wangersky, 1974) is effective but introduction of contamination may be a problem (Sharp, 1974). Telek and Marshall (1974) report that differential heating in a CHN analyzer is a more effective way to eliminate carbonate interference than is acidification. Because a significant fraction of particulate matter is hydrolyzable by enzymes (Gordon, 1970), care must be taken to control biological utilization of the labile fraction of POC (Zsolnay, 1973). If filters can not be analyzed immediately (Hagell and Pocklington, 1973), they should be stored frozen until analysis.

Direct comparison of analytical methods for POC determination has been hampered by variability in the POC distribution in natural waters, over even short distances (Wangersky, 1974). Reasons why higher amounts of organic carbon are retained by glass filters in comparison with silver filters should be determined. The differences may result from more efficient particle retention, adsorption of dissolved organics, or a combination of both. These problems are compounded by lack of understanding of what is meant by the working or operational definition of particulate organics. If our measurements of particulate organic matter are to usefully represent the particulate organic carbon of natural waters, we must appreciate how the methods affect what is measured, a conclusion earlier posed by Cauwet in Chapter 4.

2.2. Volatile organic carbon

The volatile component of organic matter in most marine systems is a small fraction (<10%) of the total organic matter. The distribution and cycling of volatile organic matter is poorly understood because of the lack of reliable quantitative and qualitative data. Its study has been hampered by the lack of a clear definition of the ambiguous term "volatile", usually defined by a working definition based on the method of extraction and analysis. This "volatile" fraction includes those organic compounds of high vapor pressure, low molecular weight, and low-water solubility (Giger, 1977) which can be easily purged or vaporized from water systems under natural conditions of wind and turbulence. Some organic compounds which might be classed as volatile on the basis of their high vapor pressure ($\geqslant$water) are too polar to be completely removed from water by simple purging systems.

To many workers, volatile organic matter is that fraction of the total organic matter which is lost during acidification and decarbonation steps required in present oxidation methods for determining the DOC in sea water. Wangersky (1972) assumed that in wet oxidation procedures all volatile components would be lost during removal of inorganic carbonates. However, Duursma (1961) reported only about a 5—10% loss of low molecular weight, organic acids by this procedure. Organic compounds of high vapor pressure and low solubility are likely to be lost (Van Hall et al., 1965) but probably only partially unless the decarbonation step is prolonged (MacKinnon,

1978). In dry oxidation procedures where the water sample must be taken to dryness, there should be complete loss of volatile organics with vapor pressure equal to or greater than that of water (Montgomery and Thom, 1962; Skopintsev, 1966; Gordon and Sutcliffe, 1973; MacKinnon, 1978). The effects of this loss on measurement of TOC concentrations will depend on sample origin, but in most oceanic samples thiss loss should be less than 5% (MacKinnon, 1977).

The importance and role of the volatile fraction of total organic matter can only be understood after samples from different environments have been analyzed. Both indirect and direct methods of analyses have been used. Skopintsev (1966) estimated that about 15% of TOC was volatile since measured TOC concentrations with his dry oxidation method were higher when evaporation of water samples was performed at lower temperatures. Volatile organic materials would most probably have been lost from his water samples during the drying step in this procedure irrespective of the temperature of evaporation. Other indirect methods such as the direct-injection method described by Van Hall and Stenger (1967), in which VOC is determined as a difference between total organic (TOC) before and after purging the sample, are limited to waters of high organic load.

Some of the direct methods used for the analysis of volatile organics in natural waters include distillation (Armstrong and Boalch, 1960; Lamar and Goerlitz, 1966), liquid—liquid extraction (Koyama, 1962; Kamata, 1966), head-space analysis (Corwin, 1970; Hurst, 1974), dynamic head-space analysis (Swinnerton and Linnebom, 1967; Mieure and Dietrich, 1973; Lovelock et al., 1973; Zlatkis et al., 1973; Grob, 1973; Novak et al., 1973; Bellar and Lichtenberg, 1974; Dowty et al., 1976; Zürcher and Giger, 1976), and gas-chromatographic direct-injection methods (Harris et al., 1974). These methods have been used mainly for analysis of specific organic compounds. A distillation procedure, based on chemical oxidation of the distillate, was described by Ryabov et al. (1972) for measurement of total VOC concentration in waters of high organic content, but its application to sea-water systems is questionable. At present, the most effective extraction method for analysis of volatile organics in water is the dynamic head-space procedure by which volatile organics are swept from the water with a gas and concentrated in a trap (Grob et al., 1975). This method is recommended by the U.S. Environmental Protection Agency (1977) for analysis of volatile contaminants in drinking water. Many different trapping systems (e.g., cryogenic, carbon, "Porapak Q", "Chromosorb 101", "Tenax G.C.", "Tenax G.C." + silica gel) have been used. Because of its good thermal stability and adsorption—desorption efficiency as well as its freedom from water or CO_2 interferences, the porous polymer "Tenax G.C" (2,6 diphenyl-paraphenylene oxide) is used often in trapping systems of head-space analysis methods (Mieure and Dietrich, 1973; Russell, 1975; Daemen et al., 1975; Kuo et al., 1977). With the dynamic head-space method, extraction efficiency is high for such low molecular weight compounds of low solubility and high vapor pressure as heptane (100%) but drops rapidly for compounds such as acetone (1—2%) which are more water-soluble (Games and Hayes, 1976). Quantitative recovery of even some of the more soluble organic materials, such as acetone, may be obtained by using a modified dynamic head-space method for sea-water analysis (MacKinnon, 1977). Conditions used were relatively harsh (purge 500 ml sea-water sample at 60—80°C with N_2 for up to 24 h, passing effluent through a series of "Tenax G.C." traps and a cryogenic (—78°C) trap), so that the possibility of thermally induced changes to labile organic species cannot be neglected. Dynamic head-space sampling methods are probably subject to errors from changes in the sample during the extraction procedure, but at present, it is the most efficient method for examining the volatile organics in natural waters. The organic materials trapped from water samples were desorbed into a high-temperature oxidation system. The concentrations of volatile organic material measured in different marine systems were quite low (0.01—0.07 ppm C, Table I). The VOC fraction of TOC expressed as the ratio (VOC/TOC) varied only slightly (2—6%) with area, depth, or season (Fig. 5). However, any method where VOC concentration is measured in a discrete sample can yield only "steady-state" concentration of VOC in the system sampled and cannot follow rapidly changing concentrations. Although volatile organic matter may be important in the cycling of organic matter in natural waters, its measured concentration is low and is only a small fraction of the total organic matter. Some of the volatile organics will be lost during analysis of DOC or TOC by current methods (either wet or dry oxidation procedures), but this loss should be small (<5%) and will usually be within the precision of the analytical methods.

2.3. Dissolved organic carbon

The dissolved organic matter in sea water is defined as the organic material which passes a 0.45-μm filter and which is not lost by evaporization during analytical procedures later used for its measurement. The term "dissolved" is an operational definition which will also include colloidal particles which are not completely retained during filtration (Sharp, 1973a; Cauwet, 1978). In most marine systems, particulate (POC) and volatile (VOC) organic carbon will together comprise less than 15—20% of total organic carbon (TOC). Of the remaining 80—95% of TOC, classed as "dissolved" organic carbon (DOC), only about 15% has been characterized as specific compounds. The concentration of DOC in natural waters is determined by oxidation of organic matter to CO_2 which is measured directly (usually with a non-dispersive infrared analyzer or by conductivity) or reduced to CH_4 which is measured using a flame ionization detector. While DOC results obtained with available detectors are comparable if potential interferences are eliminated, the non-dispersive infrared analyzer is the detection system most often used since interferences are easily controlled and operation is relatively trouble free.

However, the choice of a procedure for oxidation of organic matter in sea water is difficult because of the wide discrepancy of reported DOC concentrations which have been obtained by different oxidation methods. The oxidation of organic matter in a seawater sample is hindered by low concentration of organic matter (0.5—2.0 ppm C) in comparison to the high salt content (25,000—35,000 ppm) and the presence of inorganic carbonates (20—30 ppm C) which interfere with the oxidation and detection systems. Of the methods used at present for analysis of DOC, high-temperature oxidation procedures may be assumed to result in complete oxidation of organic matter in samples from natural waters (Wangersky, 1975, 1978). The DOC results from earlier dry oxidation methods for sea-water samples (Skopintsev, 1960; Gordon and Sutcliffe, 1973) were several times higher than would have been expected if wet oxidation procedures had been used (Fig. 6). It seems that contamination problems and control of blanks could probably have accounted for most of these differences (MacKinnon, 1977). With careful control over sample contamination during storage, evaporation, and handling, and with removal of interfering by-products during the high-temperature (850—900°C) oxidation of the acidified salt, a much closer agreement in TOC results has been obtained from analyses where the same seawater samples were analyzed by both wet and dry oxidation methods (MacKinnon, 1978). The DOC results obtained with the dry oxidation method averaged about 15% higher than those obtained with a modified persulfate oxidation procedure. This was statistically significant and relatively uniform for samples from various geographic areas and depths (Fig. 8). While absolute TOC values obtained by the two methods were different, precision of both methods was good. Results from either method would have led to the same conclusion about distributions of DOC in the systems sampled.

There have been few direct comparisons of oxidation methods (wet oxidation methods: persulfate, photo-oxidation; direct injection combustion methods; dry oxidation methods) used to measure DOC concentration. P.M. Williams (1969) found that results of persulfate (Menzel and Vaccaro, 1964) and photo-oxidation methods were not significantly different. Sharp (1973b) reported that DOC results obtained using his method of directly injecting a sample into a high-temperature oxidation furnace were 25% higher than those obtained using a modified persulfate oxidation method. However, his conclusion can not be completely justified because of the low correlation coefficient ($r = 0.5$) that was obtained when results from the two methods were compared (Gershey et al., 1979). In fresh-water analysis, Goulden and Brooksbank (1975) found no significant difference between DOC results obtained using silver-catalyzed persulfate oxidation, photo-oxidation, and high-temperature combustion methods.

The comparison of persulfate and dry oxidation methods described by

434

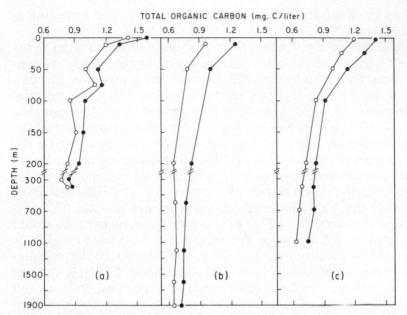

Fig. 8. Comparison of averaged TOC results for duplicate samples from different areas analyzed by the dry oxidation (●) and persulfate oxidation (○) methods (MacKinnon, 1978) a, Gulf of St. Lawrence (5, 6/75); b, Scotian shelf and slope (8/75); c, central eastern Atlantic (off Senegal) (2, 3/76).

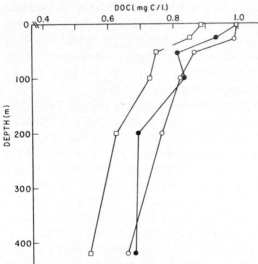

Fig. 9. Comparison of depth-averaged profiles of dissolved organic carbon (DOC) concentrations of samples from the Nova Scotian shelf measured with various oxidation methods. Data from Gershey et al. (1979). □ = persulfate oxidation (Menzel and Vaccaro, 1964); ○ = dry oxidation (MacKinnon, 1978); ● = photo-oxidation (Collins and Williams, 1977).

TABLE IV

Comparison of oxidation methods used for analysis of organic matter in sea water (based on results of comparison by Gershey et al., 1979)

	Persulfate oxidation (Menzel and Vaccaro, 1964)	Photo-oxidation (Collins and Williams, 1977)	Dry oxidation (MacKinnon, 1978)
1. Completeness of oxidation	incomplete (85—90% of dry oxidation results)	complete (>95% of dry oxidation results)	complete
2. Precision	variable	high	high
3. Blanks	high and difficult to assess accurately	low	low with proper precautions
4. Contamination	variable	low	low with proper precautions
5. Sample preparation	extensive	minimal	minimal
6. Automation of analysis	possible	yes	not likely
7. Real-time analysis	possible	almost	no
8. Ship-board operation	yes	yes	not likely
9. Changes with time in accuracy of oxidation system	no	possible — must monitor efficiency of UV lamp; simple organic standards not sufficient	no
10. Problem with particles in sample	no	yes	no
11. Loss of volatile organics	partial	partial	complete
12. Relative cost of equipment	low	high	low

MacKinnon (1978) was based on parallel samples from an extensive study of depth profiles of TOC concentrations of different water masses in different seasons. Contamination during sample handling and the biasing of results by incorrect blank evaluation were thoroughly examined. It was concluded that the observed differences (about 15%) were due to incompleteness of oxidation in the persulfate procedure. Similar results were obtained in a comparison of oxidation methods described by Gershey et al. (1979). In this study, subsamples of the same sea water (covering the expected range of natural samples from estuarine to deep-ocean waters) were analyzed for DOC by modified persulfate oxidation (Sharp, 1973a), dry oxidation (MacKinnon, 1978) and photo-oxidation methods (Collins and Williams, 1977). The DOC concentrations obtained by the photo-oxidation and dry combustion methods were not significantly different while those obtained by the persulfate oxidation method were significantly lower (about 10—15% lower) than by the other two methods (Fig. 9). However, if particles were present in samples, TOC results obtained by the photo-oxidation method were significantly lower than those obtained by the dry oxidation method. Oxidation of organic matter using reported photo-oxidation methods seems incomplete when the sample contains particles. Photo-oxidation methods have advantages for routine analysis of DOC in sea water since they are more easily automated. After questions about optimum system configuration (choice of UV lamp, lamp ageing characteristics, reaction vessel configuration, reaction time, ozone emissions) and the mechanism of oxidation of organic matter are answered, marine analysts may have access to a method which will provide real-time analysis of DOC in sea-water systems. This may provide the analyst with the means to monitor role and flux of organic matter in different marine environments.

Each of these oxidation methods which have been used for analysis of dissolved organic matter in sea water have advantages and disadvantages which are summarized in Table IV. Oxidation by the persulfate method appears to be incomplete; the photo-oxidation method seems to incompletely oxidize particles; the dry oxidation method results in loss of volatile fraction during evaporation and drying steps. However, if proper precautions are followed and sample contamination is prevented, any of these oxidation methods can probably be used with confidence to measure organic carbon in sea water and to study the distribution of DOC in sea-water systems (Fig. 9).

3. CONCLUSIONS

While organic matter is a small component of sea water, it is important to the biological, chemical and physical processes occurring in the ocean. The analysis of this organic matter in sea water is complicated by the matrix in which it is found and by the arbitrary "operational" or "working" definitions (DOC, POC and VOC) used to describe it. The oxidation methods used for analysis of organic carbon in sea water are similar and with skilled operation should lead to reasonably comparable results. Accurate and precise measurements can only be obtained when handling and preparation procedures of collected samples are controlled and monitored with the same care as the various aspects of the oxidation procedures themselves.

4. ACKNOWLEDGEMENTS

I would like to thank Drs. W.D. Jamieson and J. McLachlan for critically reading this manuscript. Also I would like to express my appreciation to Dr. P.J. Wangersky who helped formulate most of the opinions expressed in this paper during my years under his guidance.

5. REFERENCES

American Society for Testing and Materials, 1976. Annual Book of ASTM Standards, Part 31.

Andersen, N.R., 1977. Concepts in marine organic chemistry. Mar. Chem., 5: 301—638.

Armstrong, F.A. and Boalch, G.T., 1960. Volatile organic matter in algal culture media and seawater. Nature, 185: 761—762.

Armstrong, F.A. and Tibbitts, S., 1968. Photochemical combustion of organic matter in seawater, for nitrogen, phosphorous and carbon determination. J. Mar. Biol. Assoc. U.K., 48: 143—152.

Armstrong, F.A., Williams, P.M. and Strickland, J.D., 1966. Photooxidation of organic matter in seawater by ultra-violet radiation; analytical and other applications. Nature, 211: 481—483.

Armstrong, R., Goldman, C.R., Fujita, D.K., 1971. A rapid method for estimation of the carbon content of seston and periphyton. Limnol. Oceanogr., 16: 137—139.

Bada, J.L. and Lee, C., 1977. Decomposition and alteration of organic compounds dissolved in seawater. Mar. Chem., 5: 523—534.

Banoub, M.W. and Williams, P.J. LeB., 1972. Measurements of microbial activity and organic material in the western Mediterranean Sea. Deep-Sea Res., 19: 433—443.

Banoub, M.W. and Williams, P.J. LeB., 1973. Seasonal changes in the organic forms of C, N and P in seawater at E_1 in the English Channel during 1968. J. Mar. Biol. Assoc. U.K., 53: 695—703.

Baturina, M.V., Ljutsarev, S.V. and Romankevich, E.A., 1975. Fractionation of organic matter in seawater by ultrafiltration. Trans. P.P. Shirshov Inst. Oceanogr., 102: 76—79 (in Russian).

Bellar, T.A. and Lichtenberg, J.J., 1974. Determining volatile organics at microgram-per-liter levels by gas chromatography. J. Am. Water Works Assoc., 66: 739—744.

Blumer, M., 1975. Organic compounds in nature: Limits of our knowledge. Angew. Chem. Int. Edit., 14: 507—514.

Brown, W., 1975. High molecular-weight material in Baltic seawater. Mar. Chem., 3: 253—258.

Caperon, J., Harvey, W.A. and Steinhilper, F.A., 1976. Particulate organic carbon, nitrogen and chlorophyll as measures of phytoplankton and detritus standing crops in Kaneohe Bay, Oahu, Hawaiian Islands. Pac. Sci., 30: 317—327.

Carlberg, S.R., 1976. A study of the distribution of major organic constituents measured as organic carbon, oxidability and yellow substances in Baltic waters. Preprint from Symposium on "Concepts in Marine Organic Chemistry", Edinburgh, 1976.

Cauwet, G., 1978. Organic chemistry of sea water particulates. Concepts and developments. Oceanol. Acta, 1: 99—105.

Chester, R. and Stoner, J.H., 1974. The distribution of P.O.C. and N. in some surface waters of the world oceans. Mar. Chem., 2: 263–275.

Collins, K.J. and Williams, P.J. LeB., 1977. An automated photochemical method for the determination of dissolved organic carbon in sea and estuarine waters. Mar. Chem., 5: 123–141.

Copin-Montegut, C. and Copin-Montegut, G., 1972a. Chemical analyses of suspended particulate matter collected in the northeast Atlantic. Deep-Sea Res., 19: 445–452.

Copin-Montegut, C. and Copin-Montegut, G., 1972b. Comparison between two processes of determination of particulate organic carbon in seawater. Mar. Chem., 1: 151–156.

Corwin, J.F., 1970. Volatile organic matter in seawater. In: D.W. Hood (Editor), Organic Matter in Natural Waters. Inst. Mar. Sci., Alaska Occas. Publ., No. 1: 169–180.

Cranston, R.E. and Buckley, D.E., 1972. The application and performance of microfilters in analyses of suspended particulate matter. Report Series/BI-R-72-7 Bedford Institute of Oceanography, Dartmouth, N.S., 14 pp.

Croll, B.T., 1972. The determination of total organic carbon in sea water. Chem. Ind., p. 386.

Daemen, J.M., Dankelman, H.W. and Hendriks, M.E., 1975. Properties and applications of Tenax GC as a column packing material in gas chromatography. J. Chrom. Sci., 13: 79–83.

Deuser, W.G., 1971. Organic carbon budget of the Black Sea. Deep-Sea Res., 18: 995–1004.

Dowty, B.L., Green, L.E. and Laseter, J.L., 1976. Automated gas chromatographic procedure to analyze volatile organics in water and biological fluids. Anal. Chem., 48: 946–949.

Duce, R.A. and Duursma, E.K., 1977. Inputs of organic matter to the ocean. Mar. Chem., 5: 319–340.

Duursma, E.K., 1961. Dissolved organic carbon, nitrogen and phosphorus in the sea. Neth. J. Sea Res., 1: 1–148.

Duursma, E.K., 1965. The dissolved organic constituents of seawater. In: J.P. Riley and G. Skirrow (Editors), Chemical Oceanography, 1. Academic Press, New York, N.Y., pp. 433–475.

Duursma, E.K. and Marchand, M., 1974. Aspects of organic marine pollution. Oceanogr. Mar. Biol. Annu. Rev., 12: 315–431.

Eadie, B.J. and Jeffrey, L.M., 1973. $\delta^{13}C$ analyses of oceanic P.O.M. Mar. Chem., 1: 199–209.

Ehrhardt, M., 1969. A new method for the automatic measurement of dissolved organic carbon in sea water. Deep-Sea Res., 16: 393–397.

Ehrhardt, M., 1976. The automatic determination of dissolved organic carbon. In: K. Grasshoff (Editor), Methods of Seawater Analysis. Verlag Chemie, Weinheim, pp. 289–297.

Ehrhardt, M., 1977. Organic substances in seawater. Mar. Chem., 5: 307–316.

Eppley, R.W., Harrison, W.G., Chisholm, S.W. and Stewart, E., 1977. Particulate organic matter in surface waters off southern California and its relationship to phytoplankton. J. Mar. Res., 35: 671–696.

Faust, S.D. and Hunter, J.V. (Editors), 1971. Organic Compounds in Aquatic Environments. Marcel Dekker, New York, N.Y., 638 pp.

Fredericks, A.D. and Sackett, W.M., 1970. Organic carbon in the Gulf of Mexico. J. Geophys. Res., 75: 2199–2206.

Gagosian, R.B. and Stuermer, D.H., 1977. The cycling of biogenic compounds and their diagenetically transformed products in seawater. Mar. Chem., 5: 605–632.

Games, L.M. and Hayes, J.M., 1976. Isotopic and quantitative analysis of the major carbon fractions in natural water samples. Anal. Chem., 48: 130–135.

Gershey, R.M., MacKinnon, M.D. and Williams, P.J. LeB., 1979. Comparison of three oxidation methods used for the analysis of the dissolved organic carbon in seawater. Mar. Chem., 7: 289—306.

Giger, W., 1977. Inventory of organic gases and volatiles in the marine environment. Mar. Chem., 5: 429—442.

Goldberg, E.D., 1976. Strategies for Marine Pollution Monitoring. Wiley, New York, N.Y., 310 pp.

Gordon, D.C., Jr., 1970. Some studies on the distribution and composition of P.O.C. in the North Atlantic Ocean. Deep-Sea Res., 17: 233—243.

Gordon, D.C., Jr., 1971. Distribution of particulate organic carbon and nitrogen at an oceanic station in the central Pacific. Deep-Sea Res., 18: 1127—1134.

Gordon, D.C., Jr., 1977. Variability of particulate organic carbon and nitrogen along the Halifax—Bermuda section. Deep-Sea Res., 24: 257—270.

Gordon, D.C., Jr. and Sutcliffe, W.H., Jr., 1973. A new dry combustion method for the simultaneous determination of total organic carbon and nitrogen in seawater. Mar. Chem., 1: 231—244.

Gordon, D.C., Jr. and Sutcliffe, W.H., Jr., 1974. Filtration of seawater using silver filters for particulate nitrogen and carbon analysis. Limnol. Oceanogr., 19: 989—993.

Goulden, P.D. and Brooksbank, P., 1975. Automated determination of dissolved organic carbon in lake water. Anal. Chem., 47: 1943—1946.

Grasshoff, K., 1976. Methods of Seawater Analysis. Verlag Chemie, New York, N.Y., 317 pp.

Grob, K., 1973. Organic substances in potable water and its precursor. Part 1. Methods for their determination by gas-liquid chromatography. J. Chromatogr., 84: 255—273.

Grob, K., Grob, K., Jr. and Grob, G., 1975. Organic substances in potable water and in its precursor. Part III. The closed-loop stripping procedure compared with rapid liquid extraction. J. Chromatogr., 106: 299—315.

Hagell, G.T. and Pocklington, R., 1973. A seagoing system for the measurement of particulate carbon, hydrogen and nitrogen. Rep. Ser/BI-R-73, 14: 1—18.

Handa, N., 1977. Land sources of marine organic matter. Mar. Chem., 5: 341—360.

Harris, L.E., Buddle, W.L. and Eichelberger, J.W., 1974. Direct analysis of water samples for organic pollutants with G.C.-M.S. Anal. Chem., 46: 1912—1917.

Hobson, L.A., 1971. Relationships between particulate organic carbon and micro-organisms in upwelling areas off southwest Africa. Inv. Pesq., 35: 195—208.

Hobson, L.A., Menzel, D.W. and Barber, R.T., 1973. Primary productivity and sizes of pools of organic carbon in the mixed layer of the ocean. Mar. Biol., 19: 298—306.

Holm-Hansen, O., Strickland, J.D.H. and Williams, P.M., 1966. A detailed analysis of biologically important substances in a profile off southern California. Limnol. Oceanogr., 11: 548—561.

Hood, D.W. (Editor)., 1970. Proceedings of Symposium. Organic Matter in Natural Waters. Inst. Mar. Sci., Alaska, Occas. Publ., No. 1: 625 pp.

Hood, D.W. (Editor), 1971. Impingement of Man on the Oceans. Wiley—Interscience, New York, N.Y., 738 pp.

Hunter, K.A. and Liss, P.S., 1977. The input of organic materials to the oceans: air—sea interactions and the organic chemical composition of the sea surface. Mar. Chem., 5: 361—380.

Huntsman, S.A. and Barber, R.T., 1977. Primary production off northwest Africa: the relationship to wind and nutrient conditions. Deep-Sea Res., 24: 25—33.

Hurst, R.E., 1974. A method of collecting and concentrating headspace volatiles for G.C. analysis. Analyst, 99: 302—305.

Johnson, B.D., 1976. Non living organic particle formation from bubble dissolution. Limnol. Oceanogr., 21: 444—446.

Kamata, E., 1966. Aldehydes in lake and sea waters. Bull. Chem. Soc., Jap., 39: 1227—1229.

Kay, H., 1954. A micromethod for the chemical determination of organic carbon in seawater. Kieler Meeresforsch., 10: 26—35.

Khaylov, K.M. and Ayzatullin, T.A., 1972. Kinetics of transformation of proteins and polysaccharides dissolved in seawater during interaction with detritus. Oceanology, 12: 809—816.

Kinney, P.J., Loder, T.C. and Groves, J., 1971. Particulate and dissolved organic matter in the Amerasian Basin of the Arctic Ocean. Limnol. Oceanogr., 16: 132—137.

Koyama, T., 1962. Organic compounds in seawater. J. Oceanogr. Soc. Jap., 20: 563—576.

Kuo, P.R.K., Chian, E.S.K., DeWalle, F.B. and Kim, J.H., 1977. Gas stripping, sorption and thermal desorption procedures for preconcentrating volatile polar water-soluble organics from water samples for analysis by gas chromatography. Anal. Chem., 49: 1023—1029.

Lamar, W.L. and Goerlitz, D.F., 1966. Organic acids in naturally colored surface waters. Geol. Surv. Water Suppl. Pap., 1817-A.

Liss, P.S., 1975. Chemistry of the sea surface microlayer. In: J.P. Riley and G. Skirrow (Editors), Chemical Oceanography, 2. Academic Press, London, pp. 193—243.

Lovelock, J.E., Maggs, R.J. and Wade, R.J., 1973. Halogenated hydrocarbons in and over the Atlantic. Nature, 241: 194—196.

Ljutsarev, S.V., Mirkina, S.D., Romankevich, E.A. and Smetankin, A.V., 1975. Dissolved organic carbon and phosphorous in the waters of the eastern part of the Equatorial Pacific Ocean. Trans. P.P. Shirshov Inst. Oceanogr., 102: 70—75.

Loder, T.C. and Hood, D.W., 1972. Distribution of organic carbon in a glacial estuary in Alaska. Limnol. Oceanogr., 17: 349—355.

MacKinnon, M.D., 1977. The Analysis of the Total Organic Carbon in Seawater: a. Development of Methods for the Quantification of T.O.C. b. Measurement and Examination of the Volatile Fraction of the T.O.C. Thesis, Dalhousie University, Halifix, N.S., 183 pp.

MacKinnon, M.D., 1978. A dry oxidation method for the analysis of the T.O.C. in sea water. Mar. Chem., 7: 17—37.

Malcolm, R.L. and Leenheer, J.A., 1973. The usefulness of organic carbon parameters in water quality investigations. Inst. Environ. Sci., 19: 336—341.

Martinie, G.D. and Schilt, A.A., 1976. Investigation of wet oxidation efficiencies of perchloric acid mixtures for various organic substances and the identities of residual matter. Anal. Chem., 48: 70—74.

Mattson, J.S., Smith, C.A., Jones, T.J. and Gerchakov, S.M., 1974. Continuous monitoring of dissolved organic matter by UV-visible photometry. Limnol. Oceanogr., 19: 530—535.

Maurer, L.G. and Parker, P.L., 1972. The distribution of D.O.M. in the near shore waters off the Texas coast. Contrib. Mar. Sci., 16: 109—124.

Menzel, D.W., 1964. The distribution of dissolved organic carbon in the Western Indian Ocean. Deep-Sea Res., 11: 757—766.

Menzel, D.W., 1970. The role of in situ decomposition of organic matter on the concentration of non-conservative properties in the sea. Deep-Sea Res., 17: 751—764.

Menzel, D.W., 1974. Primary productivity, dissolved, particulate organic matter and the sites of oxidation of organic matter. In: M.N. Hill (Editor), The Sea, 5. Interscience, New York, N.Y., pp. 659—678.

Menzel, D.W. and Ryther, J., 1968. Organic carbon and the oxygen minimum in the South Atlantic Ocean. Deep-Sea Res., 15 (3): 327—337.

Menzel, D.W. and Ryther, J.H., 1970. Distribution and cycling of organic matter in the oceans. In: D.W. Hood (Editor), Proc. Symp. Organic Matter in Natural Waters. Inst. Mar. Sci., Alaska, Occas. Publ., No. 1: 31—54.

Menzel, D.W. and Vaccaro, R.F., 1964. The measurement of dissolved organic and particulate carbon in seawater. Limnol. Oceanogr., 9: 138—142.

Mieure, J.P. and Dietrich, M.W., 1973. Determination of trace organics in air and water. J. Chromatogr. Sci., 11: 559—570.

Minas, M., Abboud, M. and Slawyk, G., 1978. Production primaire et relations entre les paramètres de la biomasse microplanctonique dans le dome de Guinée. Symp. on the Canary Current, Las Palmas de Gran Canaria, 11—14 April.

Mitchell, D.G., Aldous, K.M. and Canelli, E., 1977. Determination of organic carbon by thermal volatilization—plasma emission spectrometry. Anal. Chem., 49: 1235—1238.

Montgomery, H.A.C. and Thom, N.S., 1962. The determination of low concentrations of organic carbon in water. The Analyst, 87: 689—697.

Moore, R.M., 1977. Trace Metals — Dissolved Organic Matter and Their Association in Natural Waters. University of Southampton, 175 pp.

Morris, R.J. and Eglington, G., 1977. Fate and recycling of carbon compounds. Mar. Chem., 5: 559—572.

Novak, J., Zluticky, J., Kubelka, V. and Mostecky, J., 1973. Analysis of organic constituents present in drinking water. J. Chromatogr. 76: 45—50.

Ogura, N., 1970a. The relation between dissolved organic carbon and apparent oxygen utilization in the Western North Pacific. Deep-Sea Res., 17: 221—231.

Ogura, N., 1970b. Dissolved organic carbon in the equatorial region of the Central Pacific. Nature, 227: 1335—1336.

Ogura, N., 1972. Rate and extent of decomposition of dissolved organic matter in surface seawater. Mar. Biol., 13: 89—93.

Ogura, N., 1974. Molecular weight fractionation of dissolved organic matter in coastal seawater by ultrafiltration. Mar. Biol., 24: 305—312.

Parsons, T.R., 1975. Particulate organic carbon in the sea. In: J.P. Riley and G. Skirrow, (Editors), Chemical Oceanography, 2. Academic Press, London, pp. 365—383.

Pocklington, R., 1973. Organic carbon and nitrogen in sediments and particulate matter from the Gulf of St. Lawrence. Rep. Ser. B1-R-73-8 Bedford Institute, 16 pp.

Pocklington, R., 1978. Organic matter in waters off Senegal and The Gambia, February—March, 1976. Rep. Ser., Bedford Institute, 17 pp.

Pocklington, R. and MacKinnon, M.D., in press. Organic matter in upwelling off Senegal and the Gambia. Spec. Vol. Rapp. Proc. Verb. ICES.

Pocklington, R., Richman, J. and Smith, S., 1978. Baffin cruise 77-030 Report. Bedford Institute.

Poirier, S.J. and Wood, J.H., 1978. A new approach to the measurement of organic carbon. Am. Lab., 10: 78—89.

Provasoli, L., 1963. Organic regulation of phytoplankton fertility. In: M.N. Hill (Editor), The Sea, 2. Interscience, New York, N.Y. pp. 165—219.

Quinn, J.G. and Meyers, P.A., 1971. Retention of dissolved organic acids in seawater by various filters. Limnol. Oceanogr., 16: 129—131.

Riley, G.A., 1970. Particulate organic matter in seawater. Adv. Mar. Biol., 8: 1—118.

Riley, J.P., 1971. Dissolved and particulate organic compounds in the sea. In: J.P. Riley and R. Chester (Editors), Introduction to Marine Chemistry. Academic Press, New York, pp. 182—215.

Russell, J.W., 1975. Analysis of air pollutants using sampling tubes and gas chromatography. Environ. Sci. Technol., 9: 1175—1178.

Ryabov, A.K., Nabivanets, B.I. and Litvinenko, Z.S., 1972. Determination of bichromate oxidizability of water with trapping of volatile organic compounds. Hydrobiol. J., 8: 97—101.

Sharp, J.H., 1973a. Size classes of organic carbon in seawater. Limnol. Oceanogr., 18 (3): 441—447.

442

Sharp, J.H., 1973b. Total organic carbon in seawater — comparison of measurements using persulfate oxidation and high temperature combustion. Mar. Chem., 1: 211—229.

Sharp, J.H., 1974. Improved analysis for "particulate" organic carbon and nitrogen. Limnol. Oceanogr., 19: 984—989.

Sheldon, R.W., 1972. Size separation of marine seston by membrane and glass-fiber filters. Limnol. Oceanogr., 17: 494—498.

Sholkovitz, E.R., 1976. Flocculation of dissolved organic and inorganic matter during the mixing of river water and seawater. Geochim. Cosmochim. Acta, 40: 831—845.

Skopintsev, B.A., 1960. Organic matter in seawater. Mar. Hydrophys. Inst., 19: 1—14.

Skopintsev, B.A., 1966. Some considerations on the distribution and state of OM in ocean water. Oceanology, 6: 361—368.

Skopintsev, B.A., 1972a. On the age of stable organic matter — Aquatic humics in the oceanic water. In: D. Dyrssen and D. Jagner (Editors), The Changing Chemistry of the Oceans. Wiley-Interscience, New York, N.Y., pp. 205—207 (Nobel Symp. 20).

Skopintsev, B.A., 1972b. A discussion of some views on the origin, distribution and composition of OM in deep ocean waters. Oceanology, 12: 471—474.

Skopintsev, B.A., 1976. On the determination of organic carbon in chloride rich water by the persulfate method. Oceanology, 16: 630—633.

Skopintsev, B.A. and Timofeyeva, S.N., 1962. Content of organic carbon in the waters of the Baltic and North Seas and in the subtropical and tropical regions of N. Atlantic. Tr. Mar. Hydrophys. Inst., 25: 110—117.

Skopintsev, B.A., Timofeyeva, S.N. and Vershinina, O.A., 1966. Organic carbon in the Near Equatorial and Southern Atlantic and in the Mediterranean. Oceanology, 6: 201—210.

Skopintsev, B.A., Romenskaya, N.N. and Sokolova, M.V., 1968. Organic carbon in the waters of the Norwegian Sea and of the Northeast Atlantic. Oceanology, 8: 178—186.

Smith, R.G., Jr., 1976. Evaluation of combined applications of ultrafiltration and complexation capacity techniques to natural waters. Anal. Chem., 48: 74—76.

Starikova, N.D., 1970. Vertical distribution patterns of DOC in seawater and interstitial solutions. Oceanology, 10: 796—807.

Starikova, N.D. and Yablokova, O.G., 1974. Organic matter in Northwestern Pacific Ocean waters (Tsugasu Strait—Wake Island section). Oceanology, 14: 833—837.

Starikova, N.D., Yablokova, O.G. and Korzhikova, L.I., 1976. Determining the molecular composition of dissolved organic matter by gel filtration. Oceanology, 16: 571 –575.

Strickland, J.D.H. and Parsons, T.R., 1972. A practical handbook of seawater analysis. Bull. Fish. Res. Board Can., No. 167: 311 pp. (2nd ed.).

Swinnerton, J.W. and Linnenbom, V.J., 1967. Gaseous hydrocarbons in seawater: Determination. Science, 156: 1—2.

Tarasova, Ye.N., 1972. Correlation of organic carbon with different kinds of oxidizability in the open waters of Lake Baikal. Hydrobiol. J., 8: 53—57.

Telek, G. and Marshall, N., 1974. Using a CHN analyzer to reduce carbonate interference in particulate organic carbon analyses. Mar. Biol., 24: 219.

U.S. Environmental Protection Agency, 1977. Sampling and analysis procedures for survey of industrial effluents for priority pollutants. 1FB No. WA-77-B133.

Vallentyne, J.R., 1957. The molecular nature of organic matter in lakes and oceans with lesser reference to sewage and terrestial soils. J. Fish. Res. Board Can., 14 (1): 33—82.

Van Hall, C.E. and Stenger, V.A., 1967. An instrumental method for rapid determination of the carbonate and total carbon in solutions. Anal. Chem., 39: 503—507.

Van Hall, C.E., Barth, D. and Stenger, V.A., 1965. Elimination of carbonates from aqueous solutions prior to organic carbon determination. Anal. Chem., 37 (6): 769—771.

Wagner, F.S., 1969. Composition of the dissolved organic compounds in seawater: A Review. Contrib. Mar. Sci., 14: 115—153.

Wangersky, P.J., 1965. The organic chemistry of sea water. Am. Sci., 53: 358—374.

Wangersky, P.J., 1972. The cycle of organic carbon in seawater. Chimica, 26: 559—564.

Wangersky, P.J., 1974. Particulate organic carbon: sampling variability. Limnol. Oceanogr., 19: 980—984.

Wangersky, P.J., 1975. The measurement of organic carbon in seawater. In: T.R.P. Gibb (Editor), Analytical Methods in Oceanography. Adv. Chem. Ser., 147. Am. Chem. Soc., Washington, D.C., pp. 148—162.

Wangersky, P.J., 1976a. The surface film as a physical environment. Ann. Rev. Ecol. Syst., 7: 161—176.

Wangersky, P.J., 1976b. Particulate organic carbon in the Atlantic and Pacific Oceans. Deep-Sea Res., 23: 457—465.

Wangersky, P.J., 1977. The role of particulate matter in the productivity of surface waters. Helgol. Wiss. Meeresunters, 30: 546—564.

Wangersky, P.J., 1978. Production of dissolved organic matter. In: O. Kinne (Editor), Marine Ecology, 4. Wiley, New York, N.Y., pp. 115—220.

Wangersky, P.J. and Hincks, A.V., 1978. The shipboard intercalibration of filters used in the measurement of particulate organic carbon. N.R.C. Canada, Mar. Anal. Chem. Standards Progr. Tech. Rep. 4. NRCC No. 16767, 35 pp.

Wangersky, P.J. and Zika, R.G., 1978. The analysis of organic compounds in Sea Water. N.R.C. of Canada, Mar. Anal. Chem. Standards Progr., Tech. Rep. 3. NRCC No. 16566, 177 pp.

West, D.L., 1964. Determination of total carbon by combustion gas-chromatography. Anal. Chem., 36: 2194—2195.

Wheeler, J.R., 1977. Dissolved organic carbon: spectral relationships in coastal waters. Limnol. Oceanogr., 22: 573—575.

Whittle, K.J., 1977. Marine organisms and their contribution to organic matter in the ocean. Mar. Chem., 5: 381—412.

Williams, P.J. LeB., 1969. The wet oxidation of organic matter in seawater. Limnol. Oceanogr., 14: 292—297.

Williams, P.J. LeB., 1975. Biological and chemical aspects of dissolved organic matter in seawater. In: J.P. Riley and G. Skirrow (Editors), Chemical Oceanography, 2. Academic Press, London, pp. 301—363.

Williams, P.M., 1968. Organic and inorganic constituents of the Amazon River. Nature, 218: 937—938.

Williams, P.M., 1969. The determination of dissolved organic carbon in seawater: a comparison of two methods. Limnol. Oceanogr., 14: 297—298.

Williams, P.M., 1971. The distribution and cycling of organic matter in the ocean. In: S.D. Faust and J.W. Hunter (Editors), Organic Compounds in Aquatic Environments. Dekker, New York, N.Y., pp. 145—163.

Williams, P.M., Oeschger, H. and Kinney, P.J., 1969. Natural radiocarbon activity of the dissolved organic carbon in the Northeast Pacific Ocean. Nature, 224: 256—258.

Wilson, R.F., 1961. Measurement of organic carbon in seawater. Limnol. Oceanogr., 6: 259—261.

Woodwell, G.M. and Pecan, E.V. (Editors), 1973. Carbon and the Biosphere. U.S.A.E.C. Conf.-720510.

Zafiriou, O.C., 1977. Marine organic photochemistry reviewed. Mar. Chem., 5: 497—522.

Zika, R.G., 1977. An Investigation in Marine Photochemistry. Thesis, Dalhousie University, Halifax, N.S., 346 pp.

Zlatkis, A., Lichtenstein, H.A. and Tichbee, A., 1973. Concentration and analysis of trace volatile organics in gases and biological fluids with a new solid absorbent. Chromatographia, 6: 67—70.

Zsolnay, A., 1973. Hydrocarbon and chlorophyll: a correlation in upwelling region off West Africa. Deep-Sea Res., 20: 923—925.

Zürcher, F. and Giger, W., 1976. Flüchtige organische Spurenkompenenten in der Glatt. Vom Wasser, 47: 37—55.

Chapter 15

THE ANALYTICAL METHODS FOR THE CHARACTERISATION
OF ORGANICS IN SEAWATER

R. DAWSON and G. LIEBEZEIT

1. INTRODUCTION

An awareness that biogeochemical processes in seawater cannot be adequately described by a consideration of merely the elemental composition of the dissolved organic fraction, but that descriptions of biochemical processes necessitate the analysis of biochemically relevant compounds, has challenged the skills of the analytical organic chemist during recent years.

The multiplication of the introductions of reports concerning organic matter in the marine environment, whereby we are repeatedly reminded that we are dealing with the detection of compounds at the part per billion level (or below) in the face of enormous quantities of inorganic salts, is testament to the difficulties encountered by workers in this field.

Progress made in developing sensitive techniques for the analysis of organic compounds in other fields of science filters only slowly through to the marine chemist, largely due to problems of adaptation of the techniques to the electrolyte solution and often to the use of sophisticated equipment on board ship (Riley, 1975).

The ultimate aim of the marine organic chemist, i.e. to be able to detect and quantify compounds by direct analysis of seawater without any pre-concentration or desalting procedure, has been realised in only a few instances. There is, however, considerable optimism that this will be accomplished for several more compound classes in the near future (Degens, 1979). Some authors whilst sharing this optimism warn of the dangers of an information explosion in describing the fine structure of organic chemicals at the expense of analysing enough samples to enable meaningful oceanographic interpretations (Wangersky and Zika, 1978). Following the doctrine of Blumer (1975), who suggested the acceptance of the "limits of ignorance" in order to maximise the return on research investment, Wangersky and Zika (1978) suggest that we may have to accept a value for "total carbohydrates as glucose equivalents" rather than attempt a complete analysis.

While we accept that such an approach at the moment appears to be some solution in appeasing the appetite for data concerning organic compounds in the marine environment, it will however direct research to the monitoring of bulk biochemical processes and may reveal little more information than elemental analysis.

Detailed analysis within compound classes may suffer from the disadvantage that the techniques require sophisticated equipment with experienced personnel. However, it may be erroneous to suggest that these techniques are either time-consuming or unsuitable for real-time analysis at sea.

It is the intention of this chapter to describe some of these techniques presently employed to characterise organics in seawater and to evaluate how these techniques may be further developed to enable direct analysis.

It is to the credit of earlier workers who, with limited analytical facilities and painstaking methods, were able to at least identify some of the more obvious organic compounds in seawater. The modern methods of analysis owe their development to their precursors and as such, a short historical review of alternative methods will be given for each class of compound considered.

The recent review *The Analysis of Organic Compounds in Seawater* (Wangersky and Zika, 1978) is not only an extensive compilation of available methods; the critical approach adopted, together with a clear division into compound classes and useful predictions for future research directions, makes this work primary reading.

Sampling and the problems of storage and contamination will not be treated separately here, the reader is referred to Riley (1975), Williams (1975a), Grasshoff (1976) and Wangersky and Zika (1978).

Although the efforts in pollution research have undoubtedly been instrumental in the refinement of the techniques for the analysis of natural organics in seawater, the proliferation of literature concerning pollutant chemicals (Duursma and Marchand, 1974) has most certainly made the reviewing of relevant literature for natural organics more difficult.

In this chapter we confine ourselves to methods of analysis for naturally occurring organic compounds. The distinction between natural and anthropogenic hydrocarbons and the analytical methods adopted has been treated by Saliot (Chapter 11). The determination of dissolved organic carbon has been described by MacKinnon (Chapter 14). Comparison of data concerning levels of individual dissolved organic compounds may, in many cases, only be meaningful if the concentrations of the components are referred to a value for DOC.

1.1. Inventories of organic compounds in seawater

The approach adopted to inventorise the organic compounds in seawater has been one of reviewing the literature and listing substances detected along with an expected concentration range in some form of class system (e.g. Duursma, 1965; Williams, 1975b and more recently the discussions on inventories by Daumas and Saliot, 1977; Giger, 1977; and Zsolnay, 1977).

The components of these inventories have for the most part been determined by the available techniques. Degens (1979) has recently suggested "that the past approach to trace organic analysis, i.e. analysis of seawater and the postulation of a feasible process to explain the presence of substance X, should be replaced by a consideration of the known biochemistry of cellular systems and attempts should be made to detect these biochemical intermediates".

Since we may logically expect to find any organic compound known to exist in marine life released into the water by various mechanisms (Gagosian and Lee, Chapter 5), our efforts should be directed to the analysis of those compounds whose functions are well defined. Carbohydrates for example are being produced in the marine environment in significant amounts yet the analysis of constituent saccharides has attracted little attention in the past and measurements of total carbohydrate, by often unspecific methods, have been favoured.

Other compounds with known important biological functions remain undetected e.g. those involved in the citric acid cycle, purines and pyrimidines and related compounds, multi-nitrogen compounds, etc. (Gagosian and Lee, Chapter 5).

The scope of this chapter is limited to a description of selected analytical techniques for the compound classes suggested to be produced in the sea, for those compound classes previously analysed, and to a description of chromatographic techniques for a more detailed analysis within the compound classes.

2. DISSOLVED FREE AND COMBINED AMINO ACIDS

There has been considerable interest in the detection of free amino acids in seawater. The efforts to analyse for these compounds began at the end of the fifties (Belser, 1959, using a bioassay technique) and since then widely varying analytical techniques have been applied to tackle the problem of low-level determinations with the inherent risk of sample contamination.

Lee (1975) clearly demonstrated that the levels of free amino acids found in seawater are inversely correlated with the refinement of the analytical technique (in most cases the year of publication).

Two approaches have been adopted in the study of amino acids in seawater: the measurement of total amounts and the determination of single amino acids after chromatography.

The attempts to use some of the more established reagents for specific analysis of protein e.g. Folin-phenol reagent or the variation of the method by Lowry et al. (1951), whereby total protein content is calculated from the content of aromatic amino acids (tryptophan, tyrosine and phenylalanine), lacks the sensitivity for determinations on seawater samples (Wangersky and Zika, 1978). Furthermore, detailed analysis of the combined amino acid fraction of seawater does not show any aromatic amino acid to be present in significant quantity (Garrasi et al., 1979).

2.1. Methods for estimation of total amino nitrogen

All locating agents commonly used in ion-exchange chromatography have been tested for use as reagents in determining total amino nitrogen in seawater.

2.1.1. Colorimetric method

Coughenower and Curl (1975) employed ninhydrin in an automatic procedure to detect amino acids in Lake Washington. The limit of detection (0.5 μM) and the salt dependence detracts from its use with oceanic samples.

2.1.2. Fluorimetric methods

The use of fluorimetric reagents generally results in an increase of sensitivity of about 2 to 3 orders of magnitude. The reagents commonly used, fluorescamine and o-phthalaldehyde, react with primary amines and are not restricted to amino acids. They therefore also react to a varying degree with oligopeptides, proteins, polyamines, ammonia, etc., and the results of the analysis can at best be given as amounts of fluorescamine or o-phthalaldehyde reactive substances referred to a model compound (e.g. glycine equivalents).

Fluorescamine (Udenfriend et al., 1972; Weigele et al., 1972) was adapted for use with seawater samples by North (1975), Packard and Dortch (1975) and Zika (1977). It may be employed to make some distinction between free amino acids and peptides or proteins by changing the reaction pH from 9.0 to 7.0, respectively. The reaction takes place at room temperature with only low ammonia interference. On the other hand, fluorescamine is unstable in aqueous solutions and the relative fluorescence intensities for the various amino acids vary over a wide range (Zika, 1977).

o-Phthalaldehyde (Roth, 1971) was first described as a fluorimetric reagent in column chromatography by Roth and Hampai (1973). The method was adapted for determination of total o-phthalaldehyde reactive substances (ORS) in seawater using an automated procedure (Josefsson et al., 1977). The reagent is stable in aqueous solution, reacts rapidly at room temperature to form a highly fluorescent product without any salt dependence and, with few exceptions, the relative fluorescence intensities of the amino acids are constant. Ammonia interferes in the determination thus necessitating the separate determination of the ammonia content of the water and subsequent correction of the measured ORS values. However, the ammonia interference in the method described below amounts to only 5% of the response of glycine (commonly used as reference substance) and correction of the ORS value is thought only to be necessary in waters with extreme ammonia contents e.g. pore waters or anoxic layers.

2.1.3. The determination of o-phthalaldehyde reactive substances

The simplicity of the method coupled with its high sensitivity and speed of analysis has caused the method to gain popularity as a shipboard technique, allowing the processing of large numbers of samples. The method suggested below is a modification of the technique described by Josefsson et al. (1977) and has been tested for widely varying samples. Although the values obtained (ORS) are somewhat ill-defined as to the nature of compound analysed, the routine monitoring of a more labile fraction of the DOC pool is certainly welcome for ecological studies.

The reagent consists of 100 mg of analytical quality o-phthalaldehyde predissolved in 10 ml absolute ethanol and added to 1 l of 0.4 M boric acid adjusted to pH 10.5 with 8 M NaOH solution.

The addition of 0.5 ml of mercaptoethanol (purified and stored under nitrogen) activates the reagent. The mixed reagent is stable over several days but it is recommended that the reagent be prepared daily since ammonia absorption may increase its reagent blank over long exposure times.

The unfiltered seawater sample is mixed with the reagent by means of a peristaltic pump with air segmentation in a sample-and-wash routine based on the autoanalyser principle (Fig. 1). After a reaction time of approximately 2 min the reaction stream is degassed and passed through a filter fluorimeter where the fluorescence is monitored (λ_{ex} 340 nm, λ_{em} 455 nm).

Since the purity of the wash water determines the sensitivity limit, doubly distilled ion-exchange water is recommended for use. The water should be stored in a vessel with a concentrated sulphuric acid trap to reduce the adsorption of ammonia. Likewise, the air for segmentation should be scrubbed of ammonia.

Samples should be analysed immediately where possible, otherwise addition of a fixing agent (e.g. dichloromethane) and deep freezing — an undesir-

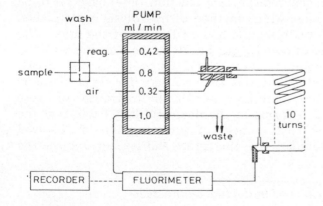

I = AUTOMATIC SAMPLER

Fig. 1. Flowscheme of autoanalyser for ORS determinations.

able alternative (Wangersky and Zika, 1978) — should be the procedure adopted.

Since the amino acid content of a single wet thumb print (Hamilton, 1965) exceeds in many cases the concentration in 1 l of seawater, special care is required in the handling of sample cups during transfer steps. The sampling cups should be rinsed thoroughly with the samples to be analysed. A stock standard solution of glycine (1 mM) made up in 0.5 M NaCl with addition of a fixing agent is used to prepare subdilutions for calibration. Glycine response is linear up to 100 μmole gly l^{-1}, the limit of detection is strongly dependent upon the wash water, the fluorimeter employed and the stability of the electrical signal on the recorder and in normal use lies typically around 0.05 μmole l^{-1}.

Ammonia response is linear up to 1000 μmole l^{-1} and corresponds to 5% of glycine response. The relative standard deviation at the 10-μM level has been found to be ±1.8% ($n = 15$); at the 1-μM level ±11.5% ($n = 14$) due to the increased signal: noise ratio.

The method is suitable not only for a rapid determination of "free" primary amino groups in seawater but can also be employed for an estimation of the combined amino nitrogen content of water samples and particulate matter. Butcher and Lowry (1976) employed o-phthalaldehyde after hydrolysis of different defined proteins and found a recovery of 70% based on the calculated amino acid content. Although when working with undefined protein-like material, the recovery is dependent upon the fluorescent response of the individual monomers released on hydrolysis, and also on the destruction losses. Maske (1979) claimed a recovery of 74% whilst using the method for plankton material after hydrolysis with 4 M HCl.

The application of the method to the determination of dissolved protein-like material in seawater is somewhat more difficult. The hydrolysis conditions required (acid strength, temperature and reaction time) have yet to be specified. It is difficult to assess which method of hydrolysis is preferable, i.e. lyophilisation of the seawater sample and hydrolysis of the dried salts or addition of large quantities of concentrated acids to the seawater with the associated increased risk of contamination (C. Garrasi, pers. comm., 1978). It is likely that very different conditions are necessary for surface waters containing "fresh" nitrogenous material than for the hydrolysis of more resistant older organic nitrogen forms in the deep sea. It is likely that the fluorimetric approach to the measurement of some fraction of dissolved organic nitrogen will gain popularity in the future and perhaps replace the standard Kjeldahl method.

2.2. Chromatographic separations of individual amino acids

The measurement of individual free dissolved amino acids in seawater has been a challenge for numerous research workers over the past two decades.

The major obstacles facing investigators in the past were those of preconcentration and desalting, and the sampling and treatment of large volumes of water to enable analysis at the sensitivity levels available.

It is now clear that such preconcentration steps and desalting procedures increased the risk of contamination and led to sample artifacts and possibly to a misinterpretation of processes in the sea involving free amino acids.

Garrasi et al. (1979) compared the results obtained when employing two of the more popular desalting/concentration procedures (Cu-chelex and cation exchange) with a recently developed technique for direct analysis and reported vast differences both quantitatively and qualitatively.

Since techniques are now available for the direct analysis of amino acids in seawater, preconcentration and desalting procedures will not be discussed in any detail and only mentioned in the historical review of the progress made in chromatography.

Even in the possession of sensitive analysers for such low-level determinations, several questions remain unanswered: whether or not to filter samples; how long may samples be stored before analysis; can we poison samples of seawater effectively and do the results of the analysis refer to "free" amino acids? The last question is one of definition. The analyst's definition of a free dissolved species does not necessarily match that of the microbiologist, where the former assumes that if he performs no hydrolysis step and analysed purely the dissolved phase, the results obtained are to be defined as a "free" compound, whereas, the latter's definition may be one of heterotrophic availability. Dawson and Gocke (1978) have pointed out the discrepancy between the results of the two approaches and it has also been suggested (Dawson and Pritchard, 1978; Dawson and Gocke, 1978) that association of amino acids with either metals (Degens and Matheja, 1967; Degens, 1970) or other organic compounds (to form peptides or other condensation products), may be the cause of the discrepancy. Such "bound" forms of amino acids may well be labile under the chromatographic conditions employed but their actual speciation in the seawater medium may render them unavailable to heterotrophs.

Certainly the use of cation exchange resins with weakly acidic eluting buffers (i.e. the standard amino-acid analyser) may detract from any distinction between free and metal-chelated ligands.

The high-performance liquid chromatographic techniques with precolumn derivatisation (Lindroth and Mopper, 1979, see below) may to some extent overcome some of these limitations. A comparative investigation of the two procedures may even give some insight into the extent of these phenomena in seawater.

2.2.1. Paper and thin-layer chromatography

The earliest methods for the detection of amino acids in seawater consisted of concentrating the amino acids from seawater by either evaporating the acidified samples to dryness and extraction of the residual salts with ethanol (e.g. Degens et al., 1964) or coprecipitation with Fe(OH)$_3$ (Tatsumoto et al., 1961). The lack of sensitivity of the chromatographic systems employed i.e. circular paper chromatography (Palmork, 1963), paper chromatography (Degens and Reuter, 1962; Degens et al., 1964; Starikova and Korzhikova, 1969), thin-layer chromatography (e.g. Riley and Segar, 1970) and thin layer chromatography after dansylation and extraction (Litchfield and Prescott, 1970), required that large volumes of seawater be desalted and concentrated.

Nevertheless, it is creditable that these earlier researchers identified up to 15 or 16 amino acids in their seawater samples. The risk of contamination during the tedious work-up procedures together with the difficulties of quantification probably led to gross overestimations of the FAA content of water and to a misinterpretation of the pattern of individual acids present.

2.2.2. Gas chromatographic determinations

The high sensitivity of the gas chromatography caused the technique to gain some popularity in the attempts to detect low concentrations of amino acids (Wangersky and Zika, 1978).

A volatile derivative of the amino acid must be prepared before analysis and desalting; concentration steps are obligatory before extraction. The techniques adopted in the past, e.g. Palmork (1969) TAB-amino-acids, Pocklington (1971, 1972) TMS-amino-acids, or electron-capture detection procedures, e.g. Hosaku and Maita, 1971 and Gardner and Lee (1973) employing halogen-containing derivatives, are not recommendable for studies on the free amino-acid pool of seawater. The arguments against the use of GC techniques for FAA arise from the increased risk of sample contamination during the multiple steps of derivative formation and the fact that combined amino acids are not separated before derivatisation employing strong acids (Dawson and Pritchard, 1978).

However, these techniques may be usefully employed for studies on the combined fraction (dissolved or particulate after hydrolysis, e.g. Siezen and Mague, 1978; employing n-heptafluorbutyryl, O-isobutyl esters). The techniques are also eminently suitable for differentiating between D/L-enantiomers (Lee, 1975; Lee and Bada, 1977).

2.2.3. Ion-exchange liquid chromatography

Following the introductions of the cation exchange technique for separation of amino acids (Moore and Stein, 1951) and the development of the first automatic amino acid analyser (Spackman et al., 1958), amino-acid analysis became routine in clinical and biological research.

The detecting reagent employed, ninhydrin can be described as being a sensitive reagent, although the limits of detection of the original analysers lay above the concentration found in seawater. Whilst the improvements in the manufacture of close-tolerance ion-exchange resins did much to speed up the chromatographic separation and higher pressure pumps enabled the use of narrow bore columns, desalting and preconcentration steps were still prerequisites for the marine chemist. The development of reagents which formed fluorescent products with amino acids, fluorescamine (Udenfriend et al., 1972) and o-phthalaldehyde (Roth, 1971) and the adaptation of the techniques for seawater analysis (Dawson and Pritchard, 1978; Gardner, 1978; Garrasi et al., 1979) drastically reduced the limits of detection to a point where direct analysis became a possibility. Since the first publication of the use of ion-exchange chromatography for separation of amino acids in seawater (Tatsumoto et al., 1961) there have been several short reviews on the subject of desalting, concentration and analysis employing both nin-hydrin and fluorescent reagents (cf. Bohling, 1970; Lee, 1975; Garrasi and Degens, 1976; Dawson and Pritchard, 1978; Garrasi et al., 1979).

Since methods now exist for the direct injection of seawater samples, these desalting and concentration methods will not be described here. Garrasi et al. (1979) have presented a detailed comparison between direct analysis and the two commonest methods, i.e. absorption onto Cu-Chelex 100 or cation-exchange resins as adopted by various authors, e.g. Siegel and Degens (1966), Webb and Wood (1967), Clark et al. (1972), Lee and Bada (1975), Daumas (1976), Garrasi and Degens (1976), Dawson and Pritchard (1978), and Gardner (1978).

The major drawbacks appear to be the introduction of artifacts during desalting, losses on desalting and evaporation to dryness (Dawson and Mopper, 1978a), contamination due to bleeding of the resins employed and the time required for sample work-up detracting from real time analyses.

The direct determination of amino acids using ion-exchange analysers. Any standard amino-acid analyser may be modified by the addition of a sensitive filter fluorimeter to allow the determination of amino acids down to a few picomoles. The reagent employed, o-phthalaldehyde, is best prepared as given in section 2.1.3 and mixed 1 : 1 with the column effluent, allowing 2 min reaction time at room temperature before passing the reacted stream through the fluorimeter cell.

Garrasi et al. (1979) describe such a system for a full amino-acid analyser using a 3-buffer stepwise elution of the acidic, neutral and basic amino acids on a 6×420 mm column filled with 12-μm cation-exchange resin. The analysis time including regeneration and equilibration was around 4 h.

Fig. 2 shows a typical standard run with 50 picomoles of each acid together with a chromatogram obtained from injection of 2 ml of seawater (filtered Black Sea water from a depth of 2160 m; total amino-acid concentration 419 nmole l^{-1}).

Whilst microbore columns (around 2 mm i.d.) offer the advantage of speed of separation (Gardner, 1978) higher system pressures are involved and

454

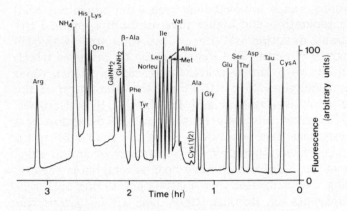

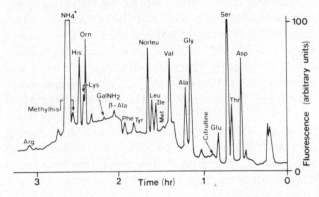

Fig. 2. Upper: Chromatogram of an amino-acid standard mixture. Each peak is 50 pmol. Lower: Chromatogram showing the dissolved free amino-acid composition after direct injection of 2.0 ml of seawater sampled from a depth of 2160 m in the Black Sea. The total concentration of DFAA is 419 nmole l^{-1}. The composition on a mole % basis is: Asp 8.3, Thr 6.5, Ser 25.7, Glu 3.6, Gly 18.6, Ala 7.4, Val 1.6, Ile 1.9, Leu 2.7, Tyr 1.9, Phe 1.7, Orn 9.4, Lys 1.0, His 5.2, Arg 3.1. Norleu, internal standard, recovery 97%. Full scale sensitivity as in standard (upper). Courtesy of Garrasi et al. (1979).

only low amounts of sample (up to 100 μl) may be injected. Garrasi et al. (1979) and Dawson and Pritchard (1978) favour the use of wider bore colums for their tolerance to larger volumes (up to 2 ml) of seawater (up to 35‰ salinity) with little loss of separation through peak broadening.

The method described by Garrasi et al. (1979) was tested on samples which had been filtered (0.45 μm), poisoned by the addition of HgCl$_2$ and stored at 4°C for some weeks. Whilst this is most commonly the fate of samples intended for analyses of organics back in the static laboratory, the authors point out that the amino-acid spectra obtained from such samples may not accurately reflect the composition at the time of sampling.

To what extent filtration, preservation and storage affects the compositions of the sample is at present difficult to assess. G. Liebezeit (unpublished results) has suggested that filtration of samples, even when carefully performed with the risks of contamination in mind, through various filter materials, increases contamination with serine and glycine.

Other postulated sources of error may include: hydrolysis of polymeric material during storage, adsorption onto the vessel walls, leakage of amino acids from organisms passing the filter and losses due to condensation of monomeric labile organic substances. Whilst the absolute amounts of total free amino acids may remain constant (if a balance between losses and gains is achieved) the composition may be significantly affected. Shipboard analyses of DFAA would certainly go a long way to overcome the problems of storage. Filtration still remains a problem. A suggested filtration procedure is as follows: samples for analysis of DFAA should be drawn from the thoroughly cleaned sampler (i.e. Niskin bottle, see Wangersky and Zika, 1978) immediately upon retrieval. A filter funnel containing a 0.2-μm filter should be previously rinsed with organic-free water and subsequently with the water from the sampler. Plastic gloves should be worn and all glassware should be chemically clean and sterilised. Since only a few millilitres of sample are required for analysis, filtration may proceed under gravity flow (with a head of only 10 cm) thus minimising the possibility of cell rupture. The filtrate of a few millilitres is caught in a precombusted test tube and analysed immediately where possible after the addition of a few microlitres of concentrated sulphuric acid to adjust the pH to between 2.0 and 2.5 (the sulphuric acid should be heated at a temperature over 300°C for several hours to remove any amino-acid contamination).

Full amino-acid analysers with multiple buffer-eluting systems are neither simple in operation nor suited for use at sea. Problems are often encountered with the baseline rises, when changing from one buffer to another, due to differences in degrees of buffer contamination when working at high sensitivities (Dawson and Gocke, 1978). At the sacrifice of the detection, of all the amino acids in seawater isocratic analysers, separating groups of amino acids (i.e. acidic and neutral or basic amino acids), overcome many of the difficulties of operation and stability of baselines, are considerably simpler in their construction and are easier to use at sea.

Fig. 3 depicts schematically such an isocratic amino acid analyser which has been successfully employed for direct analysis in the field. The chromatographic conditions for the separation of acidic and neutral acids are given in Table I. Using a similar analyser with the conditions given in Table II, basic amino acids may be separated with a single buffer. Other groups of amino acids, not covered in this range, may be separated by variations of the chromatographic parameters. Fig. 4 depicts typical chromatograms obtained by the direct injection of 2 ml unfiltered seawater, acidified with H_2SO_4 to pH 2.0—2.5, using the two analysers.

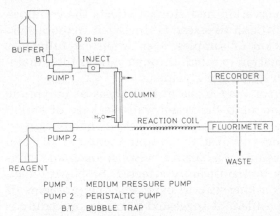

BUFFER

PUMP 1   MEDIUM PRESSURE PUMP
PUMP 2   PERISTALTIC PUMP
B.T.   BUBBLE TRAP

Fig. 3. Schematic diagram of isocratic amino-acid analyser.

The limit of detection, for quantification of the peak area above an acceptable baseline, largely a question of the sensitivity of the fluorimeter employed, lies around 5 nmole l^{-1} for an individual acid when 2 ml is injected.

The capabilities of the analytical technique are sufficient to monitor dissolved free amino acids in oceanic samples, e.g. from the Sargasso Sea, where average concentrations as low as 25 nmole l^{-1} have been reported for total DFAA (Lee and Bada, 1977; Liebezeit et al., 1980).

TABLE I

Chromatographic parameters for acidic and neutral amino acid separations

Resin	Locarte Cation Resin, 7 μm
Column	23 × 0.9 cm glass, thermostated at 53°C
Buffer	0.2N trisodium citrate, pH 3.25 ± 0.01; 50 ml ethyleneglycol monomethylether l^{-1}
Reagent	o-phthalaldehyde + mercaptoethanol (as in section 2.1.3)
Flowrates	buffer 40 ml h^{-1}; reagent 48 ml h^{-1}
Reaction temp., time	room temp.; 2 min
Detection	fluorimetric, λ_{ex} 340 nm, λ_{em} 440 nm, cell volume 45 μl
Regeneration	depending on sample composition; after 2—4 samples with 4 ml 0.2 M NaOH
Equilibration	20 min
Amino acids separated	taurine, aspartic acid, threonine, serine, glutamic acid, citrulline, glycine, alanine and cysteine (increasing run-time from 60 to 75 min allows determination of valine)

TABLE II

Chromatographic parameters for basic amino acid separations

Resin	Locarte Cation Resin, 7 μm
Column	11 × 0.9 cm glass, thermostated at 70°C
Buffer	0.35 N trisodium citrate, pH 5.96 ± 0.01; 50 ml ethanol l^{-1}
Reagent	o-phthalaldehyde + mercaptoethanol (as in section 2.1.3)
Flowrates	buffer 40 ml h^{-1}; reagent 48 ml h^{-1}
Reaction temp., time	room temp.; 2 min
Detection	fluorimetric, λ_{ex} 340 nm, λ_{em} 440 nm, cell volume 45 μl
Regeneration	unnecessary
Amino acids separated	phenylalanine, tyrosine, histidine, lysine, ornithine, ammonia and arginine

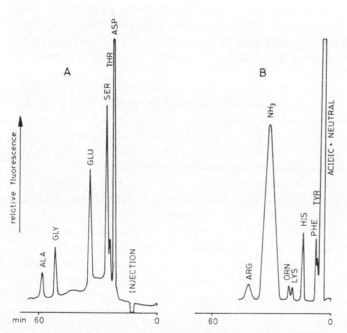

Fig. 4. Chromatograms obtained from the analysis of 2 ml of unfiltered Baltic Sea water, by direct injection into two isocratic analysers: acidic-neutral (A); basic (B). The glutamic acid peak (GLU) represents a concentration of ca. 100 nmole l^{-1}. Peak identity: ALA = alanine; GLY = glycine; GLU = glutamic acid; SER = serine; THR = threonine; ASP = aspartic acid; ARG = arginine; ORN = ornithine; LYS = lysine; HIS = histidine; PHE = phenylalanine; TYR = tyrosine. $ACIDIC + NEUTRAL$ = amino acids not separated under conditions for basic amino acids.

2.2.4. Determination of DFAA by HPLC

The high-performance liquid chromatograph (HPLC) has been suggested to be only at the very beginning of its development and had been predicted to "become one of the major analytical tools for organic materials in sea-water in the next ten years" (Wangersky and Zika, 1978).

The major drawbacks in the case of this system have been connected in the past with the unspecific nature and lack of sensitivity of the detectors employed (e.g. refractive index detection, UV monitor, flow calorimeter). These may be overcome by the use of a locating agent after the column as in ion-exchange chromatography or by the analysis of specific derivatives followed by colorimetric or fluorimetric detection. The latter, termed pre-column derivatisation, appears to be the more suitable approach since commercially available HPLC systems may be adopted without modification and the risk of peak broadening with increase in system dead volume are mini-mised.

Zika (1977) compared two precolumn reagents — fluorescamine and dansyl chloride — and found the latter to be more suitable for seawater determinations. The long reaction times and the solvent extraction step involved prior to injection, together with the bulky nature of the derivative (leading to a loss of resolution), and the fact that the reagent and the reac-tion by-product fluoresce themselves, limits the usefulness of the method.

In a recent publication, Lindroth and Mopper (1979) have presented a method for precolumn derivatisation followed by separation on reversed phase packing materials with subsequent fluorimetric detection of the amino acid-derivatives. The method has been applied for the analysis of small volumes (100 μl or less) of seawater with separation of up to 25 amino acids within 25 min at sensitivities in the femtomolar (10^{-15} M) range. The speed of analysis together with the sensitivities attainable on a routine basis and the simplicity of the analytical instrumentation and procedure, makes the technique eminently suitable for real-time analysis in the field (Garrasi et al., 1979). It is certain that this approach is an exciting development in marine analytical organic chemistry and sets the standard for discovery of equiv-alent techniques to detect and quantify other biochemical compounds.

The precolumn derivatisation is based on the formation of substituted iso-indoles (Fig. 5) on reaction of o-phthalaldehyde with primary amines

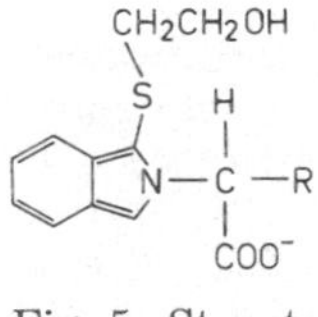

Fig. 5. Structure of fluorescent isoindole derivatives of amino acids. R = amino-acid car-bon chain.

in the presence of a reducing agent (2-mercaptoethanol) in strong alkaline medium. The derivatives exhibit the same fluorescence characteristics as those described for the post-column system (λ_{ex} 340, λ_{em} 455 nm).

The n-substituted iso-indoles are chromatographed on reversed phase silica (bonded C_{18} porous spherical silica, 5 μm), with gradient elution employing methanol-phosphate buffers. Since the buffers show no native fluorescence under the conditions employed, the resulting baselines are thus a product of electronic noise alone.

Full details of the chromatographic procedures may be found in Lindroth and Mopper (1979). A few difficulties remain when employing the method:

(1) The reagent for use with seawater samples should contain at least a thousandfold excess of o-phthalaldehyde on a molar basis compared with the content of amino acids in the sample and the reagent should be added in small quantities to the seawater, i.e. 10 μl to 1 ml of seawater sample, and should be sufficiently basic to ensure a reaction pH of least 9.0 when mixed. An "aged" reagent should be used to ensure that any fluorophor, resulting from contamination of the chemicals employed, has decayed.

(2) Since the reagent reacts with primary amines rapidly at room temperature to form fluorescent compounds with different stabilities, the derivatised sample should be injected after a precisely determined reaction interval (usually 2 min), thus limiting the apparatus in its present form to manual operation.

(3) Since the analysis of unfiltered seawater samples may sometimes yield irreproducible results (Lindroth and Mopper, pers. comm.), samples should be filtered through 0.2-μm filters. The effects of filtration on sample composition (as mentioned on p. 455) are not fully investigated for the HPLC system.

(4) The effect of the salt content on the separation efficiency and column life requires further investigation. Standards should be prepared in artificial seawater or natural seawater depleted of DFAA in order to correct for any possible "salt-effects" on the retention times, or elution order.

(5) There is some question as to which internal standard should be employed. (Lindroth and Mopper, pers. comm., suggest α-amino butyric acid for seawater samples.)

An example of a chromatogram obtained from the analysis of a 25-μl coastal seawater sample without prior clean-up is depicted (courtesy Lindroth and Mopper) in Fig. 6. The concentration of aspartic and glumatic acids represents around 10 nmole l^{-1}.

2.3. Combined amino acids (CAA) in the dissolved and particulate matter

Any of the previously described techniques for the determination of free amino acids (total, ion exchange, GC, HPLC) may be employed for estimation of CAA after a suitable hydrolysis procedure, measurement of the

460

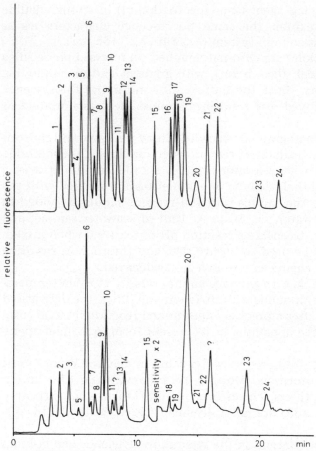

Fig. 6. HPLC separations of OPA derivatives of amino acids. Upper chromatogram shows a standard run employing gradient elution with phosphate-methanol (details given by Lindroth and Mopper, 1979). Each peak represents 30 pmol of an individual amino acid. Lower chromatogram shows the separation of amino acids from a sample of 100 μl of Baltic surface water, unfiltered and injected after derivatisation. Peak 15 (internal standard) represents 2 pmol.

Peak identification: 1 = cysteic acid; 2 = aspartic acid; 3 = glutamic acid; 4 = S-carboxymethyl cysteine; 5 = asparagine; 6 = serine; 7 = glutamine; 8 = histidine; 9 = threonine; 10 = glycine; 11 = arginine; 12 = β-alanine; 13 = tyrosine; 14 = alanine; 15 = α-amino butyric acid; 16 = tryptophan; 17 = methionine; 18 = valine; 19 = phenylalanine; 20 = ammonia; 21 = isoleucine; 22 = leucine; 23 = ornithine; 24 = lysine. Chromatograms courtesy of P. Lindroth and K. Mopper.

hydrolysate for free amino acids and subtraction of a predetermined value for DFAA.

Whilst the conditions for the hydrolysis of protein material has been a subject of discussion in the clinical or analytical fields (see Blackburn, 1978), little attention has been focused on the hydrolysis of nitrogenous compounds in seawater.

Hydrolysis with 5—6 M HCl at 100°C for 20—24 h under nitrogen (or slight modifications) appears to have been the most popular procedure. It has often been reported in the literature that hydrolysis of proteins with strong HCl results in the destructive loss of tryptophan and incomplete recoveries of methionine, threonine and serine (Blackburn, 1978). It is at present difficult to assess the extent of these losses and to evaluate the efficiency of hydrolysis when dealing with marine samples. Attempts to estimate recoveries often include the total analytical errors involved in other preconcentration desalting steps (e.g. Lee and Bada, 1975; Siezen and Mague, 1978).

Since organic matter is variable in its composition with time and space, both in the dissolved and particulate fraction, its lability or stability towards hydrolysis with strong acids is likely also to be highly variable. For hydrolysis of DCAA two approaches have been adopted:

(1) Evaporation of the seawater sample to the dried salts, addition of a small quantity of acid followed by dilution of the acid after hydrolysis before direct injection (Garrasi et al., 1979) or neutralisation of the acid hydrolysate and subsequent desalting (e.g. Lee and Bada, 1975).

(2) Mixing of equal quantities of 10 M HCl and seawater sample followed by neutralisation before direct injection (C. Garrasi, pers. comm.).

Whilst the latter avoids a lyophilisation step, larger quantities of acids and bases are required whose purity should be guaranteed beforehand.

2.4. The pattern of amino acids in seawater

In the last few years there have been considerable developments in the detection and quantification of amino acids in seawater. The pattern of amino acids found in the DFAA pool was believed in the past to be fairly uniform (see Dawson and Gocke, 1978) with a dominance of neutral amino acids (particularly glycine and serine). Fig. 7 was, in the past, considered to be a typical pattern for amino acids in seawater, including all artifacts arising from the desalting, concentration and transfer steps.

Although there is a paucity of data concerning the distribution of amino acids determined by direct injection techniques, preliminary studies (Dawson and Liebezeit, 1978; Garrasi et al., 1979; Lindroth and Mopper, 1979; Liebezeit et al., 1980) suggest the following: glycine and serine are not necessarily dominant acids; valine may be present in amounts more significant than reported; aspartic and glutamic acids may dominate the DFAA spectrum in many surface waters and porewaters; histidine concentrations may have been considerably underestimated; ornithine is present in many of the samples analysed; arginine may make up a considerable proportion of the DFAA at depth but is almost absent in surface samples; aromatic and sulphur-con-

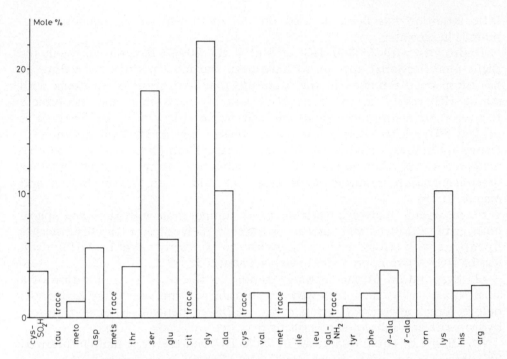

Fig. 7. Average molar % composition of free amino acids in seawater (according to Dawson and Gocke, 1978).

taining amino acids are almost undetectable in DFAA, but present in small amounts in DCAA; in the DCAA fraction, serine and valine are quantitatively less important; distinct differences are to be observed in the pattern of amino acids in samples from the euphotic zone and those from lower-lying layers.

It is likely that the advent of direct methods of analysis will also show that the absolute concentrations of amino acids in seawater have been overestimated. Few measurements exist for samples from the deep ocean and until further data are available, the lower limit for total free amino acids, as suggested by Lee and Bada (1975), whose techniques avoided any sample preservation, may be considered to be a reasonable estimate.

At the present it is estimated that the concentration of dissolved combined amino acids lies 4 to 5 times higher than that of DFAA (Garrasi et al., 1979).

2.5. Amino-acid related compounds

Urea has been reported as an excretion product of protein, purine and pyrimidine metabolism of microorganisms and zooplankton and may be regarded as being a part of the N-pool available for phytoplankton growth (Remsen, 1971). Although urea may be separated on the amino acid analyser, the weak response with OPA reagent (less than 5% of glycine response) renders the method too insensitive for analysis of marine water. The method most favoured by researchers in the field has been the colorimetric detection of the product formed on reaction of urea with biacetylmonoxime (Newell et al., 1967), an automatic version of which was developed by DeMarche et al. (1973).

Creatine, reported as being an excretion product specific to fish and not to zooplankton was determined in seawater by Whitledge and Dugdale (1972) using an automated procedure. The method was based on the reaction of creatine with alkaline ninhydrin solution after chelation of the alkaline earth elements with a complexing agent and fluorimetric detection of the reaction products. The method does not appear to have been widely adopted and the detection limit above the fluorimetric blank (around 0.10 μmole creatine l^{-1}) and the dubious stability of creatine in aqueous solution probably renders the method useful only in the laboratory.

3. CARBOHYDRATES

The most abundantly produced class of compounds in the biosphere are the carbohydrates. As the primary product of photosynthesis, free and combined sugars constitute a large fraction of aquatic organisms (up to 80% in the case of macrophytes). Between 15 and 90% of photoassimilated carbon may be released as carbohydrates by algae during growth periods (Hellebust, 1974) and even at low concentrations sugars provide a substantial food source for heterotrophic bacteria (e.g. Wright and Hobbie, 1966; Williams and Yentsch, 1976). Carbohydrates participate in several metabolic functions, serve as food storage products and are in many cases dominant structural products. Mucopolysaccharides are abundant in the marine environment, chitin the commonest bio-polymer in the sea, second only to cellulose as the most abundant polysaccharide on earth.

Despite the great interest in this class of compound and the numerous reports concerning the determination of total carbohydrates in seawater, surprisingly few determinations of individual monosaccharides have been made. Unlike the determination of free amino acids in seawater, no method for direct determination of free monosaccharides exists at present. Analogous to the developments in the detection of DFAA, the introduction of direct analysis techniques may change our concepts of the pattern of free sugars in seawater at a future date.

3.1. Methods for the determination of total carbohydrates in seawater

A number of colorimetric procedures, relying on the formation of furfural derivatives of carbohydrates in concentrated sulphuric acid, employing

various reagents, have been applied for the determination of the total carbohydrate content of seawater. Short reviews have been given by Josefsson et al.
(1972) and Burney and Sieburth (1977). Josefsson et al. (1972) have tested
five of these reagents in an automatic procedure with the conclusion that the
l-tryptophan method gives the best results. Other reagents, N-ethylcarbazole
(Zein-Eldin and May, 1958), anthrone (e.g. Antia and Lee, 1963; Walsh and
Douglass, 1966), orcinol (Larsson and Samuelson, 1967) and phenol (Handa,
1966), show either a salt dependence or variations in the responses of different sugars, suffer from interferences from non-carbohydrate compounds or
are too insensitive for oceanic samples (Josefsson et al., 1972). Bikbulatov
and Skopintsev (1974) further report an increase in the absorbance of
"yellow organic materials" on treatment with sulphuric acid, which may
interfere in the determination of total carbohydrate.

In general, all the above-mentioned methods, employing concentrated
sulphuric acid, have common drawbacks which may lead both to under- and
over-estimation of carbohydrate content. During dehydration of individual
sugars, non-detectable by-products such as γ-pyrones or benzene derivatives
may be formed. No distinction is made between mono- and polysaccharides
and since the different methods have different responses towards the various
sugars, comparison of results obtained by the different authors is difficult.

These difficulties may be overcome by the method of Johnson and Sieburth (1977), who combined several well-known reactions of carbohydrates
to gain specificity in the determination. Their spectrophotometric assay
included reduction of free monosaccharides to sugar alcohols with KBH_4,
periodate oxidation and subsequent determination of the liberated formaldehyde with 3-methyl-2-benzothiazolinone hydrazone hydrochloride. Since
only one compound is actually determined (formaldehyde), the problems
associated with varying responses of different sugars are eliminated.

In addition to supplying information about the monosaccharide content
of seawater down to low levels, the method may be used for the estimation
of the polysaccharide content of water and particulate material, after a hydrolysis step: $0.1N$ HCl for 20 h at $100°C$ (Burney and Sieburth, 1977).

The method shows no salinity dependence and there are few appreciable
interferences. Whilst the method involves a large number of procedural steps,
which renders it somewhat tedious for routine use, this sensitive and precise
procedure is probably the only recommendable "total" method to date.

The use of fluorimetric reagents generally increases the sensitivity of
detection by 1 to 2 orders of magnitude. Hirayama (1974) used anthrone for
fluorimetric determination of pentoses and 5-hydroxy-1-tetralone for hex-
oses. However, sodium chloride was found to interfere, thus necessitating an
additional desalting step. Concentrated sulphuric acid was used in both
determinations and thus the method has its drawbacks and interferences as
stated above. Furthermore, an increase in the natural fluorescence background of seawater samples may occur on heating and lead to overestimations.

Although Wangersky and Zika (1978) have suggested the use of Cu-bicin-choninate (Mopper and Gindler, 1973) for total carbohydrate estimation, its use may be limited in that it detects reducing substances and not exclusively carbohydrates. A value for reducing substances may be obtained using this method, after removal of alkaline earth metals by cation exchange.

3.2. Determination of individual monosaccharides

In comparison to the free amino acids in seawater, relatively few deter-minations of free dissolved sugars have been carried out. With the exception of a few enzymatic procedures, no direct methods for the determination of sugars in seawater exist at present and lengthy desalting concentration steps are imperative. This has certainly detracted from making shipboard mea-surements of monosaccharides.

3.2.1. Enzymatic techniques

Almost all studies of the uptake of dissolved organic compounds by heterotrophic marine microorganisms have used ^{14}C glucose as a model sub-strate since the introduction of the technique by Parsons and Strickland (1962). The actual concentration of glucose in seawater was unknown in the past and its presence in seawater was inferred from its presence in biota. The need to estimate actual glucose flux, which requires a knowledge of the *in situ* glucose concentration, together with the inadequacy of chemical methods of analysis in the sixties, were the stimuli for the development of enzymatic methods for glucose determination in seawater.

The method for the determination of glucose in seawater, described by Hicks and Carey (1968) used hexokinase as phosphorylating agent, reduc-tion of the resulting glucose-6-phosphate with glucose-6-phosphate-dehy-drogenase in the presence of NADP. The oxidised form of NADP is reduced by a dye which itself forms a fluorescent product. Since the blank values for this procedure were as high as 80%, Cavari and Phelps (1977) omitted the use of glucose-6-phosphate-dehydrogenase in order to reduce the blanks, but in doing so stated that the method would lead to an overestimation since hexokinase alone is not specific for glucose. Hanson and Snyder (1979), using a modification of the Hicks and Carey method, were able to eliminate the interference of other sugars by treatment of a duplicate sample with glucose oxidase. Thus, the difference between treated and untreated samples represents the glucose content. The enzymatic methods for glucose deter-mination have mainly been applied to freshwaters and their adaptation to saline waters was evidently not quite as straightforward as hoped. The enzymes employed may be "poisoned" by the ionic species in seawater and a salt error correction may be necessary in regions of widely varying salini-ties. A crude desalting of the major polyvalent cations and anions in seawater in a batch process with ion-exchange resins before analysis of the seawater may produce more reliable results (Saltzmann, pers. comm., 1979).

Unfortunately, no other enzymatic methods have been adopted for the determination of other sugars in seawater. A method for the determination of fructose would be of special interest since Mopper (1977), Meyer-Reil et al. (1978) and Mopper et al. (1980) have shown that this sugar may contribute significantly to the total free monosaccharide content of water. The methods used by Ruchti and Kunkler (1966) for the estimation of the glucose, saccharose and fructose contents of freshwaters may be successfully adapted for analysis in marine waters.

3.2.2. Desalting and concentration procedures prior to chromatography

Until the sensitivity of an analytical technique is improved to a point where direct analyses become feasible, a desalting and concentration step remains mandatory for the detection of free sugars in seawater.

A treatment of the methods for the isolation of organic compounds from seawater has been given by Jeffrey and Hood (1958) and more specifically for carbohydrates by Josefsson (1970).

Desalting by ion exchange, either after extracting the dried salts of a seawater evaporate with aqueous organic solvents (Degens et al., 1964), or by passing the seawater directly through a mixed cation—anion exchanger, suffers from several disadvantages: a number of transfer steps may be involved; the amounts of ion-exchange resins employed are considerable when desalting higher-salinity waters and the distilled water rinses combined with the column eluates mean that large volumes of water must be evaporated to dryness.

If desalting on ion-exchange resins is the adopted method, then the anion-exchange resin should be employed in the bicarbonate or formate form and the water percolated first through the cation resin in the H^+ form. This avoids destruction losses of sugars under strongly alkaline conditions (Mopper, 1973).

A somewhat milder procedure for removal of inorganic salts from filtered seawater samples for sugar analysis has been described by Jeffrey and Hood (1958), Josefsson (1970) and Hirayama (1974). Ion-exchange membrane electrodialysis, whereby cation and anion permeable membranes are used to separate a seawater sample from a cathode and anode supplying a d.c.-field, has been shown to be a non-destructive method for desalination of solutes of neutrally charged organic compounds. Recoveries of glucose after desalting have been shown to approach 100% (Josefsson, 1970) and the method results in no increase in the sample volume to be subsequently evaporated to dryness and few transfer are steps involved. Fig. 8 depicts an exploded view of the apparatus employed by Josefsson (1970). The volume of the seawater compartment may be varied to suit the analytical requirements. The current applied across the membranes should not exceed 10 mA cm^{-2}, the voltage applied should be adjusted to maintain the maximum cur-

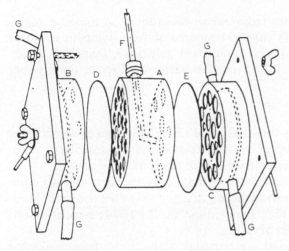

Fig. 8. Exploded view of electrodialysis apparatus. *A*, center compartment for seawater; *B* and *C*, anode and cathode compartment; *D* and *E*, ion-exchange membranes; *F*, stirrer; *G*, flow of tapwater. Courtesy of Josefsson (1970).

rent. A sample of 100 ml of 35‰ salinity seawater requires around 6 h to desalt to a conductivity of $<0.01 \ \Omega^{-1}$.

The desalted sample is then evaporated to dryness either by rotary evaporation under reduced pressure at 40°C or by lyophilisation. The addition of a small quantity of glycerine (e.g. 100 μl to 100 ml of sample) prior to reduction in volume avoids any losses of sugars by wall-induced condensation reactions or adsorption (Dawson and Mopper, 1978a). The residue is then taken up in a small volume of 20% aqueous ethanol and may be stored refrigerated if not analysed immediately by liquid chromatography.

Residues intended for derivatisation for gas chromatography are dissolved in the appopriate solvent. Using liquid chromatographic techniques described below with their present day detection limits, approximately 50—100 ml of seawater (5—10 ml interstitial water) require desalting and concentrating to 1 ml final volume.

Although the desalting and concentration steps are time-consuming, and therefore only a limited number of samples may be processed, efforts should be taken to perform these steps soon after sampling and filtration to ensure that sample composition and concentrations of constituents reflects those *in situ*. Until a direct method of determination in real time is available, the effects of sample filtration, preservation and storage will be difficult to assess. It is also difficult to judge whether the pattern of free monosaccharides reported reflects the actual *in situ* distribution.

3.2.3. Paper chromatography
The use of paper chromatography to identify sugars in seawater is now

only of historic interest and apart from earlier techniques to measure sugars in sediment (e.g. Plunkett, 1957) the first reports of free sugars in seawater came from Degens et al. (1964) who quantified galactose, mannose, rhamnose and arabinose in seawater by visual comparison of spots on a paper chromatogram.

3.2.4. Gas chromatography

Gas chromatography provides the analytical chemist with a powerful tool for rapid separations of a variety of volatile compounds. Non-volatile compounds such as sugars must be derivatised before analysis.

TMS (trimethylsilyl) derivatives of sugars may only be formed with completely dry samples, thus the method as adapted for seawater analysis (Modzeleski et al., 1971; Tesarik, 1972; Brockmann et al., 1974) requires steps for desalting and concentration to dryness.

The gas chromatograms obtained are complex since up to four peaks for the anomeric and ring forms of one simple sugar are separated. Eklund et al. (1977) achieved complete separation of the anomers using gas chromatography with high-sensitivity electron-capture detection of trifluoroacetyl (TFA) esters of nine sugars on glass capillary columns. Although only small volumes of seawater required desalting and drying, the derivatisation step with trifluoro-acetyl anhydride at 130°C cannot guarantee that only free monosaccharides are analysed.

3.2.5. Liquid chromatography

Two liquid chromatographic separation procedures have been applied for the determination of individual saccharides in seawater samples.

Partition chromatography of sugars on anion-exchange resins with ethanol-water eluant (Larsson and Samuelson, 1967), and colorimetric detection of the separated compounds with orcinol-sulphuric acid (Josefsson, 1970), tetrazolium blue (Mopper and Degens, 1972) or Cu-biconchininate (Mopper and Gindler, 1973), whilst fulfilling many of the requirements for resolution and sensitivity, has serious shortcomings when applied to the analysis of marine samples. The interference from other compounds in the residues after preconcentration, leads to a rapid deterioration in the performance of the anion-exchange columns.

Chromatography of the borate complexes of saccharides on strong anion-exchange resins in the borate form overcomes many of the difficulties associated with partition chromatography and in conjunction with a sensitive colorimetric reagent for the detection of reducing sugars, Cu-biconchininate (Mopper and Gindler, 1973), has been the method most commonly applied for the analysis of water and sediment samples (e.g. Dawson and Mopper, 1978a; Mopper, 1978a; Meyer-Reil et al., 1979; Boehm et al., 1980; Mopper et al., 1980; Liebezeit et al., 1980).

The elution behaviour of the borate complexes of saccharides on strong anion-exchange resins is dependent upon a number of factors, including medium grain size of the resin, degree of cross linkage, column length and diameter, column temperature, molarity and pH and the flowrate of the eluting buffer. Details of the influences of several of these parameters on separations may be found in Sinner et al. (1975), Simatupang et al. (1976) and Mopper (1978a). For general reviews of analytical techniques for carbohydrate analysis the reports by Jandera and Churacek (1974) and Kennedy (1974) should be consulted. The apparatus for the chromatographic separation and detection of reducing sugars is shown schematically in Fig. 9 and typical running parameters are given in Table III. The detecting reagent, Cu-bicinchoninate (Mopper and Gindler, 1973), produces a lavender-coloured product on reaction with reducing sugars after heating for 4 min at 124°C. The reaction temperature is held constant at 124°C by use of a reflux bath containing methyl cellosolve. The absorption maximum for colorimetric detection lies around 560 nm and sensitivities are such that 100 picomoles of a reducing sugar may be quantified above an acceptable baseline. Fig. 10 shows a typical chromatogram together with examples of the pattern of free sugars found in 100 ml of seawater sample, desalted and concentrated using ion-exchange electro-dialysis and lyophilisation.

Up to 19 individual sugars have been separated in one run of 3 h using the borate system: 2-deoxyribose, rhamnose, ribose, maltose, lactose, cellobiose, mannose, fructose, arabinose, fucose, galactose, tagatose, lyxose, xylose, glucose, gentiobiose, melibiose, mannoheptulose, 2-deoxygalactose. About

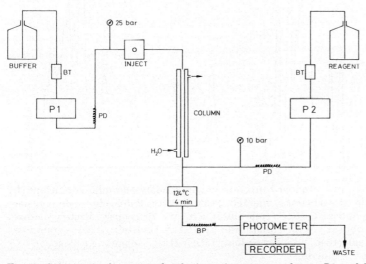

Fig. 9. Schematic diagram of colorimetric sugar analyser. *P1* and *P2* are medium pressure pumps; *BT* = bubble trap; *PD* = pulse dampeners; and *BP* = back pressure coil to suppress boiling.

470

TABLE III

Chromatographic parameters for separation of reducing saccharides

Resin	Durrum DAX-4, 20 µm anion-exchange resin borate form (regenerated as per Mopper, 1978a)
Column (a) (b)	25 × 0.32 cm, thermostated at 54°C 35 × 0.6 cm, thermostated at 78°C
Buffer (a) (b)	0.8 M boric acid adjusted to pH 8.30 with 8 M NaOH 0.65 M boric acid, pH 9.25 (NaOH)
Reagent	Cu-bicinchoninate/aspartic acid (Mopper and Gindler, 1973)
Reaction temp.; time	124°C (boiling methyl cellosolve); 4 min
Flowrates (a) (b)	36 ml h^{-1} buffer; 36 ml h^{-1} reagent 42 ml h^{-1} buffer; 42 ml h^{-1} reagent
Detection	spectrophotometric at 562 nm, optical pathlength 10 mm

Conditions (a) may be employed for rapid analysis of a limited number of sugars, whereby the total analysis time to glucose would amount to 30 min.

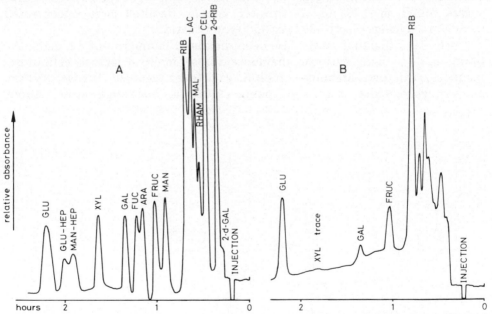

Fig. 10. A. Chromatogram of a standard mixture of sugars analysed under conditions (b) Table III. Five nanomole of each sugar injected. Peak identity: *2-d-GAL* = 2-deoxygalactose; *2-d-RIB* = 2-deoxyribose; *CELL* = cellobiose; *RHAM* = rhamnose; *MAL* = maltose; *LAC* = lactose; *RIB* = ribose; *MAN* = mannose; *FRUC* = fructose; *ARA* = arabinose; *FUC* = fucose; *GAL* = galactose; *XYL* = xylose; *MAN-HEP* = mannoheptulose; *GLU-HEP* = glucoheptose; *GLU* = glucose.
B. Chromatogram obtained from the analysis of 100 ml desalted and evaporated seawater. The free glucose concentration amounts to approximately 400 nmole l^{-1}. Chromatographic conditions as above.

10 of these have been identified in the free dissolved fraction of seawater. Recent attempts to employ fluorimetric locating agents for the detection of sugars after ion-exchange chromatography have to date not resulted in any significant lowering of detection limits. Reaction of aliphatic amines with reducing sugars under alkaline conditions to produce Schiff's base condensation products results in stable fluorescent compounds which may be monitored with a flow-through fluorimeter. Unfortunately, these compounds form only after long reaction times at high temperature and the fluorescent quantum yield (e.g. from the condensation product of ethylenediamine and glucose; Honda et al., 1974) is very low and high amplification factors are required to achieve sensitivities comparable with colorimetric methods. On the other hand, the reagent itself is non-fluorescent and therefore baselines are stable, and reacted products fluoresce above a low background.

The fluorescent yield may be increased by incorporating aromatic ring structures in the final condensation product and therefore reaction with o-phenylene diamine, for example, may produce an order of magnitude increase in sensitivity, perhaps at the expense of an elevated background fluorescence (K. Mopper, pers. comm.). Any reagent employed should be compatible with aqueous borate buffers at elevated pH, if ion-exchange techniques are adopted.

3.2.6. HPLC techniques for determination of sugars

High-performance liquid chromatographic techniques for the determination of carbohydrates have not yet been applied to the analysis of seawater samples. The separations achieved on porous silica columns, whilst being adequate for most determinations (of e.g. foodstuffs), do not match the performance of the borate ion-exchange system. The use of primary amine bonded silica, or silica packing materials saturated with Amine Modifier I (NATEC, Hamburg; Aitzetmüller, 1978) have led to some improvement in the performance of columns for separating monosaccharides. The problem of detection still remains and the refractive index detector commonly used to detect sugars in aqueous acetonitrile eluant is sensitive only to μg amounts of sugars.

A pre-column derivatisation step, as in the case of amino-acid analysis, followed by fluorimetric detection of the derivatives after separation, would be a logical step. However, fluorescent labels for sugars are more difficult to produce and a number of transfer steps are often employed during the derivatisation (e.g. dansylation) which renders the method less specific for free dissolved monosaccharides. Post-column reaction with a colorimetric locating agent after HPLC may decrease the limit of detection of a single monosaccharide to around 50 pM (ca. 250 nmole l^{-1} if 200 μl seawater was analysed). The method suggested by D'Amboise et al. (1980) and Noel et al. (1980) involving separation of sugars on primary amine-bonded silica or HPLC anion-exchange resin with acetonitrile-water eluant followed by reaction with tetrazolium blue (Mopper and Degens, 1972), may perhaps be adapted successfully to the low-level detection of a limited number of sugars in seawater.

3.3. Polysaccharides

Methods of analysis for total carbohydrate employing concentrated sulphuric acid (see section 3.1) do not distinguish between monomeric and polymeric carbohydrate forms in seawater, or furnish information as to their identity.

Since no methods exist for analysing discrete seawater polysaccharides directly, their identifications can at best only be inferred by an examination of their constituent monomers, including uronic acids and amino sugars.

A method for the determination of the total amount of polysaccharide material, which allows a distinction between monomeric and polymeric carbohydrates has been reported by Burney and Sieburth (1977). The authors employed the procedure described by Johnson and Sieburth (1977) after hydrolysis, and estimated the polysaccharides following subtraction of a value for free monosaccharides. Results indicated that the mean values for total carbohydrate in surface samples from Narrangansett Bay and Rhode Island Sound represented 9.5% of DOC (monomeric carbohydrates 3.1% and 6.4% for polysaccharides). These data probably represent our best available estimate of the actual sugar content of seawater (excluding any discussion of how representative these values may be for the open ocean) and point to the requirements for future research into the composition of this significant pool of organic carbon. An investigation of the monomeric composition of the dissolved polysaccharides in seawater requires a hydrolysis procedure, after desalting and concentration steps. Although Mopper (1973) discussed in detail the effects of hydrolysis with various strengths of acids for varying reaction times and temperature, these were limited to sediment and particulate extracts. Much milder hydrolysis procedures have been suggested by Burney and Sieburth (1977), who employed hydrolysis with $0.1N$ HCl at $100°C$ for 20 h and reported recovery of a variety of marine polysaccharides.

If individual monosaccharides are to be determined by a chromatographic procedure subsequent to hydrolysis, destruction losses of the individual compounds should be taken into account. For example, it has been reported that up to 90% of combined fructose may be destroyed under certain hydrolysis conditions (Mopper, 1973), and thus the pattern of monosaccharides found in samples after hydrolysis may not accurately reflect the original composition.

Selective hydrolysis (see Adams, 1965) at varying acid strengths may supply additional information about the nature (storage or structural) of the dissolved polysaccharides in seawater. For investigations of the composition of particulate matter, a selective extraction technique, to differentiate between water-soluble and residual polysaccharides, may be applied (Handa and Yanagi, 1969).

3.4. Uronic acids

Uronic acids occur in seawater in the particulate polysaccharide fraction as the building blocks of the cell walls of bacteria and certain algae (Percival, 1968). Algal excretions have also been shown to contain uronic acids (Hellebust, 1965) and phytoplankton and particulate organic matter have previously been analysed for total "hexuronic acid" (Williams and Craigie, 1970). Handa et al. (1972) isolated a glucuromannan from POM by extraction with 4% NaOH followed by acid hydrolysis.

Although uronic acids are produced by marine biota and the residues may be found in particulate matter and sediments, with the exception of a limited number of analyses (Mopper, 1977; Mopper and Larsson, 1978), data on the occurrence of uronic acids, in either the monomeric or polymeric dissolved fractions of seawater, are non-existent. The values reported for five identified uronic acids by the above authors are in the range between 10—100 nmole l^{-1} for individual acids in a single sample of hydrolysed North Sea surface water.

There is no standardised procedure for the hydrolysis of dissolved polymeric uronic acids. Mopper (1977) suggests much milder procedures than are used for sugars since uronic acids are much more labile towards decarboxylation or transformation reactions. The hydrolysis procedure suggested by Burney and Sieburth (1977) may therefore prove to be adequate. The use of cation-exchange resin in the H^+ form may also be effective.

Determination of uronic acids in the free dissolved fraction of seawater poses several difficulties, not least the low expected levels (<0.02 μmole l^{-1}; Mopper, 1977) and difficulties in removing inorganic salts from per se ionic organic compounds.

The chromatographic analyses described by Mopper and Larsson (1978), with an improved version described by Mopper (1978b), rely on the separation of complex uronic acid mixtures on anion-exchange columns in the acetate form with acetate buffers as eluant and are capable of detecting uronic acids in the nanomole range with Cu-bicinchoninate as locating agent.

The borate system described in Section 3.2.5 may be adopted for the separation of uronic acids using a mixed borate/chloride buffer as eluant. A separation of mannuronic acid from guluronic acid is, however, not possible with this chromatographic technique, although the analysis time may be reduced considerably.

3.5. Amino sugars

The occurrence of amino sugars in polymeric forms in marine biota, e.g. as mucopolysaccharides and glycoproteins, is undisputed, chitin (N-acetyl glucosamine polymer) being the most abundant of these. It would appear therefore that these compounds should be identifiable after hydrolysis of dissolved and particulate combined organic matter.

Past determinations of amino sugars have usually been carried out using standard amino-acid techniques with hydrolysis of polymeric material under the conditions applied for proteinaceous material and reporting of the total hexosamine content (e.g. Degens, 1970). Dawson and Mopper (1978b) have suggested that the conditions for hydrolysis should resemble those employed for polysaccharide material to avoid substantial destruction losses of amino sugars.

The above authors used a modified cation exchange based amino-acid analyser for the separation of the amino sugars glucosamine, mannosamine and galactosamine, followed by detection with Cu-bicinchoninate thus avoiding any interference from amino acids present in the hydrolysed samples. Although only surface sediments were analysed using the above system, Dawson and Liebezeit (unpublished results) have detected the presence of mainly glucosamine and galactosamine (with occasional traces of mannosamine) in hydrolysed coastal seawater samples.

The low reported levels of free amino sugars in seawater (e.g. Garrasi et al., 1979) together with the difficulties in concentrating "charged" organic species after customary desalting techniques, pose difficulties for the marine analytical chemist. Although amino sugars form pre-column derivatives with OPA-reagent, as in the case of amino acids (Lindroth and Mopper, 1979), no attempt has yet been made to separate them on HPLC reversed phase columns.

3.6. The pattern of carbohydrates in seawater

With improvements in analytical techniques for the analysis of sugars in seawater, it is likely that our concepts of the distribution of individual compounds will change analogous to the developments in amino-acid research.

As far as free sugars in seawater are concerned, our only available direct techniques for analysis are enzymatic ones. Hanson and Snyder (1979), whilst employing a technique for the assay of glucose, demonstrated that 43% (mean) of hexokinase phosphorylated sugars are not glucose. Mopper et al. (1980) discussed the surprisingly uniform pattern of free sugars in seawater with glucose and fructose being the dominant sugars. The ratio glucose/fructose was found to be fairly constant and close to 1 : 1. The authors concluded that the relatively high amounts of fructose in seawater result from abiotic isomerisation of glucose, since no significant natural sources of fructose in the sea have been reported.

The conditions employed for analysis exclude the possibility that such an isomerisation occurs during sample work up. However, little is known about the effects of filtration, fixation or storage on the absolute recoveries.

Too few analyses of combined sugars after hydrolysis have been performed to allow any interpretation of their pattern in seawater.

4. HYDROPHOBIC COMPOUNDS IN SEAWATER

Hydrophobic compounds are commonly considered to be those compounds extractable with immiscible semi- or non-polar organic solvents. Included in this group are: long-chain fatty acids, fatty acid esters and waxes, long-chain alcohols, steroids, hydrocarbons, phospholipids and anthropogenic compounds (e.g. chlorinated hydrocarbons, oil hydrocarbons, phthalate esters).

Within this inventory of naturally occurring lipids, the classes fatty acids, hydrocarbons and sterols have received most attention and the analytical techniques for their determination are now well established (Zsolnay, 1977). By adjustment of the extraction pH, more polar compounds (e.g. phenols) can be included in this category. Many of the above compounds classes may be analysed from a common extract by suitable choice of extracting solvent and extraction conditions. The extract is then subdivided into the various classes by selective or preparative chromatography prior to final separation and detection. The gas chromatograph has almost exclusively been chosen for final detection of the compounds since these are either volatile or form-volatile derivatives after simple derivatisation procedures. This, according to Wangersky and Zika (1978) has resulted in the accumulation of data describing the detailed, qualitative and quantitative composition of many marine lipids (due to the high resolution of modern GC—MS system), but that these individual compounds are in many cases reported without any reference to DOC levels.

Class reactions for many of the compounds are either few or non-existent and thus an evaluation of recoveries or reproducibility of the analyses is difficult. A review of the literature concerning natural lipids in seawater highlights the fact that detailed information has been gathered at the expense of any wide aerial surveys of the levels of these compounds in ocean waters. The concentration ranges found by various workers vary widely and in the absence of standardised methods of extraction and analysis, intercomparison of results is often impossible.

The methods for the analysis of hydrocarbons (natural or anthropogenic) have been treated separately in Chapter 11.

4.1. Fatty acids

Dissolved fatty acids and esters in seawater and interstitial waters have been reported to occur in the low μg l^{-1} range (e.g. Slowey et al., 1962; Williams, 1965; Stauffer and McIntyre, 1970; Saliot, 1975; Meyers, 1976; Boussuge et al., 1979). Reports of the occurrence of fatty acids in particulate matter (Schultz and Quinn, 1972; Saliot and Boussuge, 1978) and surface films (Garrett, 1967; Duce et al., 1972; Larsson et al., 1974; Marty and Saliot, 1974; Kattner and Brockmann, 1978) demonstrate a considerable enrichment in these phases.

Very few determinations have been made of the total fatty acid content of marine waters. Of the few methods available, a technique which is reported to be reliable is that of extraction of the free acids with chloroform at pH 2 followed by formation of a copper complex and final estimation of the complexed copper by atomic absorption spectrophotometry (Treguer et al., 1972). The free fatty acid contents are reported as palmitic acid equivalents. The method, however, has not yet been widely adopted for routine use.

The fluorescent tag proposed by Dünges (1977, see below) although not yet applied to marine samples (Wangersky and Zika, 1978) may offer a sensitive method for total estimations.

Since gas chromatography has been adopted in the majority of the investigations, the occurrence of saturated and unsaturated acids with chain lengths from C_{10} to C_{32} has commonly been reported. The fatty acid composition of oceanic samples is typified by a distribution showing a dominance of the even-numbered acids (C_{12}—C_{18}) over the unsaturated and the odd-numbered compounds.

Free fatty acids, hydroxy acids and their esters may be extracted from seawater after filtration with various organic solvents in high yields after acidification to pH 2—3 or extraction at pH 8 and subsequent acidification to pH 2—3 (A. Saliot, pers. comm., 1979). Chloroform is the most commonly employed extracting solvent, three repeated extractions being sufficient to quantitatively extract 1—2 l of seawater (Treguer et al., 1972).

Following reduction in volume of the extracting solvent under nitrogen, subsequent saponification and re-esterification is required prior to chromatographic separations of the mixture. After saponification no distinction is possible between free acids and their naturally occurring esters. Only direct GC—MS of the crude extract would allow any estimations of the amounts of free esters present in the sample.

Gas chromatography of the methyl esters of fatty acids on two or more columns is most commonly used for final quantification. Liquid chromatography after derivatisation of the acids to form UV sensitive or fluorescent compounds is an alternative approach which at present has not been refined to the extent of gas chromatography but, with improvements in the performance of HPLC columns, may in future offer a rapid method for screening samples.

Examples of such visualising agents for carboxylic acids (also short-chain acids) include 2-naphthacyl esters (Cooper and Anders, 1974), phenacyl esters (Borch, 1975; Durst et al., 1975), p-nitrobenzyl esters (Knapp and Krueger, 1975), p-methoxyanilides (Hoffman and Liao, 1976), p-bromophenacyl esters (Pei et al., 1976) and 4-bromomethyl-7-methoxycoumarin (Dünges, 1977). Hydroperoxy and hydroxy derivatives may be formed for unsaturated fatty acids (Aoshima, 1978).

The phenacyl esters are more readily prepared than the other derivatives. The fluorescent tag 4-bromomethyl-7-methoxycoumarin may however offer increased sensitivity. Due to the bulky nature of the fluorescent products, separation factors may be poorer and allow only positive identification of a limited number of fatty acids.

Fig. 11 depicts a proposed scheme for the extraction and further treatment of samples for fatty acid analyses.

Short-chain carboxylic acids, as products of bacterial metabolism in anoxic environments (e.g. lactic, propionic, butyric, etc.) have been successfully extracted with diethylether at pH 2 and quantified after HPLC separation (M. Hines, pers. comm., 1979).

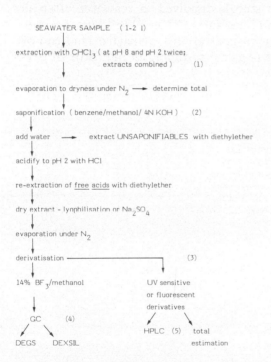

Fig. 11. Extraction and analysis of fatty acids from seawater. (1) According to A. Saliot (pers. comm.). (2) According to Boussuge et al. (1979). (3) See text for derivatives. (4) Depending on the chromatographic conditions, saturated and unsaturated acids with carbon numbers from C_{10} to C_{32} may be detected as their methyl esters. Computer interfaced GC—MS may be essential for positive identification. (5) Depending on the chromatographic conditions, mixtures of C_1 to C_{24} acids may theoretically be separated.

478

4.2. Steroids

Sterols and structurally similar molecules are present in almost all forms of marine life (Morris and Culkin, 1977). Their role as hormonal regulators of growth and reproduction and their incorporation in lipoprotein structures has been briefly discussed in a previous chapter (Gagosian and Lee, Chapter 5). Considerable interest has recently been shown in the determination of trace amounts of these biologically active substances. Due to the high diversity of their molecular structures, the steroid compounds lend themselves to use as tracers of processes involving labile organic matter.

The steroid hydrocarbon structure may incorporate functional groups such as alcohols, ketones and olefinic linkages, either in the four-ring system or on the side chain originating at C_{17}. Sterol is a term commonly used to describe all steroidal alcohols. Stenols refer to sterols with $\Delta 5$ or $\Delta 7$ double bonds, whereas stanols have a fully saturated ABCD ring system (Gagosian and Heinzer, 1979).

The concentrations of total free sterols dissolved in seawater often lies under 100 ng l^{-1} in deeper waters (Gagosian, 1975, 1976; Tusseau et al., 1978; Gagosian and Heinzer, 1979). In coastal areas receiving land run-off, some of the more important sterols (e.g. coprostanol, an indicator of possi-

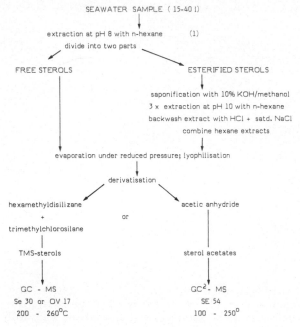

Fig. 12. Extraction and analysis of sterols from seawater. (1) At this stage, sulphur removal with an activated Cu-column may be necessary. The sample extracts may also be further purified by chromatography on silica gel (see Gagosian and Heinzer, 1979).

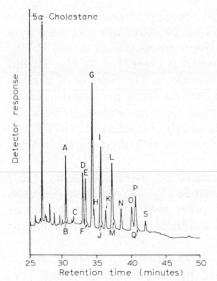

Fig. 13. Gas chromatogram of sterol acetates from 10 m sample from Black Sea. Chromatographic conditions as given by Gagosian and Heinzer (1979). Peak *O* and non-labelled peaks are unidentified non-steroidal compounds. Peak *K* is assigned as a phthalate ester. Chromatogram and peak identification list courtesy of above authors. Peak identity: *A* = 24-norcholesta-5,22-dien-3β-ol; *B* = 24-Nor-5α-cholest-22-en-3β-ol; *C* = 24-norcholest-5-en-3β-ol (norcholesterol); *C* = cholesta-5,22(Z)-dien-3β-ol; *E* = cholesta-5,22(Z)-dien-3β-ol; *F* = 5-cholest-22(E)-en-3β-ol; *G* = cholest-5-en-3β-ol; *H* - 5-cholestan-3β-ol; I_1 = cholesta-5,7-dien-3β-ol (?), cholesta-5,24-dien-3β-ol (?) (desmosterol); I_2 = 24-methylcholesta-5,22-dien-3β-ol (brassicasterol/crinosterol); *J* = 24-methyl-5-cholest-22-en-3β-ol (spongesterol); *L* = 24-methylcholesta-5,24(28)-dien-3β-ol; *M* = 24-methyl-5α-cholest-24(28)-en-3β-ol; *N* = 24-ethylcholesta-5,22-dien-3β-ol (stigmasterol/poriferasterol); P_1 = 24-ethylcholest-5-en-3β-ol (β-sitosterol/clionasterol); P_2 = 24-ethylcholest-5,24(28)(E)-dien-3β-ol (fucosterol); *S* = 24-ketocholest-5-en-3β-ol.

ble fecal pollution) may be found in higher quantities and compound specific reactions or less sensitive GC methods may be applied after concentration (Wun et al., 1976; Kanazawa and Teshima, 1978).

The low concentrations of free sterols in ocean waters require that large volumes of seawater are extracted in order to decrease the detection limit for FID detection. The structural similarity of the sterol compounds found in seawater makes demands on the chromatographic separations and identification, which may only be reasonably met with the use of the GC—MS systems (e.g. Gagosian and co-workers; Saliot and co-workers).

No class reaction exists for measurement of total sterols and hence the researcher has no method for the rapid screening of seawater and also has no reference value with which to compare the concentrations of individual compounds.

Although sterols have recently been determined by HPLC techniques

(Kiuchi et al.,1975), these have been limited to applications where either the sterol contents were comparatively high (e.g. foodstuffs) or where only a limited number of sterols were present in the extracts (Colin et al., 1979).

An extraction and analysis scheme for sterols is presented in Fig. 12 and is based on the methods of Gagosian and co-workers. The reports of individual and total free sterols include the contributions from particulate matter since filtration through various filter materials has been shown to be a significant source of analytical error and contamination (Gagosian, 1975). The above authors make a distinction between free and esterified sterols in their analytical procedures.

An example of the capability of glass-capillary gas chromatography (GC^2) in separating complex sterol mixtures is shown in Fig. 13 (courtesy Gagosian and Heinzer, 1979). Sixteen sterol acetates were identified with the aid of mass spectral data in a sample of Black Sea water. 5α-cholestane was used as internal standard.

The pattern of sterols in oceanic waters shows a predominance of cholesterol over other major sterols: 22-dehydrocholesterol, campesterol, stigmasterol, β-sitosterol and fucosterol. Based on the spectrum of sterols found in marine organisms it is likely that sterols with carbon numbers ranging from C_{21} to C_{30} (Gagosian and Stuermer, 1977) may be present in the dissolved or sedimentary phases of the marine environment.

4.3. Phenols

The toxic nature of phenolic compounds has aroused interest in their detection at low concentrations in natural waters and industrial effluents. Whilst anthropogenic sources of phenols may have an impact in polluted inshore waters, significant contributions of phenols are probably entering the seawater environment by the decomposition of attached algae and phytoplankton. To distinguish between the several sources of phenols, information is required on the kinds of phenols present and the concentration ranges to be expected.

The assumed involvement of phenolic compounds in formation of marine humus led early workers (e.g. Degens et al., 1964) to tentatively identify and quantify a few phenols, notably syringic, vanillic and p-hydroxybenzoic acids, in seawater at the low μg l^{-1} level.

The methods of extraction of phenols from water include steam distillation (Goulden et al., 1973), liquid—liquid extraction (Afghan et al., 1974) and adsorption onto macroporous anion-exchange resin (Chriswell et al., 1976). Distillation is suggested by Afghan et al. (1974) to introduce contamination when working under the 5-μg l^{-1} range and difficulties often arise in cleaning-up macroporous resins to a quality required for low-level work.

Extraction of seawater, acidified to pH 2, with either n-butyl acetate or iso-amyl acetate has been shown to be an efficient method for extraction of phenols (Afghan et al., 1974). Further selectivity is achieved in their method by back extracting the phenols from the organic phase with 1.5 M NaOH solution.

Wangersky and Zika (1978) have reviewed the commonly applied techniques for the analysis of phenols in natural waters. Spectrophotometric methods lack the sensitivity required for low-level determinations of individual phenols. Class reactions have been investigated and discussed by Afghan et al. (1974); the detection limit of the spectrophotometric 4-aminoantipyrine reagent lies around 5 μg l^{-1}. The elegant method (involving few transfer steps and reagents), differential UV spectrophotometry of the dissolved phenols in acidic and basic media, may only be usefully applied when sample concentrations are higher than 5 μg l^{-1}. This technique has advantages over the colorimetric methods since para-blocked compounds are also detected. A spectrofluorimetric determination based on the measurement of the native fluorescence of certain phenols at 305 nm when excited at 275 nm (Murray, 1949), is capable of detecting as little as 0.1 μg l^{-1}. The Raman emission of water interferes at these wavelengths, however, measurements performed on dried extracts would eliminate this blank.

A logical step would be HPLC separation of the extracts and measurement of the native fluorescence after chromatography. Enhanced sensitivity in the detection of the individual phenols may be achieved by chromatography of dansyl derivatives (Cassidy et al., 1974) in which case ng quantities of phenolic compounds were estimated in physiological fluids.

Gas chromatography (Chriswell et al., 1976) or GC—MS determinations (Kunte and Slemrova, 1975) of phenols have been limited to pollution applications in freshwaters receiving industrial effluents (e.g. Dietz and Traud, 1978).

5. COMPOUNDS RELATED TO BIOLOGICAL ACTIVITY

The compounds included in this section have mainly been detected in the particulate fractions of seawater and are suggested to correlate with biological activity. The oldest determined compounds in this class are the photosynthetic pigments, mainly the chlorophylls. The methods for the determination of the plant pigments have been standardised by international agreements (UNESCO, 1966) and the published methods of Strickland and Parsons (1972). A review of the methods for the determination of both total and individual pigments has been presented by Rai (1973).

Several vitamins have been detected and estimated in seawater samples with the aid of highly sensitive (although time-consuming) bioassay techniques. Whilst these compounds are of obvious interest as bioactive substances, the methods for their determination are more biological than chemical in nature.

The determination of ATP is a routinely adopted method which has reached a high degree of methodological refinement. With improvements in column chromatography systems, further information is likely to come from research into other nucleotides in seawater particulates. Over the last decade there have been some interesting achievements in determining the activity of specific enzymes in the marine environment.

5.1. ATP (ADP, AMP)

ATP has been shown to be present in all living organisms so far investigated, and serves the general function as an energy source in endothermic metabolic processes. As a universal component of the living cell, ATP determinations have been used for the indirect determination of biomass (Holm-Hansen and Booth, 1966), although the ratio of cell carbon to ATP may in fact vary between 50 to 5000 (Cavari, 1976; Skjoldal and Bamstedt, 1977). An average accepted value for microorganisms lies around 250 : 1 (Holm-Hansen and Booth, 1966). A modern application area for the determination of adenosine nucleotides is that of the calculation of energy charge, whereby from the ratio of the different phosphates, a measure of the physiological condition of the cells (or populations) may be derived (e.g. Atkinson, 1969).

The commonly accepted method for the analysis of ATP is based on enzymatic determination employing the lantern firefly extract luciferin—luciferase; a detailed review has been presented by Deluca (1977). Methods for the analysis of ATP, ADP and AMP may be found in the reports by Skjoldal and Bamstedt (1977), Karl and Holm-Hansen (1978), Witzel (1979) and Graf and Lagaly (1980). In the majority of investigations, the seawater particulates and the microporous filters used for collection were extracted with boiling Tris buffer (0.02 M, pH 7.8) for periods between 1 and 5 min. To avoid "stress" on the cells, the volume filtered should be kept to a minimum (50—200 ml). The extracts obtained may be stored deep frozen for several months prior to analysis back in the laboratory.

Hodson and Azam (1977) have reported that up to 20% of the total ATP in seawater occurs in the free form. It is, however, difficult to say whether this finding was a product of sample manipulation. At least in bacterial cultures, it has been shown that nucleotides may be released into the medium (Chapmann et al., 1971). On the other hand, it is questionable whether ATP is stable enough to exist in solution over longer periods owing to its low stability constant or the possibility of complexation with divalent cations and fulvic acids (Tetas and Lowenstein, 1963; Hulett, 1970; Bulleid, 1978).

The results of numerous analyses show that ATP concentrations occur within a wide range from 10 to 2000 ng l^{-1} for seawater with almost an order of magnitude higher values for the particulates from interstitial waters. These levels lie well above the absolute detection limit of the luciferin—

luciferase method, nominally given as around 200 pg l⁻¹.

Among the various other nucleotides involved in the metabolism of marine organisms, only guanosinetriphosphate (GTP) and uraciltriphosphate (UTP) have been analysed for in the marine environment (Karl, 1978). HPLC separations of the extracts from marine particulates seems a logical next step in the investigation of cell processes involving high-energy phosphates.

5.2. Nucleic acids

The nucleic acid content of particulate matter in seawater has been employed as a parameter to describe the bioactive potential of this material. A knowledge of the DNA content of seawater may provide information on the growth of planktonic organisms and may further serve as a biomass indicator (Holm-Hansen et al., 1968), since there is some suggestion of a high logarithmic correlation of DNA to carbon content per cell (Holm-Hansen, 1969). This point is at present disputed (Skjoldal and Bamstedt, 1976).

Holm-Hansen and co-workers adapted the method previously employed by Kissane and Robins (1958) for DNA determinations in mammalian tissues, using the fluorescent 2-d-ribose: 3,5-diaminobenzoic acid complex for quantitative estimations. Pillay and Ganguly (1970) absorbed nucleic acids onto homogeneously precipitated BaSO₄ and after extraction with NaCl and subsequent hydrolysis, the DNA content was determined from either phosphate or 2-d-ribose contents. Although nothing is yet known about the presence of other 2-deoxysugars in particulates, it should be borne in mind that the presence of these would interfere in determinations by the Kissane—Robins method.

Derenbach (1970) made a distinction between soluble and combined RNA of planktonic material from the different behaviour of these compounds after different extraction procedures. He and others (see Derenbach, 1970 for references), employed the orcinol/hydrochloric acid reagent according to Ceriotti (1955). Since a variety of other sugars may interfere (see section 3.1; orcinol/sulphuric acid has also been used as a reagent for total carbohydrate determinations) correction factors or preferably chromatographic separations of the extracts are required for quantitative results.

Iwamura et al. (1970) obtained DNA and RNA fractions after several extraction and purification steps to remove lipids, pigments and proteins (also characterised from the common extract).

DNA was then determined by the Kissane—Robins method; RNA was estimated from the difference in optical densities at 260 and 320 nm.

Since total methods yield only a limited amount of information, Breter et al. (1977) after precipitation of the polyanionic fraction and subsequent hydrolysis, separated the extracted DNA and RNA bases by HPLC. A complete run including adenine, cytosine, guanine, thymine and uracil lasted approximately 130 min, although this figure may easily be reduced by use

TABLE IV

Determinations of enzyme activity of marine samples

Enzyme	Substrate	Detection	Source [1]	Literature
Phosphatase	calcium phosphate culture medium phenyldisodium orthophosphate	solubilisation zones phenol	sed	Ayyakkannu and Chandromohan (1971)
Esterase	p-nitrophenyl acetate	p-nitrophenol	film-forming bacteria	Corpe and Winters (1972)
Phosphatase	p-nitrophenyl phosphate	p-nitrophenol		
β-D-glucosidase	p-nitrophenyl-β-D-glucoside	p-nitrophenol		
Proteinase	Azocoll (protein-dye-complex)	spectrophotometry		
Alkaline phosphatase	3-O-methylfluorescein phosphate	fluorimetry	sw	Perry (1972)
Deoxyribonuclease	highly polymerised calf thymus DNA	2-d-ribose:3,5-di-amino benzoic acid complex/fluorimetry	sed sw	Maeda and Taga (1973)
Arylsulfatase	tripotassiumphenolphthalein disulfate	spectrophotometry	sed	Chandromohan et al. (1974)
Alkaline phosphatase	p-nitrophenyl phosphate	p-nitrophenol iodometry;	sw	Kim and ZoBell (1974)
Amylase	starch amylopectinazure	spectrophotometry		
Urease	urea	ammonia		
Succinate dehydrogenase	sodium succinate	nitro-blue tetrazolium		

Enzyme	Substrate	Product/method	[1]	Reference
Dehydrogenase	NADH2	2(p-...) (p-NO₂-phenyl)-5-phenyl-tetrazolium chloride	sed	West and Ison (1978)
L-glutaminase	L-glutamine	ammonia	sed	Dharmaraj et al. (1977)
L-asparaginase	L-asparaginase	ammonia	sed	Selvakumar et al. (1977)
β-D-glucosidase	p-nitrophenyl-β-D-glucoside	p-nitrophenol	sw	Morrison et al. (1977) Bobbie et al. (1978)
α-D-mannosidase	p-nitrophenyl-α-D-mannoside	p-nitrophenol		
β-D-galactosidase	p-nitrophenyl-β-D-galactoside	p-nitrophenol		
Alkaline phosphatase	p-nitrophenol disodium orthophosphate	p-nitrophenol		
Phosphodiesterase	bis-p-nitrophenyl phosphate	p-nitrophenol	lw	Wunderlich (1977)
C_x-cellulase	carboxy methylcellulose	viscosity changes		
C_l-cellulase	cellulose	photometry of residue	sw	Taga and Kobori (1978)
Alkaline phosphatase	sodium-β-glycero phosphate	phosphate	lw	Little et al. (1979)
	3-O-methylfluorescein phosphate	fluorimetry		L.A. Meyer-Reil (unpublished results)
Proteolytic enzymes	hide powder azure	spectrophotometry	sed	

[1] sw = seawater; sed = sediment; lw = lakewater.

of gradient elution. For example, separations of up to 28 nucleotides, bases, nucleosides, aromatic amino acids and metabolites have recently been achieved for samples of human serum using reversed-phase HPLC and gradient elution (Hartwick et al., 1979).

5.3. Enzymes

The universal association of enzymes with life and the occurrence of organisms in all areas of the marine environment has evoked interest in the study of this class of compound. Besides the determination of heterotrophic activity and knowledge of the presence of particular strains of bacteria, the measurement of enzymatic activity may give further insight into degradation processes of *in situ* produced organic matter.

Although suspected to be present in the ocean since 1925 (Harvey, 1925) and known to occur in freshwaters (Overbeck and Babenzien, 1964), the determination of enzyme activity has gained popularity only since the beginning of the last decade.

Two types of enzymes can be distinguished: (a) extracellular, cell-free or exoenzymes, formed either by active excretion or as a result of cell lysis, and (b) enzymes closely associated with living cells. Whereas the former seem generally to be present, the second type may only by induced by addition of a suitable substrate. This requires that two different approaches be adopted for their measurement.

Removal of bacterial cells and other particulate material is necessary before estimation of exoenzymes by the addition of a substrate and measurement of the released product(s). In this case, the "actual" activity will be the determined parameter. In the case of substrate induced enzymes, homogenised sediment or water samples are incubated after substrate addition and thus the method can only yield a measure of the "potential" activity.

The added substrate should meet two requirements: the released product should be easy to detect and the chemical nature of the substrate (i.e. the stability of the bond to be cleaved) should be comparable with that of the available natural substrate. The effects of temperature, pH and possible inhibitors should be taken into account. Petterson and Jansson (1978) tested the activity of enzymes derived from freshwaters in comparison with commercially available alkaline phosphatases using different artificial substrates and found that the factors affecting the results included temperature, pH and phosphate concentration.

Normally, natural substrates are present in addition to added substrates, thus parallel determinations of these natural concentrations are required for accurate estimations of the "real" activity. A similar approach has been adopted for the determination of actual uptake rates of heterotrophs using radio-labelled substrates (e.g. Meyer-Reil, 1978).

Table IV presents a review of the types of enzymes determined to date, either in the sedimentary or water phase of the marine environment. It is interesting to note that Goodrich and Morita (1977) failed to detect any chitinase activity in estuarine sediments, in spite of the fact that chitin is considered to be one of the more abundant marine polymers. Considerable chitinase activity has, however, been detected in the stomachs of some fish species investigated by the above authors.

6. ACKNOWLEDGEMENTS

Numerous people have assisted us in the preparation of this chapter, particularly our colleagues from the SFB 95 of the University of Kiel. Dr. G. Graf supplied the information concerning ATP and Dr. L.A. Meyer-Reil assisted in the review of enzymatic methods. We particularly wish to thank Drs. Smetacek, Boelter, Mopper and Rumohr for their support throughout. We are indebted to Fritz Bohde for his painstaking efforts in collecting literature at the expense of a great deal of patience and shoe leather, and to Brigitte Lohmann-Wörner for all manner of technical and moral support. Preparation of the manuscripts would have been intolerable without the support of Hedy Sladowich, Siggi Radzewitz and Anneke van de Kasteele. Barbara Heywood demonstrated almost extreme patience in proof-reading our efforts and the Bermuda Biological Station provided an excellent hide-away for the major part of the writing.

The research involved was supported by funding from the Sonderforschungsbereich 95 of the Deutsche Forschungsgemeinschaft, and carries the contribution No. 265.

7. REFERENCES

Adams, G.A., 1965. Complete acid hydrolysis. In: R.L. Whistler (Editor), Methods of Carbohydrate Chemistry, 5. Academic Press, New York, N.Y., pp. 269—276.

Afghan, B.K., Belliveau, P.E., Larose, R.H. and Ryan, J.F., 1974. An improved method for determination of trace quantities of phenols in natural waters. Anal. Chim. Acta, 71: 355—366.

Aitzetmüller, K., 1978. Sugar analysis by high-performance liquid chromatography using silica columns. J. Chromatogr., 156: 354—358.

Antia, A.L. and Lee, C.T., 1963. Studies on the determination and differential analysis of dissolved carbohydrate in sea water. Fish. Res. Board Can., Manuscript Rep. Ser. 168.

Aoshima, H., 1978. Analysis of unsaturated fatty acids and their hydroperoxy and hydroxy derivatives by high-performance liquid chromatography. Anal. Biochem., 87: 49—55.

Atkinson, D.E., 1969. Regulation of enzyme function. Annu. Rev. Microbiol., 23: 47—68.

Ayyakkannu, K. and Chandromohan, D., 1971. Occurrence and distribution of phosphate solubilizing bacteria and phosphatase in marine sediments at Porto Novo. Mar. Biol., 11: 201—205.

Belser, W.H., 1959. Bioassay of organic micronutrients in the sea. Proc. Natl. Acad. Sci. U.S.A., 45: 1533—1542.

Bikbulatov, E.S. and Skopintsev, B.A., 1974. Determination of the total amount of dissolved carbohydrates in natural waters in the presence of humic substances. Gidrokhim. Mater., 60: 179—185 (in Russian).

Blackburn, M., 1978. Amino Acid Determination — Methods and Techniques. Marcel Dekker, New York, N.Y., 367 pp.

Blumer, M., 1975. Organic compounds in nature: limits of our knowledge. Angew. Chem. Int. Ed., 14: 507—514.

Bobbie, R.J., Morrison, S.J. and White, D.C., 1978. Effects of substrate biodegradability on the mass and activity of the associated estuarine microbiota. Appl. Environ. Microbiol., 35: 179—184.

Boehm, E.L., Dawson, R., Liebezeit, G. and Wefer, G., 1980. Monosaccharides for particle identification in sediments. Sedimentology, 27: 167—177.

Bohling, H., 1970. Untersuchungen über freie gelöste Aminosäuren in Meerwasser, Mar. Biol., 6: 213—225.

Borch, R.F., 1975. Separation of long chain fatty acids as phenacyl esters by high pressure liquid chromatography. Anal. Chem., 47: 2437—2439.

Boussuge, C., Goutx, M., Saliot, A. and Tissier, M.J., 1979. Acides gras et hydrocarbures aux interface eau de mer—sediment et eau interstitielle — sediment en Atlantique tropical Est. In: Géochimie organique des sédiments marins profonds, Orgon III, Mauretanie, Sénégal, Isles de Cap Vert. C.N.R.S., Paris, 189 pp.

Breter, H.-J., Kurelec, B., Müller, W.E.G. and Zahn, R.K., 1977. Thymine content of seawater as a measure of biosynthetic potential. Mar. Biol., 40: 1—8.

Brockmann, U.H., Eberlein, K., Junge, H.D., Maier-Reimer, E., Siebers, D. and Trageser, H., 1974. Entwicklung natürlicher Planktonpopulationen in einem Outdoor-Tank mit nährstoffarmen Meerwasser. II. Konzentrationsveränderungen von gelösten neutralen Kohlenhydraten und freien Aminosäuren. Ber. Sonderforsch. Meeresforsch., SFB 94, Hamburg, Heft 6.

Bulleid, N.C., 1978. An improved method for the extraction of ATP from marine sediment and seawater. Limnol. Oceanogr., 23: 174—178.

Burney, C.M. and Sieburth, J. McN., 1977. Dissolved carbohydrates in seawater II. A spectrophotometric procedure for total carbohydrate analysis and polysaccharide estimation. Mar. Chem., 5: 15—28.

Butcher, E.C. and Lowry, O.H., 1976. Measurement of nanogram quantities of protein by hydrolysis followed by reaction with ortho-phthalaldehyde or determination of glutamate. Anal. Biochem., 76: 502—523.

Cassidy, R.M., LeGau, D.S. and Frei, R.W.,.1974. Analysis of phenols by derivatization and high-speed liquid chromatography. J. Chromatogr. Sci., 12: 85—89.

Cavari, B., 1976. ATP in Lake Kinneret: Indicator of microbial biomass or of phosphorus deficiency. Limnol. Oceanogr., 21: 231—236.

Cavari, B. and Phelps, G., 1977. Sensitive enzymatic assay for glucose determination in natural waters. Appl. Environ. Microbiol., 33: 1237—1243.

Ceriotti, G., 1955. Determination of nucleic acids in animal tissues. J. Biol. Chem., 214: 59—68.

Chandromohan, D., Devendran, K. and Natarajan, R., 1974. Arylsulfatase activity in marine sediments. Mar. Biol., 27: 89—92.

Chapman, A.G., Fall, L. and Atkinson, D.E., 1971. Adenylate energy in *Escherichia* coli during growth and starvation. J. Bacteriol., 108: 1072—1086.

Chriswell, C.D., Kissinger, L.D. and Fritz, J.S., 1976. Chromatographic determination of phenols in water. Anal. Chem., 47: 1325—1329.

Clark, M.E., Jackson, G.A. and North, W.J., 1972. Dissolved free amino acids in Southern California coastal waters. Limnol. Oceanogr., 17: 749—758.

Colin, H., Guiochon, G. and Siouffi, A., 1979. Comparison of various systems for the separation of free sterols by high performance liquid chromatography. Anal. Chem., 51: 1661—1666.

Cooper, M.J. and Anders, M.W., 1974. Determination of long chain fatty acids as 2-napthacyl esters by high pressure liquid chromatography and mass spectrometry. Anal. Chem., 46: 1849—1852.

Corpe, W.M. and Winters, H., 1972. Hydrolytic enzymes of some periphytic marine bacteria. Can. J. Microbiol., 18: 1483—1490.

Coughenower, D.D. and Curl, H.C. Jr., 1975. An automated technique for total dissolved free amino acids in seawater. Limnol. Oceanogr., 20: 128—131.

D'Amboise, M., Noel, D. and Hanai, T., 1980. Characterization of bonded-amine packing for liquid chromatography and high-sensitivity determination of carbohydrates. Carbohydr. Res., 78: in press.

Daumas, R.A., 1976. Variations of particulate proteins and dissolved amino acids in coastal seawater. Mar. Chem., 4: 225—242.

Daumas, R.A. and Saliot, A., 1977. The inventory in marine organic chemistry. Mar. Chem., 5: 417—427.

Dawson, R. and Gocke, K., 1978. Heterotrophic activity in comparison to the free amino acid concentrations in Baltic Seawater samples. Oceanol. Acta, 1: 45—54.

Dawson, R. and Liebezeit, G., 1978. Bestimmung gelöster Aminosäuren an verschiedenen Substraten. In: G. Wefer, and G. Hempel (Editors), Rep. Sonderforschungsber. 95, No. 48: 21—25.

Dawson, R. and Mopper, K., 1978a. A note on the losses of monosaccharides, amino sugars and amino acids from extracts during concentration procedures. Anal. Biochem., 84: 186—190.

Dawson, R. and Mopper, K., 1978b. An automatic analyzer for the specific determination of amino sugars. Anal. Biochem., 84: 191—195.

Dawson, R. and Pritchard, R.G., 1978. The determination of α-amino acids in seawater using a fluorimetric analyser. Mar. Chem., 6: 27—40.

Degens, E.T., 1970. Molecular nature of nitrogenous compounds in seawater and recent marine sediments. In: D.W. Hood (Editor), Organic Matter in Natural Waters. Mar. Sci. Inst. Univ. Alaska, Occas. Publ., No. 1: pp. 77—106.

Degens, E.T., 1979. Carbon in the sea. Nature, 279: 191—192.

Degens, E.T. and Matheja, J., 1967. Molecular mechanisms of interactions between oxygen co-ordinated metal polyhedra and biochemical compounds. Woods Hole Oceanogr. Inst. Tech. Rep. Ref. No. 67-57: 312 pp.

Degens, E.T. and Reuter, J.H., 1962. Analytical techniques in the field or organic geochemistry. Proc. Int. Meet. Org. Geochem., Sept. 10—12, 1962, Milan.

Degens, E.T., Reuter, J.H. and Shaw, K.N.F., 1964. Biochemical compounds in offshore California sediments and seawater. Geochim. Cosmochim. Acta, 28: 45—66.

Deluca, M., 1977. Recent experiments on firefly and bacterial luciferase. 2nd Biannual ATP Methodology Symposium, San Diego, SAI Technol. Comp. Calif., pp. 1—26.

DeMarche, J.M., Curl. H. Jr. and Coughenower, D.D., 1973. Automated analysis for urea in seawater. Limnol. Oceanogr., 18: 686—689.

Derenbach, J.B., 1970. Zur Bestimmung der Ribonucleinsäure in planktischem Analysenmaterial. Kieler Meeresforsch., 26: 79—84.

Dharmaraj, K., Selvakumar, N., Chandromohan, D. and Natarajan, R., 1977. L-Glutaminase activity in marine sediments. Indian J. Mar. Sci., 6: 168—170.

Dietz, F. and Traud, J., 1978. Zur Spurenanalyse von Phenolen, insbesondere Chlorphenolen in Wässern mittels Gaschromatographie-Methoden und Ergebnisse. Vom Wasser, 51: 235—257.

Duce, R.A., Quinn, J.G., Olney, C.E., Piotrowicz, S.R., Ray, B.J. and Wade, T.L., 1972. Enrichment of heavy metals and organic compounds in the surface microlayer of Narragansett Bay, Rhode Island. Science, 176: 161—163.

Dünges, W., 1977. 4-Bromomethyl-7-methoxycoumarin as a new fluorescence label for fatty acids. Anal. Chem., 49: 442—445.

Durst, H.D., Milano, M., Kikta, E.J. Jr., Connelly, S.A. and Grushka, E., 1975. Phenacyl esters of fatty acids via crown ether catalyts for enhanced ultraviolet detection in liquid chromatography. Anal. Chem., 47: 1797—1801.

Duursma, E.K., 1965. The dissolved organic constituents of sea water. In: J.P. Riley and G. Skirrow (Editors), Chemical Oceanography, 1. Academic, Press, London, pp. 433—475.

Duursma, E.K. and Marchand, M. 1974. Aspects of organic marine pollution. Oceanogr. Mar. Biol. Annu. Rev., 12: 315—431.

Eklund, G., Josefsson, B. and Roos, C., 1977. Gas—liquid chromatography of monosaccharides at the picogram level using glass-capillary columns, trifluoroacetyl derivatization and electron-capture detection. J. Chromatogr., 142: 575—585.

Gagosian, R.B., 1975. Sterols in the western North Atlantic Ocean. Geochim. Cosmochim. Acta, 39: 1443—1454.

Gagosian, R.B., 1976. A detailed vertical profile of sterols in the Sargasso Sea. Limnol. Oceanogr., 21: 702—710.

Gagosian, R.B. and Heinzer, F., 1979. Stenols and stanols in the oxic and anoxic waters of the Black Sea. Geochim. Cosmochim. Acta, 43: 471—486.

Gagosian, R.B. and Stuermer, D.H., 1977. The cycling of biogenic compounds and their diagenetically transformed products in seawater. Mar. Chem., 5: 605—632.

Gardner, W.S., 1978. Sensitive fluorimetric procedure to determine individual amino acids in marine waters. Mar. Chem., 6: 15—26.

Gardner, W.S. and Lee, G.F., 1973. Gas chromatographic procedure to analyze amino acids in lake waters. Environ. Sci. Technol., 7: 719—724.

Garrasi, C. and Degens, E.T., 1976. Analytische Methoden zur säulenchromatographischen Bestimmung von Aminosäuren und Zuckern im Meerwasser und Sediment. Berichte aus dem Projekt DFG-De 74/3: "Littoralforschung-Abwasser in Küstennähe", DFG-Abschlusskolloquium, Bremerhaven.

Garrasi, C., Degens E.T. and Mopper, K., 1979. The free amino acid composition of seawater obtained without desalting and preconcentration. Mar. Chem., 8: 71—85.

Garrett, W.D., 1967. The organic chemical composition of the ocean surface. Deep-Sea Res., 14: 221—227.

Giger, W., 1977. Inventory of organic gases and volatiles in the marine environment. Mar. Chem., 5: 429—442.

Goodrich, T.D. and Morita, R.Y., 1977. Incidence and estimation of chitinase activity associated with marine fish and other estuarine samples. Mar. Biol., 41: 349—353.

Goulden, P.D., Brooksbank, P. and Day, M.B., 1973. Determination of submicrogram levels of phenol in water. Anal. Chem., 45: 2430—2433.

Graf, G. and Lagaly, G., 1980. Interaction of clayminerals with adenosine-5-phosphates. Clay Clay Min., 28: 12—18.

Grasshoff, K., 1976. Methods of Seawater Analysis. Verlag Chemie, Weinheim, 317 pp.

Hamilton, P.B., 1965. Amino-acids on hands. Nature, 205: 284—285.

Handa, N., 1966. Distribution of dissolved carbohydrates in the Indian Ocean. J. Oceanogr. Soc. Jap., 22: 16—22.

Handa, N. and Yanagi, K., 1969. Studies on water-extractable carbohydrates of the particulate matter from the northwest Pacific Ocean. Mar. Biol., 4: 197—207.

Handa, N., Yanagi, K. and Matsunga, K., 1972. Distribution of detrital materials in the western Pacific Ocean and their biochemical nature. Mem. Ist. Ital. Idrobiol., 29 (Suppl.): 53—71.

Hanson, R.B. and Snyder, I., 1979. Enzymatic determination of glucose in marine environments: improvement and note of caution. Mar. Chem., 7: 353—362.

Hartwick, R.A., Assenza, S.P. and Brown, P.R., 1979. Identification and quantification of nucleosides, bases and other UV absorbing compounds in serum using reversed phase high performance liquid chromatography. I. Chromatographic methodology. In: A. Zlatkis (Editor), Advances in Chromatography. Lausanne Symp., pp. 725—736.

Harvey, H.W., 1925. Oxidation in seawater. J. Mar. Biol. Assoc. U.K., 18: 953—969.

Hellebust, J.A., 1965. Excretion of some organic compounds by marine phytoplankton. Limnol. Oceanogr., 10: 192—206.

Hellebust, J.A., 1974. Extracellular products. In: W.D.P. Stewart (Editor), Algal Physiology and Biochemistry. Botanical Monographs, 10. Blackwell, Oxford, pp. 838—863.

Hicks, S.E. and Carey, R.G., 1968. Glucose determination in natural waters. Limnol. Oceanogr., 13: 361—363.

Hirayama, H., 1974. Fluorimetric determination of carbohydrates in seawater. Anal. Chim. Acta, 70: 141—148.

Hodson, R.E. and Azam, F., 1977. Determination and biological significance of dissolved ATP in seawater. 2nd Biannual ATP Methodology Symposium, San Diego, SAI Technol. Comp. Calif., pp. 127—139.

Hoffman, N.E. and Liao, J.C., 1976. High pressure liquid chromatography of p-methoxyanilides of fatty acids. Anal. Chem., 48: 1104—1106.

Holm-Hansen, O., 1969. Determination of microbial biomass in ocean profiles. Limnol. Oceanogr., 14: 740—747.

Holm-Hansen, O. and Booth, Ch.R., 1966. The measurement of adenosine triphosphate in the ocean and its ecological significance. Limnol. Oceanogr., 11: 510—519.

Holm-Hansen, O., Sutcliffe, W.H. Jr. and Sharp, J., 1968. Measurement of deoxyribonucleic acid in the ocean and its ecological significance. Limnol. Oceanogr., 13: 507—514.

Honda, S., Kakimoto, K., Sudo, K., Kakehi, K. and Takiura, K., 1974. Fluorimetric determination of reducing sugars with ethylenediamine sulfate. Anal. Chim. Acta, 70: 133—139.

Hosaku, K. and Maita, Y., 1971. Gas-liquid chromatographic determination of amino acids and vertical distribution of proteinaceous substances in sea water. J. Oceanogr. Soc. Japan, 27: 27—33.

Hulett, H.R., 1970. Non-enzymatic hydrolysis of adenosine phosphates. Nature, 225: 1240—1249.

Iwamura, T., Nagai, H. and Ichimura, S., 1970. Improved methods for determining contents of chlorophyll, protein, ribonucleic acid and deoxyribonucleic acid in planktonic populations. Int. Revue Ges. Hydrobiol., 55: 131—147.

Jandera, P. and Churacek, J., 1974. Ion-exchange chromatography of aldehydes, ketones, ethers, alcohols, polyols and saccharides. J. Chromatogr., 98: 55—104.

Jeffrey, L.M. and Hood, D.W., 1958. Organic matter in seawater: an evaluation of various methods for isolation. J. Mar. Res., 17: 247—271.

Johnson, K.M. and Sieburth, J.McN., 1977. Dissolved carbohydrates in seawater. I. A precise spectrophotometric analysis for monosaccharides. Mar. Chem., 5: 1—13.

Josefsson, B.O., 1970. Determination of soluble carbohydrates in seawater by partition chromatography after desalting by ion-exchange membrane electrodialysis. Anal. Chim. Acta, 52: 65—73.

Josefsson, B.O., Upsström, L. and Östling, G., 1972. Automatic spectrophotometric procedure for the determination of the total amount of dissolved carbohydrates in seawater. Deep-Sea Res., 19: 385—395.

Josefsson, B.O., Lindroth, P. and Östling, G., 1977. An automated fluorescence method for the determination of total amino acids in natural waters. Anal. Chim. Acta, 89: 21—28.

Kanazawa, A. and Teshima, S., 1978. The occurrence of coprostanol, an indicator of faecal pollution, in sea water and sediments. Oceanol. Acta, 1: 39—44.

492

Karl, D.M., 1978. Occurrence and ecological significance of GTP in the ocean and in microbial cells. Appl. Environ. Microbiol., 36: 349—355.

Karl. D.M. and Holm-Hansen, O., 1978. Methodology and measurement of adenylate energy charge ratios in environmental samples. Mar. Biol., 48: 185—197.

Kattner, G.G. and Brockmann, U.H., 1978. Fatty acid composition of dissolved and particulate matter in surface films. Mar. Chem., 6: 233—241.

Kennedy, J.F., 1974. Methodology and instrumentation in the chromatography of carbohydrates on ion-exchange resins. Biochem. Soc. Trans., 2: 54—64.

Kim, J. and Zobell, C.E., 1974. Occurrence and activity of cell-free enzymes in oceanic environments. In: R.R. Colwell and R.Y. Morita (Editors), Effect of the Ocean Environment on Microbial Activities. University Park Press, Baltimore, Md., pp. 368—385.

Kissane, J.M. and Robins, E., 1958. The fluorimetric measurement of deoxyribonucleic acid in animal tissues with special reference to the central nervous system. J. Biol. Chem., 233: 184—188.

Kiuchi, K., Ohta, T. and Ebine, H., 1975. High-speed liquid chromatographic separation of glycerides, fatty acids and sterols. J. Chromatogr. Sci., 13: 461—466.

Knapp, D.R. and Krueger, S., 1975. Use of O-p-nitrobenzyl-N,N'-diisopropylisourea as a chromogenic reagent for liquid chromatographic analysis of carboxylic acids. Anal. Lett., 8: 603—610.

Kunte, H. and Slemrova, J., 1975. Gas chromatographic and mass spectrometric identification of phenolic substances from surface waters. Z. Wasser—Abwasser-Forsch., 8: 176—182.

Larsson, K., Odham, G. and Södergren, A., 1974. On lipid surface films on the sea I. A simple method for sampling and studies of composition. Mar. Chem., 2: 49—57.

Larsson, L.T. and Samuelson, O., 1967. An automated micromethod for the separation of monosaccharides by partition chromatography. Microchim. Acta, 2: 328—332.

Lee, C.L., 1975. Biological and Geochemical Implications of Amino Acids in Sea Water, Wood and Charcoal. Thesis, University of California, San Diego, Calif., 178 pp.

Lee, C.L. and Bada, J.L., 1975. Amino acids in Equatorial Pacific ocean water. Earth Planet. Sci. Lett., 26: 61—68.

Lee, C.L. and Bada, J.L., 1977. Dissolved amino acids in the Equatorial Pacific, the Sargasso Sea and Biscayne Bay. Limnol. Oceanogr., 22: 502—510.

Liebezeit, G., Boelter, M., Brown, I.F. and Dawson, R., 1980. Dissolved free amino acids and carbohydrates at pycnocline boundaries in the Sargasso Sea and related microbial activity. Oceanolog. Acta, 3: 357—362.

Lindroth, P. and Mopper, K., 1979. High performance liquid chromatographic determination of subpicomole amounts of amino acids by precolumn fluorescence derivatization with o-phthaldialdehyde. Anal. Chem., 51: 1667—1674.

Litchfield, C.D. and Prescott, J.M., 1970. Analysis by dansylation of amino acids dissolved in marine and freshwaters. Limnol. Oceanogr., 15: 250—256.

Little, J.E., Sjogren, R.E. and Carson, G.E., 1979. Measurement of proteolysis in natural waters. Appl. Environ. Microbiol., 37: 900—908.

Lowry, O.H., Rosebrough, N.J., Lewis-Farr, A. and Randall, R.J., 1951. Protein measurement with the Folin phenol reagent. J. Biol. Chem., 193: 265—275.

Maeda, M. and Taga, N., 1973. Deoxyribonuclease activity in seawater and sediment. Mar. Biol., 20: 58—63.

Marty, J.C. and Saliot, A., 1974. Étude chimique comparée du film de surface et de l'eau de mer sous-jacente: acides gras. J. Rech. Atmos., 13: 563—570.

Maske, H., 1979. Das Wachstum von Skeletonema costatum (Grev.) Cleve in ammoniaklimitierten, kontinuierlichen Kulturen — Experimentelle Daten und Modellergebnisse. Thesis, University of Kiel, Kiel, 173 pp.

Meyer-Reil, L.-A., 1978. Uptake of glucose by bacteria in the sediment. Mar. Biol., 44: 293—298.

Meyer-Reil, L.-A., Dawson, R., Liebezeit, G. and Tiedge, H., 1978. Fluctuations and interactions of bacterial activity in sandy beach sediments and overlying waters. Mar. Biol., 48: 161—171.

Meyer-Reil, L.-A., Bölter, M., Liebezeit, G. and Schramm, W., 1979. Short-term variations in microbiological and chemical parameters. Mar. Ecol. Progr. Ser., 1: 1—6.

Meyers, P.A., 1976. Dissolved fatty acids in seawater from a fringing reef and a barrier reef at Grand Cayman. Limnol. Oceanogr., 21: 315—319.

Modzeleski, J.E., Laurie, W.A. and Nagy, B., 1971. Carbohydrates from Santa Barbara Basin sediments; Gas chromatographic—mass spectrometric analysis of trimethylsilyl derivatives. Geochim. Cosmochim. Acta, 34: 825—838.

Moore, S. and Stein, W.H., 1951. Chromatography of amino acids on sulfonated poly-styrene resins. J. Biol. Chem., 192: 663—681.

Mopper, K., 1973. Aspects of the biochemistry of carbohydrates in aquatic environments. Thesis, Woods Hole Oceanographic Institute, Woods Hole, Mass., 221 pp.

Mopper. K., 1977. Sugars and uronic acids in sediment and water from the Black Sea and North Sea with emphasis on analytical techniques. Mar. Chem., 5: 585—603.

Mopper, K., 1978a. Improved chromatographic separations on anion exchange resins. III. Sugars in borate medium. Anal. Biochem., 87: 162—168.

Mopper, K., 1978b. Improved chromatographic separations on anion-exchange resins. II. Separation of uronic acids in acetate medium and detection with a noncorrosive reagent. Anal. Biochem., 86: 597—601.

Mopper, K. and Degens, E.T., 1972. A new chromatographic sugar autoanalyzer with a sensitivity of 10^{-10} moles. Anal. Biochem., 45: 147—153.

Mopper, K. and Gindler, E.M., 1973. A new non-corrosive dye reagent for automatic sugar chromatography. Anal. Biochem., 56: 440—442.

Mopper, K. and Larsson, K., 1978. Uronic and other organic acids in Baltic Sea and Black Sea sediments. Geochim. Cosmochim. Acta, 42: 153—163.

Mopper, K., Dawson, R., Liebezeit, G. and Ittekot, V., 1980. The monosaccharide spectra of natural waters. Mar. Chem., 10: in press.

Morris, R.J. and Culkin, F., 1977. Marine lipids: sterols. Oceanogr. Mar. Biol. Annu. Rev., 15: 73—102.

Morrison S.J., King, J.D., Bobbie, R.J., Bechtold, R.E. and White, D.C., 1977. Evidence for microfloral succession on allochthonous plant litter in Apalachicola Bay, Florida, USA. Mar. Biol., 41: 229—240.

Murray, M.J., 1949. Total phenols in gasolines and in cresylic acids — spectrophotometric determination. Anal. Chem., 21: 941—945.

Newell, B.S., Morgan, B. and Cundy, J., 1967. The determination of urea in seawater. J. Mar. Res., 25: 201—202.

Noël, D., Hanai, T. and D'Amboise, M., 1980. Systematic liquid chromatographic separa-tion of poly-, oligo- and monosaccharides. J. Liquid Chromatogr., in press.

North, B.B., 1975. Primary amines in California coastal waters: utilization by phyto-plankton. Limnol. Oceanogr., 20: 20—27.

Overbeck, J. and Babenzien, H., 1964. Über den Nachweis von freien Enzymen in Ge-wässern. Arch. Hydrobiol., 60: 107—114.

Packard, T.T. and Dortch, Q., 1975. Particulate protein-nitrogen in North Atlantic sur-face waters. Mar. Biol., 33: 347—354.

Palmork, K.H., 1963. Studies of the dissolved organic compounds in the sea. Nor. Fish. Mar. Invest., 13: 120—125.

Palmork, K.H., 1969. Amino acids in seawater by gas chromatography. Int. Counc. Explor. Sea, Hydrogr. Comm., C.M., 1969/C 17.

Parsons, T.R. and Strickland, J.D.H., 1962. On the production of particulate organic car-bon by heterotrophic processes in the sea. Deep-Sea Res., 8: 211—222.

Pei, P.T.-S., Kossa, W.C., Ramachandran, S. and Henly, R.S., 1976. High pressure reverse phase liquid chromatography of fatty acid p-bromo-phenacyl esters. Lipids, 11: 814—816.

Percival, E., 1968. Marine algal carbohydrates. Oceanogr. Mar. Biol. Annu. Rev., 6: 137—161.

Perry, M.J., 1972. Alkaline phosphatase activity in subtropical Central North Pacific waters using a sensivitive fluorimetric method. Mar. Biol., 15: 113—119.

Petterson, K. and Jansson, M., 1978. Determination of phosphatase activity in lake water — a study of methods. Verh. Int. Verein. Limnol., 20: 1226—1230.

Pillai, T.N.V. and Ganguly, A.K., 1970. Nucleic acids in the dissolved constituents of seawater. Curr. Sci., 39(22): 501—504.

Plunkett, M.A., 1957. The qualitative determination of some organic compounds in marine sediments. Deep-Sea Res., 5: 259—262.

Pocklington, R., 1971. Free amino acids dissolved in North Atlantic Ocean waters. Nature, 230: 374—375.

Pocklington, R., 1972. Determination of nanomolar free amino acids dissolved in North Atlantic Ocean waters. Anal. Biochem., 45: 403—421.

Rai, H., 1973. Methods involving the determination of photosynthetic pigments using spectrophotometry. Verh. Int. Ver. Theor. Angew. Limnol., 18: 1864—1875.

Remsen, C.D., 1971: The distribution of urea in coastal and oceanic waters. Limnol. Oceanogr., 16: 732—740.

Riley, J.P., 1975. Analytical chemistry of sea water. In: J.P. Riley and G. Skirrow (Editors), Chemical Oceanography, 3. Academic Press, London, pp. 193—514.

Riley, J.P. and Segar, D.A., 1970. The seasonal variation of the free and combined dissolved amino acids in the Irish Sea. J. Mar. Biol. Assoc. U.K., 50: 713—720.

Roth, M., 1971. Fluorescence reaction for amino acids. Anal. Chem., 43: 880—882.

Roth, M. and Hampai, A., 1973. Column chromatography of amino acids with fluorescent detection. J. Chromatogr., 83: 353—356.

Ruchti, J. and Kunkler, D., 1966. Enzymatische Bestimmung von Glucose, Fructose und Saccharose in Gewässern. Schweiz. Z. Hydrol., 28: 62—68.

Saliot, A., 1975. Acides gras, stérols et hydrocarbures en milieu marin: inventaire, applications géochimiques et biologiques, Thesis University of Paris VI, Paris, 234 pp.

Saliot, A. and Boussuge, C., 1978. Inventaire et dynamique des lipides a l'interface eau de mer-sédiment II. — Acides gras de l'eau de mer et de l'eau interstitielle. In: A. Combaz and R. Pelet (Editors), Géochimie organique des sédiments marins profonds, Orgon II, Atlantique — N.-E. Brésil. C.N.R.S., Paris, pp. 239—251.

Schultz, D.M. and Quinn, J.G., 1972. Fatty acids in surface particulate matter from the North Atlantic. J. Fish. Res. Board. Can., 29: 1482—1486.

Selvakumar, N., Chandromohan, D. and Natarajan, R., 1977. L-Asparaginase activity in marine sediments. Curr. Sci., 46: 287—291.

Siegel, A. and Degens, E.T., 1966. Concentration of dissolved amino acids from saline waters by ligand-exchange chromatography. Science, 151: 1089—1101.

Siezen, R.J. and Mague, T.H., 1978. Amino acids in suspended particulate matter from oceanic and coastal waters of the Pacific. Mar. Chem., 6: 215—231.

Simatupang, M.H., Hoffmann, P. and Augustin, H., 1976. Einfluss der Versuchsbedingungen bei der ionenaustauschchromatographischen Bestimmung neutraler Zucker in Holz- und Zellstoffhydrolysaten. Holz Roh Werkst., 34: 289—293.

Sinner, M., Simatupang, M.H. and Dietrichs, H.H., 1975. Automated quantitative analysis of wood carbohydrates by borate complex ion exchange chromatography. Wood Sci. Technol., 9: 307—322.

Skjoldal, H.R. and Bamstedt, U., 1976. Studies on the deep water pelagic community of Korfjorden, Western Norway: Adenosine phosphates and nucleic acids in Meganyctiphanes norvegica (Euphasiacea) in relation to the life cycle. Sarsia, 61: 1—14.

495

Skjoldal, H.R. and Bamstedt, U., 1977. Ecobiochemical studies on the deep-water pelagic community of Korfjorden, Western Norway. Adenine nucleotides in zooplankton. Mar. Biol., 42: 197—211.
Slowey, J.F., Jeffrey, L.M. and Hood, D.W., 1962. The fatty-acid content of ocean water. Geochim. Cosmochim. Acta, 26: 607—616.
Spackman, D.H., Stein, W.H. and Moore, S., 1958. Automatic recording apparatus for use in the chromatography of amino acids. Anal. Chem., 30: 1190—1206.
Starikova, N.D. and Korzhikova, R.I., 1969. Amino acids in the Black Sea. Okeanologiya, 9: 509—518 (in Russian).
Stauffer, T.B. and MacIntyre, W.G., 1970. Dissolved fatty acids in the James River Estuary, Virginia, and adjacent ocean waters. Chesapeake Sci., 11: 216—220.
Strickland, J.D.H. and Parsons, T.R., 1972. A practical handbook for sea water analysis. Fish. Res. Board Can. Bull., 167: 1—310.
Taga, N. and Kobori, H., 1978. Phosphatase activity in eutrophic Tokyo Bay Mar. Biol., 49: 223—229.
Tatsumoto, M., Williams, W.T., Prescott, J.M. and Hood, D.W., 1961. On the amino acids in samples of surface sea water. J. Mar. Res., 19: 89—95.
Tesarik, K., 1972. The separation of some monosaccharides by capillary column gas chromatography. J. Chromatogr., 65: 295—302.
Tetas, M. and Lowenstein, J.M., 1963. The effect of bivalent metal ions on the hydrolysis of ADP and ATP. Biochemistry, 2: 350—357.
Treguer, P., Le Corre, P. and Courtot, P., 1972. A method for determination of the total dissolved free fatty-acid content of sea water. J. Mar. Biol. Assoc. U.K., 52: 1045—1055.
Tusseau, D., Barbier, M. and Saliot, A., 1978. Inventaire et dynamique des lipides à l'interface eau de mer sédiment III. — Stérols de l'eau de mer et de l'eau interstitielle. In: A. Combaz and R. Pelet (Editors), Géochimie organique des sédiments marins profonds. Orgon II, Atlantique — N.-E. Brésil. C.N.R.S., Paris, pp. 253—261.
Udenfriend, S., Stein S., Böhlen, P., Dairman, W., Leimgruber, W. and Weigele, M., 1972. Fluorescamine: A reagent for assay of amino acids, peptides, proteins and primary amines in the picomole range. Science, 178: 871—872.
UNESCO, 1966. Determination of photosynthetic pigments in sea water. Monogr. Oceanogr. Methodol. (UNESCO), 1: 1—66.
Walsh, G.E. and Douglass, J., 1966. Vertical distribution of dissolved carbohydrate in the Sargasso Sea off Bermuda. Limnol. Oceanogr., 11: 406—408.
Wangersky, P.J. and Zika, R.G., 1978. The Analysis of Organic Compounds in Seawater. Natl. Res. Counc. Can., Rep. 3, NRCC No. 16566, 177 pp.
Webb, K.L. and Wood, L., 1967. Improved techniques for analysis of free amino acids in sea water. In: N.B. Scova et al. (Editors), Automation in Analytical Chemistry, Technicon Symposium, 1966. Mediad, New York, N.Y., 1: 440—444.
Weigele, M., De Bernardo, S.L., Tengi, J.P. and Leimgruber, W., 1972. A novel reagent for the fluorimetric assay of primary amines. J. Am. Chem. Soc., 94: 5927—5928.
Whitledge, T.E. and Dugdale, R.C., 1972. Creatine in sea water. Limnol. Oceanogr., 17: 309—314.
Wieser, W. and Zech, M., 1976. Dehydrogenases as tools in the study of marine sediments. Mar. Biol., 36: 113—122.
Williams, P.J. LeB., 1975a. Determination of Organic Components. In: J.P. Riley and G. Skirrow (Editors), Chemical Oceanography, 3. Academic Press, London, pp. 453—514.
Williams, P.J. LeB., 1975b. Biological and chemical aspects of dissolved organic material in sea water. In: J.P. Riley and G. Skirrow (Editors), Chemical Oceanography, 2. Academic Press, London, 2nd ed., pp. 301—363.
Williams, P.J. LeB. and Yentsch, C.S., 1976. An examination of the photosynthetic production, excretion of photosynthetic products and heterotrophic utilization of dis-

solved organic compounds with reference to results from a coastal subtropical sea. Mar. Biol., 35: 31—40.

Williams, P.M., 1965. Fatty acids derived from lipids of marine origin. J. Fish, Res. Board Can., 22: 1107—1122.

Williams, P.M. and Craigie, J.S., 1970. Microdetermination of uronic acids and related compounds in the marine environment. In: D.W. Hood (Editor), Organic Matter in Natural Waters. Mar. Sci. Inst., Univ. Alaska, Occas. Publ., No. 1: 509—519.

Witzel, K.P., 1979. The adenylate energy-charge as a measure of microbial activities in aquatic ecosystems. Arch. Hydrobiol. Beih. Ergebn. Limnol., 12: 146—165.

Wright, R.T. and Hobbie, J.E., 1966. Use of glucose and acetate by bacteria and algae in aquatic ecosystems. Ecology, 47: 447—464.

Wun, C.K., Walker, R.W. and Litsky, W., 1976. The use of XAD-2 resin for the analysis of coprostanol in water. Water Res., 10: 955—959.

Wunderlich, M., 1977. Free and dissolved cellulases in natural waters and their ecological significance. Arch. Hydrobiol., 81: 65—105.

Zein-Eldin, Z.P. and May, B.Z., 1958. Improved n-ethylcarbazole determination of carbohydrates with emphasis on sea water samples. Anal. Chem., 30: 1935—1941.

Zika, R.G., 1977. An investigation in Marine Photochemistry. Thesis, Dalhousie University, Halifax, N.S., 346 pp.

Zsolnay, A., 1977. Inventory of nonvolatile fatty acids and hydrocarbons in the oceans. Mar. Chem., 5: 465—475.

Chapter 16

STATE OF THE ART

R. DAWSON and E.K. DUURSMA

1. INTRODUCTION

The most spectacular developments in the field of organic chemistry of sea water are linked with the detection of specific compounds directly in sea water without any major chemical pretreatment or concentration procedures. Whether the findings, when employing these methods on a wider scale, fit in with earlier chemical oceanographic theories for the formation, distribution and decomposition of organic matter, remains to be seen.

A few of the major problem areas requiring investigation are dealt with briefly below: production — particulate organic matter and dissolved organic matter production; distribution — horizontal, vertical and temporal distribution; humification; age; molecular weight distribution; and chelation.

2. PRODUCTION

2.1. Particulate organic matter

A major step in biological oceanography was the introduction of the ^{14}C method for measuring primary production (Steemann-Nielsen, 1952). The discussion on various aspects with regard to procedure and accuracy still continues, even though measurements have been made in the remotest parts of the World Ocean. What exactly ^{14}C incorporation means is demonstrated in an autoradiographic print (Fig. 1). Cells containing ^{14}C are marked with black spots. These findings, combined with ^{14}C measurements by radioactive counting and quantitative microscopic determination of the plankton spectrum, reveal that ^{14}C uptake shows wide variations depending on the composition, physiological state and environment of the phytoplankton populations. The percentage uptake per species does not depend on its abundance (Fig. 2). The above observations are from *in situ* measurements (Van Iwaarden, 1979) and should reflect normal marine conditions.

Equally problematic are the estimates of primary production in terms of total organic matter. Terms such as net and gross, "new" and "regenerated" production (Eppley and Peterson, 1979) are employed but their meanings are approximations derived from a series of mathematical equations with more unknowns than equations.

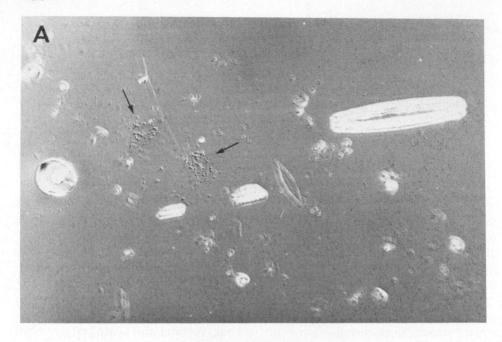

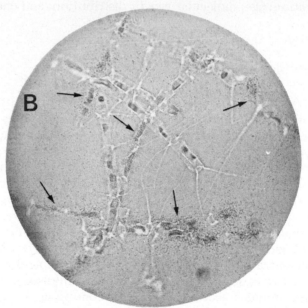

Fig. 1. A. Autoradiograph of natural phytoplankton collected from the saline Lake Greve-lingen and exposed to $^{14}CO_2$ at its sampling depth (3 m) for 5 h. Only flagellates (arrows) were productive. B. Autoradiograph of $^{14}CO_2$ enriched culture of *Chaetoceros* sp. Only parts were productive (arrows). Figure reproduced with permission of the author (Van Iwaarden, 1979).

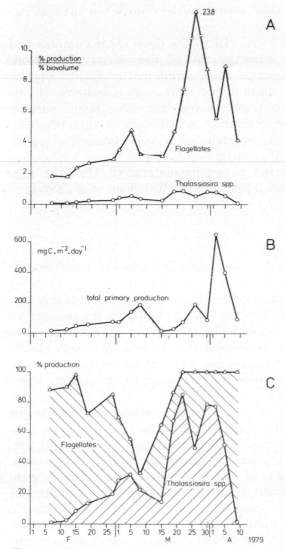

Fig. 2. A. Production ratio (% production/% biovolume) of natural phytoplankton of the saline Lake Grevelingen. The two dominant taxa were flagellates and *Thalassiosira* spp. (diatom). B. The total primary production of the same water. C. The contribution of the dominant taxa to the total primary production. (After Bakker, Van Iwaarden, Vegter and Visscher, unpublished). Figure reproduced with permission of the authors.

A point to be considered is the purpose for the determination of production. Is it to make estimates of the available particulate material for the marine food cycle or is it to make estimates of particular compounds which have other functions in the sea (or perhaps no function at all)? The answers may differ and this should be taken into consideration when interpreting bio- and geochemical cycles.

The composition of particulate material in the deep sea is complex and little investigated. Microscopic investigations of deep-water suspensions reveal an array of aggregates, flocculates of combined inorganic—organic nature, together with identifiable intact biological remains (Chesselet, 1979).

Differentiation between dissolved, macromolecular colloidal and particulate is arbitrary and based on our difficulties in retrieving samples (Cauwet, Chapter 4). Information concerning the detailed organic chemical composition of suspended material is only recently developing with the application of deep-sea collection techniques for particulate material (S. Honjo, J. Farrington and coworkers, Woods Hole, Mass., U.S.A.; Wakeham et al., 1980).

2.2. Dissolved organic matter

The production of organic matter has been investigated from various angles; for example as a significant energy source for aquatic organisms, or the other extreme, in the formation of inert compounds outside any biochemical cycle. Many controversial theories exist not only on the production of DOC (e.g. excretion) but also on its value as food for aquatic organisms. On this subject it is worthwhile to re-read the reviews by Krogh (1931, 1934), who experimented and philosophized extensively on these aspects. Many of his remarks are still up to date and pertinent to present-day scientific investigations.

A tendency in the past has been to look to exudation/excretion as a major mechanism leading to the production of DOC, whereas probably lysis, exoenzymatic activity leading to the breakdown of polymeric material and, perhaps over longer time scales, in situ hydrolysis are the dominant processes. Monomers are likely to be products of cell exudation. These probably turnover quickly by secondary biotic or abiotic processes and the variations in composition may correlate with primary biological activity. Polymeric material, on the other hand, will undergo transformation by various mechanisms and the released lower molecular-weight products may bear no resemblance to primary synthesized material.

An essential discussion on the excretion of organic matter by phytoplankton has been held between Sharp (1977), Fogg (1977), Aaronson (1978) and again Sharp (1978). The consensus of this discussion is that, although excretion may exist as a functional process (particularly when algal cells are "stressed"), the evidence presented to date in support of any extensive excretion by phytoplankton is probably based on methodological artifacts.

A functional excretion product is for example glycollate (Al-Hasan et al., 1975), which is considered to play a role in the bicarbonate balance of algal cells (cf: also Tolbert and Zill, 1956):

$$\frac{(glycollate^-)\ cell}{(glycollate^-)\ medium} = k\ \frac{(HCO_3^-)\ cell}{(HCO_3^-)\ medium}$$

It is unfortunate, however, that the analytical method for the determination of glycolic acid (Shah and Fogg, 1973) is far too insensitive and perhaps not reproducible enough to detect low-level excretion *in situ*. It is also unlikely that information may be derived from studies involving the determination of bulk parameters (e.g. DOC) leading to an estimate of the extent of natural excretion processes.

Several other biochemical compounds can be envisaged to be excreted by algal cells (e.g. vitamins, enzymes, chelating agents) whose function, when released, may be to "condition" the environment to the competitive advantage of the algal cell (cf. Chapter 8).

Recent work, whereby the spectrum and concentrations of dissolved free amino acids were monitored by direct analysis of seawater from a fixed station over the period of a day at hourly intervals (K. Mopper, pers. comm., 1979), showed pronounced variations in both composition and concentration. Particularly striking was the sharp increase in basic amino acids (low C : N) at the expense of a decrease in acidic-neutral acids (high C : N) outside the daylight hours. An explanation can at present only be speculative, but the results suggest that the phenomenon is directly linked with photosynthetic activity. This suggests, therefore, that the approach to study excretion *in situ* requires the detection of bioactive compounds at their natural levels, which generally lie several orders of magnitude below those of bulk parameters.

These numerous facets of DOC production will certainly continue to stir up further emotions among chemical oceanographers. A point which should not be overlooked, however, is that experimental cultures rarely reflect the natural environment and the methodologies employed have a great impact on the results.

3. DISTRIBUTION

3.1. Horizontal

Since the early sixties abundant data have been collected describing the levels of DOC and POC in the oceans. Variations with season and distance from shore are relatively well understood. Less is known about the distribution of individual constituent compounds, although in this book, full credit is given to the more modern findings based on relatively few samples. Lacking is the combination of elementary analysis of organic matter (particulate, dissolved and colloidal) together with the determination of specific compounds. DON and DOP are essentially more functional than DOC, although DOC is most widely determined, since automatic methods exist

502

and have been extensively investigated to assess their accuracies (MacKinnon, Chapter 14). The same is, however, not the case for DON and DOP measurements.

The process of breakdown of organic compounds in the euphotic zone is equally poorly understood. A general (but wrong) opinion is that coastal zones should have higher DOM concentrations than open ocean areas. In fact bacterial decomposition is favoured at high suspension loads. Superimposed on this are the seasonal variations and thus an estimation of the input of organics from terrestrial sources and coastal zones to the oceans is problematic and may even sometimes be negative.

3.2. Vertical

Very few data exist on DOC/DON ratios of deep waters even though DOC has been routinely measured. Can such ratios (on atomic basis) really be as low as 2.3 : 1 as suggested in Fig. 3 for North Atlantic Deep Water? If so, what is the composition of the nitrogenous components; uracil (C/N 2 : 1), urea (C/N 1 : 2) creatinine (C/N 4 : 3), guanine (C/N 1 : 1) and glycine (C/N 2 : 1)?

Modern analytical techniques are now capable of identifying the majority of amino group-containing compounds, both before and after controlled hydrolysis, and analysis of freshly sampled water onboard ship may reveal a part of the mystery. The incorporation of basic amino acids, polyamines and even urea (which may have been underestimated in seawater and which has been suggested to form stable multifunctional polymers; Degens, 1970) into polymeric forms in deep-water samples would serve to lower the C/N ratio. Although only a few deep-water measurements of DOC/DON ratios have been made, it has been suggested (Duursma, 1965) that the low ratios were not directly related to "old", deep water. Deep water from the Norwegian

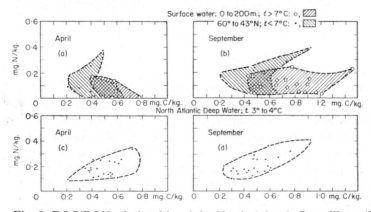

Fig. 3. DOC/DON relationship of the North Atlantic Deep Water (Duursma, 1965).

Sea for example is supposed to be "older" than North Atlantic Deep Water south of Greenland but nonetheless, had DOC/DON ratios between 6 : 1 and 8 : 1 by weight (7 : 1 and 9.3 : 1 atomic ratio).

In reality, here exists a blank field of research, requiring an input in the form of new analytical techniques for the estimation of DON and DOP together with determination of constituent compounds in a multidisciplinary effort to characterise watermasses. This coincides with the strong future research recommendations made by Gagosian et al. (1978).

3.3. Temporal

In addition to gross-seasonal variations in production and decomposition of organic matter (mainly in the euphotic zone) and the spatial patchiness of organic matter when sampling with commonly adopted methods, the gathering of representative data is further complicated by the possibility of short-term variations of biological and chemical parameters. Biological rhythms have long been taken into consideration in the study of land plants, however, the difficulties in sampling have detracted from any major investigation of marine organisms *in situ*. Several investigators have observed large deviations of even conservative parameters such as DOC in coastal areas, which, after allowing a generous margin for analytical error, suggested either diurnal or more complicated hourly cycles. Ferguson and Palumbo (1979) found significant variations in bacterial biomass and substrate uptake over a 24-h period in Long Island Sound. Meyer-Reil et al. (1979) and K. Mopper (unpublished results) have also demonstrated considerable variations in free glucose, fructose and individual amino acids, which were well correlated with fluctuations in heterotrophic activity of Baltic Sea water samples collected over 24-h periods. Such short-term variations of chemical and microbiological parameters may exceed the variations in these parameters with season (Cavari and Hadas, 1979; for freshwaters)!

4. HUMIFICATION

In principle it is wrong to speak of humic substances in seawater. Humus is the term reserved to describe degraded soil organics which improve the quality of agricultural or forested ground; a traditional term with an intuitive rather than chemical meaning. Although Chapter 6 of this book is devoted to humification, the term will probably be used less in the future when our knowledge of specific compounds increases.

The generally accepted meaning of the term humification is polymerisation or condensation of low molecular weight compounds resulting in the formation of organic substances which are supposedly more stable in the marine environment. It becomes gradually clear (although already suggested by Kalle, 1966) that the majority of these polycondensates are of marine rather than terrestrial origin. Their stability is questionable until this term is properly defined. The definitions may include such distinctions as resistant to UV and persulphate oxidation, unattractive to bacterial decomposition or simply "old".

The amounts of humic or fulvic acids (distinguished solely by their solubilities in aqueous acid or base), recovered from seawater, strongly depend on the extraction procedure employed; these extraction procedures have a pronounced influence on the types of compounds recovered.

Most researchers support the theory that "humic substances" are derived from natural products of biological production and that we may logically expect to find amino acid, sugar, amino sugar, lipid and pigment moieties in the condensed structures (Gagosian and Stuermer, 1977). Surprisingly low amounts of these building blocks have been recovered on hydrolysis of the polymeric material. Whether this is simply a methodological difficulty remains to be seen.

It may be misleading to think of humics as being "old" and a clear division may be necessary to distinguish between condensed products in the euphotic zone and those in deeper layers. Basic amino acids (arginine, lysine, histidine, etc.), were found to be incorporated in high molecular weight material recovered on Amberlite XAD-2 (R. Dawson, unpublished results). The absence of these high-energy amino acids in free dissolved form in the euphotic zone (Garrasi et al., 1979), suggests that the incorporation into high molecular weight structures proceeds at a rate comparable with microbial utilisation.

The reaction of amino acids with sugars to form Schiff's base-type compounds has been often suggested to be a feasible process in the formation of coloured condensed material in seawater, particular in the presence of clay minerals (Hedges, 1978). It is also worthy to note that natural free sugars and amino acids when heated in seawater form fluorescent compounds whose excitation and emission maxima correspond with those formed on reaction of amines with aldehydes (Honda et al., 1974). These experiments were performed at high temperatures (G. Liebezeit, R. Dawson and K. Mopper, unpublished results) and there is some suggestion that the apparent disappearance of amino acids on standing in seawater at room temperature, may be connected in part to this abiotic process.

In the natural environment photolytic reactions leading to the production of high-energy unsaturated compounds (see Zika, Chapter 10) may play a role in the production of polymeric material in the euphotic zone, while on the other hand fluorescent substances may be degraded by sunlight excitation (Kramer, 1979). In deeper layers, adsorption onto colloids and further condensation of macromolecules may be determining processes.

Many of these points are open to criticism and a great deal of research is required to elucidate any one of the possible pathways. It is advisable, however, that we realise the inconsistencies in using the term stability in this context and employ a more suitable definition for the substances we are attempting to describe.

5. AGE

Since Williams et al. (1969) determined the average age of dissolved organic matter in the oceans from $^{14}C/^{12}C$ ratios, the figure of 3400 years has been cited in extenso. The problem remains, however, to determine what exactly this number means. Is the bulk of the dissolved organic matter in the deep sea of that age or is just a small fraction of it very old, while the majority of the material turns over rapidly.

As discussed in the introduction, this turnover cannot be too large, otherwise the oxygen consumption in the deep sea would have been much higher and old watermasses would have become anoxic. This leads to the suggestion that a large proportion of organic matter is of similar, average age, although we have no data concerning statistical variations. Since ocean circulation is more rapid than 3400 years (Gross, 1967), a part of the "old" organics must pass at least once through the euphotic zone to arrive at a later date (and other location) again in the deep sea.

A further question is whether age may be correlated with stability. Unsolved remain the problems of distinguishing between potential and actual stability of organic substances in seawater, and of estimating the role of particulate matter in destabilising DOM. The adsorption of DOM to particles would provide favourable boundary conditions for bacterial growth and since bacterial activities are higher at higher suspension loads, it would be interesting to have estimates of the differences in age of DOM from coastal and open ocean areas.

6. MOLECULAR WEIGHT DISTRIBUTION

Excretion products of phytoplankton, either functional or produced from maltreatment of the cells during filtration, have been shown by ^{14}C techniques to comprise up to 95% of compounds with molecular weights less than 3500 (Derenbach and Williams, 1974). It is clear that this primary material must undergo further polymerisation to account for the high proportions (up to 45%) of macromolecules with molecular weights over 100,000 (Ogura, 1977, and references therein).

Relatively few determinations have been made of molecular size distributions of organic matter in seawater, this being probably associated with the difficulties in processing samples. The distinction between "dissolved" macromolecular and colloidal material is very vague; the lower the cut-off limit of the ultrafilters employed to retain colloids, the higher are the proportions of DOM recovered (e.g. Sharp, 1973).

The problem areas of research into high molecular weight compounds have been highlighted by Ogura (1977): Although ultra filtration appears to be the best method for concentration and isolation of the components, alteration (aggregation) may occur during the process. The author suggests that selective enzymatic hydrolysis may give insights into the chemical composition of the recovered macromolecules, whilst mild extraction conditions should be employed in studies of the resistance of the material towards microbial degradation. The mechanisms whereby macromolecules are transformed into particles and the association of trace metals with the different molecular weight fractions are problem areas warranting further study.

7. CHELATION

Many myths are associated with the study of organic matter in seawater. Chelation may be one good example. Chelation involves the formation of chemical bonds which sterically form a claw-like structure; EDTA-chelation is the commonest example given. This type of bonding implies the formation of a specific organic molecule to which, for example, a metal is attached by a kind of covalent bonding with two carboxylate groups and two bridges with amine groups. Various combinations are allowed as long as the steric structure is maintained.

Chelation usually supposes an equilibrium between the organic molecule and the metal in solution, which can be defined with a stability constant. The reactions are relatively rapid and reversible and the equilibrium concentrations of a metal chelate are determined by the concentrations of the components and those of competing metals in seawater. This is demonstrated by the example with EDTA (Fig. 4).

Complexation is usually employed in a wider sense, although the term also includes chelation. It is possible to determine stability constants in the case of chelation, whereas for some complexation this is supposed not to be the case.

Few examples of metals being chelated in reversible form with natural dissolved organics exist (Mantoura, Chapter 7). Reports of metal complexation are more common, e.g. nickel in vitamin B_{12} or the complexed forms of zinc (Bernhard and Zattera, 1969). Many of these complexes have been discovered by employing careful oxidation procedures prior to metal detection (Duinker and Kramer, 1977).

A number of detectable organic substances have complexing or metal-binding properties and a complete field of research is open for the determination of the stability constants of compounds known to exist in seawater. Since the methods for the determination of trace metals have been perfected over many years of research, it has been tempting to use the data from metal analyses alone to estimate the extent of complexation. With the improvements in the analytical techniques for the determination of at least constituents of the monomeric pool of organics, it may become possible to consider

complexation from the ligand's point of view. High performance reversed-phase chromatography is likely to become a powerful tool capable of separating "intact" complexes with post-column detection of the eluted compounds with atomic absorption spectrophotometry. These techniques may at first only be feasibly employed with more concentrated extracts (sediments, biological samples, particulates or seawater concentrates) but may serve to detect important complexes involved in cell-wall transport and metabolic processes in biota.

8. ACKNOWLEDGEMENTS

We are particularly grateful to G. Liebezeit, V. Smetacek and B. Knoppers for their useful comments and help with the manuscript.

9. REFERENCES

Aaronson, S., 1978. Excretion of organic matter by phytoplankton in vitro. Limnol. Oceanogr., 23: 838.

Al-Hasan, R.H., Coughlan, S.J., Pant, A. and Fogg, G.E., 1975. Seasonal variations in phytoplankton and glycollate concentrations in the Menai Straits, Anglesey. J. Mar. Biol. Assoc. U.K., 55: 557—565.

Bernhard, M. and Zattera, A., 1969. A comparison between the uptake of radioactive and stable zinc by a marine unicellular alga. In: J.N. Nelson and D. Evans (Editors), Second Natl. Symp. on Radioecology, Ann Arbor, Mich., pp. 389—398.

Cavari, B.Z. and Hadas, O., 1979. Heterotrophic activity, glucose uptake and primary productivity in Lake Kinneret. Freshwater Biol., 9: 329—338.

Chesselet, R., 1979. Modes of settling and organic input to the sediment seawater interface. Colloque International C.N.R.S. Biogeóchimie de la matière organique à l'interface eau—sédiment marin, Marseille (extended abstract).

Degens, E.T., 1970. Molecular nature of nitrogenous compounds in seawater and recent marine sediments. In: D.W. Hood (Editor), Organic Matter in Natural Waters. Inst. Mar. Sci., Univ. Alaska, Occas. Publ., No. 1: 77—106.

Derenbach, J.B. and Williams, P.J.LeB., 1974. Autotrophic and bacterial production: fractionation of plankton population by differential filtration of samples from the English Channel. Mar. Biol., 25: 263—269.

Duinker, J.C. and Kramer, C.J.M., 1977. An experimental study on the speciation of dissolved zinc, cadmium, lead and copper in river Rhine and North Sea water, by differential pulsed anodic stripping voltametry. Mar. Chem., 5: 207—228.

Duursma, E.K., 1965. The dissolved organic constituents of seawater. In: J.P. Riley and G. Skirrow (Editors), Chemical Oceanography, 1. Academic Press, London, 1st ed., pp. 433—475.

Duursma, E.K., 1970. Organic chelation of ^{60}Co and ^{55}Zn by leucine in relation to sorption by sediments. In: D.W. Hood (Editor), Organic Matter in Natural Waters. Inst. Mar. Sci., Univ. Alaska, Occas. Publ., No. 1: 387—397.

Eppley, R.W. and Peterson, B.J., 1979. Particulate organic matter flux and planktonic new production in the deep ocean. Nature, 282: 677—680.

Ferguson, R.L. and Palumbo, A.V., 1979. Distribution of suspended bacteria in neritic waters south of Long Island during stratified conditions. Limnol. Oceanogr., 24: 697—705.

Fogg, G.E., 1977. Excretion of organic matter by phytoplankton. Limnol. Oceanogr., 22: 576—577.

512

Gagosian, R.B. and Stuermer, D.H., 1977. The cycling of biogenic compounds and their diagenetically transformed products in seawater. Mar. Chem., 5: 605—632.

Gagosian, R.B., Ahmed, S.I., Farrington, J.W., Lee, R.F., Mantoura, R.F.C., Nealson, K.H., Packard, T.T. and Reinhart, K.L., 1978. Future research problems in marine organic chemistry. Mar. Chem., 6: 375—382.

Garrasi, C., Degens, E.T. and Mopper, K., 1979. The free amino acid composition of seawater obtained without desalting and preconcentration. Mar. Chem., 8: 71—85.

Gross, M.G., 1967. Oceanography. Merrill, Columbus, Ohio, 135 pp.

Hedges, J.I., 1978. The formation and clay mineral reactions of melanoidins. Geochim. Cosmochim. Acta, 42: 69—76.

Honda, S., Kakimoto, K., Sudo, K., Kakehi, K. and Takiura, K., 1974. Fluorimetric determination of reducing sugars with ethylenediamine sulphate. Anal. Chim. Acta, 70: 133—139.

Kramer, C.J.M., 1979. Degradation by sunlight of dissolved fluorescing substances in the upper layers of the Eastern Atlantic Ocean. Neth. J. Sea. Res., 13: 325—329.

Kalle, K., 1966. The problem of the Gelbstoff in the sea. Ocean. Mar. Biol. Annu. Rev., 4: 91—104.

Krogh, A., 1931. Dissolved substances as food of aquatic organisms. Biol. Rev., 6: 412—442.

Krogh, A., 1934. Conditions of life in the oceans. Ecol. Monogr., 4: 421—429 (Part I); 430—439 (Part II).

Meyer-Reil, L.-A., Boelter, M., Liebezeit, G. and Schramm, W., 1979. Short-term variations in microbiological and chemical parameters. Mar. Ecol. Progr. Ser., 1: 1—6.

Ogura, N., 1977. High molecular weight organic matter in seawater. Mar. Chem., 5: 535—549.

Shah, N.M. and Fogg, G.E., 1973. Determination of glycolic acid in seawater. J. Mar. Biol. Assoc. U.K., 53: 321—324.

Sharp, J.H., 1973. Size classes of organic carbon in seawater. Limnol. Oceanogr., 18: 441—447.

Sharp, J.H., 1977. Excretion of organic matter by marine phytoplankton: do healthy cells do it? Limnol. Oceanogr., 22: 381—389.

Sharp, J.H., 1978. Reply to comment by S. Aaronson. Limnol. Oceanogr., 23: 839—840.

Steemann-Nielsen, E., 1952. The use of radioactive carbon (C-14) for measuring organic production in the sea. J. Cons. Int. Explor. Mer., 18: 117—140.

Tolbert, N.E. and Zill, L.P., 1956. Excretion of glycolic acid by algae during photosynthesis. J. Biol. Chem., 222: 895—906.

Van Iwaarden, A.J.W., 1979. Micro-autoradiographical Determination of the Primary Productivity of Different Phytoplankton Species During the Spring Bloom (1979) of the Lake Grevelingen. Delta Institute, Yerseke, Stud. Rep. D9-1979, 85 pp.

Wakeham, S.G., Farrington, J.W., Gagosian, R.B., Lee, C., DeBaar, H., Nigrelli, G.E., Tripp, B.W., Smith, S.O. and Frew, N.M., 1980. Organic matter fluxes from sediment traps in the equatorial Atlantic Ocean. Nature, 286: 798—800.

Williams, P.M., Oeschger, H. and Kinney, P.J., 1969. Natural radiocarbon activity of the dissolved organic carbon in the Northeast Pacific Ocean. Nature, 224: 256—258.

SUBJECT INDEX

adsorbed film, 271
adsorption, 272, 354
advection, 92, 97, 110, 111
aerosols, 152, 153, 294
aggregates, 77, 79, 80, 83, 85, 141, 153, 159
aggression substances, 243, 244
alarm, pheromones involved in, 239
albuminoid-N, 270
alcohols, 264, 269
aldehyde, 91, 98, 321
alicyclic hydrocarbons in bacteria, 350, 351
— — — benthic algae, 350
— — — phytoplankton, 351
— — — sea-surface microlayer, 360
— — — seawater, 358
— — — sediment, 363
alkali earth metals, 194—196
— metals, 194—196
alkanes, 91, 95, 157, 342—345
alkenes, 91, 94, 95, 347—349
alkyl sulphonates, 270
allelochemics, 226
alloiohormones, 226
allomones, attractant, 242
—, repulsive or allelopathic, 243
Alteromonas, production of antibiotics, 229
amines, 96, 107
— as trophic attractants, 242
— — behavioural inhibitors, 244
amino-acid analyser, 452—457
amino acids, 4, 91, 93, 94, 96, 103—105, 112, 141, 144, 259, 260, 290, 501, 504
— —, aromatic, 447, 461
— — combined, hydrolysis of, 447, 450, 459—461
— — determination of, 450—459
— — free dissolved, 447
— — occurrence in seawater, 400
— — pattern of, 461, 462
— — production, 399
— — sulphur containing, 398, 399
amino sugars, 91, 473, 474
ammonia, interference of, 448—450
ammonification, 126, 127, 128, 137

Anacystis marina, 376
anaerobic experiments, 133, 143—145
anaerobiosis, 136
anoxic basin, 92, 102, 103
antagonism between phytoplankton and zooplankton, 236
anthopleurin, 239
antibiotics from bacteria, 229, 230, 233
— — marine animals, 235, 236
— — phytoplankton, 230—233
— — seaweeds, 232
aplysiatoxin, 377
aromatic hydrocarbons in algae, 352
— — — analysis, 338, 339
— — — bacteria, 352
— — — seawater, 288, 358, 359
— — — sea-surface microlayer, 360
— — — sediment, 364
arthropodin, 243
Asparagopsis spp., 379
atmospheric transport, 91, 107, 108
ATP, 32, 57, 58, 95, 295
—, determination of, 482, 483
autotrophs, 93
autoxidation, 313

Babylonia japonica, 384
bacteria, 93, 94, 96, 103, 130, 139, 145, 151, 157, 165, 167
—, alicyclic hydrocarbons in, 350, 351
—, aromatic hydrocarbons in, 352
—, biomass, 55, 56, 57
—, hydrocarbons in, 342, 345—347, 350—352, 355, 356
—, isoprenoids in, 346
—, n-alkanes in, 342, 345
—, n-alkenes in, 347, 350
—, open water, 54
—, production, 55—57
—, sediments, 56, 57
bacterial decomposition, 502, 503, 505, 506
bacteriolytic enzymes, 229
barium, 195
behavioural inhibition, 244
benthic algae, alicyclic hydrocarbons in, 350